Bayesian Networks for Probabilistic Inference and Decision Analysis in Forensic Science

STATISTICS IN PRACTICE

Series Advisors

Human and Biological Sciences
Stephen Senn
CRP-Santé, Luxembourg

Earth and Environmental Sciences
Marian Scott
University of Glasgow, UK

Industry, Commerce and Finance
Wolfgang Jank
University of Maryland, USA

Founding Editor
Vic Barnett
Nottingham Trent University, UK

Statistics in Practice is an important international series of texts which provide detailed coverage of statistical concepts, methods and worked case studies in specific fields of investigation and study.

With sound motivation and many worked practical examples, the books show in down-to-earth terms how to select and use an appropriate range of statistical techniques in a particular practical field within each title's special topic area.

The books provide statistical support for professionals and research workers across a range of employment fields and research environments. Subject areas covered include medicine and pharmaceutics; industry, finance and commerce; public services; the earth and environmental sciences, and so on.

The books also provide support to students studying statistical courses applied to the above areas. The demand for graduates to be equipped for the work environment has led to such courses becoming increasingly prevalent at universities and colleges.

It is our aim to present judiciously chosen and well-written workbooks to meet everyday practical needs. Feedback of views from readers will be most valuable to monitor the success of this aim.

A complete list of titles in this series appears at the end of the volume.

Bayesian Networks for Probabilistic Inference and Decision Analysis in Forensic Science

Second Edition

Franco Taroni

University of Lausanne, Switzerland

Alex Biedermann

University of Lausanne, Switzerland

Silvia Bozza

University Ca' Foscari of Venice, Italy

Paolo Garbolino

University IUAV of Venice, Italy

Colin Aitken

University of Edinburgh, UK

WILEY

This edition first published 2014
© 2014 John Wiley & Sons, Ltd

Registered office

John Wiley & Sons Ltd, The Atrium, Southern Gate, Chichester, West Sussex, PO19 8SQ, United Kingdom

For details of our global editorial offices, for customer services and for information about how to apply for permission to reuse the copyright material in this book please see our website at www.wiley.com.

The right of the author to be identified as the author of this work has been asserted in accordance with the Copyright, Designs and Patents Act 1988.

Wiley also publishes its books in a variety of electronic formats. Some content that appears in print may not be available in electronic books.

Designations used by companies to distinguish their products are often claimed as trademarks. All brand names and product names used in this book are trade names, service marks, trademarks or registered trademarks of their respective owners. The publisher is not associated with any product or vendor mentioned in this book.

Limit of Liability/Disclaimer of Warranty: While the publisher and author have used their best efforts in preparing this book, they make no representations or warranties with respect to the accuracy or completeness of the contents of this book and specifically disclaim any implied warranties of merchantability or fitness for a particular purpose. It is sold on the understanding that the publisher is not engaged in rendering professional services and neither the publisher nor the author shall be liable for damages arising herefrom. If professional advice or other expert assistance is required, the services of a competent professional should be sought.

Library of Congress Cataloging-in-Publication Data

Taroni, Franco, author.
  Bayesian networks for probabilistic inference and decision analysis in forensic science / Franco Taroni,
Alex Biedermann, Silvia Bozza, Paolo Garbolino, Colin Aitken. – Second edition.
     pages cm
  Includes bibliographical references and index.
  ISBN 978-0-470-97973-0 (hardback)
  1. Bayesian statistical decision theory – Graphic methods.   2. Uncertainty (Information theory) – Graphic
methods.   3. Forensic sciences – Graphic methods.   I. Title.
  QA279.5.T37 2014
  363.2501′519542 – dc23

                                                                                              2014014247

A catalogue record for this book is available from the British Library.

ISBN: 978-0-470-97973-0

Set in 10/12pt, TimeLTStd by Laserwords Private Limited, Chennai, India

1   2014

To our families

Bayes Nets and Influence Diagrams [...] will be the core elements of new computer technologies that will make the 21st century the Century of Bayes. [...] Bayes Nets and Influence Diagrams are not just important design tools; they also represent a major enhancement of the understanding about how important intellectual tasks typically performed by people should and can be performed.

– Ward Edwards

Hailfinder: tools for and experiences with Bayesian normative modeling.
American Psychologist 53 (1998) 416–428.
Reprinted in: J.W. Weiss and D.J. Weiss (Eds),
A science of decision making–the legacy of Ward Edwards.
Oxford University Press, Oxford (2009) 134–150 at 134.

Contents

Foreword

I became a forensic scientist in 1966, training as a document examiner at the Home Office Forensic Science Laboratory, Cardiff. I recall that, after I had been in the job for 2 or 3 years, one of my colleagues travelled up to London for a Friday evening meeting of the British Academy of Forensic Science: a visit to London was something of an adventure in those days and attendance at such a meeting was regarded as a perk to be savoured. On the following Monday, I asked my colleague how the meeting had gone and he replied that it was not bad as meetings went, but it would not have interested me very much. However, he did give me a photocopy that had been used as a handout at the meeting. My colleague was right in the sense that I had not paid much attention to the theory of evidence interpretation up to that point, but he did me an enormous favour because the handout was a prepublication version of a paper by two Americans called Finkelstein and Fairley[1] (1970). This was my first introduction to Bayesian reasoning, and I must confess that it took me several years to come to terms with it.

Later, having qualified as a statistician, I devoted my efforts largely to the interpretation of glass refractive index measurements – addressing what we would now call source-level propositions. This work was entirely frequentist, but again I was fortunate when I learned of a new paper by Dennis Lindley (1975). This, of course, was also Bayesian and what made it particularly exciting for me was that it addressed what we would now call offence-level propositions. In time, I took Dennis' analysis and modified it to address what we would now call activity-level propositions, though my first attempt to publish the results gained scathing criticism from a reviewer and rejection by the journal! The analysis took account of the phenomena of transfer and persistence and also the background distribution of extraneous material on clothing. At a conference in Sacramento in 1982 and later in Vancouver in 1987, I heard David Stoney's talk about the notion of relevance, and I realized that here was a further factor to be built in to our interpretation. My work in this area culminated in a paper that was published in the *Journal of the Royal Statistical Society* (Evett 1984). Although I was comfortable with the theory, I realized that it was far too mathematical for practical use as it stood so it was a great delight for me when, while I was visiting at Victoria University, Wellington, Tony Vignaux and Bernard Robertson introduced me to Bayes' nets (Tony and Bernard published a paper on Bayes' nets back in 1993). The issues that I had been trying to address by means of complicated and relatively inaccessible algebra could be captured simply by diagrams such as

---

[1] Many years later, I met both Michael Finkelstein and Will Fairley at a conference in Arizona – it was a great pleasure to inform them that they were two of my heroes.

those in Figure 6.4 that follows. This was, for me, greatly exciting and the prospects seemed almost limitless.

During the 1990s, much of my work was with DNA statistics, but I was enormously fortunate to be invited to be part of a project called 'Case Assessment and Interpretation' (CAI). This was devoted to pursuing the notion of delivering forensic science in a manner that best fulfilled the needs of the criminal justice system (CJS). So much came out of this project that clarified our view of the forensic scientist, and I recall still today the meeting at FSS Headquarters in Birmingham at which my colleague and friend, Graham Jackson, first used the phrase 'hierarchy of propositions'. Using the notion of the hierarchy, we ranged across forensic science disciplines looking at issues such as the combination of evidence types, the two-trace problem, and transfer from scene to offender. On reflection, we did not employ Bayes' nets nearly as much as we could have done in the early days. However, we were later joined by a specialist in that field, Roberto Puch-Solis, and he was able to develop software for implementing several of the CAI initiatives within a Bayes' nets framework.

The work of the authors of this book shows with clarity how Bayes' nets clearly establish inferential structure, for example: Figures 3.1 to 3.4 for the two-trace problem; Figure 5.8 for cross-transfer of evidence; and Figure 6.10 for raising inference up from source-level to crime-level propositions. One of the important new developments from the CAI project was the notion of carrying out a pre-assessment of a case to establish the strategy that has the best expectation of meeting the needs of the CJS. Chapter 10 demonstrates how a Bayes' nets approach clarifies thinking and provides structure to this process: Figure 10.1 shows how the essential structure may be graphically displayed.

It might be an exaggeration to claim that DNA profiling has led to a revolution in forensic science. Whether or not that is the case, it is inarguable that there have been many new inferential challenges over the last 20 years. Here too, it is possible to become submerged in complex mathematics, but again, Bayes' nets enable the core issues to be exposed diagrammatically. For example, dealing with what is sometimes called the 'brother's defence' is something that I handled algebraically back in the 1990s, but here the authors present a more transparent analysis by means of the Bayes' net in Figure 7.7. A major area for debate, research and development is the interpretation of DNA mixtures, again providing challenges that are peculiar to the field. Much elegant work has been done by a number of different groups, and an introduction is given in Section 7.8, with Figure 7.14 showing how a Bayes' net captures the issues of a two-person mixture. Also because of my work on DNA interpretation, I found myself drawn into the fascinating world of inference of kinship ('paternity testing' being just one aspect of this complex pursuit). It is amazing how complex a kinship analysis can become in, for example, a case of identification of human remains where comparison samples come from a range of different possible relatives of the deceased. Later, in 2002, I was fortunate to be present at the Fifth International Conference on Forensic Inference and Statistics in Venice when Robert Cowell demonstrated the power of Bayes' nets for kinship analyses: Figure 7.15 shows how a simple paternity analysis may be represented in this way.

And what of the future? A naïve view of forensic inference that held some sway in the late twentieth century was that statistical methods, allied with the power of modern computers, would replace human judgement in due time. However, we have gradually come to see that the knowledge, judgement and understanding of the forensic practitioner will always be at the core of evidence interpretation. The notion of the 'real LR' is an illusion – any calculation invokes a model and the validity and parameters of that model rest on the knowledge of the practitioner who provides the court with an LR that reflects his/her personal judgement.

However, no one would wish to see human judgement to be unconstrained, and it is essential that we have systems in place to calibrate personal probabilities and LRs. There is much good work currently being carried out in the field of calibration of forensic opinions and, it seems to me, that work naturally complements the structured approach to logical inference that flows from the application of Bayes' nets. Bayesian reasoning provides the logical framework for bringing together all of the personal probabilities necessary for a given inferential problem: it is the calibration process that considers the robustness of those probabilities and the Bayes' net that enables the exploration of the sensitivity of the output, given the calibration data. Pure synergy: this, for me, is the future of forensic science.

Finally, I wish to compliment the authors on the painstaking scholarly work that has led to the creation of this book, which should, I believe, have a place on the bookshelf of every forensic scientist who cares about the science of evidence interpretation.

Ian Evett,
January 2014

Preface to the second edition

Suppose that you are a forensic scientist, facing a large quantity of information coming from various observations, data or, more generally, findings related to a case under investigation. Your task is to help express a probabilistic conclusion on the joint value of such a quantity items of information or to assist a court of justice in expressing a belief on a judicial question of interest, typically expressed in terms of a proposition, compared to a particular alternative. How should you proceed? Ten years ago, Professor Dennis Lindley wrote in his foreword for another book of two of us (Aitken and Taroni 2004, p. 24):

> A problem that arises in a courtroom, affecting both lawyers, witnesses and jurors, is that several pieces of evidence have to be put together before a reasoned judgement can be reached: as when motive has to be considered along with material evidence. Probability is designed to effect such combinations but the accumulation of simple rules can produce complicated procedures. Methods of handling sets of evidence have been developed: for example Bayes nets (...). There is a fascinating interplay here between the lawyer and the scientist where they can learn from each other and develop tools that significantly assist in the production of a better judicial system.

Indeed, during the past three decades, the so-called Bayesian networks have gradually become a centre of attention for researchers from several academic fields. Whenever complicated inference problems involving uncertainty as a characterizing feature need to be captured and approached in a coherent way, that is using the normative framework of probability, their clarity of formulation and thorough computational architecture can provide a level of assistance that in many fields is unprecedented, in particular when there is a need to associate a reasoning process with a wider context of decision analysis and decision making.

As pointed out in forensic science and judicial literature, the merit of the Bayesian network graphical probability environment goes well beyond a purely descriptive account that focusses on the translation of a reasoner's view of a particular inference problem of interest. On the one hand, Bayesian networks support the concise description of challenging practical problems and the communication of their essential features so as to favour their understanding amongst discussants. On the other hand, Bayesian networks extend to a dynamic dimension that provides a means for belief computations; that is, the revision of a reasoner's belief structure as a result of knowing the truth or otherwise of one or more propositions that are part of the description of the overall problem. One of the very strengths of Bayesian networks is that their users can concentrate their efforts on eliciting sensible network structures and probability assignments, while leaving the computational burden to computerized implementations of Bayesian network models. However, there is no claim here of a 'true' model: indeed, different analysts may come up with different models for the same problem. Definitions of basic entities, the specification of their relationships and probability assignments may naturally differ because different analysts may hold different background information and may have different views of how a particular problem ought to be understood. However, this is not a drawback of the Bayesian network modelling language; it is one of its very strengths to make such differences explicit and provide a transparent framework for exploring the nature and extent of these differences. With respect to the theory of Bayesian networks, this amounts to applied research, but with respect to forensic science, such research is fundamental because it can provide original and innovative insights.

Inference and decision analysis, supported by Bayesian networks, should help us to acquire a better understanding of the problem we face, in terms of the target propositions of interest, our uncertainties about their true state and the way in which new items of information ought to affect our view. This better understanding can help to place scientists in a more secure position when they are required to advise other participants in the legal process on issues concerning the evaluation of forensic results. Typical questions include, but are not limited to: What is the bearing this finding has on this proposition, as compared to a given alternative proposition? If so, to what extent can we affirm degree of support? Should we attempt to acquire further information? If so, which other information?

One point that is clear from these introductory thoughts is that there are no pre-defined solutions. Bayesian networks are an abstract concept, and besides some aspects of definition that prescribe particular modelling constraints, there is nothing in the concept as such to tell us how to define sensible Bayesian network structures. This places forensic scientists in a responsible position: they need to make up their minds seriously and invoke further argument to justify particular model structures and their relevance for particular contexts of application. Like probability, Bayesian networks are both a very strict and a very liberal concept. They require the analyst to observe a few general principles of probabilistic reasoning, but beyond this, there are no prescriptions of as to how the basic terms ought to be interpreted. This highlights the personal nature of the approach, for which Bayesian analysis in general is so well known.

At the same time, this paradigm leads directly to one of the main motivations for a book on Bayesian networks for forensic science. It is driven by the question of how forensic scientists may use Bayesian networks meaningfully in their work. The idea thus is to offer the reader a guided introduction to the use of Bayesian networks for analysing forensic inference problems that arise in connection with various types of traces. To convince the reader that Bayesian networks can be specified in a defensible way, it is useful to point out that they can capture and illustrate the rationale behind particular probabilistic solutions, notably likelihood ratio formulae described in existing literature, which are now generally accepted as a measure of probative value. This is illustrated through various examples given throughout this book with reference to the original literature. The aim is to clarify the logic of generic structures of inferential networks that readers may transfer to their own contexts of application. Often, original literature provides numerical examples based on scenarios inspired by real cases that will allow the reader to track particular numerical output (Evett et al. 1998b).

Descriptions and analyses of entire real cases demand a substantial amount of additional discussion and explanation, in particular with respect to numerical specification. This would have clearly exceeded the space available in this book. The subtlety of real case analysis is illustrated, for example, by a whole book by Kadane and Schum (1996) devoted to the Sacco and Vanzetti case and papers covering selected case studies by Biedermann et al. (2011b); Evett et al. (2002). For a book with chapters focusing on selected practical applications from different fields –not necessarily forensic – see, for example, Pourret et al. (2008).

The second edition of this book on Bayesian networks features a series of changes. The theoretical introduction offered by Chapters 1 and 2 on probability and inference has been extended with material related to decision theory and its application. The reason for this is that scientists, but most importantly Courts of Justice, must reach decisions on the basis of particular items of information. In this context, Bayesian decision networks allow one to describe a general framework for logical decision analysis (and, hence, decision making) and how graphical models can support coherent decision making. In addition, aspects of terminology related to object-oriented Bayesian networks, a concept to support advanced graphical modelling, have been added.

Chapters 3–6 lay out the logic of forensic evaluation given the established levels of propositions known as *source*, *activity* and *crime (or offense)*, respectively. Each level has its own particular features, although there are connections between them, and Bayesian networks are an excellent means by which these can be made explicit. The discussion with respect to the various levels of propositions is kept separate in order to ease the understanding. This structure allows the reader to see the impact of an increased number of variables and their effect on inferential tasks. A note on the use of standard statistical distributions to define node tables, as offered by some Bayesian network software, is also included.

Evaluation of DNA profiling results is presented in Chapter 7, with new material on database searching and ways to account for the probability of (laboratory) error. Chapter 8 relates to the challenging topic of the joint value of multiple items of evidence. It covers material on the foundational aspects of such assessment given by Professor David Schum's pioneering works [Schum (1994)]. In turn, Chapter 9 deals with the use of continuous variables for Bayesian network construction, including both continuous and mixed networks with examples of applications.

Chapter 10 on 'Pre-assessment' introduces new sections on consignment inspection (i.e. sampling) and gunshot residues particles, followed by an entirely new Chapter 11, focusing on how Bayesian networks can be logically extended to incorporate questions of decision making. Bayesian decision networks, also sometimes called *influence diagrams*, contain nodes with lists of possible actions and nodes for utilities or losses to account for the valuation of consequences of decisions. This discussion relies on theory presented in Chapters 1 and 2.

Chapter 12 on object-oriented Bayesian networks is also an entirely new addition. Like Chapter 11, it relies on theory introduced at the beginning of the book (Chapter 2). The utility of object-oriented networks can be particularly well illustrated with examples focussing on DNA profiling results, because their evaluation often involves analogous and repetitive inferential patterns. It is for this reason that Chapter 12 will invoke network fragments introduced in Chapter 7 and show how they can be reused efficiently using the notion of class networks. This introduces a hierarchical construction process, utilizing more local and modular network fragments as building blocks.

Chapter 13 relates to qualitative and sensitivity analyses and can be concisely summarized by words written by Professor Schum in his Foreword of the first edition of the book (Taroni et al. 2006a, p. 18): 'Those of us who have employed Bayesian network analyses for various purposes come to recognize their virtues in allowing us to tell different stories about some complex process depending on what probabilistic ingredients we will include in them.' The aim in this chapter thus is to investigate the insight in particular inferential questions based on the way in which we may make various component assessments. This should help us recognize those parts of a model that have a critical impact on a given line of reasoning and that may thus require particular attention. Such insight has the potential to help with the allocation of resources for model construction and analysis in an informed way.

Bayesian networks, including their extension to decision networks, represent a wide area of research, including additional aspects that have not been included in this book, partly because they are technicalities that are not immediately necessary for a basic understanding of Bayesian networks. Examples of such technicalities are probabilistic and structural learning. These aspects focus on the automated, or partially automated, construction of Bayesian networks based on large datasets, which is a point of interest in domains such as artificial intelligence (Korb and Nicholson 2011; Pourret et al. 2008). This idea is not pursued in this book because, in the main, evaluative and decisional questions in forensic science refer to single cases with a distinct set of circumstances that cannot realistically be conceptualized as an instance of a large collection of cases. Indeed, the perspective adopted in this book is that of considering the scientist as the person in charge of specifying both the network structure and the numerical assignments for probabilities and utilities or losses. This is a subjective approach which may be informed or assisted by data, but not reduced to this exclusively. Notable exceptions are contexts where large datasets may be available and used for investigative purposes, such as inference about general donor features (e.g. handedness and gender) based on general ridge skin features (e.g. Champod 2010). Another example is inference about general offender characteristics based on descriptors of the crime and/or the victim (Aitken et al. 1996a,b), as described in the first edition of our book.

The material presented in this book is the fruit of a longstanding collaboration between the authors, which was supported, at various instances, by the *Swiss National Science Foundation* (FT, AB, SB, PG, CGGA), the *Fondation du 450e anniversaire de l'Université de Lausanne* (AB, PG), the *Société Académique Vaudoise* (AB), and by the *Italian Ministry of Education, University and Research* (PG, Grant no. PRIN 2011RP5RNM). We also express our appreciation to the *Université de Lausanne*, the *Università Ca' Foscari di Venezia*, the *Università IUAV di Venezia* and the *University of Edinburgh* for support during the writing of the book.

Franco Taroni, Alex Biedermann, Silvia Bozza, Paolo Garbolino and Colin Aitken
Lausanne, Venezia and Edinburgh

Preface to the first edition

Forensic scientists may have hardly ever been able to gather and offer as much information, analytical or otherwise, as is possible today. Owing to advances made in science and technology, today's forensic scientists can choose amongst a broad scope of methods and techniques, applicable to various kinds of evidence even in their remotest quantities. However, despite the potentially abundant amount of available information, there now is an increased awareness amongst a significant part of the forensic community – including legal scholars – that there are risks associated with these tendencies. A sense of some sort of overconfidence, for instance.

Scientific evidence as encountered in the real world is always incomplete to some degree, thus uncertainty is a prevalent element with which forensic scientists have to deal. Evidence does not tell anything as for itself; its significance needs to be elucidated in the light of competing propositions and given background knowledge about the case at hand. There is a great practical necessity for forensic scientists of advising their clients, be they lawyers, prosecutors, jurors or decision makers at large, of the significance of their findings. Forensic scientists are required to qualify and, where possible, quantify their states of knowledge and to be consultants in assessing uncertainties associated with the inferences that may be drawn from forensic evidence.

For this task, forensic scientists should consider probability theory as the fundamental concept to govern their reasoning. The aim of this book will be to show that the practical application of probabilistic reasoning in forensic science can be assisted and its rationale substantially clarified if performed in a graphical environment, in particular, using a formalism known as *Bayesian networks*.

Thus, the idea for a book on *Bayesian networks and probabilistic inference in forensic science* is guided by a series of questions currently asked by forensic scientists and other participants in the criminal justice system. The aim is to offer theoretical and practical elements to help solve the following questions.

- What are the relationships amongst a set of (usually unobservable) causes and a set of (observable) scientific evidence?

- What are the structural relationships amongst arguments based on different kinds of evidence in an inference to one or more propositions of interest?

- How can we construct coherent, credible and defensible arguments in reasoning about the evidential value of scientific evidence?

- Given that the known set of evidence is not sufficient to 'diagnose' the cause(s) of their origin with certain degrees of certainty, that is inductive inferences being risky, what additional information should be obtained?

- What is the value of each of these additional pieces of information?

- Can we build expert systems to guide forensic scientists and other actors of the criminal justice system in their decision making about forensic evidence?

- How can one collect, organize, store, update and retrieve forensic information (hard data or linked with expert judgement) in expert systems?

The current state of the art in forensic science, notably in scientific evidence evaluation, does not allow scientists to cope adequately with the problems caused by the complexity of evidence (typically the combination of evidence, especially if contradictory) even if such complexity occurs routinely in practice.

Methods of formal reasoning have been proposed to assist the forensic scientist to understand all of the dependencies which may exist between different aspects of evidence and to deal with the formal analysis of decision making. Notably, graphical methods, such as Bayesian networks, have been found to provide valuable assistance for the representation of the relationships amongst characteristics of interest in situations of uncertainty, unpredictability or imprecision. Recently, several researchers (mainly statisticians) have begun to converge on a common set of issues surrounding the representation of problems which are structured with Bayesian networks. Bayesian networks are a widely applicable formalism for a concise representation of uncertain relationships amongst parameters in a domain (in this case, forensic science). The task of developing and specifying relevant equations can be made invisible to the user, and the arithmetic can be almost completely automated. Most importantly, the intellectually difficult task of organizing and arraying complex sets of evidence to exhibit their dependencies and independencies can be made visual and intuitive. Bayesian networks are a method for discovering valid, novel and potentially useful patterns in data where uncertainty is handled in a mathematically rigorous, but simple and logical, way. A network can be taken as a concise graphical representation of an evolution of all possible stories related to a scenario. However, the majority of scientists are not familiar with such new tools for handling uncertainty and complexity.

Thus, attention is concentrated here on Bayesian networks essentially because they are relatively easy to develop and – from a practical point of view – they also allow their user to deduce the related formulae (expressed through likelihood ratios) for the assessment of scientific evidence. Examples of the use of Bayesian networks in forensic science have already been presented in several papers. In summary, the use of Bayesian networks has some key advantages that could be described as follows:

- the ability to structure inferential processes, permitting the consideration of problems in a logical and sequential manner;

- the requirement to evaluate all possible narratives;

- the possibility to calculate the effect of knowing the truth of one proposition or piece of evidence on the plausibility of others;

- the communication of the processes involved in the inferential problems to others in a succinct manner, illustrating the assumptions made;

- the ability to focus the discussion on probability and underlying assumptions.

A complete mastery of these aspects is fundamental to the work of modern forensic scientists.

The level of the book is the same as that of Aitken and Taroni (2004), namely those with a modest mathematical background. Undergraduate lawyers and aspiring forensic scientists attending a course in evidence evaluation should be able to cope with it though it would be better appreciated by professionals in law or forensic science with some experience of evaluating evidence.

The aim of the authors is to present a well-balanced book which introduces new knowledge and challenges for all individuals interested in the evaluation and interpretation of evidence and, more generally, the fundamental principles of the logic of scientific reasoning. These principles will be set forth in Chapter 1. Chapter 2 will show how they can be operated within a graphical environment – through Bayesian networks – with the reward of being applicable to problems of increased complexity. The discussion of the logic of uncertainty is then continued in the particular context of forensic science (Chapter 3) with studies of Bayesian networks for dealing with general issues affecting the evaluation of scientific evidence (Chapter 4). Later chapters will focus on more specific kinds of forensic evidence, such as DNA (Chapter 5) and transfer evidence (Chapter 6). Bayesian network models studied so far will then be used for the analysis of aspects associated with the joint evaluation of scientific evidence (Chapter 7). In Chapter 8, the discussion will focus on case pre-assessment, where the role of the forensic scientists consists of assessing the value of expected results *prior* to laboratory examination. Here, Bayesian networks will be constructed to evaluate the probability of possible outcomes in various cases together with their respective weight. Chapter 9 will emphasize the importance of the structural dependencies amongst the basic constituents of an argument. It will be shown that qualitative judgements may suffice to agree with the rules of probability calculus and that reasonable ideas can be gained about a model's properties through sensitivity analyses. This book concludes with a discussion of the use of continuous variables (Chapter 10) and further applications including offender profiling and Bayesian decision analysis (Chapter 11).

An important message of this book is that the Bayesian network formalism should primarily be considered an aid to structure and guide one's inferences under uncertainty, rather than a way to reach 'precise numerical assessments'. Moreover, none of the proposed models is claimed to be, in some sense, 'right'; a network is a direct translation of one's subjective viewpoint towards an inference problem, which may be structured differently according to one's extent of background information and knowledge about domain properties. It is here that a valuable property of Bayesian networks comes into play: they are flexible enough to accommodate readily structural changes whenever this is felt to be necessary.

The authors believe that their differing backgrounds (forensic science, statistics and philosophy of science) have common features and interactions amongst them enable the production of results that none of the disciplines could produce separately. A book on this topic aims to offer insight not only for forensic scientists but also for all persons faced with uncertainty in data analysis.

We are very grateful to Glenn Shafer for his permission to use his own words as the heading of Section 1.1.1, to Michael Dennis for help with Chapter 10 and to Silvia Bozza for commenting on Section 11.2. We thank Hugin Expert A/S who provided the authors with discounted copies of the software to enable them to develop the networks described throughout the book. Other software used throughout the book are R, a statistical package freely available at www.r-project.org, and XFIG, a drawing freeware running under the X Window System and available at www.xfig.org.

One of us, Paolo Garbolino, has been supported by research grant COFIN 2003 2003114293, 'Probabilistic judgments and decision processes' of the Italian Ministry of Education, University and Research.

We would like to express our appreciation to the Universities of Lausanne and Edinburgh and the IUAV University of Venice for support during the writing of the book.

Franco Taroni, C.G.G. Aitken, Paolo Garbolino, Alex Biedermann
Lausanne, Edinburgh, Venezia, Lausanne

1

The logic of decision

1.1 Uncertainty and probability

1.1.1 Probability is not about numbers, it is about coherent reasoning under uncertainty

The U.S. Federal Rule of Evidence 401 says that

> 'Relevant evidence' means evidence having any tendency to make the existence of any fact that is of consequence to the determination of the action more probable or less probable than it would be without the evidence. (Mueller and Kirkpatrick 1988, p. 33)

The term *probable* here means the degree of belief the fact finder entertains that a certain fact occurred. If it is not known whether the fact occurred, only a degree of belief less than certainty may be assigned to the occurrence of the fact, and there can then be discussion about the strength of this degree. Sometimes, we are satisfied with speaking loosely of 'strong' or 'weak' beliefs; sometimes, we would prefer to be more precise because we are dealing with important matters. A way to be more precise is to assign numerical values to our degrees of belief and use well-defined rules for combining them together.

People are usually not very willing to assign numbers to beliefs, especially if they are not actuaries or professional gamblers. In this book, we shall ask our readers to assign numbers, but these numbers are not important by themselves: what really matters is the fact that numbers allow us to use powerful rules of reasoning which can be implemented by computer programmes. It is not really important that the numbers be 'precise', whatever the meaning of 'precision' may be in reference to subjective degrees of belief based on personal knowledge. What is really important is that we are able to use sound rules of reasoning to check the logical

Bayesian Networks for Probabilistic Inference and Decision Analysis in Forensic Science, Second Edition.
Franco Taroni, Alex Biedermann, Silvia Bozza, Paolo Garbolino and Colin Aitken.
© 2014 John Wiley & Sons, Ltd. Published 2014 by John Wiley & Sons, Ltd.

consequences of our propositions, and that we are able to answer questions like[1]: 'What are the logical consequences with respect to the degree of belief in A of assuming that the degree of belief in B is high, let us say x, or between x and y?', 'How the degree of belief in A does change, if I lower the degree of belief in B by, let us say, z?', 'What is the best decision I can make if my degree of belief in A is x?', 'It is worthwhile to conduct this test T, before I take a decision D?'

If you are willing to take seriously the task of making up your mind to quantify, as best as you can, your degrees of belief, then the reward will be the possibility of using the laws of probability to answer questions like those formulated above. As a distinguished scholar of the logic of uncertainty, Glenn Shafer once said: 'Probability is not really about numbers; it is about the structure of reasoning'.[2] The mathematical theory of probability can be interpreted as a logical calculus for *doing inferences* and *making decisions* under uncertainty. It provides not only inference rules but also decision rules: as the laws of deductive logic can be used to define formal notions of coherence for beliefs entertained with certainty and provide constraints to deductive reasoning by means of rules of inference, so the laws of probability can be used as *a standard of coherence for beliefs and actions*. If your degrees of belief are not coherent according to the probabilistic standard, and you wish to use them as a guide for action, you will be acting in a way that will bring you consequences that are worse than they might have been if your degrees of belief had been coherent. If your *preferences* amongst actions with uncertain consequences are coherent according to the probabilistic standard, then the values (utilities) these consequences have for you can be measured by numerical values in the range zero to one. Section 1.4, it is shown that these 'utilities' obey, like degrees of beliefs, the laws of probability calculus.

As noted by Lindley (1990, at p. 55)

> [t]he Bayesian paradigm is a complete recipe for appreciation of the world by You, and for Your action within it. The central concept is probability as the sole measure of uncertainty and as a means of expressing Your preferences through (expected) utility. Properly appreciated, both measures are adequate for inference and decision making.[3]

Throughout this chapter, readers will learn what the 'Bayesian paradigm' is and why it is 'adequate' for making inferences and decisions.

1.1.2   The first two laws of probability

In order to be able to 'measure' your degree of belief that a certain fact occurred, it is necessary to be precise about what 'a degree of belief' is. In this book, it will be defined as a *personal degree of belief* that a proposition of a natural language, describing that fact, is true. Information and data, more generally also termed *evidence* throughout the first two chapters of this book, bearing on that proposition are expressed by means of other propositions. Therefore, it shall be said, on first approximation, that a proposition B is *relevant*, according to an

---

[1] Some textbooks on probability theory address the readers by using the second person. This is a rhetorical artifice to help them to keep in mind that probabilities are their own degrees of belief based on information they have. We shall follow this usage throughout this chapter but only in this chapter.

[2] This is a hearsay. Glenn Shafer told one of the authors that he heard it by Judea Pearl.

[3] Note that 'You' is capitalized in this quote to emphasize the point made in footnote 1.

opinion, for another proposition A if, and only if, knowing the truth (or the falsity) of B would change the degree of belief in the truth of A.

Having defined what is to be measured, it is next necessary to choose a function that assigns numbers to propositions. There are several possibilities available and, according to the choice, there are different rules for combining degrees of belief. In this book, a *probability function* is chosen, denoted by the symbol $Pr()$, where the () contains the event or proposition, the probability of which is of interest. Numerical degrees of belief must satisfy, for any propositions A and B, the laws of the mathematical theory of probability.

The first two laws can be formulated as follows. Degrees of belief are real numbers between zero and one: $0 \leq Pr(A) \leq 1$. This is the first law. The second law states that if A and B are mutually exclusive propositions, that is, they cannot be both true at the same time, then the degree of belief that one of them is true is given by the sum of their degrees of belief, taken separately: $P(A \text{ or } B) = Pr(A) + Pr(B)$. This is also sometimes called the *addition law*. The addition law can be extended to any number n of exclusive propositions. If no two of $A_1, A_2, \dots, A_n$ can be both true at the same time, then

$$Pr(A_1 \text{ or } A_2 \text{ or } \dots A_n) = Pr(A_1) + Pr(A_2) + \dots + Pr(A_n).$$

Satisfying the first law means that if it is known that proposition A is true, then the degree of belief should take the maximum numerical value, that is, $Pr(A) = 1$. The degrees of belief in propositions not known to be true, for example, the proposition A, 'this coin lands heads after it is tossed', are somewhere between the certainty that the proposition is true and the certainty that it is false. A straightforward consequence of the probability laws is that when the degree of belief for heads is fixed, then it is necessary to assign to the proposition B, 'this coin lands tails after it is tossed', the degree of belief $Pr(B) = 1 - Pr(A)$, assuming that pathological results such as the coin balancing on its edge do not occur.

This is the simplest example of how probability calculus works as a *logic for reasoning under uncertainty*. The logic places constraints on the ways in which numerical degrees of belief may be combined. Notice that the laws of probability require the degrees of belief in A and B to be such that they are non-negative and their sum is equal to one. Within these constraints, there is not an obligation for A to take any particular value. Any value between the minimum (0) and the maximum (1) is allowed by the laws.

This result holds, in general, for any two propositions A and B that are said to be mutually exclusive and exhaustive: one and only one of them can be true at any one time and together they include all possible outcomes. Thus, given that a proposition and its logical negation are mutually exclusive and exhaustive, the degree of belief in the logical negation of any proposition A is $Pr(\text{not} - A) = Pr(\bar{A}) = 1 - Pr(A)$, where $\bar{A}$ is read as 'A-bar'. The logical negation of an event or proposition is also known as its *complement*.

1.1.3   Relevance and independence

A proposition B is said to be *relevant* for another proposition A if and only if the answer to the following question is positive: if it is supposed that B is true, does that supposition change the degree of belief in the truth of A? A judgement of relevance is an exercise in hypothetical reasoning. There is a search for a certain kind of evidence[4] because it is known in advance

---

[4] Note that the term *evidence* is used here in an entirely general sense of a given item of information or data point. It can, but need not, refer to the idea of 'evidence' as admitted before court.

that it is relevant. If someone submits particular findings maintaining that they constitute relevant evidence, a hypothetical judgement has to be made as to whether or not to accept the claim. In doing that, a distinction has to be drawn, not only between the *hypothesis A* and *evidence B* for the hypothesis *A* but also between that particular evidence *B* and whatever else is known.

When evaluating one's degree of belief in a proposition of interest, there is always exploitation of available *background information*, even though such information is not explicit. An assessment of the degree of belief in the proposition 'this coin lands heads after it is tossed' is made on the basis of some background information that has been taken for granted: if the coin looks like a common coin from the mint, and there is no reason for doubting that, then it is usually assumed that it is well balanced. Should it be realized, after inspection, that the coin is not a fair coin, this additional information is 'evidence' that changes the degree of belief about that coin, even though it is still believed that coins from the mint are well balanced. A *relevant proposition* is taken to mean a proposition that is not included in the background information. The distinction between 'evidence' and 'background information' is important, because sometimes it has to be decided that certain propositions are to be considered as evidence, whereas others are to be considered as part of the background information.

For example, suppose that the results of DNA analyses are to be evaluated. Assume that all scientific theories that support the methodology of the analysis are true, that the analysis has been conducted correctly, and that the chain of custody has not been broken. These assumptions all form part of the background information. Relevant evidence is only those propositions that describe the results of the examinations, *plus* some other propositions reporting statistical data about the reliability of this type of scientific result. Alternatively, propositions concerning how the analyses have been conducted, and/or the chain of custody can also be taken to be part of the evidence whilst scientific theories are still left in the background. Therefore, it is useful to make a clear distinction between what is considered, in a particular context, to be 'evidence', and what is considered to be 'background'. For this reason, background information is introduced explicitly, from time to time, in the notation.

Let $Pr(A|I)$ denote 'the degree of belief that proposition A is true, given background information I', and let $Pr(A|B, I)$ denote 'the degree of belief that proposition A is true, given that proposition B is assumed to be true, *and* given background information I'. Then, B is relevant for A if and only if the degree of belief that A is true, given that B is assumed to be true, is different from (greater or smaller than) the degree of belief that A is true, given background information I only. Thus,

$$Pr(A|B, I) \neq Pr(A|I). \tag{1.1}$$

It is assumed that for the evaluation of what is referred to as *evidence* in forensic science applications, the distinction between the background knowledge, of the court or of the expert, and the findings submitted for judgement is clear from the context and the distinction between 'background information' and 'relevant information' shall not be dwelt on further.

One more remark is pertinent here, however. The definitions of *epistemic relevance* and, implicitly, *irrelevance* have been given in terms of *probabilistic dependence* and *probabilistic independence*, respectively. 'Probabilistic independence' is a subtle concept that must be handled with care, always making up one's mind about what counts, personally, as background information. For example, imagine that a coin is to be tossed twice: is the outcome of the first toss relevant for the belief in the outcome of the second toss? That is, does the degree of belief in the outcome of the second toss change, knowing the outcome of the first one? The answer is 'It depends'. If it is known that the coin is well balanced, and that it is a coin from the mint, then the answer is 'No'. But, if it is not known if the coin is well balanced, then the answer is 'Yes'. Surely, one toss is not enough to influence beliefs above a significant psychological threshold, but if a sequence of many tosses is considered, the difference between the two situations can be seen. If it is not known whether the coin is fair, a sequence of tosses is just an experiment to test the hypothesis that the coin is fair, and the outcomes of the tosses are not independent.

It is not necessary to go into the subtleties of the concept of probabilistic independence. The purpose here is only to emphasize the point that *probability is always conditional on knowledge*. It is obvious that personal beliefs depend on the particular knowledge one has. If the choice is made to represent degrees of belief by means of probabilities, then it must be kept in mind that it will always be the case that probabilities are relative to the knowledge available *and* the assumptions made. Statisticians say that probability is always relative to a *model*. The assumption that the coin is well balanced is a model. An alternative assumption that, possibly, the coin is not well balanced is another model or, better, is equivalent to the postulate of a cluster of models. For this example, a continuous probability distribution may be used to represent this set of models. Continuous probability distributions will be introduced in Chapter 9.

1.1.4   The third law of probability

Another constraint on one's degrees of belief is given by the so-called multiplication law of probability. It states that for any propositions A and B, the degree of belief that A and B are both true, given background information I, is equal to the degree of belief that A is true, given background information I, times the degree of belief that B is true, given that one assumes that A is true: $Pr(A, B|I) = Pr(A|I) \times Pr(B|A, I)$.

In what follows, $Pr(A, B|I)$ will denote the degree of belief that propositions A and B are both true. The mathematical law is commutative, so one's numerical degrees of belief must satisfy commutativity as well. One can calculate the degree of belief in A and B in two different ways, but the final result must be the same:

$$Pr(A, B|I) = Pr(A|I) \times Pr(B|A, I) = Pr(B|I) \times Pr(A|B, I). \tag{1.2}$$

It can be checked that probabilistic dependence is a symmetrical relationship. If B is relevant for A, then A is also relevant for B: if $Pr(A|I)$ is different from $Pr(A|B, I)$, then $Pr(B|I)$ must also be different from $Pr(B|A, I)$, for (1.2) has to be satisfied.

Consider again the example of coin tossing, with B and A denoting, respectively, the propositions 'the outcome of the first toss is heads' and 'the outcome of the second toss is heads'. If one knows that the coin is fair, then B is probabilistically independent from A: the degree of belief in the outcome of the second toss would not change, if one were able to know in advance the outcome of the first toss. Therefore, in this case, the multiplication law reads: $Pr(A, B|I) = Pr(B|I) \times Pr(A|I)$.

Just as for the addition law, the multiplication law can be extended to any number of independent propositions $A_1, A_2, \ldots, A_n$:

$$Pr(A_1, A_2, \ldots, A_n|I) = Pr(A_1|I) \times Pr(A_2|A_1, I)$$
$$\times \ldots \times Pr(A_n|A_1, A_2, \ldots, A_{n-1}, I).$$

Let $A_1, A_2, \ldots, A_n$ be the propositions 'the outcome of the first toss is heads', 'the outcome of the second toss is heads' and so on. If one knows that the coin is fair, then the outcomes of the tosses 1 to $(n - 1)$ are not relevant for the degree of belief about the outcome of the toss n. Hence, $Pr(A_1, A_2, \ldots, A_n|I) = Pr(A_1|I) \times Pr(A_2|I) \times \cdots \times Pr(A_n|I)$.

1.1.5   Extension of the conversation

A quite straightforward theorem of probability calculus, which follows immediately from the second law plus propositional calculus, puts another constraint on one's degrees of belief and is an all-important rule of the logic of reasoning under uncertainty. Dennis Lindley has underlined its role by dubbing it the *extension of the conversation rule* (Lindley 1985, p. 39). Thus, the theorem states that for any propositions A and B, the degree of belief that A is true, given background information I, is given by the sum of the degree of belief that A and B are both true, given background information I, and the degree of belief that A is true and B is false, given background information I: $Pr(A|I) = Pr(A, B|I) + Pr(A, \bar{B}|I)$.

Its importance comes from the fact that it is easier to assign a degree of belief to a proposition if it is thought about it in a context, extending the conversation to include other propositions judged to be relevant. Applying the multiplication law, the formula reads

$$Pr(A|I) = [Pr(A|B,I) \times Pr(B|I)] + [Pr(A|\bar{B},I) \times Pr(\bar{B}|I)]. \tag{1.3}$$

For example, let A be the proposition 'the outcome of the first toss is heads'. Assume that there are two, and only two hypotheses about the state of the coin: either it is fair (proposition B) or it has two heads (proposition $\bar{B}$). Then, one's overall degree of belief in A can easily be calculated via the formula. We can extend the conversation to any number of propositions, provided they are exclusive and exhaustive. In the example of coin tossing suppose, one knows that only three propositions B_1, B_2 and B_3 are possible, for the hypotheses that 'the coin is fair', 'the coin has two tails' 'the coin has two heads', respectively, then the overall degree of belief in A is given by

$$Pr(A|I) = [Pr(A|B_1,I) \times Pr(B_1|I)] + [Pr(A|B_2,I) \times Pr(B_2|I)]$$
$$+ [Pr(A|B_3,I) \times Pr(B_3|I)].$$

It is convenient to have a more compact notation in order to deal with more than two alternative propositions, and, following mathematical usage, the symbol $\sum_{i=1}^{n}$, which reads 'add the results for each value of i from 1 to n', shall be used. If $B_1, B_2, \ldots, B_n$ are mutually exclusive and exhaustive propositions, then, for any proposition A:

$$Pr(A|I) = \sum_{i=1}^{n} Pr(A|B_i,I) \times Pr(B_i|I).$$

Finally, notice that the multiplication law, under the form (1.3), implies that definition (1.1) of relevance is equivalent to saying that B is relevant for A if and only if

$$Pr(A|B,I) \neq Pr(A|\bar{B},I).$$

Indeed, suppose that B is not relevant for A, that is, $Pr(A|B,I) = Pr(A|\bar{B},I)$. Then, given that $Pr(\bar{B}|I) = 1 - Pr(B|I)$, this implies that the right-hand side of (1.3) is identically equal to the left-hand side, independent of the value of $Pr(B|I)$.

1.1.6   Bayes' theorem

So far, it has been said that a relevant proposition B, with respect to another proposition A, is a proposition such that the supposition if it were true would change our degree of belief in A.

The rule according to which the degree of belief in A must change has not yet been provided. This rule is given by an elementary manipulation of (1.2). Namely, for any propositions A and B, the degree of belief that A is true, given that one assumes that B is true, is equal to the degree of belief that A and B are both true, given background information I, divided by the degree of belief that B is true, given background information I, provided that $Pr(B|I) > 0$:

$$Pr(A|B, I) = \frac{Pr(A, B|I)}{Pr(B|I)}. \tag{1.4}$$

Equation (1.4) is called *Bayes' theorem* because it is the algebraic version of the formula first proved in the second half of the eighteenth century by the Reverend Thomas Bayes, a member of the English Presbyterian clergy and Fellow of the Royal Society (Bayes 1763). The importance of Bayes' theorem is due to the fact that it is a rule for *updating degrees of belief on receiving new evidence*.

The process of evidence acquisition may be modelled as a two-step process in time. At time t_0, it is planned to look for evidence B because B is believed to be relevant for hypothesis A. At time t_0, the change in degree of belief in A if it were to be discovered that B were true may be calculated by use of Equation (1.4). Denote the degrees of belief at time t_0 by $Pr_0()$. Then, by use of the multiplication rule and the extension of the conversation rule, Equation (1.4) may be rewritten as

$$Pr_0(A|B, I) = \frac{Pr_0(B|A, I)Pr_0(A|I)}{[Pr_0(B|A, I)Pr_0(A|I)] + [Pr_0(B|\bar{A}, I)Pr_0(\bar{A}|I)]}. \tag{1.5}$$

The probability $Pr_0(A|B, I)$ is also called the *probability* of A, *conditional on B* (at time t_0). $Pr_0(A|I)$ and $Pr_0(\bar{A}|I)$ are also called *prior*, or *initial*, probabilities. $Pr_0(B|A, I)$ is called the *likelihood* of A given B (at time t_0). Analogously, $Pr_0(B|\bar{A}, I)$ is the *likelihood* of $\bar{A}$ given B (at time t_0). This use of a likelihood is another exercise in hypothetical reasoning. Assessment of the likelihood requires consideration of the question: supposing that hypothesis A is true (or false), what is the degree of belief that B is true?

At time t_1, it is discovered that B is true. Denote the degree of belief at time t_1 by $Pr_1()$. What, then, is the degree of belief in the truth of A at time t_1, that is $Pr_1(A|I)$? A reasonable answer seems to be that, if it has been learned at time t_1 that B is true, *and nothing else*, then knowledge that B is true has became part of the background knowledge at time t_1. Therefore, the overall degree of belief in A at time t_1 is equal to the degree of belief in A, conditional on B, at time t_0:

$$Pr_1(A|I) = Pr_0(A|B, I). \tag{1.6}$$

The situation is the same as when there is information at time t_1 that a proposition B is true that had not been thought of before time t_1. Assessment of the relevance of B for A, and the effect of this relevance on the degrees of belief, requires thought about likelihoods and initial probabilities as if one were at time t_0: what would have been the degree of belief that B were true, given the background knowledge less the information that B is true, and supposing that hypothesis A is true, or false? What would have been the prior probability for A, given the background knowledge less the information that B is true?

1.1.7 Probability trees

It may be helpful in clarifying the general mechanism underlying probabilistic updating to consider (1.6) in another way. Consider a possible scenario that can be derived from the

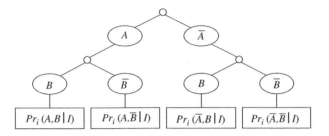

Figure 1.1 Probability tree for the uncertain propositions A and B at time t_i (for $i = \{0, 1\}$), given background information I.

combination of two logically compatible propositions A and B with a probability assignment Pr_0 at time t_0. This scenario can be represented graphically by means of a *probability tree* as shown in Figure 1.1.

A probability tree is a type of graphical model that consists of a series of branches stemming from nodes, usually called *random nodes*, which represent uncertain events. At every random node, there are as many branches starting there as the number of the possible outcomes of the uncertain event. In this context, outcomes of uncertain events are described by propositions and branches containing more than one node, which correspond to the logical conjunction of as many propositions as the number of nodes. Branches have associated with the probabilities of the corresponding conjunctions of propositions calculated via the multiplication rule, and the probability of each proposition is given by the sum of the probabilities of the branches containing it (extension of the conversation). The sum of the probabilities of all the branches must add to one (addition rule).

Then, at time t_1, proposition B is known to be true. The probability tree still has the structure shown in Figure 1.1, but for $i = 1$. As B is known to be true, only two scenarios are possible in the new state of knowledge, and thus two of the probability nodes in Figure 1.1 are zero, which will be verified later.

Again, the sum of the probabilities of the branches in Figure 1.1 must add to one. The original probabilities of the branches must be amended in such a way that their sum turns out to be one. This can be done by multiplying them by the constant factor $1/Pr_0(B|I)$; hence,

$$Pr_1(A, B|I) = Pr_0(A, B|I) \times \frac{1}{Pr_0(B|I)} = \frac{Pr_0(A, B|I)}{Pr_0(B|I)};$$

$$Pr_1(\bar{A}, B|I) = Pr_0(\bar{A}, B|I) \times \frac{1}{Pr_0(B|I)} = \frac{Pr_0(\bar{A}, B|I)}{Pr_0(B|I)}. \tag{1.7}$$

Indeed, addition of the terms in (1.7) gives the desired result.

$$Pr_1(A, B|I) + Pr_1(\bar{A}, B|I) = \frac{Pr_0(A, B|I)}{Pr_0(B|I)} + \frac{Pr_0(\bar{A}, B|I)}{Pr_0(B|I)}$$

$$= \frac{Pr_0(B|I)}{Pr_0(B|I)} = 1.$$

The new probability of A is now identically equal to the probability of the branch containing both A and B. Since it is known that B is true, $Pr_1(A, \bar{B}|I) = 0$; hence, $Pr_1(A|I) = Pr_1(A, B|I)$.

Therefore, substitution of this result in (1.7) gives (1.6):

$$Pr_1(A|I) = \frac{Pr_0(A,B|I)}{Pr_0(B|I)}; \quad Pr_1(\bar{A}|I) = \frac{Pr_0(\bar{A},B|I)}{Pr_0(B|I)}.$$

Result (1.7) demonstrates that the use of Bayes' theorem for probability updating is equivalent to the redistribution of probabilities amongst possible scenarios in a symmetric way. Given that the state of information has changed on learning only that B is true, and nothing else, there is no reason to make a change biased for or against certain particular scenarios. This important point is mentioned again in Section 1.2.6. In what follows, time subscripts will be omitted unless strictly necessary.

1.1.8   Likelihood and probability

In this book, likelihoods will play an important role. In some less technical talks, the expressions 'it is likely' and 'it is probable' are used interchangeably so it is worthwhile to emphasize the difference between the *likelihood* of A, given B, and the *probability* of A, conditional on B.

As an illustration of the importance of the difference between likelihoods and probabilities, consider the following example. Mr. Jones has been seen by an eyewitness running away from the house where a crime has been committed, at approximately the time of the crime. Let B be the proposition 'Mr. Jones was running away from the scene of the crime at the time when it was committed' and A be the proposition 'Mr. Jones committed the crime'. It is reasonable to believe that the likelihood $Pr(B|A,I)$ is high, but not necessarily $P(A|B,I)$ is high as well. If Mr. Jones actually committed the crime, it is expected that he would want to hasten away from the scene of the crime. The hypothesis of culpability provides a good account of the evidence. However, the fact that he was running away from the house does not, by itself, make it very probable that he committed the crime. There are many other possible explanations of that fact.

The confusion of the likelihood of hypothesis A, given evidence B, with the probability of the same hypothesis, conditional on the same evidence, is known as the *fallacy of the transposed conditional* (Diaconis and Freedman 1981). The fallacy occurs when, from the fact that if A has occurred, then B occurs with a high probability, it is erroneously concluded that if B has occurred, then A occurs with high probability. The best antidote against the fallacy is to remember Bayes' theorem. Inspection of (1.5) shows that the two numbers are not necessarily close. In the second term of the sum in the denominator, the likelihood of $\bar{A}$, given B, could be high as well: there can exist another good explanation of evidence B even in the case that A is false. Moreover, the degree of belief of A, given the background knowledge only, has also to be taken properly into account.

The conversation can obviously be extended to n exclusive and exhaustive propositions $(A_i, \ i = 1, \ldots, n)$, which represent possible alternative explanations of B, and in the general case, for any A_i, Bayes' theorem reads

$$Pr(A_i|B,I) = \frac{Pr(B|A_i,I)Pr(A_i|I)}{\sum_{i=1}^{n} Pr(B|A_i,I)Pr(A_i|I)}. \tag{1.8}$$

1.1.9 The calculus of (probable) truths

A meaningful proposition can be true or false. Twentieth-century logicians have introduced the 'trick' of associating to any proposition, say A, a function f, called its *truth-function*, that assigns to A either the value $f(A) = 1$ if A is true or the value $f(A) = 0$ if A is false. In this way, a one-to-one correspondence can be established between the truth-function of a proposition A and a *random variable*, say X_A, whose value is uncertain. Therefore, $Pr(X_A = 1|I)$ may be written instead of $Pr(f(A) = 1|I)$ to denote the probability that A is true, given information I, that is $Pr(A|I) = Pr(X_A = 1|I)$. In order to simplify notation, explicit mention of the background information I will be omitted when using random variables.

The concept of a random variable is especially helpful in dealing with continuous quantities. Continuous random variables will be considered in Chapter 9. At present, only discrete quantities like truth-values that can take only two values, 0 and 1, are considered. In general, a *discrete random variable X* is an uncertain quantity that can take a discrete number of mutually exclusive and exhaustive values x with probabilities $Pr(X = x)$, such that

$$0 \le Pr(X = x) \le 1 \quad \text{and} \quad \sum_x Pr(X = x) = 1.$$

Note the use of the abbreviated notation $\sum_x$ that reads 'add the results for each value x of the variable X'. The *probability distribution* of a random variable X is the set of probabilities $Pr(X = x)$ for all possible values x of X. In the special case of the random variable associated to a proposition A, this set contains only two values, $\{Pr(X_A = 1), Pr(X_A = 0)\}$ and $Pr(X_A = 1) + Pr(X_A = 0) = 1$.

Let $\{X_1, X_2, \ldots, X_n\}$ be a set of random variables. The *joint distribution* of these variables is the set of probabilities $Pr((X_1 = x_1), (X_2 = x_2), \ldots, (X_n = x_n))$ for all possible values x_i of X_i ($i = 1, 2, \ldots, n$). For instance, for any pair of propositions A and B, the elements of the *joint probability distribution* of the random variables X_A and Y_B are the probability that A and B are both true, the probability that A is true and B is false and so on. That is, more formally, $Pr((X_A = 1), (Y_B = 1))$, $Pr((X_A = 1), (Y_B = 0))$, $Pr((X_A = 0), (Y_B = 1))$ and $Pr((X_A = 0), (Y_B = 0))$.

Through the multiplication law, the joint distribution of two random variables X and Y can be expressed as a product of two other distributions, one being the distribution of the single variable X and the other being the *conditional distribution* of the single variable Y, conditional on X:

$$Pr((X = x), (Y = y)) = Pr(X = x) \times Pr(Y = y|X = x)$$

$$= Pr(Y = y) \times Pr(X = x|Y = y).$$

In this book, conditional distributions for propositions are tabulated by means of conditional probability tables like Table 1.1 where, by abuse of language, the symbols denoting propositions are used instead of the corresponding random variables (t stands for 'true' and f for 'false').

In this context, the probability distribution of a variable X_A can be obtained as the sum over y of its joint distribution with Y_B:

$$Pr(X = x) = \sum_y Pr(X = x, Y = y).$$

This operation is called *marginalization*, and it can be seen that it is the same operation as 'extending the conversation'. The probability $Pr(X = x)$ is referred to as the *marginal*

Table 1.1 Illustrative conditional distribution of B given A.

A:		t	f
B:	$Pr(B = t\|A)$	0.4	0.1
	$Pr(B = f\|A)$	0.6	0.9

distribution of X and in Chapter 2 shorthand notation of the following type:

$$Pr(A) = \sum_B Pr(A, B) = Pr(A, B) + Pr(A, \bar{B})$$

$$Pr(A) = \sum_{BC} Pr(A, B, C)$$

$$= Pr(A, B, C) + Pr(A, B, \bar{C}) + Pr(A, \bar{B}, C) + Pr(A, \bar{B}, \bar{C})$$

$$Pr(A, B) = \sum_C Pr(A, B, C) = Pr(A, B, C) + Pr(A, B, \bar{C}),$$

shall be used for the marginalization of probabilities of propositions.

A way of representing a joint distribution is to draw the *table of marginals*. Assume that the distribution of variable A is $Pr(A) = (0.8, 0.2)$ and the conditional distribution of B, given A, is as given in Table 1.1. Then, the table of marginals for the joint distribution of A and B is Table 1.2, where the bottom row with the column totals provides the marginal distribution of A, and the rightmost column, with the row totals, provides the marginal distribution of B. For example

$$Pr(A = t, B = t) = Pr(A = t) \times Pr(B = t|A = t) = 0.8 \times 0.4 = 0.32.$$

Calculations of conditional probabilities via marginal probability tables are straightforward. In Table 1.3, the (rounded off) entries are calculated from Table 1.2. For example,

$$Pr(A = t|B = t) = \frac{Pr(A = t, B = t)}{Pr(B = t)} = \frac{0.32}{0.34} = 0.94.$$

All these definitions and operations generalize to three or more random variables. The joint distribution of several random variables may be decomposed via the multiplication law into a product of distributions for each of the variables individually.

Table 1.2 Illustrative joint and marginal distributions of A and B.

A:		t	f	Marginal
B:	t	0.32	0.02	0.34
	f	0.48	0.18	0.66
	Marginal	0.80	0.20	1.00

Table 1.3 Conditional distribution of
A, given B, derived from Table 1.2.

B:	t	f
A: $Pr(A = t\|B)$	0.94	0.73
$Pr(A = f\|B)$	0.06	0.27

1.2 Reasoning under uncertainty

1.2.1 *The Hound of the Baskervilles*

The following is a passage from the Devon County Chronicle of May 14th, 189 … :

The recent sudden death of Sir Charles Baskerville, whose name has been
mentioned as the probable Liberal candidate for Mid-Devon at the next election,
has cast a gloom over the county. (…) The circumstances connected with the
death of Sir Charles cannot be said to have been entirely cleared up by the
inquest, but at least enough has been done to dispose of those rumours to which
local superstition has given rise. There is no reason whatever to suspect foul play,
or to imagine that death could be from any but natural causes. (…) In spite of his
considerable wealth he was simple in his personal tastes, and his indoor servants
at Baskerville Hall consisted of a married couple named Barrymore, the husband
acting as butler and the wife as housekeeper. Their evidence, corroborated by that
of several friends, tends to show that Sir Charles's health has for some time been
impaired, and points especially to some affection of the heart (…) Dr. James
Mortimer, the friend and medical attendant of the deceased, has given evidence
to the same effect.

The facts of the case are simple. Sir Charles Baskerville was in the habit every
night before going to bed of walking down the famous yew alley of Baskerville
Hall. The evidence of the Barrymores shows that this had been his custom. On
the fourth of May (…) he went out as usual for his nocturnal walk (…) He never
returned. At twelve o'clock Barrymore, finding the hall door still open, became
alarmed, and, lighting a lantern, went in search of his master. The day had been
wet, and Sir Charles's footmarks were easily traced down the alley. Halfway down
this walk there is a gate which leads out on to the moor. There were indications
that Sir Charles had stood for some little time here. He then proceeded down the
alley, and it was at the far end of it that his body was discovered. One fact which
has not been explained is the statement of Barrymore that his master's footprints
altered their character from the time he passed the moor-gate, and that he appeared
from thence onward to have been walking upon his toes. (…) No signs of vio-
lence were to be discovered upon Sir Charles's person, and though the doctor's
evidence pointed to an almost incredible facial distortion (…) it was explained
that that is a symptom which is not unusual in cases of dyspnoea and death from
cardiac exhaustion. This explanation was borne out by the post-mortem examination, which showed long-standing organic disease, and coroner's jury returned a
verdict in accordance with the medical evidence.

Many readers will have recognized a passage from Conan Doyle's novel *The Hound of the Baskervilles* (Conan Doyle 1953, pp. 676–677). The newspaper report is read by Dr. Mortimer himself to Sherlock Holmes and his friend John Watson, whose counsel he is asking for because certain circumstances he knows make the case less simple than the public facts let suppose to be. So he relates to Holmes the strangest of those circumstances [*The Hound of the Baskervilles* (Conan Doyle 1953, at p. 679)]:

> I checked and corroborated all the facts which were mentioned at the inquest. I followed the footsteps down the yew alley, I saw the spot at the moor-gate where he seemed to have waited, I remarked the change in the shape of the prints after that point, I noted that there were no other footsteps save those of Barrymore on the soft gravel, and finally I carefully examined the body, which has not been touched until my arrival. (…) There was certainly no physical injury of any kind. But one false statement was made by Barrymore at the inquest. He said that there were no traces upon the ground round the body. He did not observe any. But I did - some little distance off, but fresh and clear.
>
> 'Footprints?'
>
> 'Footprints.'
>
> 'A man's or a woman's?'
>
> (…) 'Mr. Holmes, they were the footprints of a gigantic hound!'

The readers will remember that at the core of the plot there was the legend of the devilish hound which punished for his sins, a wicked ancestor of the Baskervilles, information which is part of Holmes' and Watson's background knowledge, given that they have been told by Dr. Mortimer. The beginning of this famous story is used as an example to show how the logic of probability works and how inferences licenced by probabilistic reasoning do agree with common-sense inferences, as far as the problem is simple enough to be tackled by intuitive reasoning. But Conan Doyle's story will also offer a good example of what was claimed before, namely, that probability is always relative to a model and that seemingly 'simple' problems might be less simple than they look like.

Readers' common sense, besides that of Holmes and Watson, will conclude, from the public and private facts told by Dr. Mortimer, that the official inquest verdict is highly credible but that there is some ground, although very thin, to arouse curiosity about the circumstances of Sir Charles Baskerville's death. Dr. Mortimer will relate some more facts which will make Holmes cry that 'it is evidently a case of extraordinary interest, and one which presented immense opportunities to the scientific expert' (Conan Doyle 1953, p. 680), but consideration here is limited to two pieces of evidence as mentioned above, namely, medical evidence and the footmarks.

The goal is now to show how well old Dr. John Watson, who is not naturally endowed with the genius and the prodigious insight of Sherlock Holmes but is learned about Bayes' theorem, can build up a probabilistic argument on the basis of these two pieces of evidence. First, reasoning is done step by step, taking into account the coroner's report, and then the footmarks. Afterwards, reasoning is done by taking into account all the evidence at once.

1.2.2   Combination of background information and evidence

Given Watson's background information *only*, that is, given what he knew before considering the coroner's report, three mutually exclusive, and exhaustive, hypotheses were possible:

Sir Charles Baskerville died by natural causes (H_1), Sir Charles Baskerville died by a criminal act (H_2) and Sir Charles Baskerville committed suicide (H_3).

It is assumed that a nineteenth-century positivist held to be impossible, given his background information, the hypothesis that Baskerville's death was caused by a ghost. It would have not been considered admissible by a Court in Sherlock Holmes' times, and in our times either. This only reflects the fact that background information is historically given: some centuries ago, such a fourth hypothesis might have been taken seriously into account by a Court. It is not the task of logic alone to decide what counts as background information.

Suppose Watson has to formulate his opinion knowing only the coroner's report. This report is such that Watson is warranted to believe that the following proposition is true: the proximate cause of Sir Charles Baskerville's death was heart attack (R).

What is the effect of evidence R on his hypotheses? Proposition R, of course, is evidence *for* the hypothesis of accidental death and *against* the other two hypotheses, according to common sense. We know from Section 1.1.6 that the question may be reformulated in probabilistic terms to read: what are the probabilities of the hypotheses, conditional on R?

Watson can make some reasonable judgements about prior probabilities. First of all, the three hypotheses cover all the logical possibilities, and hence the sum of their probabilities is one. Moreover, he could think that the probability of H_1 is greater than both the probabilities of H_2 and H_3 and also greater than their logical disjunction, given only background knowledge about occurrences of causes of death:

$$Pr(H_1|I) > Pr(H_2|I);$$
$$Pr(H_1|I) > Pr(H_3|I);$$
$$Pr(H_1|I) > Pr(H_2 \text{ or } H_3|I) = Pr(H_2|I) + Pr(H_3|I). \qquad (1.9)$$

Notice that Watson has expressed only comparative probability judgements so far, he does not need to assign precise numerical probabilities. As regards the likelihoods, he considers first the suicide hypothesis. It is *a priori* possible, but its likelihood, given that the proximate cause of death was a heart attack, can be safely assumed to be practically equal to zero.

He does not have reasons to exclude the crime hypothesis *a priori*. According to his background knowledge, a person, especially an aged person with bad health, can be scared to death. Surely, that would be quite a complicated way to kill a person, but this means that his prior probability for such an hypothesis will be very, very low but not equal to zero. Of course, it is quite impossible to give a particular assignment of the likelihood of the hypothesis that a criminal act has been committed, given that the proximate cause of death is a heart attack, but, again, it is not necessary to provide a precise assignment.

The likelihood of the accident hypothesis is, of course, much higher than the crime hypothesis, and Watson could even try to assign a value for it using statistical data, if he has any, about the death rate by heart attack in the class of people in Baskerville's range of age and health conditions. Therefore, the following judgements seem reasonable:

$$Pr(R|H_1, I) > Pr(R|H_2, I);$$
$$Pr(R|H_3, I) = 0. \qquad (1.10)$$

It is true that, with comparative probability judgements like (1.9) and (1.10), it is not possible to calculate the numerical value of the denominator in formula (1.5), but it is not necessary to do that to check that Bayes' theorem does yield a result in agreement with common sense. It is sufficient to notice that, using (1.5) and (1.10):

$$Pr(H_3|R,I) = 0 \qquad (1.11)$$

and hence Bayes' formula (1.4) for hypotheses H_1 and H_2 may be rewritten as

$$Pr(H_1|R,I) \times Pr(R|I) = Pr(R|H_1,I) \times Pr(H_1|I),$$

$$Pr(H_2|R,I) \times Pr(R|I) = Pr(R|H_2,I) \times Pr(H_2|I). \qquad (1.12)$$

Using (1.9) and (1.10), it follows that the probability of the accident hypothesis and that for Watson was already greater than the probability of the crime hypothesis, given background knowledge only, it is even higher knowing the coroner's report:

$$Pr(H_1|R,I) > Pr(H_2|R,I). \qquad (1.13)$$

It is worthwhile to notice that (i) exact numerical probabilities are not always needed to draw probabilistic inferences and (ii) conclusion (1.13) is a *deductive inference*, being a necessary consequence of premises (1.9) and (1.10) according to Bayes' theorem. Moreover, in those cases, where the truth of a particular hypothesis is not logically compatible with observing some particular evidence, then observation of that evidence falsifies the hypothesis. In this example, proposition R would have been false if proposition H_3 had been true (premiss (1.10)). Therefore, the truth of R implies the falsity of H_3 according to Bayes' theorem (conclusion (1.11)).

1.2.3   The odds form of Bayes' theorem

The combination of the two formulae that appear in (1.12) gives the so-called odds form of Bayes' theorem:

$$\frac{Pr(H_1|R,I)}{Pr(H_2|R,I)} = \frac{Pr(R|H_1,I)}{Pr(R|H_2,I)} \times \frac{Pr(H_1|I)}{Pr(H_2|I)}. \qquad (1.14)$$

The left-hand side is the odds in favour of H_1, conditional on R, called the *posterior odds* in favour of H_1. In the Right-hand side, the first term is the *likelihood ratio*, also called *Bayes factor*, and the second term is the *prior odds* in favour of H_1. The effect of evidence on hypotheses can be calculated by multiplying the likelihood ratio by the prior odds.

The likelihood ratio, say V, has been suggested as a measure of the probative value of the evidence with respect to two alternative hypotheses (Aitken and Taroni 2004; Robertson and Vignaux 1995; Schum 1994). If $V > 1$ in (1.14), we shall say that the probative value of evidence R is in favour of H_1; if $V < 1$, we shall say that it is in favour of H_2; and if $V = 1$, we shall say that R is not relevant for the hypotheses in question or that the evidence is 'neutral' with respect to them.

It is necessary, again, to emphasize that the probative value of a piece of evidence with respect to two hypotheses is to be distinguished from conditional degrees of belief for the same hypotheses. As shown clearly in (1.14), the probabilities of the hypotheses conditional on that particular piece of evidence on not only depend the posterior odds but also on the prior odds, that is, on the background knowledge and all other relevant evidence that is known.

1.2.4   Combination of evidence

Now consider Dr. Mortimer's evidence. For the sake of simplicity, it shall be taken that Watson can believe the following proposition to be true: A gigantic hound was running after Sir Charles Baskerville (F). Now, Watson has to evaluate the probability:

$$Pr(H_1|F,R,I) = \frac{Pr(F|H_1,R,I)Pr(H_1|R,I)}{Pr(F|R,I)}.$$

He can argue that likelihoods are against the accident hypothesis and in favour of the crime hypothesis. The probability that a big dog was in that place of the moor at that time is smaller than the probability that this event would have happened if an intentional scheme was at work:

$$Pr(F|H_1,R,I) < Pr(F|H_2,R,I). \tag{1.15}$$

At this step, the initial probabilities are (1.13), and they are in favour of H_1. So, what is the overall effect of evidence F? Consider the odds form of Bayes' rule.

$$\frac{Pr(H_1|F,R,I)}{Pr(H_2|F,R,I)} = \frac{Pr(F|H_1,R,I)}{Pr(F|H_2,R,I)} \times \frac{Pr(H_1|R,I)}{Pr(H_2|R,I)}.$$

From this form, it can be seen immediately that, given that the initial odds ratio at this point is in favour of H_1, the posterior odds can be reversed only if the likelihood ratio in favour of H_2 is greater than the initial odds ratio. $Pr(H_2|F,R,I) > Pr(H_1|F,R,I)$ if and only if

$$\frac{Pr(F|H_2,R,I)}{Pr(F|H_1,R,I)} > \frac{Pr(H_1|R,I)}{Pr(H_2|R,I)}. \tag{1.16}$$

Even if Watson is not able to quantify precisely the relevant likelihoods, he can surely say that the left-hand ratio in (1.16) cannot be greater than the right-hand one. Therefore, Watson concludes that the probability of natural death is higher by far, even though the probability of a criminal scheme has been raised by the information given by the footmarks, and this conclusion is reached by comparative probability judgements only, which Watson is able to make.

1.2.5   Reasoning with total evidence

Would the conclusion change if Watson does apply Bayes' rule to both pieces of evidence at the same time, instead of building up the argument in two steps? The probability Watson must evaluate is now the following:

$$Pr(H_1|F,R,I) = \frac{Pr(H_1,F,R,I)}{Pr(F,R,I)}.$$

Given that the multiplication law is commutative, Watson knows that the denominator can be factorized both ways and that the final result must be the same. Also

$$Pr(H_1,F,R,I) = Pr(H_1|I) \times Pr(R|H_1,I) \times Pr(F|H_1,R,I)$$
$$= Pr(H_1|I) \times Pr(F|H_1,I) \times Pr(R|H_1,F,I).$$

According to which factorization he chooses, the overall effect of the evidence will be calculated using either the likelihood ratio:

$$\frac{Pr(R|H_1, I)}{Pr(R|H_2, I)} \times \frac{Pr(F|H_1, R, I)}{Pr(F|H_2, R, I)} \tag{1.17}$$

or the likelihood ratio:

$$\frac{Pr(F|H_1, I)}{Pr(F|H_2, I)} \times \frac{Pr(R|H_1, F, I)}{Pr(R|H_2, F, I)}. \tag{1.18}$$

Watson is free to choose the formula that is more convenient for him, in the sense that he might be better able to assign probabilities in one factorization rather than in the other. We can notice that the conditioning (1.17) follows the temporal order by which evidence has been acquired, whereas path (1.18) follows the temporal order by which the hypothetical scenarios occurred.

If Watson relies on his judgements (1.10) and (1.15), then formula (1.17) looks like the easier way for him to follow. Suppose, however, that Watson wishes to give a comparative evaluation of the likelihoods in (1.18). He could sensibly say that the likelihood of H_2, given F, is higher than the likelihood of H_1, given F:

$$Pr(F|H_2, I) > Pr(F|H_1, I). \tag{1.19}$$

Then, he might argue that, given that a big dog was running after Sir Charles Baskerville, the probabilities he would die from a heart attack are the same, the reason for the presence of the dog does not matter. Therefore,

$$Pr(R|H_1, F, I) = Pr(R|H_2, F, I). \tag{1.20}$$

Following (1.18), with the seemingly reasonable premisses (1.19) and (1.20), the conclusion is reached that the total body of evidence is in favour of the crime hypothesis. Indeed, given (1.20), evidence R is 'neutral' with respect to the choice of H_1 versus H_2, whereas, given (1.19), evidence F is in favour of H_2:

$$\frac{Pr(F|H_2, I)}{Pr(F|H_1, I)} > \frac{Pr(R|H_1, F, I)}{Pr(R|H_2, F, I)} = 1.$$

This does not mean that posterior odds will be in favour of H_2, for the calculus of posterior odds includes the prior odds, and these are overwhelmingly in favour of H_1:

$$\frac{Pr(H_1|F, R, I)}{Pr(H_2|F, R, I)} = \frac{Pr(F|H_1, I)}{Pr(F|H_2, I)} \times \frac{Pr(R|H_1, F, I)}{Pr(R|H_2, F, I)} \times \frac{Pr(H_1|I)}{Pr(H_2|I)}.$$

A correct way for Watson to report his inference would be to say that, although the evidence is in favour of the crime hypothesis, the prior credibility of that hypothesis is so low that the accident hypothesis is still the more credible. To be more precise, Watson should try to assign numerical values to his opinions (1.9), (1.19) and (1.20). It is shown in Section 2.1.11 that it is easier to assign numerical probabilities following the temporal order by which the possible scenarios occurred rather than the temporal order of evidence acquisition.

1.2.6   Reasoning with uncertain evidence

Suppose that the coroner's report is such that there is still some uncertainty left about the proximate cause of Baskerville's death. Assume, for the sake of argument, that Watson's degree of belief in the truth of proposition R (The proximate cause of Sir Charles Baskerville's death was heart attack) at time t_1, after learnt the contents of the report, is higher than his initial degree of belief at time t_0, but it falls short of certainty: $1 > Pr_1(R|I) > Pr_0(R|I)$. What is the effect of this *uncertain evidence* upon the hypotheses?

The problem is that Watson *cannot* take the probability of H_1 conditional on R as his new degree of belief because *he does not know R for certain*. One might say that Watson could make some explicit proposition, say E, and that he knows for sure and is such that he can extend the conversation taking into account the probability of R conditional on E at time t_0. For the hypothesis H_1, for instance, this would require the calculation of

$$Pr_0(H_1|E, I) = Pr_0(H_1, R|E, I) + Pr_0(H_1, \bar{R}|E, I)$$

$$= \{Pr_0(H_1|R, E, I) \times Pr_0(R|E, I)\}$$

$$+ \{Pr_0(H_1|\bar{R}, E, I) \times Pr_0(\bar{R}|E, I)\}.$$

This suggestion has two drawbacks. The first is that Watson would have to evaluate the likelihoods of H_1 given R and E, which might not turn out to be an easy task. The second, and more important, is that it might be difficult for Watson to formulate explicitly such a proposition E that he knows for certain, considering that this proposition would be the logical conjunction of many other propositions, each one of which should be known for certain. Indeed, if there is any uncertainty left in the report, the best that Watson can do is to assess directly the effect of information on his degrees of belief and, indeed, this is all that Watson is reasonably asked to do. A probabilistic rule can be formulated that allows him to update directly on uncertain evidence.

Consider, for simplicity, the reduced probability tree of Watson's problem given in Figure 1.2, that is a probability tree containing only propositions H_1 and R. The difference with the tree in Figure 1.1 is that the probabilities of all the four scenarios at time t_1 are now greater than zero. The problem facing Watson is: how can he redistribute the probabilities of scenarios in such a way that they add up to one?

There are many ways of doing the redistribution but considering that the only change in Watson's state of information that has occurred is a change in the probability of R, and no information has been given about the probability ratio of different scenarios, then the least biased way is to leave unchanged that ratio. Thus, a reasonable answer is that the probabilities

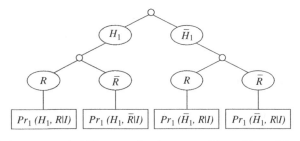

Figure 1.2   The probability tree for the propositions H_1 and R at time t_1.

are redistributed such that the ratio between the new and the old probabilities of the scenarios is the same as the ratio between the new and the old probabilities of R:

$$\frac{Pr_1(H_1, R | I)}{Pr_0(H_1, R | I)} = \frac{Pr_1(R | I)}{Pr_0(R | I)}. \tag{1.21}$$

From (1.21), the rule for calculating the new probabilities for any branch of the tree is then derived:

$$Pr_1(H_1, R | I) = Pr_0(H_1, R | I) \times \frac{Pr_1(R | I)}{Pr_0(R | I)}. \tag{1.22}$$

Formula (1.22) obeys the same principle as (1.7). The constant factor now is $Pr_1(R | I) / Pr_0(R | I)$ instead of $1/Pr_0(B | I)$, but the probabilities have been redistributed amongst the possible scenarios in a symmetric way as before. Given that the state of information has changed on learning only the new probability $Pr_1(R | I)$, and nothing else, there are no reasons to make a change biased for or against certain particular scenarios. Formula (1.22) is a straightforward generalization of Bayes' theorem, known in the philosophical literature under the name of *Jeffrey's rule*, because it was the philosopher of science Richard Jeffrey who first argued it was a reasonable general updating rule (Jeffrey 1983).

1.3 Population proportions, probabilities and induction

1.3.1 The statistical syllogism

In this book, a normative point of view about the logic of uncertain reasoning is adopted. It is not assumed that forensic scientists, and people at large, satisfy the laws of probability calculus in their common-sense reasoning. It is recommended that forensic scientists *should* use probabilities to express their degrees of beliefs, and the laws of probability to draw inferences, when they evaluate the probative force of forensic findings. This normative point of view is the core of a research programme known as *Bayesian subjectivism*, which ranges from statistical theory to artificial intelligence.

The evaluation of scientific findings in court is the expression of one's personal degree of conviction. This usually requires a combination of the knowledge about population proportions for reference classes together with a personal knowledge of circumstances for a particular case. Probability is commonly thought about with respect to proportion data (e.g. the relative frequency of a particular feature within a certain number of observations), and the use of probability as the quantitative language of personal knowledge provides a unified and powerful framework to deal with the combination of 'objective' statistical data and 'subjective' evaluation of the facts of the case.

Subjective Bayesians consider 'objective' probability judgements to be judgements about which it is possible to achieve an inter-subject agreement. For example, a degree of belief in a particular case based on the knowledge of relevant statistical data can be deemed more 'objective' than a degree of belief in a particular case not based on the knowledge of statistical data, in the sense that reasonable people tend to agree that, if statistical data are known, personal degrees of belief should take into account those data. This does not amount to saying that a probability assignment based on the knowledge of statistical data is 'objective' in the sense that there exists some kind of 'intrinsic' link between population proportions and probabilities that is independent of all information not related to proportion data.

Any probability judgement in a particular case, even when the judgement relates to a population proportion, has a component based on personal knowledge. A singular probability judgement is subsumed under a statistical law by an argument, which is sometimes called the *statistical syllogism*. The scheme of the argument is as follows, where (1)–(3) are the premisses and (4) is the conclusion:

Statistical syllogism

1. In the population R, the property Q is present with the proportion γ.

2. a_i is an individual in the population R ($i = 1, 2, \ldots, n$).

3. Nothing else is known about a_i which is relevant with respect to possession of property Q by a_i.

4. The probability that a_i has property Q is γ.

An example of a statistical syllogism is the following, where 'the probability that a particular individual has DNA profile Q' is to be understood as the *profile probability* as distinct from the so-called *random match probability* (Aitken and Taroni 2004, p. 408):

1. The proportion γ of the population R possesses the DNA profile Q.

2. The unknown criminal is an individual in the population R.

3. Nothing else is known about the criminal that is relevant with respect to possession of profile Q.

4. The probability that the criminal has profile Q is γ.

The third premiss shows how anyone who follows the argument takes into account his personal state of knowledge in addition to knowledge of statistical data. The argument goes traditionally under the name of 'syllogism' but, as it is stated, it is not really a deductive argument. It cannot be said that the conclusion follows necessarily from the premisses, even though it is the conclusion that would be accepted by most people. Conversely, the subjective Bayesian version of the statistical syllogism *is* a deductive inference.

Bayesian statistical syllogism

1. The proportion γ of the population R possesses the DNA profile Q.

2. a_i is an individual in the population R ($i = 1, 2, \ldots, n$).

3. a_i has, for You, the same probability of possessing property Q as any other individual in the population R.

4. The probability, for You, that a_i has property Q is γ.

To see that the conclusion of this schema follows necessarily from the premisses, we shall introduce the concept of 'expectation'. In addition, note that in many practical applications, the population proportion is not actually known because one cannot typically investigate each and every member of the population and note its feature. Moreover, a population is rarely stable enough for taking a 'fixed' picture (e.g. human populations may be affected by births and deaths). What is done, instead, is that an estimate of the population proportion is constructed on the basis of a survey of some of the members of the target population.

1.3.2   Expectations and population proportions

Let X be a discrete random variable which can take the values x with probabilities $Pr(X = x)$. Even when you do not know the value of X, you can calculate a representative value known as the *expectation* or *expected value* of X, denoted $E(X)$ and given by

$$E(X) = \sum_x xPr(X = x). \tag{1.23}$$

Now suppose that X_i is a random variable that takes the values $\{1, 0\}$ (respectively, 'true' and 'false'), associated with the proposition A_i: 'The individual a_i has property Q', where i belongs to a population R of n individuals. Using (1.23), it immediately follows that the expectation of this random variable is the probability that the proposition is true:

$$E(X_i) = \{Pr(X_i = 1) \times 1\} + \{Pr(X_i = 0) \times 0\} = Pr(X_i = 1).$$

Given that expectation is additive, the sum of probabilities of propositions A_i is equal to the expected value of the sum of random variables X_i:

$$Pr(X_1 = 1) + \cdots + Pr(X_n = 1) = E(X_1) + \cdots + E(X_n)$$
$$= E(X_1 + \cdots + X_n). \tag{1.24}$$

But the sum of random variables is the number of individuals who are Q, amongst the n individuals in the population R, that is the population proportion for feature Q in R (also sometimes referred to as the *frequency* of Q amongst R).

Dividing each side of (1.24) by n and applying the linearity of E, that is $E(X_i)/n = E(X_i/n)$, we have that average probability is equal to expected population proportion (or expected relative frequency):

$$\frac{Pr(X_1 = 1) + \cdots + Pr(X_n = 1)}{n} = E\left(\frac{X_1 + \cdots + X_n}{n}\right). \tag{1.25}$$

Suppose the distribution $Pr(X_i = x_i)$ is your subjective probability. Suppose, also, that you know the population proportion γ. Then, γ may be substituted on the right-hand side of (1.25) and, if you do not know which $n\gamma$ of R are Q, and you believe that all of them have the same probability of being Q, then it immediately follows that

$$Pr(X_i = 1) = \gamma \quad (i = 1, \ldots, n).$$

Therefore, if probabilities are given a subjective interpretation, population proportions are known and the probabilities of membership of Q for all members of R are taken to be equal, then population proportions (or, relative frequencies) determine the individual probabilities.

On the other hand, if there is evidence that individual a_i belongs to a sub-population S of R, with different characteristics that are relevant with respect to Q, then the probability that a_i is Q would not be equal to the probability that an individual in the general population R is Q. You should change your premises to obtain a new, valid, argument of the same form, by substituting S for R and a new value, say γ', for γ, corresponding to the proportion of Q's in S, if it is known (see for example, Section 3.3.3).

1.3.3   Probabilistic explanations

Kadane and Schum in their reference book *A Probabilistic Analysis of the Sacco and Vanzetti Evidence* synthesized as follows the differences between three basic forms of reasoning (Kadane and Schum 1996, p. 39):

> As far as concerns the process of discovery and the generation of hypotheses, evidence, and arguments linking them, there is an important issue being debated these days. The issue is: Can we generate new ideas by the processes of deduction and induction alone? The American philosopher Charles S. Peirce believed that there was a third form of reasoning according to which we generate new ideas in the form of hypotheses. He called this form of reasoning *abduction* or *retroduction* (...). Deduction shows that something is *necessarily* true, *induction* shows that something is *probably* true, but abduction shows that something is *possibly*, or *plausibly* true. Most human reasoning tasks, such as those encountered by the historian and criminal investigator, involve mixtures of these three forms of reasoning.

Abduction is classically considered the first step of an investigation when a particular event has been considered, and it is desired to *explain the occurrence of that event*. Therefore, in order to understand what 'abduction' is, it has to be understood what is meant by 'explanation'.

Deductive–Nomological (D–N) explanations are the simplest case of sound explanations, and their schema is as follows. Statements (1) and (2) constitute what is called the *explanans*, (3) constitutes what is called the *explanandum*, and the explanandum is assumed to be true. If premiss (2) is not actually true, but only hypothetically true, then we have a potential explanation.

Deductive–nomological explanation

1. The statement of a scientific law saying that if events of type B and $C_1, C_2, \ldots, C_n$ occur, then an event of type A occurs.

2. The statement that a particular event is of type B and $C_1, \ldots, C_n$.

3. The statement that a particular event is of type A.

A straightforward $D–N$ explanation of the event that can be described by the proposition 'Mr. Smith's blood specimen and the blood stain from the crime scene share the same DNA profile' is

1. If a stain of organic liquids comes from a person and it has not been in contact with extraneous organic material, then the stain shares the DNA profile of that person.

2. The blood stain found on the crime scene comes from Mr. Smith and it has not been in contact with extraneous organic material.

3. The blood stain found on the crime scene shares Mr. Smith's DNA profile.

The explanans contain here a *common-sense generalization* instead of a scientific law or, if you wish, a shorthand for a set of scientific laws too large for the enumeration of all of them. Authors such as Schum, Twining and Anderson have underlined the role of common sense-generalizations in evidential reasoning, and we shall turn on this important issue in the Chapter 2 (Anderson and Twining 1998; Schum 1994).

D–N explanations are only one amongst many kinds of 'explanations'. In social science, biological sciences, medicine and, last but not least, forensic science, the scientific laws available are, in general, statistical laws, and, whatever the philosophical view of the world is, it cannot be helped to make use of *probabilistic explanations*, namely, explanations that subsume particular facts under statistical laws. The schema of the *Inductive-statistical (I-S)* explanation originally proposed by Hempel (1965) was as follows:

Inductive-statistical explanation

1. The statement of the probability γ of occurrence of events of type Q in the reference class of events of type R.

2. The statement that a particular event is of type R.

3. The statement that a particular event is of type Q with probability γ.

If it is not known whether premiss (2) is actually true, then it is a *potential I-S* explanation that is being used. In order to establish the first premiss, the argument presupposes the use of the statistical syllogism (Section 1.3.1), as acknowledged by Hempel himself, who introduced the proviso that reference classes should be chosen on the basis of all relevant knowledge available prior to the explanandum. This requirement has been criticized for relativizing statistical explanations to the knowledge of scientists at a given time. Subjective Bayesians acknowledge this fact not as a shortcoming of statistical explanations but as an unavoidable matter of fact.

Suppose it is desired to explain the particular event 'the blood stain from the crime scene shares DNA profile Q with the suspect' by means of an *I-S* explanation:

1. The probability that a stain of organic liquids coming from a member of the population R, different from the suspect, has DNA profile Q is γ.

2. The blood stain found on the crime scene comes from a member of the population R different from the suspect.

3. The probability that the blood stain found on the crime scene has DNA profile Q is γ.

Here, the probability γ is the 'conditional profile probability' based on knowledge of not only the population proportion but also of the fact that another person, the suspect, has the profile: the 'conditional profile probability' γ can be different from the population proportion (Aitken and Taroni 2004, p. 404); see also Sections 6.1 and 7.1. This means that the third premiss of the statistical syllogism should be modified as follows: 'nothing else is known about the member of the population which is relevant with respect to the possession of profile Q but the fact that another member of the same population shares the same profile'.

Is the above explanation a legitimate *I-S* explanation? The answer is negative because Hempel's *I-S* explanations required a high probability γ for the explanandum, a probability at least greater than 0.5. This requirement for a high probability for the explanandum means that in the context of forensic DNA profiling we cannot use *I-S* explanation in conjunction with the 'fortuitous correspondence' hypothesis because the probability of fortuitous correspondence is usually very low. It is true that the 'correspondence by chance' hypothesis can be considered only an explanation by default, so to speak, of the fact that the blood stain found on the scene of the crime shares the suspect's DNA profile but, nevertheless, it is a potential alternative explanation. Moreover, a high probability is neither a sufficient nor a necessary condition for explanation, as the following example can help to make clear (Salmon 1999, p. 27):

> Suppose that Bruce Brown has a troublesome neurotic symptom. He undergoes psychotherapy and his symptoms disappear. Can we explain his recovery in terms of the treatment he has undergone? (...) If the rate of recovery for people who undergo psychotherapy is no larger than the spontaneous remission rate, no matter how large is this rate, it would be a mistake to consider treatment a legitimate explanation of recovery. (...) If, however, the recovery rate is not very large, but is greater than the spontaneous remission rate, the fact that the patient underwent psychotherapy has at least some degree of explanatory force.

The so-called *Statistical-relevance (S-R)* model of explanation (Salmon et al. 1971) has been put forward to obviate this shortcoming of Hempel's classical analysis. All that is required by this model to provide a probabilistic explanation of a particular event is that there exists a partition of the relevant population into reference classes such that the probability of the explanandum is different for any member of the partition. What it does matter are differences in probabilities, not their size. The basic idea is that there is an explanation or, at least, an approximation to a satisfactory explanation, whenever factors can be identified that make a difference for the probability of the explanandum.

A *partition* of a population R is a collection $\{S_1, \ldots, S_n\}$ of sub-populations of R such that the S_i are exclusive and exhaustive, that is, they do not have common members and they jointly contain all the members of R.

Statistical-relevance explanation

1. The statement of a partition $\{S_1, \ldots, S_n\}$ of the relevant population R such that each element S_i of the partition is probabilistically relevant for Q.

2. The statement of probabilities γ_i of occurrence of events of type Q, in the reference class of events of type S_i.

3. The statement that a particular event is of type S_k.

4. The statement that a particular event is of type Q.

In Salmon's example, the reference classes are in the population R of people who suffer the symptom, the class of people who undergo psychotherapy (S_1) and the class of people who undergo no therapies (S_2). An *S-R* explanation is not an argument; a conclusion does not follow from the premises and, in order to explain why Bruce Brown recovered (Q), all that can be done is to point out the fact that he underwent therapy and that the probability of recovery is higher for members of that class than for members of the other class.

An *S-R* explanation for DNA analysis containing in the explanans, the hypothesis of fortuitous correspondence, can be framed as follows.

1. The statement of a partition $\{S_1, \ldots, S_n\}$ of the potential perpetrator population, defined as the population of the possible perpetrators of the crime excluding the suspect, where the elements of the partition are all the sub-populations that are probabilistically relevant with respect to the property of having DNA profile Q.

2. The statement of random match probabilities (as distinct from profile probabilities) γ_i that a blood stain coming from a member of the sub-population S_i has profile Q.

3. The statement that the blood stain found on the crime scene comes from a member of the sub-population S_k.

4. The statement that a blood stain found on the crime scene has profile Q.

Sub-populations $S_1, \ldots, S_n$ are identified by a conjunction of the factors that provide information relevant to the allele proportions, such as ethnic group membership and degrees of relatedness with the suspect. For example, a coarse partition (that could be refined) of the suspect population is

$\{S_1 = \text{Relatives}, S_2 = \text{Not-relatives and members of ethnic group } X_1, S_3 = \text{Not-relatives and members of ethnic group } X_2, \ldots, S_n = \text{Not-relatives and members of ethnic group } X_{n-1}\}$.

The sub-population S_k is identified by the hypothesis proposed by the defence: it is the *relevant population* as defined by Aitken and Taroni (2004, p. 281). The magnitude of the probability γ_k does not matter for an *S-R* explanation: what does matter is that it is different from all the other probabilities.

1.3.4   Abduction and inference to the best explanation

It is now possible to give the general schemata of abductive and inductive arguments. For the sake of simplicity in what follows, the proposition describing the particular event of type Q that is to be explained is denoted by E, and the proposition describing the particular event of type R or S_k, mentioned in the explanans of *potential D-N or S-R* explanations, is denoted by H. The particular event described by H is defined as an *explanatory fact*. Then, the general schema of *abductive arguments* is

Abduction

1. E is observed.

2. H is an explanatory fact in a potential explanation of E.

3. H is possibly true.

In the contemporary debate about the term *abduction*, it is often taken to mean what the philosopher Gilbert Harman has called the rule of *inference to the best explanation* (Harman 1965). In summary, the rule says that, given some evidence E and alternative explanatory facts H_1 and H_2, H_1 should be inferred rather than H_2 if, and only if, H_1 belongs to a better potential explanation of E than H_2. The two concepts must be carefully distinguished, as philosopher Paul Thagart pointed out (Thagart 1988, pp. 75, 143):

> Mechanisms such as abduction and conceptual combination can lead to the formation of new theories (...). But we clearly do not want to *accept* a theory merely on the basis that there is something that it explains (...). Clearly, we want to accept a theory only if it provides the best explanation of the relevant evidence (...). Abduction only generates hypotheses, whereas inference to the best explanation evaluates them.

The abductive schema presented above serves to generate hypotheses and not to evaluate them, because no alternative explanations are mentioned. As Holmes pointed out in the short story *The Adventure of Black Peter* (Conan Doyle 1953, p. 567):

> One should always look for a possible alternative and provide against it. It is the first rule of criminal investigation.

In order to 'provide against' a potential alternative explanation, or against many of them, some philosophers have suggested we should reason by inference to the best explanation.

Inference to the best explanation

1. E is observed.

2. H_1 and H_2 are explanatory facts in two potential alternative explanations of E.

3. The explanation containing H_1 is overall better than the explanation containing H_2.

4. H_1 is provisionally accepted.

Hence, 'abduction' by itself is only a part of an inference to the best explanation. It is what introduces the second premiss of the argument. An important proviso must be added: the 'best explanation' must always be the best *overall* explanation of all available evidence. A critical point, and a long debated one, is the meaning of the 'best overall explanation'. The Bayesian answer is given in the following section.

1.3.5   Induction the Bayesian way

In the Bayesian approach, generation and evaluation of the prior probabilities of scenarios, similar to those exemplified in Figure 1.1, correspond to the abductive step. In order to qualify as an explanatory fact, H must satisfy two necessary conditions. That is, H is an explanatory fact for E only if (i) the probability of H, given background knowledge only, is greater than zero and (ii) the likelihood of H, given E and background knowledge, is greater than zero. More formally, this writes (i) $Pr(H|I) > 0$ and (ii) $Pr(E|H,I) > 0$.

In order to define the best overall explanation in probabilistic terms, it is first necessary to consider when an explanation is better than another. An explanation of E containing H_1 is better than an explanation containing H_2 if, and only if, the likelihood of H_1, given E, is greater than the likelihood of H_2, given E: $Pr(E|H_1,I) > Pr(E|H_2,I)$.[5]

A rule can now be given to decide which one, between two alternative hypotheses, provides an explanation that is overall better for E. This rule states that an explanation of E containing H_1 is overall better than an explanation containing H_2 if, and only if, the ratio of the likelihoods, given E, is greater than the reciprocal of the ratios of their probabilities, given background knowledge only:

$$\frac{Pr(E|H_1,I)}{Pr(E|H_2,I)} > \frac{Pr(H_2|I)}{Pr(H_1|I)}.$$

It follows immediately from the odds form of Bayes' theorem (1.14) that the following schema, to be called *Bayesian inference*, is valid.

---

[5] Note that at this point the terms *likelihood* and *probability* are not used as synonyms, as explained earlier in Section 1.1.8.

Bayesian inference

1. E is observed.

2. H_1 and H_2 are explanatory facts in two potential alternative explanations of E.

3. The explanation containing H_1 is overall better than the explanation containing H_2.

4. H_1 is more probable than H_2, conditional on E.

It can be recognized that this scheme of inference has been applied in the discussion of the Baskerville's case: see (1.12) and (1.16). Note the following very important point. A hypothesis can be the best explanation of a given piece of evidence, given the competing hypotheses, without being the best *overall* explanation. It is not sufficient that the likelihood of H_1 is higher, given E, to conclude that it is also more credible, unless it is *also* assumed that H_1 and H_2 have the *same prior* probabilities. Such an assumption, for example, is made in the traditional Essen–Möller calculations for paternity, but it is not always true that the knowledge is such to justify this assumption (Taroni and Aitken 1998).

An important practical consequence to be stressed that follows from the Bayesian analysis is that the inductive step is to be made by the Court, and not by the expert, for it is only the Court which masters the total evidence of the case and which has the viewpoint to evaluate which is the best overall explanation. As correctly pointed out by Henri Poincaré in the 'expert opinion', he gave during the second appeal of the notorious Dreyfus trial, on 1904 [English translation of Darboux et al. (1908, p. 504), quoted from Taroni et al. (1998, p. 192)]:

> Since it is absolutely impossible for us [the experts] to know the a priori probability, we cannot say: this coincidence proves that the ratio of the forgery's probability to the inverse probability has that particular value. We can only say: following the observation of this coincidence, this ratio becomes X times greater than before the observation.

It has already been noted that conclusions which are only probably true are obtained deductively by premisses which are probably true, via the laws and theorems of the mathematical theory of probability plus the updating rule (1.22). This new viewpoint on 'inductive logic' constitutes one of the major philosophical achievements of the twentieth century, insofar as it has made it possible to give a constructive answer to David Hume's sceptical challenge to induction, and it has provided solid grounds to the Artificial Intelligence quest for mechanizing uncertain reasoning. Hume was right: the traditional idea that ampliative induction is possible, starting from a collection of bare facts, was an illusion. People always face observations carrying with them their background knowledge and prior beliefs. But if this fact is recognized, then there is no obligation, as Hume thought, to trust a mere innate psychological habit; a powerful logical tool is now available. As the philosopher and logician Frank P. Ramsey wrote in his seminal essay, written in 1926 (Ramsey 1931, pp. 182, 189):

> (...) a precise account of the nature of partial beliefs reveals that the laws of probability are laws of consistency, an extension to partial beliefs of formal logic, the logic of consistency. (...) We do not regard it as belonging to formal logic to say what should be a man's expectation of drawing a white or black ball from an urn; his original expectations may within the limits of consistency be any he likes, all we have to point out is that if he has certain expectations he is bound

in consistency to have certain others. This is simply bringing probability into line with ordinary formal logic, which does not criticize premises but merely declares that certain conclusions are the only ones consistent with them.

The same idea was expressed at the same time by de Finetti without knowing Ramsey's paper and from a different philosophical background, but sharing with him a common pragmatist attitude [English translation of de Finetti (1930a, p. 259) quoted from Aitken and Taroni (2004, p. 154)]:

> Probability calculus is the logic of the probable. As logic teaches the deduction of the truth or falseness of certain consequences from the truth or falseness of certain assumptions, so probability calculus teaches the deduction of the major or minor likelihood, or probability, of certain consequences from the major or minor likelihood, or probability, of certain assumptions.

Therefore, according to this opinion, 'inductive logic' is just a matter of deducing complex, may be non-intuitive, probabilistic beliefs from premises containing simpler, and more intuitive, probabilistic beliefs, and this is the reason why, in the last talk, he gave before retirement in 1976 at the University of Rome, Bruno de Finetti declared that the expression 'Bayesian induction' is as likely to be said to be redundant, as it would be to say 'Pythagoric arithmetic' is redundant (de Finetti 1989, p. 165).

1.4   Decision making under uncertainty

1.4.1   Bookmakers in the Courtrooms?

The argument that has been usually used in the philosophical and statistical literature to justify the core of the Bayesian research programme makes appeal to a notion of pragmatic coherence. It has been known for centuries that degrees of belief can be numerically expressed by betting quotients, and the idea that numerical probabilities are fair betting quotients can be traced back to the founders of the theory in the middle of seventeenth century, Blaise Pascal and Christian Huygens. But, it was only around 1930 that the mathematician Bruno de Finetti rigorously proved the so-called Dutch Book theorem (de Finetti 1937). The theorem shows that if your numerical degrees of belief do not satisfy the probability laws, then there exists a set of bets which would be *prima facie* acceptable to you, in the sense that you would consider the odds to be fair, and such that you will be doomed to lose, no matter what the outcome is of the event in which the bets are made. The acceptance of a bet is essentially the choice of an action with an uncertain outcome as an alternative to the choice of another action, that is a refusal of the bet and adherence to the status quo. The fact that betting in ordinary life is a voluntary activity should not conceal the fact that

> we are all faced with uncertain events and have to act in the reality of that uncertainty. (…) In this sense all of us 'gamble' every day of our lives (…) The essential concept is action in the face of uncertainty. (Lindley 1985, p. 19)

The 'Dutch Book' argument can be convincing or not, and often it is not for 'unbelievers', but the most forceful argument for accepting the Bayesian programme lies in the fact that the programme does work in practice. The best argument for the application of Bayesian theory in forensic science is to show that the theory agrees with personal intuitions, when inference and decision problems are simple and intuitions are reliable, and that it helps to go beyond them, when problems become complicated and intuitions are not so reliable. The proof of the pudding is in the eating and readers will be able to judge by themselves whether the Bayesian programme is convincing.

1.4.2   Utility theory

In a decision problem, choices are to be made amongst alternative courses of actions (decisions) with uncertain consequences. Uncertainty may depend on observable or unobservable circumstances. For example, the proportion of items that contain something illegal is unknown unless the entire consignment is inspected. It is uncertain, and unobservable, whether the defendant is the offender, whilst it is uncertain, but observable, what the outcome will be of future DNA analyses. Uncertainty cannot be suppressed, but scientists and decision makers have to take decisions in front of it which will have different consequences according to what is the true state of nature.

The basic elements of a decision problem can be formalized in terms of the following three elements. First, there is a set of *feasible decisions*, whose elements will be denoted d, whilst the set of all possible decisions, also called the *decision space*, will be denoted D. Second, there is a set of *uncertain events*, also called states of nature, denoted θ. The set of all possible states of nature will be denoted Θ. Third, there is a set C of *consequences*, where $c(d, \theta) \in C$ denotes the consequence of having taken decision d when event θ occurs.

Example 1.1 (*Forensic identification/individualization*) *Suppose material (a crime mark) is collected at a crime scene and an individual is apprehended. A pair of uncertain events are defined as 'The suspect is the origin of the crime mark' (θ_1) and 'Someone else is the origin of the crime mark' (θ_2). The set Θ of these states of nature is discrete, $\{\theta_1, \theta_2\}$. The process of 'identifying' (sometimes also called* individualizing*) an individual as being the source of a crime mark can be interpreted as a decision (d_1). Alternative decisions may cover statements such as 'inconclusive' (d_2) or 'exclusion' (d_3). This decision problem is illustrated in Table 1.4. The outcome of an 'identification' ('exclusion') statement can be favourable if the suspect is truly (is truly not) the origin of the crime mark. Consequences are then listed as 'correct identification' and 'correct exclusion', respectively. In contrast, the outcome of an 'identification' ('exclusion') statement can be unfavourable if the suspect is truly not (is truly) the origin of the crime mark. Consequences are then listed as 'false identification' and 'false exclusion', respectively. The decision 'inconclusive' does not convey any information that tends to associate or otherwise the suspect with the issue of the source of the crime mark. Therefore, the respective consequences are listed as 'neutral'.*

You are interested in making the best possible decision, given the basic elements of the decision problem you have figured out. Therefore, you need a decision rule that assesses the consequences of alternative courses of action, allows a comparison of different decisions and avoids irrational actions.

The steps to formulate such a decision rule involved in the formal decision model are the following. First, it is assumed that the decision maker can express preferences amongst possible consequences. Preferences are measured with a function called a *utility function* denoted $U(\cdot)$ that associates a utility value to each one of the possible consequences, specifying their desirability on some numerical scale. Second, your uncertainty about the states of nature is

Table 1.4 Decision table for an identification problem with d_i, $i = 1, 2, 3$, denoting decisions; θ_j, $j = 1, 2$, denoting states of nature; and C_{ij} denoting the consequence of taking decision d_i when θ_j turns out to be the true state of nature.

	States of nature	
Decisions	θ_1: suspect is donor	θ_2: some other person is donor
d_1: identification	C_{11}: correct identification	C_{12}: false identification
d_2: inconclusive	C_{21}: neutral	C_{22}: neutral
d_3: exclusion	C_{31}: false exclusion	C_{32}: correct exclusion

Source: Biedermann et al. (2008a).

measured with a probability mass function $Pr(\theta|I)$ when θ is discrete, and a probability density function $f(\theta|I)$ when θ is continuous, describing the plausibility of these states given the specific information I available when the decision must be taken. For the sake of simplicity, the letter I will be omitted. Third, the desirability of available actions is measured by their *expected utility*, denoted by EU:

$$EU(d) = \begin{cases} \sum_\Theta U(d, \theta)Pr(\theta), & \text{if } \Theta \text{ is discrete, and} \\ \int_\Theta U(d, \theta)f(\theta)d\theta, & \text{if } \Theta \text{ is continuous.} \end{cases}$$

Note that $U(d, \theta)$ represents the utility of taking decision d when θ turns out to be the true state of nature. We see that expected utility is the same function defined in Section 1.3.2 as the expectation of the value of random variable, where the random variable is a utility function $U(\cdot)$. The decision rule is to take the decision with the *maximum expected utility*.

Personal evaluations about the consequences of several actions will lead the decision maker to have preferences amongst consequences in any set C. When any two consequences $(c_1, c_2) \in C$ are compared, the notation $c_1 \prec c_2$ indicates that c_2 is strictly preferred to c_1. Notation $c_1 \sim c_2$ indicates that c_1 and c_2 are equivalent (or equally preferred), whilst $c_1 \preceq c_2$ indicates that c_2 is preferred to c_1 (when $c_1 \preceq c_2$, then either $c_1 \prec c_2$ or $c_1 \sim c_2$ holds). The decision maker is asked to be able to order the consequences in such a way as to satisfy the following conditions:

A.1 For any couple of consequences $(c_1, c_2) \in C$, it must be possible to express a preference or equivalence amongst them (one of the following relations must hold: $c_1 \prec c_2, c_2 \prec c_1, c_1 \sim c_2$); the preference pattern is transitive, that is for any $(c_1, c_2, c_3) \in C$, if $c_1 \preceq c_2$ and $c_2 \preceq c_3$, then $c_1 \preceq c_3$; not all the consequences are equivalent to each other, that is, for at least a pair of consequences $(c_1, c_2) \in C$, either $c_1 \prec c_2$ or $c_2 \prec c_1$.

A.2 There are not infinitely desirable (or bad) consequences.

A.3 The ordering of preferences is invariant with respect to *gambles*. For any couple of consequences $(c_1, c_2) \in C$, such that $c_1 \preceq c_2$, then, for any other consequence $c_3 \in C$, and any probability α, the gamble that offers probability α of winning c_2, and probability $(1 - \alpha)$ of winning c_3 is preferred (or it is equivalent) to the gamble that offers probability α of winning c_1 and probability $(1 - \alpha)$ of winning c_3.

Denote by $(c_i, c_j; \alpha, 1 - \alpha)$, the gamble that offers the consequence c_i with probability α and the consequence c_j with probability $(1 - \alpha)$. Then, Axiom (A.3) can be formulated this way: $c_1 \preceq c_2$ if and only if $(c_1, c_3; \alpha, 1 - \alpha) \preceq (c_2, c_3; \alpha, 1 - \alpha)$, for any $\alpha \in [0, 1]$ and any $c_3 \in C$.

Now, consider any consequences c, satisfying $(c_1, c_2 \in C)$, such that $c_1 \prec c_2$ and $c_1 \preceq c \preceq c_2$: if Axioms (A.1)–(A.3) are satisfied, then there exists a unique number $\alpha \in [0, 1]$ such that

$$c \sim [\alpha c_1 + (1 - \alpha)c_2], \tag{1.26}$$

and it can be proved that

$$U(c) = \alpha U(c_1) + (1 - \alpha)U(c_2). \tag{1.27}$$

U is a utility function such that for any $(c_1, c_2) \in C$, then $c_1 \preceq c_2$ if and only if $U(c_1) \leq U(c_2)$, and the order relation in C is preserved. Utility functions are invariant under linear transformations, which means that if $U(c)$ is a utility function, then for any $a > 0$, $aU(c) + b$ is also a utility function preserving the same pattern of preferences.

There are several ways to proceed with the construction of the utility function. A simple way is to start with a pair of consequences $(c_1, c_2) \in C$, which are not equivalent, and assign them a utility value. In this way, the origin and the scale of the utility function is fixed, and each consequence $c \in C$ will be compared with c_1 and c_2. Since utility functions are invariant under linear transformation, the choice of c_1 and c_2 and the choice of the scale of the utility are not relevant, but they are generally identified with the worst and the best consequences, respectively. It is assumed for simplicity that $U(c_1) = 0$ and $U(c_2) = 1$ and the utilities of the remaining consequences are computed using Equation (1.27).

Example 1.2 (*Forensic identification/individualization, continued (Taroni et al. 2010, at p. 94)) Consider again the identification problem illustrated in Table 1.4. Suppose your preference ordering amongst consequences is the following:*

$$C_{12} \prec C_{31} \prec C_{21} \sim C_{22} \prec C_{32} \sim C_{11}. \tag{1.28}$$

That is, the most preferred consequences are a correct identification (C_{11}) and a correct exclusion (C_{32}), and the worst consequence is a false identification (C_{12}). The construction of the utility function starts with the assignment of the minimum utility value to the worst consequence and the maximum value to the best consequence (in this case the pair C_{32} and C_{11}): $U(C_{12}) = 0$ and $U(C_{11}) = U(C_{32}) = 1$. Consider now an assignment of a utility value to C_{21} (the 'neutral' consequence). Given that the preference ranking outlined above states that

$$C_{12} \prec C_{21} \prec C_{32} \sim C_{11},$$

the utility of C_{21} can be quantified using results (1.26) and (1.27). If there exists, for You, a unique number such that the consequence C_{21} is equivalent to a hypothetical gamble where the worst consequence is obtained with probability α and the best consequence is obtained with probability $(1 - \alpha)$:

$$C_{21} \sim [\alpha C_{12} + (1 - \alpha)C_{11}], \tag{1.29}$$

then the utility of C_{21} can be computed as

$$U(C_{21}) = \alpha \underbrace{U(C_{12})}_{0} + (1 - \alpha)\underbrace{U(C_{11})}_{1} = 1 - \alpha.$$

Let us notice that the utility of consequence C_{21}, and of any consequence strictly not preferred to the best consequence, turns out to be the probability $(1 - \alpha)$ of finishing with the best consequence in the set of all possible consequences. Note, in particular, the equivalence of utility and probability in the latter sentence. Determination of such an α is the most difficult part: what would make one indifferent between a neutral consequence and a situation in which a false identification might occur?

First of all, one must be careful not to use as values for α, the assigned probabilities for the hypotheses of the case at hand, namely, $Pr(\theta_1)$ and $Pr(\theta_2)$. The number α is really a threshold: you are asked to figure out what would be, in general, the highest probability of running the risk of making the worse mistake you are willing to exchange with the consequence of giving an 'inconclusive' statement. Suppose you say zero: you never want to run such a risk. But this means that, no matter how high is the probability of a correct identification $Pr(\theta_1)$, for you a neutral conclusion is as good as a correct identification.

Given that $C_{32} \sim C_{11}$, one can substitute C_{32} for C_{11} in (1.29), so obtaining another hypothetical gamble:

$$C_{21} \sim [\alpha C_{12} + (1 - \alpha)C_{32}].$$

In order to be coherent, you should consider again a neutral conclusion as good as a correct exclusion, no matter how high is the probability of a correct exclusion $Pr(\theta_2)$. Therefore, if it were true that the highest probability of running the risk of making the worse mistake you are willing to exchange with the consequence of giving an 'inconclusive' statement is strictly zero, then this belief cannot be coherent with the preference ordering (1.28), and it can be coherent only with the following one (note the change in the fourth preference sign from the left):

$$C_{12} \prec C_{31} \prec C_{21} \sim C_{22} \sim C_{32} \sim C_{11}. \tag{1.30}$$

Looking at Table 1.4, a preference ordering like (1.30) implies that, if θ_1 is true, then both decisions d_1 and d_2 are better than decision d_3, and, if θ_2 is true, then both decisions d_2 and d_3 are both better than decision d_1. Therefore, decision d_2 is the best decision overall, because, if θ_1 is true, it is better than d_3, and, if θ_2 is true, it is better than d_1. Thus, if the highest probability of running the risk of making the worst mistake you are willing to exchange with the consequence of giving an 'inconclusive' statement is strictly zero, you should always take the decision 'inconclusive'. Given that, as a matter of fact, people do not behave that way, there must exist for them unique numbers α such that the hypothetical gamble (1.29) does make sense, although it cannot be easy to fix the threshold α. Consider, for the sake of argument, that $\alpha = 0.001$ is felt to be correct. Then,

$$U(C_{21}) = \alpha U(C_{12}) + (1 - \alpha)U(C_{11}) = 1 - \alpha = 0.999.$$

Likewise, the utility of C_{31} will be quantified in comparison with $U(C_{12})$ and $U(C_{11})$,

$$U(C_{31}) = \alpha^* U(C_{12}) + (1 - \alpha^*)U(C_{11}) = 1 - \alpha^*.$$

Again, the number α^* is a threshold value: it is the highest probability of running the risk of making the worse mistake you are willing to exchange with the consequence of giving a 'false exclusion' statement. For rational behaviour, the value of this threshold α^* must necessarily be higher than the previous value $\alpha = 0.001$, since you are facing, on the right-hand side, a gamble with the same consequences as before and, on the left-hand side, a less preferred consequence ($C_{31} \prec C_{21}$). If $\alpha = 0.01$ is felt to be correct, then $U(C_{31}) = 0.99$.

1.4.3   The rule of maximizing expected utility

In Section 1.4.2, it has been shown that the plausibility of the states of nature and the value of uncertain outcomes of alternative actions can be quantified numerically, in terms of probabilities and utilities, respectively. In this section, it will be shown that the desirability of each possible decision d can be measured in terms of expected utility, $EU(d)$. A rational behaviour requires a person to choose an action to maximize his personal expected utility $EU(d)$.

Assume decision d is taken and θ turns out to be the true state of nature, so that the outcome is $c(d, \theta)$. Equation (1.26) can be used to show that a value α can be found such that the consequence $c(d, \theta)$ is equivalent to a hypothetical gamble where the worst consequence c_1 occurs with probability α and the best consequence c_2 occurs with probability $(1 - \alpha)$

$$c(d, \theta) \sim [\alpha c_1 + (1 - \alpha)c_2], \qquad c_1 \le c(d, \theta) \le c_2.$$

Then, Equation (1.27) allows one to calculate the utility $U(d, \theta)$ of the consequence $c(d, \theta)$:

$$U(d, \theta) = \alpha \underbrace{U(c_1)}_{0} + (1 - \alpha) \underbrace{U(c_2)}_{1} = 1 - \alpha.$$

This means that, for any d and any θ, taking decision d is equivalent to assigning a probability $U(d, \theta) = 1 - \alpha$ to the outcome of the most favourable consequence. This hypothetical gamble can always be played, under any circumstances, and in particular can be played after that decision d has been taken and the state of nature θ occurred. $U(d, \theta)$ can be understood as the *conditional probability* of obtaining c_2, given decision d has been taken and the state of nature θ occurred: $Pr(c_2|d, \theta) = U(d, \theta)$. Note that, by extending the conversation (Section 1.1.5),

$$Pr(c_2|d, \theta) = \sum_{\theta \in \Theta} Pr(c_2|d, \theta)Pr(\theta). \qquad (1.31)$$

Therefore, (1.31) can be rewritten as

$$Pr(c_2|d, \theta) = \sum_{\theta \in \Theta} U(d, \theta)Pr(\theta),$$

namely, the expected utility that quantifies the probability of obtaining the best consequence once decision d is taken (Lindley 1985). The strategy of taking the decision that maximizes the expected utility is optimal because it is the decision that has associated with it the highest probability of obtaining the most favourable consequence.

Example 1.3 *(Forensic identification/individualization, continued) Suppose utility values are those shown in Table 1.5. Consider then the computation of the expected utilities of the alternative courses of action starting, with the decision d_1, and denoting by E_1 the results (or scientific information) on which you have assessed your probabilities of θ_1 and θ_2:*

$$EU(d_1) = U(C_{11})Pr(\theta_1|E_1) + U(C_{12})Pr(\theta_2|E_1).$$

It is readily seen that the expected utility of decision d_1 reduces to

$$EU(d_1) = Pr(\theta_1|E_1).$$

Table 1.5 Illustrative values for utilities $U(C_{ij}) = U(d_i, \theta_j)$ in an identification scenario (θ_1 : 'suspect is donor of the crime stain'; θ_2 : 'some other person is donor of the crime stain'.

	Uncertain events	
Decisions	θ_1	θ_2
d_1: identification	1	0
d_2: inconclusive	0.999	0.999
d_3: exclusion	0.99	1

Source: Taroni et al. (2010).

Analogously, the expected utilities of decisions 'inconclusive' (d_2) and 'exclusion' (d_3) can be found as follows:

$$EU(d_2) = U(C_{21})Pr(\theta_1|E_1) + U(C_{22})[1 - Pr(\theta_1|E_1)]$$
$$= U(C_{21}) = U(C_{22}).$$
$$EU(d_3) = U(C_{31})Pr(\theta_1|E_1) + U(C_{32})[1 - Pr(\theta_1|E_1)]$$
$$= U(C_{31})Pr(\theta_1|E_1) + [1 - Pr(\theta_1|E_1)].$$

The optimal decision depends on the relative magnitude of $Pr(\theta_1|E_1)$, $U(C_{21})$ and $U(C_{31})$.

It has been said in Section 1.4.2 that the probability α in (1.27) is really a threshold value: it is the highest probability of running the risk of making the worse mistake one is willing to exchange with the consequence of giving an 'inconclusive' statement. Intuitively, this implies that if the probability of false identification in the case at hand, namely the probability of θ_2 is higher than this threshold, then 'identification' cannot be the best decision. Let us check that this is the case using the utility values given in Table 1.5. Suppose that the probability of false identification is 0.0011, slightly greater than α:

$$Pr(\theta_2|E_1) = 0.0011 > \alpha = 0.001.$$

Therefore, the expected utility values for decision table (Table 1.5) are as follows:

$$EU(d_1) = (1 \times 0.9989) = 0.9989$$
$$EU(d_2) = (0.999 \times 0.9989) + (0.999 \times 0.0011) = 0.999$$
$$EU(d_3) = (0.999 \times 0.99) + (1 \times 0.0011) = 0.990011.$$

Therefore, $EU(d_2) > EU(d_1) > EU(d_3)$ and 'inconclusive' is better than 'identification' ('inconclusive' is the best decision).

1.4.4   The loss function

Each pair (d, θ) gives rise to a consequence $c(d, \theta)$ whose desirability is measured by the utility function $U(d, \theta)$. In statistical inference, it is often convenient to work in terms of non-negative loss functions. Suppose that information is available about the true state of nature, say θ. Then, the decision maker will choose, in the column of the decision table corresponding to the true state of nature θ, that decision to which the highest utility is associated, that is, $\max_{d \in D} U(d, \theta)$. The loss function can be simply derived from the utility function (Lindley 1985) as

$$L(d, \theta) = \max_{d \in D} U(d, \theta) - U(d, \theta)$$

and measures the penalty for choosing the wrong (i.e. not optimal) action, that is the true amount lost if the most favourable situation does not occur. Press (1989, p. 26–27) noted that loss indicates 'opportunity loss', that is the difference between the utility of the best consequence that could have been obtained and the utility of the actual one received. Note that the loss cannot by definition be negative since $U(d, \theta)$ will be lower or at best equal to $\max_{d \in D} U(d, \theta)$. The expected loss, denoted by $EL(d)$, measures the undesirability of each possible action and can be quantified as follows:

$$EL(d) = \begin{cases} \sum_{\Theta} L(d, \theta) Pr(\theta), & \text{if } \Theta \text{ is discrete, and} \\ \int_{\Theta} L(d, \theta) f(\theta) d\theta, & \text{if } \Theta \text{ is continuous.} \end{cases}$$

The best strategy, which is that of taking the decision that maximizes the expected utility, becomes that of choosing the action that *minimizes the expected loss EL(d)*. One might argue that the assumption of non-negativity of the loss function is too stringent. It is observed, however, that the loss function represents the error due to a bad choice. Therefore, it makes sense that even the most favourable action will induce at best a null loss.

Example 1.4 *(Forensic identification/individualization, continued) Table 1.6 summarizes the loss function derived from the utility functions computed in Table 1.5. The loss associated with each possible consequence is determined, for each possible θ, by subtracting the utility of the consequence at hand from the utility of the most favourable situation.*

1.4.5 Decision trees

Decision problems can be represented by graphical models called *decision trees*. A discrete decision tree has two types of nodes. The first type is chance nodes, which represents discrete random variables, with a finite set of mutually exclusive states. The other type is decision

Table 1.6 Illustrative values for losses $L(d_i, \theta_j)$ in an identification scenario (θ_1 : 'suspect is donor of the crime stain'; θ_2 : 'some other person is donor of the crime stain'.

Decisions	Uncertain events	
	θ_1	θ_2
d_1: identification	0	1
d_2: inconclusive	0.001	0.001
d_3: exclusion	0.01	0

Source: Taroni et al. (2010).

nodes, which is represented by square boxes and has a finite set of feasible alternative decisions that are mutually exclusive and exhaustive. Each decision node represents a specific point in time under the model of the problem domain where the decision maker has to make a decision.

The construction of a decision tree proceeds in two stages. First, the tree is written out in a chronological order, usually growing horizontally from left to right and moving along all the branches from left to right, decision nodes and chance nodes alternate, beginning always with a decision node and ending with a chance node. Each branch is labelled: all those springing from a decision node with a decision and all those from chance nodes with a possible state of nature. In a second step, probabilities are attached to the branches springing from chance nodes, calculated according to the rules of probability calculus. In turn, utilities (or losses) are attached to the terminal branches.

The solution to a decision tree is a *strategy* that specifies a decision for each decision node. The *optimal strategy* is the strategy with maximum expected utility (minimum expected loss), that is the strategy (not necessarily unique) for each decision node specifies the decision with maximum expected utility (minimum expected loss). Such an optimal strategy is computed proceeding back from the terminal branches to the root, calculating expected utilities (or expected losses) at chance nodes and choosing the decision with maximum expected utility (or minimum expected loss) at decision nodes. This procedure is also known as *averaging-out and folding-back*.

Example 1.5 *(Forensic identification/individualization, continued) The decision tree for the identification problem of Table 1.6 is constructed in Figure 1.3, with losses at the terminal branch, expected losses calculated at chance nodes and minimum expected losses calculated at decision nodes. In chronological order, the root of the tree is the decision node, denote it by R (short for 'preliminary analysis'), whose set of possible decisions is 'do analysis', or 'identification', 'inconclusive' and 'exclusion', that is proceeding immediately to the final decisions, without doing a preliminary analysis. In the last case, 'exclusion', the chance nodes representing the possible state of nature θ_1 and θ_2 have given prior probabilities $Pr(\theta_j|E_1, I)$, where $j = \{1, 2\}$, E_1 represents previously considered items of information and I is the framework of circumstances (omitted hereafter for shortness of notation).*

Following the branch 'do analysis', we arrive at the chance node representing the possible outcomes of the DNA analysis, and relative probabilities have to be assigned. This is done with the following extension of the conversation:

$$\sum_{j=1}^{n=2} Pr(E_2|\theta_j, E_1)Pr(\theta_j|E_1).$$

Following each one of the branches representing the possible outcomes of the analyses, that is E_2 and $\bar{E}_2$, we arrive at decision nodes representing again the alternatives 'identification', 'inconclusive' and 'exclusion', from which we choose after knowing the result of the analyses. Therefore, the probabilities of the chance nodes are the posterior probabilities of θ_1 and θ_2, calculated given the observed outcomes, E_2, or its complement, $\bar{E}_2$. Table 1.7 represents the necessary probability calculations in an easy way . Posterior probabilities in the last column are obtained by dividing the terms of the sum in the third column by the total (an application of Bayes' theorem). Note that, for the purpose of this example, the likelihood ratio associated with the finding E_2 is assigned as $Pr(E_2|\theta_1, E_1)/Pr(E_2|\theta_2, E_1) = 1/0.0001 = 10^4$. Finally, losses are assigned to the terminal branches.

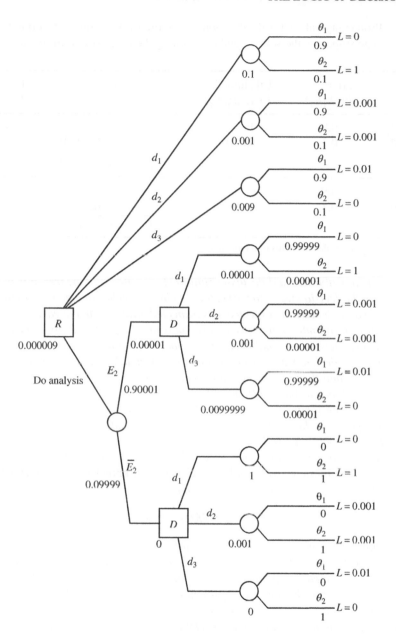

Figure 1.3   Decision tree for the identification problem with d_1, d_2 and $d_3 \in D$ denoting the available actions (i.e. expert conclusions) 'individualize', 'decide inconclusive' and 'exclude'. The possible states of nature are θ_1 (The suspect is the source) and θ_2 (Someone else is the source). Numbers below nodes represent expected losses, whereas the numbers below rectangles represent minimum expected losses. Events E_2 and $\bar{E}_2$ represent the possible outcomes 'observed correspondence' and 'observed difference', respectively, during a preliminary analysis, denoted R. Event E_1 does not appear in the tree because it is background information. Probabilities associated with states of nature and observed outcomes of the preliminary analysis are displayed below the branches. Losses L associated with particular combinations of a state of nature and a decision are shown at end of each branch.

Table 1.7   Probability values for the decision tree in Figure 1.3 with θ_1 denoting the proposition 'The suspect is the source' and θ_2 denoting the proposition 'Someone else is the source'.

		Prior $Pr(\theta_j|E_1)$	Likelihood $Pr(E_2|\theta_j,E_1)$	Product	Posterior $Pr(\theta_j|E_1,E_2)$
E_2:	θ_1	0.9	1	0.9	0.99999
	θ_2	0.1	0.0001	0.00001	0.00001
				$Pr(E_2|E_1) = 0.90001$	
$\bar{E}_2$:	θ_1	0.9	0	0	0
	θ_2	0.1	0.9999	0.09999	1
				$Pr(E_2|E_1) = 0.09999$	

Variables E_1 and E_2 denote distinct items of information.

Once all the probabilities and the losses (utilities) are entered in the tree, it is ready to be solved. To solve the decision tree, start from the terminal branches and go back to the root. At each random node, calculate the expected loss (expected utilities), and at each decision node, choose that branch with minimum expected loss (maximum expected utility). The decision of proceeding immediately to the final decision, without doing the analysis, has expected loss equal to 0.001 (i.e. decision d_2 has this minimum expected loss), whereas the decision of doing the analysis, and then taking the final decision, whatever the outcome of the analysis, has expected loss equal to 0.000009. Therefore, the optimal strategy is doing the forensic examination, and then taking decision d_1 (individualize) if the result of the analysis is an observed correspondence between the compared items (E_2) and deciding d_3 (exclusion) if the result is that the compared items do not correspond ($\bar{E}_2$).

The difference between the minimum expected loss deciding without knowledge of infor-mation E_2 *and the* minimum expected loss deciding with knowledge of information E_2 *is called the expected value of partial information E_2:*

$$(0.001 - 0.000009) = 0.00099.$$

It measures how much valuable is this piece of information for You, given Your beliefs and your preferences. If the 'cost' of the 'experiment' in terms of losses (utilities) is known, it must be added (subtracted) to the losses (utilities) in each one of the terminal branches of the sub-tree starting from the chance node representing the 'experiment'. In this example, if the 'cost' of the experiment, measured in terms of expected utility, was greater than 0.00099, then the best decision would be to not conduct the DNA analysis. Considering that your threshold for exchanging the risk of making the worse mistake with an 'inconclusive' statement is 0.001 and 0.00099 < 0.001, this calculation might offer a reason to conduct the 'experiment' beyond considerations of monetary cost.

1.4.6   The expected value of information

Information is not usually cost free. Even when there is no explicit monetary price to be paid, it takes time to obtain new data, and loss of time is a cost that sometimes can be quantified. If information was costless, and there were no deadlines for the taking of decisions, then it would always be rationale to decide to search for new information. However, this is not the case, and a typical problem scientists face is to decide whether it is worth having new information or data, if it is worth performing a new 'experiment', like another laboratory analysis, or asking police to search for a new witness. Given that this decision has to be taken before the data are available, the task is to calculate the expected gain from these new data, so that the gain can be compared with the cost of the search. Provided that the utilities, or the losses, of the outcomes of our decisions may be quantified in such a way that they can be compared to the cost of the experiment, Bayesian decision theory explains how the *expected value of information* may be calculated. Suppose you are the scientist who has assessed the loss values in decision table (Table 1.6). You know that the best decision is that one for which the expected loss is the minimum: you should take the decision $d_i (i = 1, 2, 3)$ as the one corresponding to the i for which

$$\sum_{j=1}^{2} L(d_i, \theta_j) Pr(\theta_j | E_1) \tag{1.32}$$

is minimized.

You must decide whether to perform an 'experiment' whose possible result is denoted by proposition E_2. For example, you must decide whether to perform a DNA analysis, taking biological control material from the suspect and comparing it with a biological stain found at crime scene. Your problem is to know how much should be paid for such an 'experiment'. As a first step towards the solution of the problem, it shall be shown how to calculate the *expected value of perfect information (EVPI)*, that is how much it would be worth to know with certainty which hypothesis is true. If you knew the true state of nature, you want to make the decision that minimizes the loss, that is, the decision with the smallest loss in the column corresponding to that hypothesis. Therefore, to calculate the *expected loss with perfect information (ELPI)*, you must multiply the minimum loss for each hypothesis by the probability of that hypothesis and sum all these products for n hypotheses:

$$ELPI = \sum_{j=1}^{2} \min_{i} L(d_i, \theta_j) Pr(\theta_j | E_1). \tag{1.33}$$

In this example, the calculus is very easy, for the minimum for each hypothesis is zero, so that the ELPI is zero. Your choice before knowing the truth was given by formula (1.32). Therefore, the difference between (1.32) and (1.33) is the measure of the reduction in the expected loss or, equivalently, the increase in the expected gain that could be obtained with perfect information. In other words, it is the measure of the EVPI (with the sign inverted, given that we are using losses instead of utilities):

$$EVPI = \min_{i} \sum_{j=1}^{2} L(d_i, \theta_j) Pr(\theta_j | E_1) - \sum_{j=1}^{2} \min_{i} L(d_i, \theta_j) Pr(\theta_j | E_1).$$

This is also the *maximum price that you should be willing to pay* for having that perfect information. Notice that the EVPI will always be greater than zero. Indeed, whatever decision is taken without perfect information, the value of (1.32) will be greater than the value of (1.33), since every loss $L(d_i, \theta_j)$ in the former is replaced by a loss equal to $\min_i L(d_i, \theta_j)$ in the latter which cannot be greater than any $L(d_i, \theta_j)$, though it will be equal to one of these values. This does not mean, of course, that perfect information always reduces the expected loss of the best decision (or, equivalently, raises the expected utility of the best decision). It can reduce your expectation, and it can be 'good' or 'bad' news, but it always has an informative value because 'bad' news also increases your body of knowledge. If your 'experiment' were of such a kind to tell you the truth, and all the truth, this would be the end of the story. Unfortunately, it will provide only partial information, changing your probabilities for θ_1 and θ_2 without letting them to go to one or zero. The DNA analysis you can perform has a conditional genotype probability greater than zero. The best decision after having observed the result of the 'experiment', where E_2 denotes an observed correspondence and $\bar{E}_2$ non-corresponding profiles, will be the decision that minimizes the expected loss calculated using the posterior probabilities:

$$\min_i \sum_{j=1}^{2} L(d_i, \theta_j) Pr(\theta_j | E_1, E_2). \tag{1.34}$$

Equation (1.34) is also called the *Bayes risk*. If you wish to know before performing the 'experiment' the expected gain in information in order to compare it with its 'cost', then you can proceed in this way. You do not know what will be the outcome of the experiment but you do know the likelihoods. Therefore, you can calculate the probabilities for the data by extending the conversation:

$$Pr(E_2 | E_1) = \sum_{j=1}^{2} Pr(E_2 | \theta_j, E_1) Pr(\theta_j | E_1).$$

Now, given that you should chose the action that minimizes the loss for any possible result of the 'experiment', your *ELPI* (note the lower case letter p to denote 'partial' as compared to the previous notation P denoting 'perfect') is calculated by multiplying the minimum expected loss for each possible result of the 'experiment' (Equation 1.34) by the probability of that result and sum all these products:

$$\sum_{k=t,f} \min_i \sum_{j=1}^{2} L(d_i, \theta_j) Pr(\theta_j | E_1, E_2 = k) Pr(E_2 = k | E_1). \tag{1.35}$$

Application of the formula of Bayes theorem in (1.35) gives

$$\sum_{k=t,f} \min_i \sum_{j=1}^{2} L(d_i, \theta_j) \left[\frac{Pr(E_2 = k | \theta_j, E_1) Pr(\theta_j | E_1)}{Pr(E_2 = k | E_1)} \right] Pr(E_2 = k | E_1).$$

Cancelling out $Pr(E_2 = k | E_1)$ then leads to

$$ELpI = \sum_{k=t,f} \min_i \sum_{j=1}^{2} L(d_i, \theta_j) Pr(E_2 = k | \theta_j, E_1) Pr(\theta_j | E_1). \tag{1.36}$$

Again, your choice before the 'experiment' was given by formula (1.32). Therefore, the difference between (1.32) and (1.36) is the measure of the *expected value of partial information* (*EVpI*) (with the sign inverted):

$$EVpI = \min_i \sum_{j=1}^{2} L(d_i, \theta_j) Pr(\theta_j | E_1)$$

$$- \sum_{k=t,f} \min_i \sum_{j=1}^{2} L(d_i, \theta_j) Pr(E_2 = k | \theta_j, E_1) Pr(\theta_j | E_1). \tag{1.37}$$

This is also the maximum price that you should be willing to pay for having this partial information. Thus, the original problem has been solved. If you can meaningfully compare the value of (1.37) with a given cost of the 'experiment', you will be able to decide whether it is worthwhile to conduct the 'experiment'.

Example 1.6 *(Forensic identification/individualization, continued) In this example, the expected loss is a probability, the ELPI is zero, and also the expected value of partial information will be a probability, and this allows at least a qualitative judgement to be made of the order of magnitude of the reduction in the expected loss provided by the forensic result or finding, even though the reduction in the expected loss cannot be directly compared to a quantitative cost of the 'experiment'. The calculations are given in Table 1.8.*

The decision losses $L(d_i, \theta_j)$ for each decision $d_i \in D$, $i = \{1, 2, 3\}$, states of nature θ_j ($j = \{1, 2\}$), the prior probabilities $Pr(\theta_j)$ and the likelihoods (from top to bottom) are entered in the first two numerical columns. For any decision d_i and 'experimental' result E_2, part of formula (1.37), namely,

$$\sum_{j=1}^{n-2} L(d_i, \theta_j) Pr(E_2|\theta_j, E_1) Pr(\theta_j|E_1),$$

may then be calculated as follows. Consider each of the first two columns (i.e. each hypothesis θ_j) and multiply together the values for $L(d_i, \theta_j)$, the prior probability $Pr(\theta_j|E_1)$, and the likelihoods for E_2, and sum the products over the two hypotheses. For d_1 and E_2, we have

$$(0 \times 0.9 \times 1) + (1 \times 0.1 \times 0.0001) = 0.00001.$$

For d_1 and $\bar{E}_2$, we have

$$(0 \times 0.9 \times 0) + (1 \times 0.1 \times 0.9999) = 0.09999.$$

Table 1.8 Illustrative values for losses in an identification scenario with states of nature θ_1 ('Suspect is donor') and θ_2 ('Some other person is donor'). Events E_2 and $\bar{E}_2$ represent, respectively, the possible outcomes 'observed correspondence' and 'observed difference' during a preliminary (DNA) analysis.

Decisions	θ_1	θ_2	Expected losses for decisions d_i and outcomes	
	Losses $L(d_i, \theta_j)$		E_2	$\bar{E}_2$
d_1: identification	0	1	**0.00001**	0.09999
d_2: inconclusive	0.001	0.001	0.00090001	0.00009999
d_3: exclusion	0.01	0	0.009	**0**
	Prior probabilities			
$Pr(\theta_j|E_1)$	0.9	0.1		
	Likelihoods			
$Pr(E_2|\theta_j, E_1)$	1	0.0001		
$Pr(\bar{E}_2|\theta_j, E_1)$	0	0.9999		

Variable E_1 refers to any considered information and data prior to information E_2. Values in bold in the two columns on the right-hand side indicate the minimum expected loss with, respectively, outcomes E_2 and $\bar{E}_2$.
Source: This tabulation method for calculating expected losses is taken from Lindley (1985, p. 130).

For d_2 and E_2, we have

$$(0.001 \times 0.9 \times 1) + (0.001 \times 0.1 \times 0.0001) = 0.00090001.$$

For d_2 and $\bar{E}_2$, we have

$$(0.001 \times 0.9 \times 0) + (0.001 \times 0.1 \times 0.9999) = 0.00009999.$$

For d_3 and E_2, we have

$$(0.01 \times 0.9 \times 1) + (0 \times 0.1 \times 0.0001) = 0.009.$$

For d_3 and $\bar{E}_2$, we have

$$(0.01 \times 0.9 \times 0) + (0 \times 0.1 \times 0.9999) = 0.$$

These values are entered on the right-hand side of the table in the row corresponding to d_i and in the columns labelled E_2 and $\bar{E}_2$. Then, the smallest values *of columns E_2 and $\bar{E}_2$ (shown in bold in Table 1.8) are selected and added together to give the ELpI (1.36). This value is $(0.00001 + 0) = 0.00001$.*

Given that the minimum expected loss without the information provided by the DNA analysis is 0.001, the expected value of this partial information is, following Equation 1.37, $(0.001 - 0.00001) = 0.00099$. One can immediately check that this result agrees with the decision tree analysis made above (Figure 1.3).

1.5 Further readings

The 'subjectivist interpretation' of probability defended in this chapter can be traced back to the origins of the modern mathematical theory of probability: see Daston (1988) and the classical study of Ian Hacking (1975), in particular his Chapter 11, on Christian Huygens' book *De Ratiociniis in Aleae Ludo*, published in 1657, and his concept of 'expectation'. Huygens' expectation is the same concept used here and that is explained in Section 1.3.2. Blaise Pascal, as well, defined probabilities in terms of expectations in his essay on the arithmetical triangle, written in 1654: see Edwards (1987). At the beginning of his essay, Thomas Bayes (1763) defined probability in terms of betting odds. Modern Bayesianism originates in the papers that the English logician and philosopher Frank P. Ramsey and the Italian mathematician Bruno de Finetti wrote independently of each other (Ramsey 1931; de Finetti 1930a; de Finetti 1937). The approach began to be widely known with the work of Savage (1972), first published in 1954. Modern classics are Lindley (1965) and de Finetti (1975). Related, but different, recent viewpoints about the foundations of probability may also be mentioned: the game-theoretic approach of Shafer and Vovk (2001), the 'objective Bayesianism' of Jaynes (2003) and the 'prequential' approach of Dawid and Vovk (1999). An accessible introduction to the different interpretations of the concept of probability and relationship about probability and inductive logic is Galavotti (2005).

In the framework of the Bayesian programme, the tendency to base degrees of belief on proportions (relative frequencies), as discussed in Section 1.3.1, is not only fully acknowledged as 'reasonable' but takes the form of a mathematical theorem, de Finetti's *representation theorem* (de Finetti 1930b; de Finetti 1937). The theorem says that the convergence of one's personal probabilities towards the values of observed proportions, as the number of observations increases, is a logical consequence of Bayes' theorem if a condition called *exchangeability* is satisfied by our degrees of belief, prior to observations.

A sequence of observations, for which non-Bayesian statisticians would say that it is governed by a 'true' but unknown probability distribution P (one in which observations are *conditionally independent and identically distributed*, given P), is a sequence of observations that subjective Bayesians would say is *exchangeable*.

De Finetti's theorem and other so-called convergence theorems, which have been proved lately demonstrating a generalization of de Finetti's exchangeability, enable it to be said that a probability assignment situated around the value of an empirical proportion (relative frequency) is 'objective' in the sense that several persons, whose *a priori* probabilities were different, would converge towards the same *posterior* probabilities, were they to know the same data and share the same likelihoods. This usually happens in the conduct of statistical inference where likelihoods are provided by the choice of appropriate probability distributions, that is statistical models, so that they are the same for any observer who agrees on the choice of the model. For a survey of convergence theorems in Bayesian statistics, one can see Bernardo and Smith (2000) and Schervish (1995).

With regard to the likelihood ratio as a measure of the probative value of scientific findings (Section 1.2.3), the philosopher Charles Peirce, at the end of nineteenth century, was the first to suggest the use of the *logarithm* of the likelihood ratio as a measure of the information contained in the observations, calling it the *weight of evidence*, a suggestion made anew by the mathematician Alan Turing, when he was working to break the *Enigma* code during the Second World War (Good 1985). The choice is motivated by the fact that the logarithm has many of the properties a measure of information should have. For example, if the value, $V = 1$, the logarithm of V is zero, and insofar as the information is inconclusive, that is it does not falsify one of the alternative hypotheses, the logarithm is different from zero. Another desirable intuitive property of such a measure is that the overall information content of two pieces of information, which are probabilistically independent from each other, should be given by the sum of the information. With the logarithmic measure, if two pieces of information are probabilistically independent, then Bayes' factors multiply and weights of evidence add.

Jeffrey's rule (Section 1.2.6) has been discussed at length in the philosophical literature on induction and probability, after it was put forward by Jeffrey (1983). As a matter of fact, the rule was proposed by the English astronomer William Donkin at the middle of nineteenth century (Donkin 1851). The rule is another form of Fisher's factorization theorem (Diaconis and Zabell 1982), and equivalent versions of it have been put forward, independently by Jeffrey, in the field of artificial intelligence by Lemmer and Barth (1982) and, with far, reaching consequences, by Lauritzen and Spiegelhalter (1988). The formal equivalence between Jeffrey's rule and a *minimum principle* has also been demonstrated by many authors (Domotor et al. 1980; Williams 1980). The equivalence can be stated as follows: (1.22) operates the minimal change in Probabilities, which is needed to take into account the new information $Pr_1(R|I) \neq Pr_0(R|I)$ *and only this information*, in the sense that it minimizes the 'distance' between old $Pr_0()$ and new $Pr_1()$ as measured by the *Kullback–Leibler measure of relative entropy* (Kullback 1959).

The Kullback–Leibler measure $H(Pr_0, Pr_1)$ of the distance between $Pr_0()$ and $Pr_1()$ is defined as

$$H(Pr_0, Pr_1) = \sum_e \log \left[\frac{Pr_1(e)}{Pr_0(e)} \right],$$

where all the events e together form a partition of the space of events (e.g. the sample space); the set $\{R, \bar{R}\}$ in (1.22) is such a partition.

Relative entropy $H(Pr_0, Pr_1)$ is only one amongst several possible measures of 'nearness' for probability distributions, and a *symmetry* argument has been provided by the philosopher of science van Fraassen (1986, 1989) for supporting this particular choice. He has shown that (1.22) is the only updating rule for probabilities, which satisfies a very general invariance requirement that can be simply formulated using Edwin Jaynes' words: 'In two problems where we have the same state of knowledge, we should assign the same probabilities' (Jaynes 1983, p. 284). Tikochinsky et al. (1984) have put forward another argument, which is less general than van Fraassen's, in that it makes use of the concept of independent repetitions of the same experiment.

The Bayesian approach to induction outlined briefly in Section 1.3.5 has been defended by many philosophers as a valid unifying 'rational reconstruction' of scientific reasoning (Bovens and Hartmann 2003; Horwich 1982; Howson and Urbach 1996; Jaynes 2003; Jeffrey 2004; Jeffrey 1992; Salmon 1990; Swinburne 2002). Without dwelling on the merits (or demerits) of the Bayesian approach in general, it is worth noting that some of the criticisms raised against the claim that Bayesian inference is a good 'rational reconstruction' of scientific inference do not apply in the context of forensic inference. First, it has been observed that, in the case where H_1 is a high-level scientific theory, it is very often impossible to specify an alternative theory H_2 in such a way that the likelihood $Pr(E|H_2)$ would turn out to be something different from 'anybody's guess' (Earman 1992, p. 164). In the case of forensic inference, however, the competing 'theories' are well defined: they are the prosecution hypothesis and the defence hypothesis. Second, Bayesian inference cannot give any rational guidance when the body of evidence for H_1 is different from the body of findings for H_2 for, obviously, the odds form of Bayes' theorem can be applied only when the findings, E, are the same for both theories. However, in the case of forensic evidence, the body of evidence accepted by the Court is the same for prosecution and defence. Third, when both hypotheses are deterministic and entail the evidence E, then the ratio of their probabilities conditional on E reduces to the ratio of their prior probabilities. In the case of forensic evidence, rarely is this the case. In summary, it has been shown that Bayesian inference is a good 'rational reconstruction' of forensic inference as a particular instance of scientific reasoning.

Formal proofs of the expected utility theorem (Section 1.4.2) were given for the first time by Ramsey (1931), and von Neumann and Morgenstern (1953) and Marschak (1950) provided the axiomatization in terms of 'gambles'. Classical textbooks were written by Savage (1954), Luce and Raiffa (1958), Raiffa (1968) and DeGroot (1970), and more recent texbooks are written by Berger (1988) and Robert (2007).

2

The logic of Bayesian networks and influence diagrams

2.1 Reasoning with graphical models

2.1.1 Beyond detective stories

Sherlock Holmes was a lucky guy. He had Conan Doyle at his disposal to arrange the plot in such a way that the mechanism of eliminative induction worked neatly, the very mechanism simple-minded Watson was apparently unable to appreciate, as Holmes remarked in the novel *The Sign of Four* (Conan Doyle 1953, p. 111).

> 'You will not apply my precept', he said, shaking his head. 'How often have I said to you that when you have eliminated the impossible, whatever remains, *however improbable*, must be the truth?'

Holmes' friend did not really deserve such a rebuke: the 'precept' is not easy to apply. The philosopher of science John Earman has observed, with respect to Holmes' *dictum*, that

> (...) the presupposition of the question can be given a respectable Bayesian gloss, namely, no matter how small the prior probability of the hypothesis, the posterior probability of the hypothesis goes to unity if all of the competing hypotheses are eliminated. This gloss fails to work if the Bayesian agent has been so unfortunate as to assign the true hypothesis a zero prior. (Earman 1992, p. 163)

In different words, the gloss fails to work if the agent has been so unfortunate as to choose the wrong model. This does not usually happen in literary fiction.

Bayesian Networks for Probabilistic Inference and Decision Analysis in Forensic Science, Second Edition.
Franco Taroni, Alex Biedermann, Silvia Bozza, Paolo Garbolino and Colin Aitken.
© 2014 John Wiley & Sons, Ltd. Published 2014 by John Wiley & Sons, Ltd.

> The classic English detective story is the paradigm of eliminative induction at work. The suspects in the Colonel's murder are limited to the guests and servants at a country estate. The butler is eliminated because two unimpeachable witnesses saw him in the orangery at the time the Colonel was shot in the library. Lady Smyth-Simpson is eliminated because at the crucial moment she was dallying with the chauffeur in the Daimler, *etc.* Only the guilty party, the Colonel's nephew, is left. (Earman 1992, pp. 163–164)

In real life, it can happen that the true hypothesis is not considered. This is neither the only problem in applying the precept of the down-to-earth Holmes nor the biggest. If the model does not include the true hypothesis, this is not a shortcoming of probability theory. Probability rules merely deduce probable consequences from probable premises: if the premises are wrong, it is not the fault of the theory. The real problem is different; it is the very application of Bayesian machinery to real-life problems that seems to be practically challenging. Besides background knowledge, the application of Bayes' theorem to a new piece of information requires the totality of the evidence already considered, with the omission of nothing, for the logic of probability is non-monotonic: the probability of B, conditional on A, can be different from the probability of B, conditional on A and C. For example, the probability that footprints are left in the Meadow, given that a certain person walked through it is different, according as to whether the soil was wet or not.

Denote by E the totality of the evidence less B. Evaluation of the probability of a proposition H, $Pr(H|B, E, I)$ requires account to be taken of the fact that E is a logical conjunction of many components. As the number of those components increases, the task will rapidly grow to an intractable level of complexity. At the dawn of the twentieth century, Watson and Holmes might have known Bayes' theorem, but they would have stopped before the wall of that complexity. At the dawn of the twenty-first century, Bayesian networks have come to the rescue.

2.1.2   Bayesian networks

A Bayesian network is a type of graphical model whose elements are nodes, arrows (also called *arcs*) between nodes and probability assignments. A finite set of nodes together with a set of arrows (i.e. directed links) between nodes forms a mathematical structure called a *directed graph*. If there is an arrow pointing from node X to node Y, it is said that X is a parent of Y and Y is a child of X. In Figure 2.1(i), for example, node A is parent of nodes B and E, E

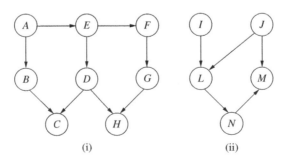

(i)   (ii)

Figure 2.1   (i) A directed acyclic graph and (ii) a directed graph with one cycle.

is a parent of D and F and B and D are parents of C. A node with no parents is called a *root node*. Nodes A and I in Figure 2.1(i) and (ii), respectively, are root nodes.

A sequence of consecutive arrows connecting two nodes X and Y, independently from the direction of the arrows, is known as a path between X and Y. For example, in Figure 2.1(i), there are three paths linking nodes A and C: one through the intermediate node B, the second through the intermediate nodes E and D and the third is the path $A - E - F - G - H - D$; in Figure 2.1(ii), the nodes I and J are linked by the paths $I - L - N - M - J$ and $I - L - J$.

Nodes can have ancestors and descendants: node X is an ancestor of node Y (and Y is a descendant of X) in the case where there is a unidirectional path from X to Y, linking intermediate nodes. In Figure 2.1(i), A is a parent of B and E and an ancestor of every other node in the graph. In Figure 2.1(ii), node L is an ancestor of itself. Hereafter, only directed graphs that do not contain cycles and that are connected will be considered. A cycle is said to exist if a node is an ancestor, and hence descendant, of itself, and a graph is connected if there exists at least one path between every two nodes. Both graphs in Figure 2.1 are connected. A connected directed graph with no cycles is called a *directed acyclic graph (DAG)*.

A formal definition of Bayesian networks is given in Section 2.1.8. Meanwhile, consider a Bayesian network as a DAG in which there are two main elements, nodes and arcs. Nodes represent random variables where the random variable may be either discrete or continuous. The discrete nodes may be categorical (qualitative) or quantitative taking integer values with a finite set of mutually exclusive states.

Arcs represent direct relationships amongst variables. For each variable X with parents $Y_1, Y_2, \ldots, Y_n$, there is an associated conditional probability table $Pr(X|Y_1, Y_2, \ldots, Y_n, I)$, where I denotes, as usual, the background knowledge, all the relevant knowledge that does not appear explicitly as a node in the graph. If X is a root node, then its table reduces to probabilities $Pr(X|I)$, unconditional on other nodes in the graph.

Let X be a variable with n mutually exclusive and exhaustive states $x_1, \ldots, x_n$. A state is a possible value associated with a random variable. If X is a root node, then the (un)conditional probability table $Pr(X|I)$ will be an n-table (a table with n entries) containing the probability distribution $\{Pr(X = x_i), i = 1, \ldots, n\}$, with $\sum_{i=1}^n Pr(X = x_i) = 1$. For notational simplicity, explicit mention of background information I has been omitted. Also, when the context is sufficiently clear that there will be no confusion in so doing, the subscript i is omitted from the notation and one writes $Pr(X = x)$ and $\sum Pr(X = x) = 1$.

Let Y be a variable with m states y (using the abbreviated notation introduced above). If Y is a parent of X, then the conditional probability table $Pr(X|Y)$ will be an $n \times m$ table containing all the probability assignments $Pr(X = x|Y = y)$. For example, suppose that variables B and D in Figure 2.1(i) each have two possible states and variable C has three. Then, the conditional probability table for C (Table 2.1) will contain 12 entries $Pr(C = c_i|B = b_j, D = d_k) = p_{ijk}$, with $(i = 1, 2, 3; j = 1, 2; k = 1, 2)$.

Notice that the notation for the states of the variables as used here may vary, in the course of the book, with the domain of application. It may happen, for example, that the name of a variable, say E, also characterizes a state of that variable (Section 6.1). On other occasions, the states of a variable may be described quite differently from the name of the variable. For instance, an individual's genotype, denoted gt, has states describing the components of that individual's DNA profile (i.e. genotypes or individual allele numbers, as discussed in Section 7.2). Sometimes, names and states of variables will be denoted in a way which conforms to the original literature in which the topic of interest was first described (e.g. Section 6.2 on nodes with indexed lower case letters).

Table 2.1 Conditional probability table for child C with parents B and D where B has two states (b_1 and b_2), C has three states (c_1, c_2 and c_3) and D has two states (d_1 and d_2).

		B :	b_1		b_2	
		D :	d_1	d_2	d_1	d_2
C :	$Pr(C = c_1 \mid B = b_j, D = d_k)$		p_{111}	p_{112}	p_{121}	p_{122}
	$Pr(C = c_2 \mid B = b_j, D = d_k)$		p_{211}	p_{212}	p_{221}	p_{222}
	$Pr(C = c_3 \mid B = b_j, D = d_k)$		p_{311}	p_{312}	p_{321}	p_{322}

Bayesian networks have a built-in computational architecture for computing the effect of information − in the context typically referred to as *evidence* − on the states of the variables. The term *evidence*, in a technical sense, is used largely throughout this chapter and is to be distinguished from the notion of 'evidence' as used in discussions about proceedings before trial. Upon learning new evidence, the Bayesian network architecture updates probabilities of the states of the variables. Further, it utilizes probabilistic relationships, both explicitly and implicitly represented in the graphical model, to make computation more efficient.

No algorithms are available to cope with feedback cycles, and this is the reason of the requirement that the network must be a DAG. Computational tractability has been one of the points raised against the use of probability theory in the development of artificial reasoning systems. It was claimed that probability would be not good in artificial intelligence (AI) because the complexity of probability calculations grows exponentially with the number of variables, and neither the human mind nor the most powerful computer imaginable can execute exponential-time calculations.

It is known today that computation in Bayesian networks is, in general, very hard, in the sense that it is at least as complex as a particular class of decision problems, called *NP-problems* (acronym for *Non-polynomial*), for which no algorithm has yet been found that can solve them in polynomial time, but it is also known that, although the general problem is NP-hard, for certain types of graphs, called *junction trees*, there are efficient algorithms for computing the probability distributions of nodes exactly, apart from rounding errors (the type of graph for which this is possible is defined in Section 2.2.3). These algorithms allow the users to handle problems with both discrete and continuous random variables that are already too complex for unaided common sense reasoning (discrete random variables have been defined in Section 1.1.9; continuous random variables are discussed in Chapter 9).

The human mind is good in selecting which features of reality are important but poor at aggregating those features. Human experts are good in building the model, but they are not so good in reasoning through the model. A computer program is not good in building the model, but it is very good in performing calculations. It is acknowledged that Bayesian networks do not describe how the human mind works. It is claimed only that, in simple cases, they provide intuitively reasonable answers and that they are better than human minds in performing some more complex reasoning tasks. The goal thus is not to replace human experts, and the goal is to help them.

2.1.3   A graphical model for relevance

A type of graphical model, namely, probability trees (Figure 1.1), for representing the possible alternative scenarios that can be built from a given set of propositions has already been discussed in Section 1.1.7. This kind of graphical representation can be used only for very small sets of propositions because the number of branches grows exponentially. For n binary variables, the tree has 2^n branches. Moreover, relevance relationships amongst propositions are not imbedded in the structure of the graph. Numerical assignments to the branches are needed to show the relevance relationships amongst the propositions. In Section 1.2.6, for instance, the relevance of H_1 for R can be indicated only by the fact that the probability assigned to the branch labelled R that follows the branch labelled H_1 is different from the probability assigned to the branch labelled R that follows the branch labelled $\bar{H}_1$. Directed graphs provide an economic and powerful model for visualizing dependencies between variables.

From Section 1.1.9, it is known that a proposition can be associated with a binary random variable. With an abuse of notation, nodes of the graph which stand for propositions will be labelled with the same letters as used for corresponding binary random variables. It is interesting to know that the person who first foresaw the possibility of transforming propositions into numerical values, Gottfried Wilhelm Leibniz, was a scholar of mathematics and law. In a paper written in 1669, when he was only 23 years old, dealing with the issue of 'conditional rights', that is, rights which hold only if certain conditions happen to be true, he proposed to assign the number 0 to *jus nullum* (a person has no rights to the possession of a certain asset), the number 1 to *jus purum* (a person has unconditional rights to the possession of a certain asset), and a fractional number to *jus conditionale* (a person has rights to the possession of a certain asset that depends on the occurrence of certain conditions) (Leibniz 1930, p. 420).

Consider two nodes H_1 and R, interpreted as propositions. It is judged that proposition H_1 is relevant for R so a directed arc may be drawn from H_1 and R. The tree in Figure 1.2 can thus be shown in the DAG of Figure 2.2(i) with associated conditional probability values p, q and r, summarised in Table 2.2. The reader can appreciate how parsimonious a directed graph is, compared with a probability tree, even for the simplest possible case. Instead of three random nodes and six arcs, we have only two nodes and one arc and, above all, the relevance of H_1 for R is clearly visualized.

Another advantage of representing relevance by directed graphs is that a node can also represent a cluster of propositions. For example, a variable H, 'Hypothesis', may be defined which has as its possible values the triplet composed of the truth values of the propositions H_1 (Sir Charles Baskerville died by accident), H_2 (Sir Charles Baskerville died by a criminal act) and H_3 (Sir Charles Baskerville committed suicide). The possible states of the variable H will

                       (i)   (ii)

Figure 2.2 (i) A Bayesian network for the probability tree shown in Figure 1.2 where H_1 and R each have two values, t ('true') and f ('false'). (ii) A Bayesian network for the coroner's report.

Table 2.2 Probability tables of the nodes H_1 (left) and R (right) in the Bayesian network shown in Figure 2.2(i).

H_1:	t	f		H_1:	t	f
$Pr(H_1)$	p	$1-p$	$R:$	$Pr(R=t\|H_1)$	q	r
				$Pr(R=f\|H_1)$	$1-q$	$1-r$

Both H_1 and R are binary with states t ('true') and f ('false').

Table 2.3   Probability tables for nodes H (left) and R (right) in Figure 2.2(ii).

H:	H_1	H_2	H_3		H:	H_1	H_2	H_3
$Pr(H = H_i)$	p_1	p_2	p_3	R : $\quad Pr(R = t\|H)$		q_1	q_2	0
				$Pr(R = f\|H)$		$1 - q_1$	$1 - q_2$	1

be $\{1,0,0\}$ (H_1 is true), $\{0,1,0\}$ (H_2 is true) or $\{0,0,1\}$ (H_3 is true). The Bayesian network shown in Figure 2.2(ii) depicts Watson's beliefs (1.9) and (1.10) of the preceding chapter, where the conditional probabilities (summarised in Table 2.3) are simplified by taking, with an abuse of notation, H_1, H_2 and H_3 as the possible states of H.

2.1.4   Conditional independence

The fundamental concept in building Bayesian networks is conditional independence. It usually happens that it must be judged whether a given proposition B is relevant for another proposition A in a context that includes more than background knowledge. For instance, in the Baskerville case, Watson has to judge if the footmarks are relevant to the crime hypothesis, given knowledge of the coroner's report. It is convenient to formulate the concept of conditional independence in terms of variables.

Let X, Y and Z be random variables and let $Pr(X|Y,Z,I)$ denote 'the degree of belief that a particular value of variable X is observed, given that a particular value of variable Y is observed, *and* given that a particular value of variable Z is observed, plus background information I'. Then, X is conditionally independent of Y given Z if and only if $Pr(X|Y,Z,I) = P(X|Z,I)$ for all the states of X, Y and Z.

Statisticians use the concept of conditional independence to build computationally efficient statistical models. Here is a simple example. It is required to assess the outcomes of three successive tosses of a coin that will be thrown by another person. The result will be made available. The coin can be a mint (fair) coin, a coin with two tails or a coin with two heads. Denote the cluster of hypotheses with the variable H and the possible outcomes with binary variables X_1, X_2 and X_3, which can take the values $X_i = h$ (the outcome of toss number i is heads) and $X_i = t$ (the outcome of toss number i is tails). It is obvious that the outcome of each toss depends on which hypothesis is true, and it is also believed that the outcome of each toss in the sequence is relevant for the degree of belief in the outcome of the next toss. These dependence relationships can be represented by means of a directed graph, such as shown in Figure 2.3(i).

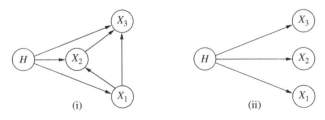

Figure 2.3   (i) A DAG and (ii) a parsimonious DAG for the coin-tossing problem.

This figure may be simplified by taking advantage of the conditional independence structure of the problem domain. Clearly, if it were known which hypothesis was true, the outcome of each toss would be not relevant for the degree of belief concerning the next one. Absence of a direct link between two nodes means that there is no direct dependency between them: they are not directly relevant to each other. The arcs between the X_i variables may be dropped.

The graph in Figure 2.3(ii) depicts explicitly all the direct dependency relationships and only them. This graph is a faithful representation of the dependence structure of the problem, in the sense that all the indirect dependency relationships are implicitly represented. This can be demonstrated by an intuitive analysis.

The outcome of one toss is relevant for the expectation of the outcome of the next toss because it changes the relative degree of belief about the hypotheses, and this, in turn, changes the expectations for the other tosses. Suppose, for instance, that the outcome of the first toss is 'heads': this information changes the expectation of the next two tosses because it falsifies the hypothesis 'two tails'. The effect of the information acts directly on node H, and only indirectly on nodes X_2 and X_3, through the direct influence that node H exerts on them. Imagine that Figure 2.3(i) and (ii) is physical nets where nodes are computing units and information can go from one unit to the others only by travelling along the arcs. Then, information from unit X_1 is transmitted to units X_2 and X_3 by the computations executed by unit H, and channels $X_1 \to X_2, X_1 \to X_3$ and $X_2 \to X_3$ are superfluous. On the other hand, if the value of the variable H is fixed, then the flow of information amongst X_1, X_2 and X_3 through H is blocked. Therefore, the arcs $H \to X_1, H \to X_2$ and $H \to X_3$ are sufficient to represent faithfully the flow of information that travels through the net.

These considerations can be generalized to cover all the possible ways by which information can be transmitted through a variable in a directed graph, obtaining a graphical decision rule to check, for any two variables, whether they are independent given knowledge of another variable or set of variables of the graph.

2.1.5 Graphical models for conditional independence: *d*-separation

There are only three possible connections by which information can travel through a variable in a directed graph, that is diverging, serial, and converging connections [Figure 2.4 (i–iii)].

The case of diverging connections has already been discussed in the coin-tossing problem (Section 2.1.4). Basic statistical models can be represented graphically by diverging connections. A diverging connection [Figure 2.4(i)] is an appropriate graphical model whenever it is believed that Z is relevant for both X and Y and that X and Y are conditionally independent given Z. This means that if the state of Z is known, then knowledge also of the state of X does

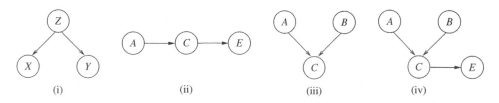

Figure 2.4 (i) A diverging connection, (ii) a serial connection and (iii) a converging connection. (iv) A DAG for 'one-trace' scenarios with binary variables A (Suspect is the offender), B (Blood stain at crime scene comes from offender), C (Blood stain at crime scene comes from suspect) and E (Suspect's blood and crime stain share same DNA profile).

not change the belief about the possible states of Y (and vice versa). It is also appropriate if it is believed that if the state of Z is not known, then knowledge of the state of X provides information about the possible states of Y (and vice versa). Thus, information may be transmitted through a diverging connection unless the value of the middle variable is fixed. It is then said that X and Y are d-separated (directionally separated) given Z.

Serial connections typically apply to 'chain' reasoning. A Markov Chain is a classical example. A serial connection [Figure 2.4(ii)] is an appropriate graphical model whenever it is believed that A is relevant for C, C is relevant for E and A and E are conditionally independent given C. This means that if the state of C is known, then knowledge also of the state of A does not change the belief about the possible states of E (and vice versa), and it also means that if the state of C is not known, then knowledge of the state of A provides information about the possible states of E (and vice versa). Information thus may be transmitted through a serial connection unless the state of the middle variable is known and it is then said that A and E are d-separated given C.

For example, let A in Figure 2.4(ii) be the proposition 'The suspect is the offender', C the proposition 'The blood stain found on the crime scene comes from the suspect' and E the proposition 'The suspect's blood sample and the blood stain from the crime scene share the same DNA profile'. Then, A is relevant for C and C for E, but there might be an explanation of the occurrence of the event described by C different from the occurrence of the event described by A and, given that C occurred, then the observed correspondence in DNA profiles does not depend on which explanation of C is true.

Converging connections describe a slightly more sophisticated kind of reasoning. The ability to cope with this kind of reasoning is a real asset of Bayesian networks. A converging connection [Figure 2.4(iii)] is an appropriate graphical model whenever it is believed that A and B are both relevant for C, A is not relevant for B, but it does become relevant if the state of C is known. In other words, it is believed that A and B are unconditionally independent but conditionally dependent given C. This means that if the state of C is known, then knowledge of the state of A provides information about the possible states of B (and vice versa). But, it also means that if the state of C is not known, then knowledge of the state of A provides no information about the possible states of B (and vice versa). The flow of information is blocked if the state of the middle variable is unknown.

Consider the following example. Let A and C in Figure 2.4(iii) be the propositions 'The suspect is the offender' and 'The blood stain found on the crime scene comes from the suspect', respectively, and B the proposition 'The blood stain found on the crime scene comes from the offender'. Knowledge that either the event described by A or the event described by B occurred does not provide information about the occurrence of the other. The fact that the source of the blood stain is the offender has no bearing at all, taken alone, on the culpability of the suspect. But, if C is true, then A and B become obviously related.

Converging connections have another peculiarity. To open the channel, it is not necessary that the state of the middle variable is known for certain but it suffices that there is some information, even though it is not such as to achieve certainty. It is customary in the literature about Bayesian networks to call information that fixes the state of a variable 'hard evidence', otherwise it is called 'soft evidence'. In other words, hard evidence bearing on a given variable allows its true state to be known, whereas soft evidence allows only the assignment of probability values lower than one (the value associated with certainty) to the states of the variable. Bayes' theorem deals with hard evidence, whereas formula (1.22), the so-called Jeffrey's rule, deals with soft evidence. For converging connections, it can be said, thus, that information may only be transmitted if either there is some evidence (hard or soft) bearing on the middle variable or there is some evidence bearing on one of its descendants.

The last condition can be easily understood by an example. Put together Figure 2.4(ii) and (iii) with the nodes interpreted as above, and a directed graph for the 'one-trace' scenario (Garbolino and Taroni 2002, p. 151), depicted by Figure 2.4(iv), is obtained. A formal development of the likelihood ratios associated with this Bayesian network is presented later in Section 6.1. Knowing that the DNA profile of the blood stain corresponds to the profile of the blood of the suspect, that is, knowing the state of child E of C, opens the channel between A and B because this information, in turn, changes the degree of belief in C, even though this degree is still lower than certainty, allowing for the possibility of a correspondence by chance.

2.1.6 A decision rule for conditional independence

Following the rules given in the preceding paragraph, it is possible to build up step by step a Bayesian network for a given set of variables such that if variables X and Y are conditionally independent given Z, then X and Y are d-separated given Z in the network. With the same rules, it is also possible to decide for any pair of variables X and Y in a given Bayesian network whether they are independent given another variable Z. Thus, a decision rule for d-separation can be formulated in general for subsets of variables in a DAG (Pearl 2000, pp. 16–17). That is, let $\mathbf{X}$, $\mathbf{Y}$ and $\mathbf{Z}$ be disjoint subsets of variables in a *DAG*. Then, $\mathbf{X}$ and $\mathbf{Y}$ are d-separated given $\mathbf{Z}$ if and only if every path between a variable in $\mathbf{X}$ and a variable in $\mathbf{Y}$ contains either (i) a serial connection $\rightarrow Z \rightarrow$, or a diverging connection $\leftarrow Z \rightarrow$, such that the middle node Z belongs to $\mathbf{Z}$, or (ii) a converging connection $\rightarrow W \leftarrow$ such that the middle node W does not belong to $\mathbf{Z}$ and no descendant of W belongs to $\mathbf{Z}$.

As noted above, nodes A and B are not d-separated given $\mathbf{Z} = \{E\}$ in Figure 2.4(iv) because E is a descendant of C. In Figure 2.1(i), nodes B and D are d-separated given $\mathbf{Z} = \{A, E\}$ because there are two paths going between B and D, namely, $B \leftarrow A \rightarrow E \rightarrow D$ and $B \rightarrow C \leftarrow D$. In the first, there is a diverging connection at A (and a serial connection at E). In the second, there is a converging connection at C and neither is it the case that C belongs to the set $\{A, E\}$ nor any descendant of C belongs to $\{A, E\}$ (in fact, C has no descendants). On the other hand, B and D are not d-separated given $\mathbf{Z} = \{A, C\}$ because, in this case, the middle node C of the converging connection belongs to the set $\mathbf{Z} = \{A, C\}$.

It can be proved for Bayesian networks that if sets of nodes $\mathbf{X}$ and $\mathbf{Y}$ are d-separated given $\mathbf{Z}$, then they are conditionally independent given $\mathbf{Z}$. Therefore, it is not necessarily the case that the graphical structure represents all the independence relationships that hold amongst the variables, but the graphical structure ensures that, at least, no false relationship of independence can be read from the graphical structure and that all the dependencies are represented.

2.1.7 Networks for evidential reasoning

Both graphs in Figures 2.3(ii) and 2.4(iv), we can distinguish 'hypothesis nodes' and 'observational (evidence) nodes'. In the statistical model, the hypothesis node is $\mathbf{H}$, the 'parameter' variable, and the observational nodes are the variables X_i describing the possible results of empirical observations. In the 'one-trace' scenario, the hypothesis nodes are A, B and C, namely the propositions that describe singular events whose occurrence is only hypothesized and the evidence node is E, the proposition describing an observed event. The arcs in these graphs are said to go top-down, from hypotheses to results (or findings).

Relevance is a symmetric relation (Section 1.1.4). One may wonder whether graphs with the direction of arcs reversed, going bottom-up, from observations to hypotheses, would not provide models for dependency, which are as good as those in Figures 2.3(ii) and 2.4(iv). The answer is that it is usually easier to construct top-down graphs. This is because Bayesian inference goes top-down, from hypotheses to evidence (Section 1.3.5): top-down arcs go from explanans to explanandum.

Scholars of evidence such as David Schum, Terence Anderson and William Twining have argued that what is to be sought in evidential reasoning are justified or warranted beliefs. As Schum (1994, p. 81, note 4) wrote:

> (...) people whose opinions we are trying to influence will wish to know why we believe we are entitled to reason from one stage to another (...) For each reasoning stage or step identified, we have to make an assertion about what we believe provides the ground for or gives us license to take each one of these steps. Amongst some logicians and legal scholars such assertions are called generalizations. Other writers refer to such assertions as warrants. Whether we call them generalizations or warrants, the intent is the same; they are assertions we make about why we believe we are entitled to reason from one stage to another.

The question 'Why do we believe we are entitled to make an inference from observation E to hypothesis H?' is not to be confused with the question, 'Why do we believe, to a certain degree, that H is true?' The answer to the latter question is: 'Because we have observed that E is true and E is relevant for the truth of H'. However, the former question asks 'Why do we believe that E is relevant for H?' The answer to this question is: 'Because H is a possible explanatory fact of E'.

This is an important point related to the issue of 'subjective' versus 'objective' probabilities, which has been already raised in Section 1.3.1. Acknowledgement in the evaluation of evidence that one has to trust one's personal beliefs does not mean that one is left merely with personal feelings that something is relevant for someone else. One is asked to explain *why* it is believed that an observed event E is relevant for another, unobserved, event H. The answer points out to a potential explanation of E whose explanans contains H plus scientific laws and common sense generalizations that provide the required warranties of the inference. Scientific laws and common sense generalizations are 'objective' in the sense that there is a widespread agreement about them. Stated otherwise, E is 'objectively' relevant for H if H is an explanatory fact in a potential explanation of E.

Anderson and Twining associate 'probandum' with 'explanans' and 'probans' with 'explanandum'. The question to be addressed by the 'proponent of evidence' is, according to them (Anderson and Twining 1998, p. 71):

> *Does the evidentiary fact point to the desired conclusion* (not as the only rational inference, but) *as the inference* (or explanation) *most plausible or most natural out of the various ones that are conceivable*? Or (to state the requirement more weakly), is the desired conclusion (not, the most natural, but) a natural or plausible one amongst the various conceivable ones? (...) How probable is the *Probandum* as the explanation of this *Probans*? [Italics by the authors]

In addition, the task of the 'opponent of evidence' is to provide an alternative explanation (Anderson and Twining 1998, p. 73):

> That fact has been admitted in evidence, but its force may now be diminished or annulled by showing that some explanation of it other than the proponent's is the true one. Thus every sort of evidentiary fact may call for treatment in a second aspect, by the opponent, viz.: *What are the other possible inferences which are available for the opponent as explaining away* the force of the fact already admitted. [Italics by the authors]

The quest for explanations breaks the symmetry of relevance relationships assigning a vantage point to top-down directed graphs: the probans is relevant for the probandum and vice-versa, but it is only the latter that explains the former. In the coin-tossing problem of Figure 2.3(ii), the state of X_i is relevant for the possible states of H but it is the state of H that explains the state of X_i.

Consider again the 'one-trace scenario' of Figure 2.4(iv). The appropriate conditional probability tables are given in Tables 2.4 and 2.5. Notice that some of the relationships in the network are deterministic, that is they have assigned probability values of one or zero. Consider the link between the variables C and E. The state of E provides information about the possible states of C, but it is the state of C that explains the state of E. Also intermediate hypotheses are to be explained by means of high-level hypotheses. Node C is the intermediate hypothesis whose state can be explained by the states of A and B. Node B represents the factor that has been called *relevance* (Stoney 1991) because its probability $Pr(B|I)$ encodes the information available for the belief that the blood stain is relevant for the case. In the conditional probability tables (Table 2.5), taken from Garbolino and Taroni (2002, p. 151), assignments of probabilities 1 and 0 are straightforward, probability γ is the probability of correspondence by chance and p the probability that the stain would have been left by the suspect even though he was innocent is to be assigned on the basis of the alternative explanation proposed by the defence and the information related to that hypothesis. If the defence agrees that C is true, then, of course, this scenario is no longer appropriate.

Table 2.4 Unconditional probabilities assigned to, respectively, the nodes A (left) and B (right) of Figure 2.4(iv).

A:	t	f	B:	t	f
$Pr(A)$	a	$1-a$	$Pr(B)$	r	$1-r$

Table 2.5 Conditional probability tables for the node E (left) and for the node C (right) of Figure 2.4(iv).

					A:	t		f	
	C:	t	f		B:	t	f	t	f
E:	$Pr(E=t|C)$	1	γ	C:	$Pr(C=t|A,B)$	1	0	0	p
	$Pr(E=f|C)$	0	$1-\gamma$		$Pr(C=f|A,B)$	0	1	1	$1-p$

As an aside, note that the notion 'explanation' used throughout this section refers to a reasonable configuration of affairs. In forensic contexts, it has been mentioned that the term *explanations* is sometimes used in an abusive way for the sole purpose of constructing an alternative scenario (e.g. for the denominator of the likelihood ratio) that 'explains' the results as well as the hypothesis of primary interest (i.e. as considered in the numerator of the likelihood ratio), so that the ensuing likelihood ratio would be one (Evett et al. 2000b). Often, such ad hoc explanations are either a statement of the obvious or fanciful or speculative. They are to be distinguished from genuine alternative hypotheses that offer reasonable explanations as understood in the discussion here.

2.1.8   The Markov property

The multiplication law of probability theory allows us to decompose a joint probability distribution with n variables as a product of $n - 1$ conditional distributions and a marginal distribution:

$$Pr(X_1, \ldots, X_n) = \left[\prod_{i=2}^{n} Pr(X_i | X_1, \ldots, X_{i-1}) \right] Pr(X_1).$$

The fundamental property, known as the *Markov property*, required of a DAG, with a joint probability distribution $Pr()$ over its variables, for it to be a Bayesian network is the following. For every variable X and every set $\mathbf{Y}$ of variables that exclude all members of the set $\mathbf{DE}(X)$ of descendants of X in the DAG, X is conditionally independent of $\mathbf{Y}$ given the set $\mathbf{PA}(X)$ of the parents of X, that is

$$Pr(X|\mathbf{PA}(X), \mathbf{Y}) = Pr(X|\mathbf{PA}(X)).$$

Example 2.1 *(Scenario 'Jack loved Lulu') Suppose that the hypothetical character Lulu, a young girl, has been found murdered at her home with many knife wounds. The knife has not been found. Some blood stains have been recovered at the scene of the crime which do not share Lulu's DNA profile. A friend of hers, Jack, has been seen by John near Lulu's house around the time of the murder. There is some evidence that Jack was madly in love with Lulu. This information is considered as background information for the evaluation of the probability of J, a root node defined in Figure 2.5. John has stated that he also was in love with Lulu. Control blood has been taken from Jack. The fact that John bore witness as he did is an event that must be explained, and the explanatory facts for it are that the event John related occurred and that John is a reliable witness, under the common sense generalization that 'reliable witnesses usually say the truth'. It is not known for certain that Jack loved Lulu, otherwise that would be part of the background information also and it would not be a node of the graph. Instead, it is a hypothesis based on information which is not totally reliable and a hypothesis that can furnish an alternative explanation of Jack's presence near Lulu's house. It is also assumed that the fact that John himself fell in love with Lulu can prejudice his reliability as a witness. The Bayesian network modelling this example is shown in Figure 2.5. The network could be enlarged by taking into account the testimony, and its reliability, of some witness that John was jealous of Jack and also by taking into account the analysis of eyewitness reliability through consideration of the variables involved in the evaluation of testimonial evidence (Schum 1994, pp. 100–114; 324–344). In this case, the network shown in Figure 2.5 would become very complex. In Section 2.2.5, we shall show how the graphical analysis can be simplified using object-oriented Bayesian networks.*

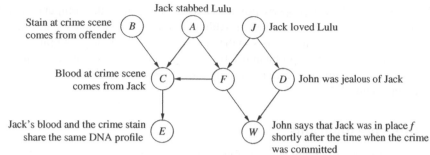

Figure 2.5   The 'Lulu' Bayesian network with node descriptions displayed beside each node.

The scenario in the example here is admittedly hypothetical, but not unrealistic as such. For a practical example of a scenario involving a victim of a stabbed and DNA profiling results, although in a different configuration, see the case of the murder of British student Meredith Kercher in Perugia (Italy) in 2007, also known as the Amanda Knox case (Vuille et al. 2013). The joint probability distribution associated with the DAG in Figure 2.5 is

$$Pr(A, B, C, D, E, F, J, W), \tag{2.1}$$

and it is easy to check that the Markov property holds. For example, B is d-separated from $\{A, F, J, D, W\}$ *given the null set* $\emptyset$ *(root nodes have the null set as the only parent), because the path to A contains a converging connection at C and C does not belong to* $\emptyset$. *On the other hand, B is not d-separated from E, given* $\emptyset$, *because the connection at C is serial. Consider two more examples: F is d-separated from* $\{B, D\}$ *given* $\{A, J\}$ *because there is a diverging connection at A and a converging connection at C, and neither C nor E belong to* $\{A, J\}$; *D is d-separated from* $\{A, B, C, E, F\}$ *given* $\{J\}$ *because there is a diverging connection at J and a converging connection at W.*

From the Markov property, the so-called chain rule for Bayesian networks follows immediately. It states that a Bayesian network can be factorized as the product, for all variables in the network, of their probabilities conditional on their parents only:

$$Pr(X_1, \ldots, X_n) = \prod_{i=1}^{n} Pr(X_i | \mathbf{PA}(X_i)).$$

For example, the joint distribution (2.1) can thus be factorised as

$$Pr(A)Pr(B)Pr(C|A, B, F)Pr(E|C)Pr(F|A, J)$$
$$Pr(J)Pr(D|J)Pr(W|D, F). \tag{2.2}$$

The task of computing a given joint probability distribution becomes very rapidly intractable because its complexity increases exponentially with the number of variables. In a simple problem with only eight binary variables like the 'Lulu' network, the joint distribution contains

$2^8 = 256$ values. The Markov property provides a method by which the computational problem of probability calculus may be handled. Although it does not guarantee a tractable task, the method is very efficient. The key lies in the observation that, although the full joint probability distribution of the 'Lulu' problem needs memory space for 2^8 values, the computation of the biggest factor in Equation (2.2) needs only memory space for 2^4 values. The method is exemplified in Section 2.2.1.

2.1.9   Influence diagrams

Decision trees are a useful graphical way of performing all the calculations needed to choose the best action in face of uncertainty, taking account also of the possibility of collecting new information, if there is such a possibility, as we have seen in Section 1.4.5. But, for complex decisions, a decision tree grows in a way that quickly became quite cumbersome. In the 1980s, a more compact representation was devised, called *influence diagrams* (Howard and Matheson 1984; Shachter 1986).

A *discrete influence diagram* is a DAG whose set of nodes **X** contains three types of nodes. One is a set $\mathbf{X_C}$ of chance nodes, representing discrete random variables, with a finite set of mutually exclusive states, with a joint probability distribution $Pr()$ over the set $\mathbf{X_C}$ of chance nodes. Another set is $\mathbf{X_D}$, with decision nodes, represented by square boxes, which have a finite set of feasible alternative decisions. Utility nodes form a third set, $\mathbf{X_U}$, represented by diamond boxes, which have no children and have no states, with a set of utility functions $U()$, or a set of loss functions $L()$, associated to the set $\mathbf{X_U}$ of utility nodes.

There is only one utility function $U(\mathbf{PA}(u))$, or a loss function $L(\mathbf{PA}(u))$, for each node $u \in \mathbf{X_U}$, where $\mathbf{PA}(u)$ is the subset of **X** constituted by the parents of u. Each local utility function $U(\mathbf{PA}(u))$ or loss function $L(\mathbf{PA}(u))$ represents an additive contribution to the total utility function $U()$ or total loss function $L()$. The total utility (loss) function is the sum of the utility (loss) functions in the influence diagram. Decision nodes have no probability tables associated with them because it is not meaningful to assign probabilities to variables under the control of the decision maker. Decision nodes can be arguments of the conditional probability table of chance nodes, that is a possible state of nature can be influenced not only by other states of nature but also by decisions formerly taken by the decision maker.

There are three types of links in an influence diagram. They are commonly represented in the same way, that is by continuous arrow like usual links in Bayesian networks, but they are conceptually different. In this chapter, for didactical purposes, we distinguish them with different names and different notation: conventional probabilistic links, denoted by continuous arrows, informational links, denoted by broken arrows (i.e. with a dashed line), and functional links, denoted by dotted arrows. These notational conventions will not be followed, for the sake of simplicity, in the following chapters, when all the links in influence diagrams will be denoted by continuous arrows, according with the standard usage.

A link from a chance node X to another chance node Y is a *probabilistic link*, representing a relation of direct probabilistic dependence amongst the nodes. In turn, a link from a chance node X to a decision node D is an *informational link*, representing the fact that knowledge of the state of random variable X has an influence on the choice of the alternative decisions represented by decision node D. When there are links from a chance node X, and from a decision node D, to a utility node u, then one refers to them as *functional links* because they represent the functional dependence of the possible consequences of D on the possible states of nature and the choice amongst possible actions in D. A link from a decision node D_i to another decision node D_j is an *informational* link, representing the fact that decision D_i is taken before decision D_j and the consequence of D_i is known before taking decision D_j. A link from a decision node D to a chance node X is considered to be a *probabilistic* link, even though the state of a random variable that represents possible states of nature is not (usually) probabilistically dependent on the choice of the decision maker. The decision maker can be able to bring about an event, although she is not able to influence the result.

For example, the decision maker can decide to conduct an 'experiment', but she is not able to influence the outcome of the 'experiment'. This situation is dealt with by assigning to a chance node X, which has a decision node D amongst its parents, a conditional probability table with some *ad hoc* values, as it will be shown in the following example. In an influence diagram, one must be able to determine which variables are known before making each decision. This means that there can be only one sequence in which the decisions are made. For building influence diagrams, the so-called *no-forgetting assumption* is usually made (Kjærulff and Madsen 2008, p. 67). Perfect recall is assumed of all 'experiments' and decisions made in the past, that is, the decision maker knows all past decisions and remembers all previously known information. This assumption implies that a decision variable and all its parents are informational parents of all successive decision variables, and, under this assumption, it is not necessary to connect directly one decision variable to the next one in the decision sequence: there need only be a *directed path* from the first decision variable in the sequence to the last. It is possible to relax the no-forgetting assumption by using *limited memory influence diagrams* (LMID) (Lauritzen and Nilsson 2001). The difference lies in the fact that, in influence diagrams with perfect recall, some informational links between decision nodes are only implicitly present in the graph, whereas in LMID, one can assume that the decision maker knows only some of the past decisions and some of the previously made observations, so that all the informational links that are assumed to be known by the decision maker must be explicitly depicted in the graph (Kjærulff and Madsen 2008, pp. 72–73).

Example 2.2 *(Forensic identification, continued) The influence diagram for the identification problem is shown in Figure 2.6. The decision node R represents the decision to perform a DNA analysis. The state spaces of the variables are as follows:* $\Theta = \{\theta_1$ *(the suspect is the source),* θ_2 *(some other person is the source)}*; $E_2 = \{true$ *(corresponding DNA profiles), false (no corresponding DNA profiles)}*; $R = \{yes$ *(perform analysis), no (do not perform analysis)} and* $D = \{d_1$ *(identification),* d_2 *(inconclusive),* d_3 *(exclusion)}. The decision sequence is* $\{R, D\}$ *and there is only one directed path from the decision node R to the decision node D so that, under the no-forgetting assumption, there is no need to add a direct link between R and D. It can be seen, comparing Figure 2.6 and Figure 1.3, that the influence diagram is simpler than the equivalent decision tree. Following previous discussion of this scenario (e.g. Example 1.6), the values 0.9 and 0.1 are assigned to the states 'the suspect is the source' and 'some other person is the source', respectively, of the node* Θ. *Two alternative options for assigning ad hoc values to the conditional probability table for node* E_2, *which has the decision node R amongst its arguments, are shown in Table 2.6. One option is to assign a uniform distribution to* $Pr(E_2|R, \Theta)$ *for situations in which the 'experiment' is not performed. That is, any possible truth value of* E_2 *does not affect the decision maker's beliefs about* Θ, *a result that is obtained by assigning equal likelihoods to all the states of* Θ, *as shown in the two rows on the right-hand side in Table 2.6, and the first two rows containing numerical values. The second option is to introduce another possible state in* E_2, *denoted 'no result'. The corresponding distribution is shown in rows three to five in Table 2.6. If the 'experiment' is not performed, then node* E_2 *is instantiated to 'no result'. The second option is semantically more clear than the first*

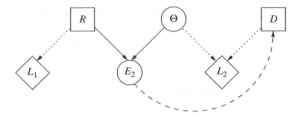

Figure 2.6 Influence diagram for the identification problem (Example 2.2). Arcs with continuous lines represent probabilistic links, arcs with dashed lines $(--)$ are informational links and arcs with dotted lines $(\cdots)$ are functional links. Node D is a decision variable with the set of actions d_1, d_2 and $d_3 \in D$ denoting, respectively, the expert conclusions 'individualise', 'inconclusive' and 'exclusion'. Node R is a further decision variable, with states 'yes' (i.e. perform analysis) and 'no' (i.e. do not perform analysis). The possible states of nature, represented by node Θ, are θ_1 ('The suspect is the source') and θ_2 ('Someone else is the source'). Node E_2 represents the possible outcomes of performing an analysis. Two options for defining the set of states for this node are given in Table 2.6. Nodes L_1 and L_2 define decision losses.

Table 2.6 Two different options for the definition of the set of states of the node E_2 and associated conditional probabilities.

	R :	Yes		No	
	θ :	θ_1	θ_2	θ_1	θ_2
First option:	$Pr(E_2 = true\|R,\theta)$	1	0.0001	0.5	0.5
	$Pr(E_2 = false\|R,\theta)$	0	0.9999	0.5	0.5
Second option:	$Pr(E_2 = true\|R,\theta)$	1	0.0001	0	0
	$Pr(E_2 = false\|R,\theta)$	0	0.9999	0	0
	$Pr(E_2 = No\ result\|R,\theta)$	0	0	1	1

Variable R has states 'yes' (i.e. perform analysis) and 'no' (i.e. do not perform analysis). The possible states of nature, represented by variable Θ, are θ_1 ('The suspect is the source') and θ_2 ('Someone else is the source').

option, for it makes clear that the variable E_2 should be instantiated to 'no result' when the 'experiment' is not performed. The loss functions for nodes L_1 and L_2 are, respectively, $L_1 = \{x, 0\}$ (for $R = \{yes, no\}$) and

$$
L_2(\Theta, D) = \begin{cases} \{0, 0.001, 0.01\}, & \Theta = \theta_1, D = \{d_1, d_2, d_3\}, \\ \{1, 0.001, 0\}, & \Theta = \theta_2, D = \{d_1, d_2, d_3\}. \end{cases}
$$

2.1.10 Conditional independence in influence diagrams

The same decision rule for d-separation that holds for Bayesian networks (Section 2.1.6) holds for influence diagrams, with the following differences. Utility nodes, with their relative functional links, and informational links are ignored because they do not represent probabilistic dependence relationships. Decision nodes are under the control of the decision maker, so they do not require prior probabilities.

With these provisos, the joint probability distribution over the set of nodes $\mathbf{X}$ of an influence diagram can be factorized as the product of the conditional probability of each chance node, given the state *only* of the chance nodes and the decision nodes which are its direct predecessors:

$$Pr(\mathbf{X}) = \prod_{X_i \in \mathbf{X}_C}^{n} Pr(X_i | \mathbf{PA}(X_i)),$$

where $\mathbf{X}_C$ is the set of chance nodes of the influence diagram.

Example 2.3 *(Forensic identification, continued) The joint probability distribution over the chance and decision nodes of the influence diagram in Figure 2.6 is*

$$Pr(\theta, E_2, R, D) = Pr(E_2 | \theta, R, D) Pr(\theta | R, D),$$

where R and D are decision nodes and do not have a prior probability and it can be factorized as follows. E_2 and D are d-separated given $\{\Theta, R\}$ because there is no path from E_2 to D. Note that $E_2 \to D$ is an informational link. Nodes Θ and $\{R, D\}$ are d-separated given $\{\emptyset\}$ because the path between Θ and $\{R, D\}$ contains a converging connection $\to E_2 \leftarrow$ such that E_2 does not belong to $\{R, D\}$ and no descendant of E_2 belongs to $\{R, D\}$. Therefore, we can simply write

$$Pr(\theta, E_2, R, D) = Pr(E_2 | \theta, R) Pr(\theta).$$

2.1.11 Relevance and causality

Explanations, especially those involving generalizations based on common sense, are often purportedly *causal explanations*. The fact that warrants (see Section 2.1.7 for discussion of the term *warrant*) for evidence are usually sought which make up scenarios or histories that follow a chronological order that leads, using common sense, to an understanding of the order as a causal order. Events that are part of the *probandum* occur *before* events that constitute the *probans* and, normally, *causes* occur before their effects. Such considerations also arise in the context of graphical modelling, as noted by Schum:

> Constructing a network representation of an inference problem is a purely subjective judgmental task, one likely to result in a different structural pattern by each person who performs it. In the contemporary literature on inference networks there is now some controversy concerning what people should and do attend to when such tasks are performed. (...) this controversy concerns whether such structuring always involves the tracking of causal relations. (Schum 1994, p.175)

Many authors, amongst them Salmon (1998) (mentioned in Section 1.3.3), consider that statistical relevance does not have a genuine explanatory interpretation and that only causal relevance has. A consequence of this position is that a probabilistic explanation would be legitimate only if it were possible to identify the causal mechanism underlying it. Pearl has forcefully argued for the advantages of building DAG models around causal relationships, claiming that they are more meaningful, more accessible, and felt by people as more reliable than mere probabilistic relevance relationships (Pearl 2009, pp. 21; 25):

(...) conditional independence judgments are accessible (hence reliable) only when they are anchored onto more fundamental building blocks of our knowledge, such as causal relationships. (...) [C]ausal relationships are more stable than probabilistic relationships. We expect such difference in stability because causal relationships are *ontological*, describing objective physical constraints on our world, whereas probabilistic relationships are epistemic, reflecting what we know or believe about the world. (...) The element of stability (of mechanisms) is also at the heart of the so-called explanatory accounts of causality, according to which causal models (...) aim primarily to provide an 'explanation' or 'understanding' of how data are generated.

In the context of evidential reasoning, it can turn out to be very difficult, if not impossible, to show genuine causal relationships. For instance, an intuitive interpretation of the path from node A to node E through node C in the graph in Figure 2.4(iv) as a 'mechanism' suggests itself easily to the mind, but the arc from node B to node C seems to represent the influence of an epistemic condition necessary to 'open' the channel from A to C rather than a causal 'mechanism' acting between B and C. One could argue that A and B fit the definition of *inus* conditions in John Mackie's regularity theory of causation (Mackie 1974) and that, therefore, there is a kind of causal explanation after all. An inus condition for some effect is an insufficient but non-redundant part of an unnecessary but sufficient condition. In our example, A and B are inus conditions for C, for (i) they are both insufficient ($Pr(C|\bar{A}, B, I) = Pr(C|A, \bar{B}, I) = 0$), (ii) they are jointly sufficient ($Pr(C|A, B, I) = 1$) and (iii) they are not necessary ($Pr(C|\bar{A}, \bar{B}, I) > 0$).

The answer is that, even though this particular explanation can be interpreted as a causal explanation according to Mackie's theory, there is no obligation to try to figure out causal explanations in the style of Mackie for any argument in evidential reasoning. Evidential analysis should take a neutral stance, as long it is possible, on the controversies about what a causal relationship is and what constitutes a causal explanation. Therefore, there is agreement with Schum's viewpoint that the capability of providing a causal explanation is not a necessary condition for justifying the existence of a relevance relationship (Schum 1994, pp. 141; 178):

> In many inference tasks the hypotheses under consideration represent possible causes for various patterns of observable effects. In such cases we may be able to trace what we regard as a causal linkage from some ultimate or major hypothesis to events or effects that are observable. If such linkage can be established, evidence of these effects would be relevant in an inference about these hypothesized causes. But there are many instances in which evidence can be justified as relevant on hypotheses of interest when links in a chain of reasoning may indicate no causal connection.

> (...) To require that the arcs of an inferential network signify direct causal influences is to hamstring the entire enterprise of applying conventional probability theory in the analyses of inferential networks.

The strategy of building up a Bayesian network following the path of Bayesian inference (Section 1.3.5) is sufficient to break the symmetry of probabilistic relevance and the only requirement to be satisfied is that the d-separation property holds. That is, if sets of variables $\mathbf{X}$ and $\mathbf{Y}$ are conditionally independent given the set $\mathbf{Z}$ in the chosen explanatory pattern, then $\mathbf{X}$ and $\mathbf{Y}$ must be d-separated given $\mathbf{Z}$ in the Bayesian network representing that pattern.

The explanatory pattern may be a causal pattern, according to one's favourite theory of causation, and it is true that 'causal' explanations which make use of 'causal' common sense generalizations are usually felt as more intuitive and therefore more 'objective' and acceptable (but this is not always the case if recourse has to be made to sophisticated scientific laws that express some 'causal' relationships that are not intuitive). It is better if causal explanations can be provided, but it is not necessary that they be so.

2.1.12   *The Hound of the Baskervilles* revisited

Doctor Watson wishes to draw a DAG that represents the explanatory inferences supporting the evidential reasoning in the Baskerville case. The propositions in question are 'Sir Charles Baskerville died by accident' (H_1), 'Sir Charles Baskerville died by a criminal act' (H_2), 'Sir Charles Baskerville committed suicide' (H_3), 'The proximate cause of Sir Charles Baskerville's death was a heart attack' (R) and 'A gigantic hound was running after Sir Charles Baskerville' (F). Therefore, the nodes will be H, with three possible states $\{H_1, H_2$ and $H_3\}$, R and F, with two possible states each [Figure 2.7(i)].

There will be an arc pointing from node H to node R because it is easy to provide a probabilistic explanation of Sir Charles Baskerville's death by heart attack: the propositions H_i constitute a partition of the general class of all the possible death causes and $Pr(R|H_i) \neq Pr(R|H_j)$ for $i \neq j$. The level of detail of the partition depends on the context in which the explanation is used and the reason why the explanation is requested.

As regards the potential explanations of F, Watson might articulate his opinion, (1.19), by arguing that the probability of F, given that an intentional action was carried out is greater than the probability of F, given that no intentional action was carried out and that the probability of F is the same in case that hypothesis H_1 is true and in case hypothesis H_3 is true. The occurrence of a heart attack does not explain the presence of a hound any more than Sir Charles Baskerville's will to commit suicide would explain it. In both cases, the only possible explanation is in terms of the occurrence of stray dogs on the moor.

Finally, according to his professional opinion, Watson believes that a stray hound running after a person with a weak heart can explain the occurrence of a heart attack. For the case, he committed suicide, there would be no connection at all between the two events. Therefore, there will be an arc pointing from node F to node R. The numerical values assigned to the various conditional probabilities (Table 2.7) are Doctor Watson's responsibility and they are coherent with the qualitative assessments (1.9), (1.19) and (1.20).

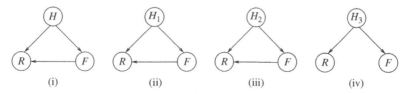

Figure 2.7   (i) A Bayesian network and (ii–iv) and local networks for the Baskerville case. Node H concerns the method by which Sir Charles Baskerville died with three states, death by accident, H_1, death by a criminal act, H_2, and death by suicide, H_3. Node F represents the event that a gigantic hound was running after Sir Charles Baskerville with two states, true, t, and false, f. Node R represents the event that the proximate cause of Sir Charles Baskerville's death was a heart attack with two states, true, t, and false, f. (ii)–(iv) The local networks for the Baskerville case.

Table 2.7 Probabilities for the node H (second row) and conditional probabilities for nodes F (rows three and four) and R (rows seven and eight) of Figure 2.7(i).

	H:	H_1	H_2	H_3
	$Pr(H)$	0.98	0.01	0.01
F:	$Pr(F = t\|H)$	0.01	0.05	0.01
	$Pr(F = f\|H)$	0.99	0.95	0.99

	H:	H_1		H_2		H_3	
	F:	t	f	t	f	t	f
R:	$Pr(R = t\|H, F)$	0.4	0.1	0.4	0.001	0	0
	$Pr(R = f\|H, F)$	0.6	0.9	0.6	0.999	1	1

Node H concerns the method by which Sir Charles Baskerville died with three states, death by accident, H_1, death by a criminal act, H_2, and death by suicide, H_3. Node F represents the event that a gigantic hound was running after Sir Charles Baskerville with two states, true, t, and false, f. Node R represents the event that the proximate cause of Sir Charles Baskerville's death was a heart attack with two states, true, t, and false, f.

This simple story contains an instance of the phenomenon called *asymmetric independence*, which occurs when variables are independent for some but not all of their values. The arc from F to R is necessary to encode the dependency of R from F, given H_1 and H_2, but the full DAG cannot represent the independence of R from F, given H_3. This asymmetric independence relationship contained in the story, at least as told by Watson, can be read only in the values of the conditional probability table. In order to be able to read it directly from a graphical model, local networks can be drawn, each representing part of the problem. Methods to deal with Bayesian networks with asymmetric independencies have been proposed in Geiger and Heckerman (1991), Friedman and Goldszmidt (1996) and Mahoney and Laskey (1999).

Following for instance the *Bayesian Multinet* method of Geiger and Heckerman, a DAG is associated with each H_i and the DAG is the Bayesian network, given that one of the values in H holds. The Bayesian Multinet of Figure 2.7(i) is the set of the local networks as given in Figures 2.7(ii)–(iv), together with a marginal probability distribution on H. Local networks represent what is also called *context-specific independence*. This example is computationally simple and it can work with the full probability distribution of the network shown in Figure 2.7(i). However, context-specific independence can be exploited to reduce the burden of model specification and improve the efficiency of the inference. To work out the example, first verify that the numerical values in the conditional probability tables (Table 2.7) are coherent with the qualitative judgment (1.10) (numbers are rounded off to the third decimal place):

$$Pr(R|H_1, I) = Pr(R|H_1, F, I)Pr(F|H_1, I) +$$
$$Pr(R|H_1, \bar{F}, I)Pr(\bar{F}|H_1, I) = 0.103,$$
$$Pr(R|H_2, I) = Pr(R|H_2, F, I)Pr(F|H_2, I) +$$
$$Pr(R|H_2, \bar{F}, I)Pr(\bar{F}|H_2, I) = 0.021,$$
$$Pr(R|H_3, I) = 0.$$

and with (1.15):

$$Pr(F|H_1, R, I) = \frac{Pr(R|H_1, F, I)Pr(F|H_1, I)}{Pr(R|H_1, I)} = 0.039,$$

$$Pr(F|H_2, R, I) = \frac{Pr(R|H_2, F, I)Pr(F|H_2, I)}{Pr(R|H_2, I)} = 0.952.$$

The likelihood ratio is slightly in favour of H_2:

$$V = \frac{Pr(F|H_1, I)}{Pr(F|H_2, I)} \times \frac{Pr(R|H_1, F, I)}{Pr(R|H_2, F, I)} = \frac{0.01}{0.05} \times \frac{0.4}{0.4} = 0.2.$$

The posterior odds, although lower than the prior odds, are still decidedly in favour of H_1:

$$V \times \frac{0.98}{0.01} = 0.2 \times 98 = 19.6.$$

Conditional on learning Dr. Mortimer's story, Watson's common sense assigns a posterior probability of about 95% to the hypothesis that poor Sir Charles Baskerville has been victim of a tragic fatality. The posterior odds are approximately 20 to 1 in favour of a tragic fatality. These odds are equivalent to a probability of 20/21 or 0.95.

2.2 Reasoning with Bayesian networks and influence diagrams

In this chapter, the 'black box' of the computational architecture of Bayesian networks is examined and it is shown that the architecture works according to the principles of the logic of uncertainty that we have outlined in Chapter 1. Formal proofs are not given of the mathematical results exploited because, for readers interested in practical applications of Bayesian networks and who are going to use commercial software already available, it is more important to understand why the answers provided by the 'black box' can be trusted, rather than to know in any detail the mathematics of the inference engine.

Example 2.4 *('Jack loved Lulu', continued) In order to be able to apply Bayes' rule in the 'Lulu' case, it is necessary to calculate the marginal distributions of the evidence nodes E and W from the conditional probabilities in (2.2). Consider, first, the task of computing the marginal probability distribution of E:*

$$Pr(E) = \sum_C Pr(E|C)Pr(C). \tag{2.3}$$

The marginal probability distribution of C is given by

$$Pr(C) = \sum_{ABF} Pr(C|A, B, F)Pr(A, B, F). \tag{2.4}$$

The variable B is d-separated from $\{A, F\}$ given $\emptyset$, therefore we can rewrite (2.4) as

$$Pr(C) = \sum_{ABF} Pr(C|A, B, F)Pr(A, F)Pr(B). \tag{2.5}$$

The final goal is the marginal probability distribution Pr(A, F),

$$Pr(A, F) = \sum_J Pr(A, F, J) = \sum_J Pr(F|A, J)Pr(A)Pr(J). \qquad (2.6)$$

A subset of nodes of a graph is called a cluster. *The determination of a value for (2.6) (also known as a sub-task) requires work with the cluster $\{A, F, J\}$ and $2^3 = 8$ probability values have to be stored. Execution of sub-task (2.5) requires work with the cluster $\{A, B, C, F\}$ and the storage of a further $2^4 = 16$ probability values. Finally, the cluster of sub-task (2.3) is $\{C, E\}$, with the requirement to store $2^2 = 4$ probability values.*

The marginal probability distribution of W may be computed in an analogous manner:

$$Pr(W) = \sum_{DF} Pr(D, F, W) = \sum_{DF} Pr(W|D, F)Pr(D, F). \qquad (2.7)$$

The next sub-task is the computation of the marginal distribution Pr(D, F). The variable D is d-separated from F given J, that is D is conditionally independent of F given J. Therefore,

$$Pr(D, F) = \sum_J Pr(D, F, J) = \sum_J Pr(D|F, J)Pr(F, J)$$

$$= \sum_J Pr(D|J)Pr(F, J). \qquad (2.8)$$

The marginal distribution P(F, J) can be obtained from the probability values that have already been stored in memory when (2.6) was executed:

$$Pr(F, J) = \sum_A Pr(A, F, J) = \sum_A Pr(F|A, J)Pr(J)Pr(A). \qquad (2.9)$$

This means that sub-task (2.9) handles the same cluster as (2.6) and that extra memory space is not needed for this computation. Sub-task (2.8) works with cluster $\{D, F, J\}$ and adds $2^3 = 8$ probability values to the store, whereas the cluster of sub-task (2.7) is $\{W, F, D\}$, and $2^3 = 8$ more values are added to the store.

Thus, $(8 + 16 + 4 + 8 + 8) = 44$ probability values have been computed instead of $2^8 = 256$, quite a remarkable saving of computational resources, and the goals have been achieved working with five clusters:

$$\{A, F, J\}, \ \{A, B, C, F\}, \ \{C, E\}, \ \{D, F, J\}, \quad \{D, F, W\}. \qquad (2.10)$$

Let call the set of clusters (2.10) the Markov domain *of the DAG in Figure 2.5.*

2.2.1   Divide and conquer

The flow of computations (2.3)–(2.9) is illustrated by Figure 2.8, where rectangular boxes are inputs and outputs, and oval boxes indicate the computations that are performed. Care has to be taken with these symbols to note the context. In other contexts, oval boxes denote discrete variables and rectangular boxes denote continuous variables (see Chapter 9).

The basic idea for the implementation of the flow of computations by an expert system can be summarised as follows. A new graph is constructed which has subsets of nodes of the original DAG as its nodes. Such a so-called *cluster graph* satisfies the two properties that (i) for every pair of clusters there exists a unique path connecting them and (ii) every node that belongs to two clusters also belongs to every cluster in the path between them.

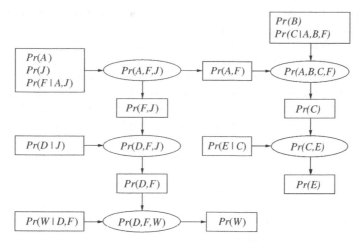

Figure 2.8 Flow diagram of the computation of $Pr(E)$ and $Pr(W)$.

These two properties together ensure a correct propagation of information because the new graph has the same topology as that depicted in Figure 2.8. There is only one path between any two oval boxes in Figure 2.8 and the oval box on the path between $Pr(A, F, J)$ and $Pr(D, F, W)$ does contain the variable F.

The cluster graph will become a permanent part of the expert system knowledge representation, a part that will change only if the original DAG changes. Each cluster stores its local joint distribution that can be used to answer queries for probabilities of any node in the cluster. In the following sections, the 'Lulu' network is used to give an overview of the procedure for the construction of the appropriate cluster graph and local propagation algorithm that passes information between pairs of adjacent clusters.

There is another desirable property the cluster graph should satisfy, beyond those already mentioned. The number of nodes contained in each cluster can be taken as a measure of computational complexity and it is desirable to use clusters with as few nodes as possible. Unfortunately, the general problem of finding clusters with minimal size is NP-hard, as is the general problem of probabilistic computation in a Bayesian network. The following procedure, *clustering* (Whittaker 1990), although it does not guarantee to find the minimal solution, is amongst the most efficient: (i) transform the DAG into an undirected graph, one with the arrowheads removed, and with links between parents and common children, the *moral graph* (the definition of which will be given in Section 2.2.2); (ii) turn the moral graph into a *triangulated graph* (the definition of which will be given in Section 2.2.2); (iii) find the clusters of the triangulated graph which are the *cliques* of the graph (the definition of which will be given in Section 2.2.2) and (iv) connect the cliques to form a *junction* or *join tree*.

A junction tree satisfies the properties of a cluster graph. The procedure is such to guarantee that the set of nodes of the junction tree, that is the set of cliques, is identical with the Markov domain of the original DAG.

2.2.2 From directed to triangulated graphs

The first step of clustering is the *moralization* of the directed graph. The name comes from the fact that parents of the same child are 'married' together. A *moral graph* is obtained by adding a link between parents of common children in a directed graph and then removing all the arrowheads. It is obvious that moralising is a mechanical procedure so that it can be executed by a computer. The second step of clustering, *triangulation*, is the crux of the matter. An undirected graph is said to be triangulated if any cycle of length four or more has at least one *chord*. A chord is a link between two nodes in a cycle that is not contained in the cycle.

The 'Lulu' moral graph [Figure 2.9(i)] is triangulated, as is the moral graph [Figure 2.9(ii)] associated with the 'one-trace' scenario [Figure 2.4(iv)]. Not every moral graph is triangulated. An example of a non-triangulated graph is shown in Figure 2.9(iii). Why should the moral graph be triangulated? There is a theorem of graph theory saying that an undirected graph has a junction tree if and only if it is triangulated. Thus, a necessary and sufficient condition for completing the procedure of construction of the junction tree is that the moral graph is triangulated.

A non-triangulated graph can be triangulated by adding chords to break the cycles. This process is called *fill-in* or *triangulation*. A cycle can be broken in several ways, so that there are different ways to triangulate a graph. Figure 2.10 shows two ways of triangulating the moral graph in Figure 2.9(iii). Readers are left to check other ways.

The problem that arises in filling-in a non-triangulated graph, with respect to the complexity issue, is that adding chords raises the size of the cliques; therefore, it is desirable to add as few chords as possible, in order to obtain clusters with the smallest size. A minimal fill-in has the minimum number of chords needed to triangulate a given graph. The fill-in in Figure 2.10(ii) is minimal but the one in Figure 2.10(i) is not because removing either the chords $A - D$ or $B - E$ will still leave a triangulated graph. The problem of finding a minimal fill-in is NP-hard, and hence the general problem of finding minimal clusters is NP-hard.

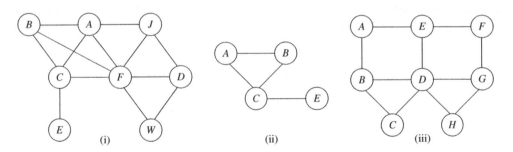

Figure 2.9 (i) 'Lulu' moral graph, (ii) moral graph of the 'one-stain' scenario DAG [Figure 2.4(iv)] and (iii) moral graph for the DAG shown in Figure 2.1(i).

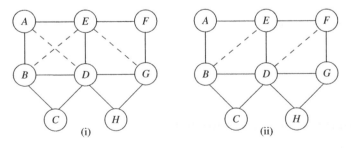

Figure 2.10 (i–ii) Two different triangulations for the moral graph depicted in Figure 2.9(iii).

We have efficient algorithms to check whether an undirected graph is triangulated and for triangulating it, if it is not, but these algorithms do not guarantee a minimum fill-in.

A simple triangulation method makes use of two algorithms, the *maximum cardinality search* (MCS), followed by the *elimination algorithm*. The MCS orders nodes of the moral graph in the following manner. First, it assigns number 1 to an arbitrary node and then, second, it assigns the next number to an unnumbered node with the largest number of numbered neighbours (the neighbours of a node X are all the nodes linked to X in an undirected graph) and breaks ties arbitrarily.

A possible ordering of nodes obtained by applying the MCS algorithm to 'Lulu' moral graph [Figure 2.9(i)] is shown in Figure 2.11. The elimination algorithm visits nodes of the moral graph in reverse numerical order and for each node X it executes the following instructions: (i) add links so that all neighbours of X are linked to each other, (ii) remove X together with all its links and (iii) continue until all nodes have been eliminated.

The original graph together with all the links that have been added executing the procedure is triangulated. A graph is already triangulated if and only if all the nodes can be eliminated without adding new links. It is clear by visual inspection that the graph in Figure 2.11(i) is triangulated without adding new links, but a computer needs a mechanical procedure to make the inspection and the elimination algorithm is such a procedure.

Starting from the highest numbered node E, its neighbour is C and node E and its link to C may be eliminated. Then move to node W: its neighbours F and D are already linked, so eliminate node W and its links to nodes F and D. Repeat the procedure with node D and so on. On arrival at node C, nodes E and B have already been eliminated, therefore, the neighbours of C at this step of the procedure are A and F, and they are linked together. Eliminate C and go to A and so on, until node J is reached. Node J can be eliminated because it has no neighbours left. Readers can check by themselves that the triangulated graph in Figure 2.10(ii) is obtained by elimination from the MCS ordering shown in Figure 2.11(ii).

2.2.3   From triangulated graphs to junction trees

There exist efficient exact probabilistic algorithms for particular kind of graphs. These graphs are a particular kind of cluster graphs, called *junction trees*, whose nodes are the cliques of a triangulated graph. A subset of nodes of an undirected graph is said to be *complete* if all its

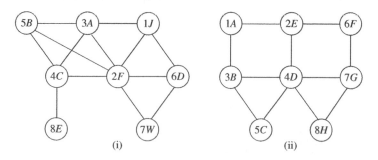

Figure 2.11   A possible ordering by MCS of (i) the 'Lulu' moral graph and (ii) the moral graph shown in Figure 2.9(iii).

nodes are pairwise linked. A complete set of nodes is called a *clique* if it is not a subset of another complete set, that is, if it is maximal.

A mechanical procedure to find the cliques of a triangulated graph is very simple, given that, for every node X of a moral graph, after step (i) of the elimination algorithm (Section 2.2.2) has been executed, it holds that X and its neighbours form a complete set.

To construct the set of cliques of a triangulated graph, the following instructions apply for every node X, after step (i) of elimination described previously in Section 2.2.2: (i) check if the complete set composed by X and its neighbours is maximal and (ii) if the answer is 'yes', add to the list of cliques.

Apply this algorithm to the 'Lulu' graph [Figure 2.11(i)]. On arrival during elimination at node C, the check for the set $\{A, C, F\}$ is negative: it is not a clique because it is a subset of the set $\{A, B, C, F\}$ already added to the list when B was eliminated. On arrival at F, the set $\{F, J\}$ is not a clique because the set $\{A, F, J\}$ was added to the list when A was eliminated. Therefore, the set of cliques of the triangulated graph in Figure 2.11(i) is (with node numbers according to MCS ordering)

$$\{C_4, E_8\}, \quad \{D_6, F_2, W_7\}, \quad \{D_6, F_2, J_1\},$$
$$\{A_3, B_5, C_4, F_2\}, \quad \{A_3, F_2, J_1\}. \tag{2.11}$$

Therefore, the set of cliques (2.11) of the triangulated graph in Figure 2.11(i) obtained by the clustering procedure is exactly the Markov domain (2.10) of the 'Lulu' DAG.

The last step of the procedure is the construction of the cluster graph having the cliques as its nodes. Observe that a connected undirected graph is called a *tree* if for every pair of nodes there exists a unique path linking them. A cluster tree is a *junction tree*, if every node that belongs to two clusters also belongs to every cluster in the path between them. This condition is called the *junction tree property*.

To guarantee that the tree of the cliques of a triangulated graph obtained via MCS and elimination is a junction tree, the following algorithm can be used: (i) order the cliques in ascending order according to their highest numbered node and (ii) go through the cliques in order, choosing for each clique **X** a clique **Y** amongst the previously numbered cliques with maximum number of common nodes and add an undirected link between **X** and **Y** and break ties arbitrarily.

In Figure 2.12, three different arrangements of the set of cliques (2.11) are given, all of them are trees but only one is a junction tree. Following the steps of the algorithm, readers can check that only the tree in Figure 2.12(i) is obtained and that trees in Figures 2.12(ii) and (iii) are not outputs of the algorithm.

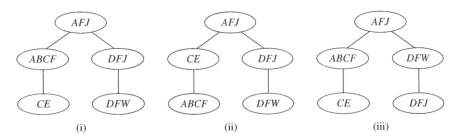

Figure 2.12   (i) Junction tree; (ii) and (iii) structures that are not junction trees.

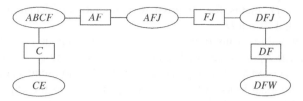

Figure 2.13   The junction tree for graph in Figure 2.11(i) with separators.

Let **X** and **Y** be adjacent cliques of a junction tree. The subset of nodes they have in common is called the *separator* for **X** and **Y** and its graphical rendering consists in associating with the link joining the cliques a box containing the nodes in the separator. Call the complement of a clique **X** with respect to the separator **S** for an adjacent clique **Y** the *residual* of **X** with respect to **S**. For example, the residual of $\{A, B, C, F\}$ with respect to $\{A, F\}$ is $\{B, C\}$, and the residual of $\{A, F, J\}$ with respect to the same separator $\{A, F\}$ is $\{J\}$. The following fact, called the *Markov property for junction trees*, holds and says that the residuals of two adjacent nodes in a junction tree are *conditionally independent given their separator*.

It can be checked in the original DAG (Figure 2.5) that $\{B, C\}$ is d-separated from $\{J\}$ given $\{A, F\}$: the path $C \leftarrow F \leftarrow J$ is serial at F and the path $C \leftarrow A \rightarrow F \leftarrow J$ is divergent at A. The residual of $\{F, D, J\}$ with respect to $\{F, J\}$ is $\{D\}$ and it is d-separated from $\{A\}$ given $\{F, J\}$: the path $A \rightarrow F \leftarrow J \rightarrow D$ is diverging at J and the path $A \rightarrow F \rightarrow W \leftarrow D$ is serial at F. Finally, $\{W\}$ is d-separated from $\{J\}$ given $\{D, F\}$ and $\{E\}$ and it is d-separated from $\{A, B, F\}$ given $\{C\}$ (Figure 2.13).

Therefore, the joint probability distribution over the junction tree can be written as the product of the conditional probability distributions of the residuals of the cliques, given their separators [left-hand side of (2.12)], and the factorization of this joint distribution [right-hand side of (2.12)] turns out to be the same as the factorization (2.2) of the joint distribution over the DAG in Figure 2.5:

$$Pr(A, F, J)Pr(B, C|A, F)Pr(D|F, J)Pr(W|D, F)Pr(E|C)$$

$$= Pr(A)Pr(J)Pr(F|A, J)Pr(B)Pr(C|A, B, F)$$

$$Pr(D|J)Pr(W|D, F)Pr(E|C). \tag{2.12}$$

2.2.4   Solving influence diagrams

A way to find a solution to an influence diagram is to transform it into a decision tree and to apply the *average-out and folding back* algorithm [Jensen and Nielsen (2007, pp. 346–353), Kjærulff and Madsen (2008, pp. 100–102)]. Under the no-forgetting assumption, each decision node and all its parents are informational parents of all successive decision nodes, and this means that, in the decision tree equivalent to an influence diagram, each decision node has the same variables in the past. This can be easily checked in the decision tree representation (Figure 1.3) of the influence diagram in Figure 2.6.

The *decision past* of a decision node D_i in influence diagram is the set of its parents and the set of the previous decisions in the decision sequence with their parents. A *policy* for a decision node D_i in an influence diagram specifies a decision for any possible configuration of the decision past of D_i, that is, for any possible combination of decisions and observations made prior to making decision D_i.

Under the no-forgetting assumptions, the decision past of a node D_i contains all the decision nodes that precede D_i in the decision sequence. In LMID, the decision past of a node D_i contains only those decision nodes that precede D_i in the decision sequence and are explicitly connected by informational links. An *optimal* policy maximizes the expected utility (minimizes the expected loss) at each decision node; an *optimal* strategy specifies the optimal policy for each decision node.

Example 2.5 *(Forensic identification, continued) Applying the average-out and folding back algorithm to the influence diagram in Figure 2.6, we compute the value of the optimal strategy as follows:*

$$\min_R \sum_{k=t,f} \min_D \sum_{j=1}^{2} Pr(\theta_j)Pr(E_2 = k|\theta_j, R)(L_1(R) + L_2(\theta_j, D)),$$

with $t = true$, $f = false$, R representing the decision to perform a DNA analysis, E_2 the result of a DNA analysis and D the terminal decision. Variable θ represents the state of nature, that is whether $(j = 1)$ or not $(j = 2)$ the suspect is the source of the analysed trace material.

We see that the probability values to be inserted in the formula are those calculated in Table 1.7 for the decision tree (see Section 1.4.5). Tables 2.8 and 2.9 (left) show the expected loss functions over R, E_2, D and over R, respectively, from which the optimal policies for D can be identified for any combination of the members R, E_2 of the decision past of D. The decision past of R is void. Table 2.9 (right) shows the optimal strategy for the identification problem. Each row of the table is an optimal policy for the decision node.

In general, it can proved that the chain rule for influence networks is

$$EU(\mathbf{X}) = \prod_{X_i \in \mathbf{X_C}} Pr(X_i|\mathbf{PA}(X_i)) \sum_{u_j \in \mathbf{X_U}} U(\mathbf{PA}(u_j)).$$

Table 2.8   The joint expected loss (EL) function over R (the decision to perform a DNA analysis), E_2 (the result of a DNA analysis) and D (the terminal decision).

R	E_2	D	EL
No	No result	Identification	0.1
No	No result	Inconclusive	0.001
No	No result	Exclusion	0.009
Yes	True	Identification	0.00001
Yes	True	Inconclusive	0.001
Yes	True	Exclusion	0.0099
Yes	False	Identification	1
Yes	False	Inconclusive	0.001
Yes	False	Exclusion	0

For decision $(R = no)$, the second option illustrated above (Table 2.6) is used, inserting 'no result'.

Table 2.9   The joint expected loss (*EL*) function over *R* (left), the decision to perform a DNA analysis and the optimal strategy for the identification problem (right).

R	EL	Optimal strategy
No	0.001	$R = yes$
Yes	0.000009	$D = identification$ if $R = yes$ and $E_2 = true$
		$D = exclusion$ if $R = yes$ and $E_2 = false$

E_2 represents the result of a DNA analysis and D the terminal decision.

An influence diagram is a compact representation of a joint expected utility (loss) function over the set of variables **X** and supports the representation and solution of sequential decision problems with multiple local utility functions. By transforming an influence diagram into a decision tree, the complexity of the decision tree representation is inherited in the solution phase that can be highly inefficient. Alternative solution methods that allow junction trees for influence diagrams to be built directly can be found in the literature [Jensen and Nielsen (2007, p. 353–367), Kjærulff and Madsen (2008, p. 102–104)].

Example 2.6 *(Forensic identification, augmented) In the decision tree in Figure 1.3 and in the equivalent influence diagram in Figure 2.6, the information provided from the first analysis, that is the analysis done on the trace collected at the crime scene, has been considered as background information. A more realistic model will consider the point of view of the decision maker prior to any observation, thus, considering also E_1 as a chance node, and considering also doing the analysis whose outcome is described by E_1 as a possible decision. The influence diagram modelling this situation is shown in Figure 2.14, where R_1 is the decision node whose possible states are $R_1 = \{yes$ (do the analysis 1), no (do not do the analysis 1)\}, R_2 is the decision node whose possible states are $R_2 = \{yes$ (do the analysis 2), no (do not do the analysis 2)\}, E_1 is a chance node whose possible states are $E_1 = \{true$ (corresponding features), false (difference)\} and L_1 and L_2 are nodes for decision loss, representing the 'cost' of the analyses. Under the 'no-forgetting' assumption, there is no need to add direct edges between the decision variables because there is a unique path from R_1 to R_2 and D.*

The (unconditional) probability assignments for the states θ_1 and θ_2 of node θ are, respectively, 0.001 and 0.999. The conditional probability table for node E_1 is given in Table 2.10. Only one of the two possible options (exemplified earlier in Table 2.6) is given, and the values are compatible with assignments $Pr(\theta_1|E_1) = 0.9$ and $Pr(\theta_2|E_1) = 0.1$, given in Table 1.7. The conditional probability table for $Pr(E_2|R, \theta)$ is the same as the second option in Table 2.6, and the loss functions for nodes L_2 and L_3 are the same as those for, respectively, L_1 and L_2 in Example 2.2. The loss function for node L_1 in the example here is $L_1 = \{y, 0\}$, for $R_1 = \{yes, no\}$.

The joint probability distribution over the influence diagram in Figure 2.14

$$Pr(\theta, E_1, E_2, R_1, R_2, D) = Pr(E_2|E_1, \theta, R_1, R_2, D)Pr(E_1|\theta, R_1, R_2, D)$$

$$\times Pr(\theta|R_1, R_2, D),$$

can be factorized as follows. The diagram with only the probabilistic links left is shown in Figure 2.15. E_2 and $\{E_1, R_1, D\}$ are d-separated given $\{\Theta, R_2\}$. E_1 and $\{E_2, R_2, D\}$ are d-separated given $\{\Theta, R_1\}$. Θ and $\{R_1, R_2, D\}$ are d-separated given $\{\emptyset\}$ because any path

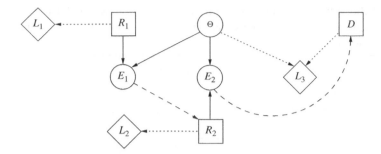

Figure 2.14 Augmented influence diagram for the identification problem (Example 2.6). Node D is a decision variable with the set of actions d_1, d_2 and $d_3 \in D$ denoting, respectively, the expert conclusions 'individualise', 'inconclusive' and 'exclusion'. Nodes R_i, $i = \{1, 2\}$, are further decision variables, with states 'yes' (i.e. perform analysis i) and 'no' (i.e. do not perform analysis i). The possible states of nature, represented by node Θ, are θ_1 ('The suspect is the source') and θ_2 ('Someone else is the source'). Nodes E_i, $i = \{1, 2\}$, represent the possible outcomes of performing analysis i. Two options for defining the set of states for these nodes are given by the second option given in Table 2.6. One option consists of using binary states 'true' and 'false', and another option uses the same states but adds a third, called no result. Nodes L_1, L_2 and L_3 define decision losses.

Table 2.10 Conditional probability table for the node E_1 in Figure 2.14.

R_1 :	Yes		No	
θ :	θ_1	θ_2	θ_1	θ_2
$Pr(E_1 = true\|R_1, \theta)$	0.95	0.001	0	0
$Pr(E_1 = false\|R_1, \theta)$	0.05	0.999	0	0
$Pr(E_1 = no\ result\|R_1, \theta)$	0	0	1	1

between Θ and $\{R, D\}$ contains a converging connection (either $\rightarrow E_1 \leftarrow$ or $\rightarrow E_2 \leftarrow$), such that E_i does not belong to $\{R, D\}$ and no descendant of E_i belongs to $\{R, D\}$. Therefore,

$$Pr(\theta, E_1, E_2, R_1, R_2, D) = Pr(E_2|\theta, R_2)Pr(E_1|\theta, R_1)Pr(\theta).$$

It is left as an exercise to draw the appropriate decision tree and to solve it, assigning appropriate utility (or loss) values.

2.2.5 Object-oriented Bayesian networks

Models of real systems are often composed of collections of identical or similar components. The framework of *object-oriented probabilistic networks* has been developed to facilitate the reuse of the same sub-model in a different part of an embedding model, to use nested sub-models in a hierarchical manner and to communicate complex models in an economical way (Bangsø and Wuillemin 2000; Koller and Pfeffer 1997; Laskey and Mahoney 1997;

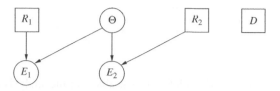

Figure 2.15   Structure underlying the influence diagram shown in Figure 2.14, showing only the probabilistic links. Node D is a decision variable with the set of actions d_1, d_2 and $d_3 \in D$ denoting, respectively, the expert conclusions 'individualise', 'inconclusive' and 'exclusion'. Nodes R_i, $i = \{1, 2\}$, are further decision variables, with states 'yes' (i.e. perform analysis i) and 'no' (i.e. do not perform analysis i). The possible states of nature, represented by node Θ, are θ_1 ('The suspect is the source') and θ_2 ('Someone else is the source'). Nodes E_i, $i = \{1, 2\}$, represent the possible outcomes of performing analysis i.

Neil et al. 2000). Chapter 12 outlines the use of this modelling framework for approaching inference problems in forensic contexts.

Example 2.7 ('*Jack loved Lulu*', continued) *The 'Lulu' Bayesian network (Figure 2.5) could be enlarged by taking into account the testimony, and its reliability, of some witnesses that Jack loved Lulu and also by taking into account the analysis of the eyewitness John, through considerations of the variables involved in the evaluation of testimony. For the sake of argument, suppose that we want to consider the reliability of John. Schum has identified three basic attributes of the credibility of human witnesses: observational sensitivity or accuracy, objectivity and veracity (Schum 1994, p. 101–114). An eyewitness X is accurate if her senses gave evidence of an event that actually occurred, and X is objective if she believes the evidence of her senses. According to this analysis, the chain of top-down reasoning between the hypothesis H that a certain event occurred, and the information E, the event that a witness X says that the event occurred, can be decomposed into two intermediate hypotheses, namely, the hypothesis S that X's senses gave evidence of H and the hypothesis B that X believes that the event occurred.*

The generalization supporting the explanation of E (Section 2.1.7) can be formulated as follows: 'any person who is an accurate eyewitness and who is objective and who is veracious will usually tell the truth' (Schum 1994, p. 101–102). This prima facie generalization can be divided into three more basic common sense generalizations: (i) if X is an accurate person, then her senses gave evidence of what she sees; (ii) if X is an objective person, then she believes the evidence of her senses and (iii) if X is a veracious person, then she says what she believes. The first attribute has to do with the physical state of the person at the time the event occurred and with the physical circumstances of its occurrence. The second attribute has to do with the psychophysical states of the person at times later than the occurrence of the event. The third attribute has to do with the intentional state of the person at the time she gave her testimony. Assuming that these attributes can be considered as independent properties, then the DAG shown in Figure 2.16 can be drawn with discrete nodes AC, O and V, each with two states, representing, respectively, the three propositions: AC = 'X is an accurate person', O = 'X is an objective person' and V = 'X is a veracious person'. The two-state discrete nodes H, S, B and E represent, respectively, the propositions H = 'event H occurred', S = 'X's senses gave evidence of H', B = 'X believes that the event H occurred' and E = 'the witness X says that the event H occurred'.

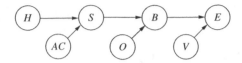

Figure 2.16   A Bayesian network with binary variables for the analysis of human witnesses reliability. Node *H* refers to an event of interest. Nodes *S*, *B* and *E* refer to, respectively, a witness's sensory perception, belief and report of event *H*. Nodes *AC*, *O* and *V* represent a witness's accuracy, objectivity and veracity, respectively.

Table 2.11   Conditional probability table for the node *S* in Figure 2.16.

H:	t		f	
AC:	t	f	t	f
$Pr(S = t|H, AC)$	1	0.5	0	0.5
$Pr(S = f|H, AC)$	0	0.5	1	0.5

A table with the same structure applies for node *B* given $\{S, O\}$, and node *E* given $\{B, V\}$.

Let the probabilities to be assigned to the states 'true' and 'false' of the nodes H, AC, O and V be denoted, respectively $\{\alpha, 1 - \alpha\}$, $\{\beta, 1 - \beta\}$, $\{\gamma, 1 - \gamma\}$ and $\{\delta, 1 - \delta\}$. Table 2.11 provides the conditional probabilities for node S. Note that a table with the same structure (i.e. conditional probabilities) applies for node B given $\{S, O\}$ and node E given $\{B, V\}$. Common sense generalizations (i)–(iii) give some of the required conditional probabilities. For conditional probabilities in case the witness is not accurate, or not objective, or not veracious, one might use a kind of 'default assumption': if the witness is not accurate, or objective, or veracious, then her answer might be considered, in a certain sense, a random answer (Bovens et al. 2002; Gärdenfors et al. 1983; Olsson 2002).

In the 'Lulu' case, knowledge of the fact that John was jealous of Jack will lower the probability that John is veracious. And one might have also information about John's accuracy and objectivity. Introducing the Bayesian network depicted in Figure 2.16 in the original 'Lulu' network of Figure 2.5 allows a finer analysis of the case but the Bayesian network grows more complicated. Object-oriented Bayesian networks can help both in reducing the visual complexity of the analysis and in making calculations easier by modularity.

A *network class* is a fragment of a Bayesian network that restricts the visibility of its nodes to the interior; that is, a node inside a class is not shown outside and vice versa. A class must be *instantiated* in order to be used. There are two kinds of possible links between classes: a node inside a class has parents outside and a node outside a class has parents inside.

The conditional probability tables of the nodes inside the class should not be changed whenever the class is instantiated, and this means that no node inside the class can have parents outside the class. In order to deal with case that a node inside a class has parents outside, we introduce a new type of node, called *input* nodes, which are placeholders for the parent nodes outside the class. Let *X* be a parent node outside the class: the *input node X** inside the class, its placeholder, has (i) the same internal definition as X, that is if X denotes a discrete random variable, then X^* has the same number of mutually exclusive states with the same values (if X denotes a continuous random variable, then X^* has the same range of values) and (ii) the same conditional probability table as X. Input nodes are also called *reference nodes* and the parents outside the class *referenced nodes*. Input nodes have a default prior probability table when no reference node is specified.

The case where a node outside has a parent inside is simpler because any node Y inside a class can be a parent of nodes outside of it. Therefore, all that we have to do is to mark Y as an *output node* (and, of course, it can become the referenced node of a reference node Y^* in another class). Nodes that are neither of type input nor of type output are called *internal* nodes. Input and output nodes are the *interface* of the class, and they are shown outside. Internal nodes are the *hidden nodes* of the class and they are not shown outside. Interface nodes may represent either decision variables or random variables, whereas internal nodes may be instantiations of other network classes, decision variables, random variables and utility functions. Instantiations of network classes that occur inside another class are represented by special nodes called *instance nodes*. Nodes that do not represent classes are called *basic nodes*. Instance nodes are represented as rectangles with arc-shaped corners, whereas input nodes are represented with a dotted outer line and output nodes are represented with a solid outer ring.

An *object-oriented Bayesian network class* $\mathbf{T}$ is a DAG with a joint probability distribution $Pr()$ over the set $\mathbf{X}$ of nodes, which contains three pairwise disjoint subsets of nodes. First, there is a set $\mathbf{I}$ of basic nodes, specified as input nodes. An input node has no parents in the class $\mathbf{T}$, and no children outside the class $\mathbf{T}$, but it can have at most one parent outside $\mathbf{T}$. Secondly, there is a set $\mathbf{O}$ of basic nodes, specified as output nodes. An output node has no parents outside $\mathbf{T}$ and it can have at most one child in the same class outside $\mathbf{T}$. Thirdly, there is a set $\mathbf{H}$ of basic nodes, specified as internal nodes. An internal node has no children or parents outside $\mathbf{T}$.

Example 2.8 (*'Jack loved Lulu', continued*) *The* network class **witness** *is shown in Figure 2.17. H is the output node, H^*, AC, O and V are the input nodes and S, B, E are the internal nodes. In this class, the input node and the output node are the same because the analysis of a witness' reliability starts with the prior probability of the hypothesis of interest as input. The end provides as an output the posterior probability of the same hypothesis. We can deal with this situation by creating two nodes H^* (the input node) and H (the output*

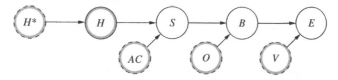

Figure 2.17   Object-oriented Bayesian network class for the analysis of the reliability of human witnesses (class **witness**). Node H, an output node, and the associated input node H^* (of which it is a copy) refer to an event of interest. Nodes S, B and E refer to, respectively, a witness's sensory perception, belief and report of event H. Input nodes AC, O and V represent a witness's accuracy, objectivity and veracity, respectively. All basic nodes are Boolean with states 'true' and 'false'.

node), which have the same internal structure, and connecting them with a probabilistic link whose conditional probability table is filled with 0 and 1 vales, as follows:

$$Pr(H|H^*) = \begin{cases} \{1,0\}, & H = true, \quad H^* = \{true, \ false\}, \\ \{0,1\}, & H = false, \quad H^* = \{true, \ false\}. \end{cases}$$

The conditional probability tables are the same in every instantiation of the class, which is as defined in Table 2.11. For any particular instantiation, there will be a different conditional probability table for the node H, and default prior distributions are assigned to the input nodes AC, O and V, unless in that particular case one has informations that allows to assign different probabilities to those nodes.

Figure 2.18(i) shows the 'Lulu' network (Figure 2.5) where node D has been substituted by an instantiation of the network class **witness**, *represented in terms of the node W because there is a witness, say Jane, who says that John was jealous of Jack, so casting doubt on the veracity of John. This can be represented with a value less than one for the internal node V of the instance node W. The Bayesian network (A, B, C, E, F, J, W) can itself be considered as a class encapsulating the class* **witness**, *where A, B, C, E, F and J are basic nodes and W is an instanced node. Figure 2.18(i) represents the network in its so-called collapsed form, that is all the nodes inside the instantiation are hidden. Figure 2.18(ii) shows part of Figure (i), but with instance node W in the so-called expanded form: only the input and output nodes are visible.*

If we do not have ancillary evidence about the reliability of John witness (witness 1), then a priori probability values can directly assigned to the input nodes AC, O and V of the instantiation W of class **witness**. *Suppose, however, that ancillary evidence is provided by other witnesses: Mary (witness 2), an ophthalmologist, says that John (witness 1) has a very poor eye sight, casting some doubts on the accuracy of John. In turn, Harry (witness 3), a physician, says that John suffers of a certain illness that impairs his psychical faculties, so*

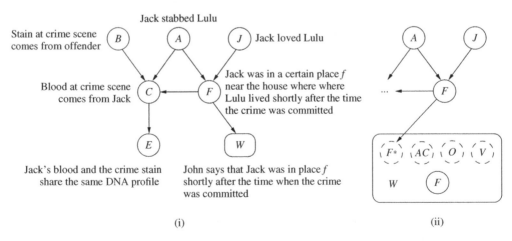

Figure 2.18   An object-oriented Bayesian network for witness about proposition *F*, with the instance node *W* in (i) collapsed and (ii) expanded forms. Nodes *AC*, *O* and *V* represent a witness's accuracy, objectivity and veracity, respectively. Node *F** is an input node (needed for technical reasons as explained in the text) and has the same definition as *F*. All basic nodes are Boolean with states 'true' and 'false'.

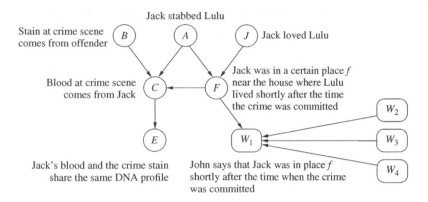

Jack stabbed Lulu

Stain at crime scene B A J Jack loved Lulu
comes from offender

Blood at crime scene C F Jack was in a certain place f
comes from Jack near the house where Lulu
lived shortly after the time
the crime was committed

W_2

E W_1 W_3

Jack's blood and the crime stain   John says that Jack was in place f W_4
share the same DNA profile   shortly after the time when the crime
was committed

Figure 2.19   Object-oriented Bayesian network in collapsed form for eyewitness evidence (W_1), and ancillary evidence for W_1, provided by witnesses W_2, W_3 and W_4 on, respectively, credibility attributes accuracy, objectivity and veracity.

*casting some doubts on his objectivity. Further, suppose that Jane (witness 4) says that John was jealous of Jack, so putting in question his veracity. Representing such ancillary evidence at the same level of hypothesis and John's witness evidence is a problem, as the adding of this ancillary evidence, or adding several witnesses for the same hypothesis (maybe all of them with their ancillary evidence) would increase the visual complexity of the Bayesian network in such a way it that would became very difficult to handle it. Schum's suggestion was to represent evidence about the credibility attributes of a witness in ancillary networks to the main network (Schum 1994, p. 188). This has been made technically possible by object-oriented Bayesian networks. In fact, there can be several instantiations of a single class, so what we have to do is add other instantiations of class **witness**. As an aside, see Hepler et al. (2007) for an example taken from the analysis of the Sacco and Vanzetti case given in Kadane and Schum (1996).*

*Figure 2.19 shows the 'Lulu' network in collapsed form with instantiations of the same network class **witness**, denoted W_1, W_2, W_3 and W_4. The node W_1 models John's direct eyewitness testimony for the hypothesis F. Nodes W_2, W_3 and W_4 model, respectively, the ancillary evidence provided by three witnesses for the attributes of accuracy, objectivity and veracity that attest to John's credibility. Figure 2.20 shows the nodes W_1, W_2 and W_3, W_4 in expanded form, displaying the interface nodes (input and output nodes), which are used to connect the instance nodes. Input and output node names are prefixed with the instance node name.*

2.2.6   Solving object-oriented Bayesian networks

For any object-oriented Bayesian class **T**, there exists an equivalent Bayesian network $BN_\mathbf{T}$, also called the *flat* or *unfolded representation of* **T**. If **T** contains instance nodes of other network classes, its representation $BN_\mathbf{T}$ is obtained by recursively unfolding all the instance nodes of **T**. The joint probability distribution of an object-oriented Bayesian network class **T** is equivalent to the joint probability distribution of its unfolded representation $BN_\mathbf{T}$:

$$Pr(\mathbf{X_T}) = \prod_{X_i \in \mathbf{X_{BN_T}}}^{n} Pr(X_i | \mathbf{PA}(X_i)),$$

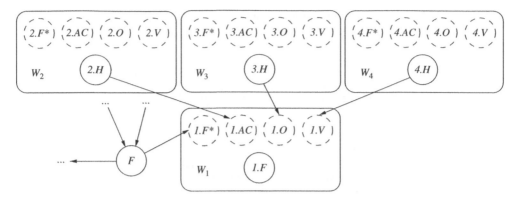

Figure 2.20 Partial representation of the object-oriented Bayesian network shown in Figure 2.19, with the expanded parts for eyewitness evidence (W_1), and ancillary evidence for W_1, provided by witnesses W_2, W_3 and W_4 on, respectively, attributes attesting to credibility of accuracy, objectivity and veracity. The nodes W_i ($i = 1, \ldots, 4$) are all instances of the class **witness**. Nodes $2.H$ and $1.AC$ denote the same proposition 'John has a very poor eye-sight'. Nodes $3.H$ and $1.O$ denote the same proposition 'John suffers from illness X'. Nodes $4.H$ and $1.V$ denote the same proposition 'John was jealous of Jack'. Nodes $i.AC$, $i.O$ and $i.V$ (for $i = 2, 3, 4$) represent, respectively, the accuracy, objectivity and veracity of, respectively, witnesses Mary (witness 2), Harry (witness 3) and Jane (witness 4).

where $\mathbf{X_T}$ is the set of nodes of the class $\mathbf{T}$ and $\mathbf{X_{BN_T}}$ is the set of nodes of its unfolded representation BN_T. The unfolded network representation of T is used for performing probabilistic inferences, the computational structure being the junction tree of the unfolded network: from the point of view of inference, an object-oriented Bayesian network is equivalent to its flat representation.

A network class $\mathbf{T}$ can be unfolded using the following algorithm: (i) add a node for all input nodes, output nodes and basic nodes of the class; (ii) add a node for each input node, output node and basic node of the instantiations contained in the class (and prefix the name of the instantiations to the node names in order to be able to distinguish different instantiations of the same class); (iii) add a link for each link in the class that is not a link between a reference node and its referenced node and (iv) for each reference node and its referenced node (there can be only one referenced node for each reference node), check that they have the same internal definition and if they do, then merge them into one node with all the parents and the children of the nodes merged, and the same conditional probability table of the referenced node.

Example 2.9 ('*Jack loved Lulu*', *continued*) *For simplicity, apply the algorithm only to a fragment of the object-oriented Bayesian network shown in Figure 2.20, composed by nodes F, W_1 and W_2. It yields the steps outlined in the following.*

First, node F is added (instance nodes W_1 and W_2 are neither input, nor output, nor basic nodes). Next, input nodes 1.F*, 2.H, i.AC, i.O and i.V, (i = 1, 2), output nodes 1.F and 2.H and internal nodes i.S, i.B, i.E (for i = 1, 2) are added. In a third step, one needs to add links to obtain the DAG shown in Figure 2.21(i), where internal nodes 1.S, 1.B, 1.E denote,

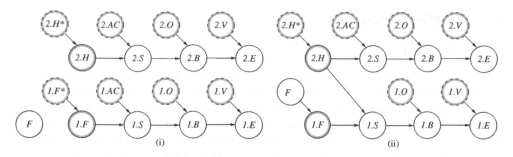

Figure 2.21   (i) State of the fragment $\{F, W_1, W_2\}$ of the object-oriented Bayesian network shown in Figure 2.19 after the third step of the algorithm for unfolding. (ii) Flat (unfolded) representation of the same network fragment.

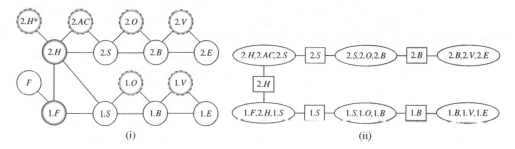

Figure 2.22   (i) The triangulated graph and (ii) junction tree of the fragment $\{F, W_1, W_2\}$ of the object-oriented Bayesian network shown in Figure 2.19.

respectively, the propositions 'John's senses gave evidence of F', 'John believes that F' and 'John says that F'. The internal nodes 2.S, 2.B and 2.E denote, respectively, the propositions 'Mary's senses gave evidence of the fact that John has a very poor eye-sight', 'Mary believes that John has a very poor eye-sight' and 'Mary says that John has a very poor eye-sight'. This step leads to the graph shown in Figure 2.21(i).

The fourth step of the algorithm involves the consideration that nodes F and 1.F* and 2.H and 1.AC are of the same type (nodes F and 1.F* denote the same proposition 'Jack was in place f', and nodes 2.H and 1.AC denote the same proposition 'John has a very poor eye-sight'). One can thus merge them to obtain the BN_T shown in Figure 2.21(ii), where node 1.F* has inherited the probability distribution of node F and node 1.AC that one of node 2.H.

It is easily seen that this Bayesian network is decomposed in such a way that the Markov property holds. The node $\{2.H\}$ d-separates the set $\{F, 1.F, 1.S, 1.O, 1.B, 1.V, 1.E\}$ from the set $\{2.H^*, 2.AC, 2.S, 2.O, 2.B, 2.V, 2.E\}$: the path $2.H^* \to 2.H \to 1.S$ is serial, and the path $1.S \leftarrow 2.H \to 2.S$ is diverging. Moralizing the graph immediately yields a triangulated graph [Figure 2.22(i)]. From the conditional probabilities $Pr(H|H^*)$ defined in Example 2.8, it follows that $Pr(1.F) = Pr(F)$ and $Pr(2.H) = Pr(2H^*)$, so that the junction tree of the triangulated graph in Figure 2.22(i) is as shown in Figure 2.22(ii).

2.3   Further readings

2.3.1   General

Jensen (2001), Jensen and Nielsen (2007) and Kjærulff and Madsen (2008) give accessible general introductions to Bayesian networks. Object-oriented Bayesian networks were first introduced in Koller and Pfeffer (1997), Laskey and Mahoney (1997) and Neil et al. (2000). A more technical treatment of the concepts and algorithms from graph theory used for Bayesian networks can be found in Castillo et al. (1997), Cowell et al. (1999) and Korb and Nicholson (2011). Practical applications of Bayesian networks can be found in Fenton and Neil (2013). The concept of d-separation is due to Judea Pearl and Thomas Verma (Pearl 1988; Pearl and Verma 1987). A good, informal, proof of the fundamental Markov property (Section 2.1.8) can be found in Charniak (1991). Interested readers can find more technical discussions of the d-separation theorem in Castillo et al. (1997), Cowell et al. (1999) and Pearl (2009). A formal proof of the fact that the joint probability distribution over the junction tree can be written as the product of the conditional probability distributions of the residuals of the cliques, given their separators (Section 2.2.3) can be found in Castillo et al. (1997, p. 228).

Junction tree propagation algorithms were first introduced by Lauritzen and Spiegelhalter (1988) and Shenoy and Shafer (1990). Readers are referred to Cowell et al. (1999) for a detailed presentation and more sophisticated examples than 'Lulu', both with discrete and continuous variables. The Hugin architecture for probability updating in junction trees has been first introduced in Jensen et al. (1990). The so-called *lazy propagation algorithm* implemented in Hugin is described in Jensen (2001) and has been introduced by Madsen and Jensen (1999). A free version of Hugin can be found at http://www.hugin.com/. Another very rich modelling environment is GeNIe and SMILE, freely available at http://genie.sis.pitt.edu/.

The propagation algorithm of Lauritzen and Spiegelhalter (1988) is also implemented in R (R Core Team 2013) through the package gRain. More generally, R is a computational environment where many developments take place. The package RHugin, for example, provides an R-interface to the Hugin software mentioned above. RHugin is available at http://rhugin.r-forge.r-project.org/ and is also described in Section 3.3.2 in Højsgaard et al. (2012). Computational aspects of the implementation of Bayesian networks in R are also addressed in Nagarajan et al. (2013).

Schum's understanding of a 'generalization' differs from the description given in Section 2.1.7 because the graphs for evidence analysis used in Schum (1994) and Kadane and Schum (1996) are not Bayesian networks but are modelled on Wigmore's reference charts (Wigmore 1937). Those graphs are bottom-up. The examples of 'generalizations' the authors put forward are also bottom-up, so to speak, as this example from Kadane and Schum (1996, p. 14) shows: 'Persons who are at the scene of a crime when it is committed are frequently (...) the ones who took part in it.' The appropriate common sense generalization is 'If a person takes part in a crime, then he is at the scene of the crime when it is committed'. This may then constitute the premiss of the following *D-N* explanation (see Section 1.3.3) that justifies the relevance of the fact that a reliable eyewitness says that they saw the individual in question at the scene of the crime (Garbolino 2001, pp. 1508–1511): (i) If a person participates in a crime, then he is at the scene of the crime when it is committed. (ii) John Smith participated in the robbery. (iii) John Smith was at the scene of the robbery when it was committed.

With regard to causality, the problem with Mackie's regularity approach (Section 2.1.11) is that (Hitchcock 2012, p. 3)

(…) the regularity approach makes causation incompatible with indeterminism: if an event is not determined to occur, then no event can be a part of a sufficient condition for that event. (…) Many philosophers find the idea of indeterministic causation counterintuitive. Indeed, the word 'causality' is sometimes used as a synonym for determinism. A strong case for indeterministic causation can be made by considering the epistemic warrant for causal claims. There is now very strong empirical evidence that smoking causes lung cancer. Yet the question of whether there is a deterministic relationship between smoking and lung cancer is wide open. The formation of cancer cells depends upon mutation, which is a strong candidate for being an indeterministic process. (…) Thus the price of preserving the intuition that causation presupposes determinism is agnosticism about even our best supported causal claims.

Hitchcock (2012) is a useful introductory survey to the important topic of probabilistic causality. Accessible introductions to the quite technical works of Pearl (2009) and Spirtes et al. (2001) can be found in Cooper (1999), Scheines (1997) and Pearl (1999). Other different viewpoints on probabilistic causation and causal modelling are offered in Dawid (2000, 2002), Williamson (2004) and Berzuini and Dawid (2012).

2.3.2   Bayesian networks and their predecessors in judicial contexts

Since the early 1990s, both legal scholars and forensic scientists have shown an increased interest in the applicability of Bayesian networks in judicial contexts. Whilst lawyers tend merely to be concerned with structuring cases as a whole, forensic scientists focus on the evaluation of selected items of trace material (and related scientific findings), such as fibres or blood. In the following discussion, relevant literature pertaining to the former topic is mentioned. A more detailed discussion of the latter topic is given in the remaining chapters of this book.

Bayesian networks have been proposed for structuring and reasoning about issues of complex and historically important cases. For example, Edwards (1991) provided an alternative analysis of the descriptive elements presented in the Collins case (People v. Collins, 68 Cal. 2d 319, 438 P. 2d 33, 66 Cal. Rptr. 497 (1968)). In this case, the prosecution assumed independence between different descriptive elements such as the colour of the getaway car or the colour of the offender's hair. The prosecution multiplied probabilities for the various descriptive elements to arrive at the result 1/12 000 000, inviting the jury to conclude that there was a chance of 1 in 12 million that the two accused were innocent. This argument has been criticized extensively in literature [e.g. Finkelstein and Fairley (1970)]. Edwards (1991) proposes various dependence relationships amongst the different descriptive elements. Bayesian networks are used as a graphical representation scheme and a means to perform probabilistic calculations.

Schum (1994), Kadane and Schum (1996) and Anderson et al. (2005) worked on a probabilistic analysis of the Sacco and Vanzetti case with an emphasis on the credibility and relevance of evidence given by human sources (i.e. testimony). Probabilistic case analysis was also proposed for the Omar Raddad case (Levitt and Blackmond Laskey 2001) and the O.J. Simpson case (Thagart 2003). Other applications in legal reasoning can be found in Fenton and Neil (2011) and Fenton et al. (2013). Object-oriented Bayesian networks have been proposed for judicial contexts by Dawid et al. (2007), Hepler et al. (2007) and Hepler and Weir (2007, 2008).

A thorough discussion of Bayesian networks for structuring and analysing legal arguments is given by Robertson and Vignaux (1993) and Vanderweele and Staudt (2011). The
authors present Bayesian networks as a contribution to the field of fact analysis and point
out the advantages of Bayesian networks over previously proposed charting methods, such as
Wigmore charts (Wigmore 1913) or route diagrams (Friedman 1986a,b). 'Wigmore charts'
are a representational scheme, intended to provide formal support for reaching and defend-
ing conclusions based on a '(...) mixed mass of evidence (...)' (Wigmore 1913, p. 79).
However, there is no incorporation of an explicit measure of uncertainty (e.g. through probability). 'Route diagrams', as proposed by Friedman, are a modification of probability trees
(Section 1.1.7), combining nodes that lead to the same final state.

3

Evaluation of scientific findings in forensic science

3.1 Introduction

Consider a legal setting. At the times of the preliminary investigation and trial, the proof of a point at issue may be assisted through information that often takes the form of results of comparative examinations and that relies on forensic science. Such scientific findings may be used to draw inferences about the existence or the nature of a criminal act (for example in forgery cases) or to help evaluators characterize relationships between elements involved in an act under investigation, such as an aggression between an offender and a victim in a sexual assault case. In the latter case, scientists may conduct comparative analyses of multiple traces (such as DNA or textile fibres) for which transfer is typically expected to occur.

Developments during the last 20 years have greatly increased the range of possible types of traces and the scope of analytical techniques that have been presented to the courts. These developments have allowed scientists to analyse a broader scope of traces, and traces of increasingly reduced size or amount. However, it is not only the mere results of analyses as such that should be presented to a court. Results need to be evaluated and interpreted in the context of the case under trial.

Scientific outcomes require considerable care in their interpretation, and it is particularly important to focus on questions of the kind 'What do the results mean in this particular case?' (Jackson 2000). Scientists and jurists should abandon the idea of absolute certainty when they approach inferential processes that aim, for example, at individualization. If it can be accepted that nothing is absolutely certain, then it becomes logical to determine the degree of belief that may be assigned to a particular proposition of interest (Kirk and Kingston 1964).

Therefore, theory, methods and applications of probability and statistics underlie the evaluation of scientific evidence. The trier of fact should engage in 'reasonable' reasoning and, thus, be familiar with probabilistic assessment so as to favour the understanding of scientific

---

Bayesian Networks for Probabilistic Inference and Decision Analysis in Forensic Science, Second Edition.
Franco Taroni, Alex Biedermann, Silvia Bozza, Paolo Garbolino and Colin Aitken.
© 2014 John Wiley & Sons, Ltd. Published 2014 by John Wiley & Sons, Ltd.

evidence (Twining 1994), which is often presented in a numerical way in front of a court of justice or in a written report to judicial authorities. Forensic scientists should give the courts an evaluation that adequately illustrates the convincing force of their results. Such an evaluation inevitably uses probability as a measure of uncertainty (Lindley 2006). Forensic science thus suggests itself as a case study for evidence evaluation and interpretation[1] as there is a broad of consensus that forensic findings should be thought about in probabilistic terms.

Scientific outcomes have − by their nature − a close link to statistical assessment (see Chapter 1), but there is a potential for misinterpretation of the value of statistical findings when, routinely, they are invoked to support scientific arguments at trial. Appropriate interpretative procedures for the assistance of both jurists and scientists have been proposed in scientific and judicial literature and applied practically to ease communication between the forensic and judicial world so as to aid in the correct interpretation of statistical information.

This chapter proposes a summary of the probabilistic approach based on Bayes' theorem and the use of the likelihood ratio for the assessment of the probative value of scientific findings. Formulae for likelihood ratios are developed and presented with a particular emphasis on the context of the case under examination and by relying on the concept of the 'hierarchy of propositions' (Cook et al. 1998a). The following chapters will translate, develop and use these equations in a graphical environment.

3.2   The value of scientific findings

The evaluation of scientific findings[2] may be thought of as the assessment of a comparison. This comparison involves material found, for example, at the scene of a crime and material found on suspects, their clothing or their environment. Let M_c denote the event of finding crime-related material on a scene, and M_s the event of collecting material on, for example, a suspect. For the remainder of this book, the events M_c and M_s will be taken to refer, by abuse of language, to the material pertaining to these events. Denote the combination by $M = (M_c, M_s)$. As an example, consider blood. The genotype of a blood stain found at a crime scene is M_c, the receptor or transferred stain form of the material, whereas M_s is the genotype of the suspect (also called the *source form of the material*). Alternatively, suppose that glass is broken during the commission of a crime. M_s would be the fragments of glass (the source form of the material) found at the crime scene, M_c would be the fragments of glass (the receptor or transferred particle form of the material) found on the clothing of a suspect. Again, M would be the two sets of fragments.

Observed qualities, such as the genotypes, or the results of measurements, such as the refractive indices of glass fragments, are taken from M. Comparisons are made of the results for the source form and the receptor form. Again, for the ease of notation, denote these by E_c and E_s, respectively, and let $E = (E_c, E_s)$ denote the combined set. Comparison of E_c and E_s

---

[1] The terms *evaluation* and *interpretation* are sometimes considered as synonyms, but it is helpful to conceive of a distinction. 'Evaluation' concentrates on deriving a value for the likelihood ratio or the Bayes' factor, whereas 'interpretation' refers to the meaning attached to such a value in the case as a whole.

[2] Note that the term *evidence* is generally used in literature and practice rather than terms such as 'finding' or 'outcome'. The term *evidence* can be conflicting, however, because in some legal contexts, it may refer to a judicial qualification of a finding. Forensic scientists are interested in the probative value of an observation before the judiciary qualifies it as 'evidence' at trial. The word 'finding' will be used from now on for the result of a forensic analysis. Note that this replacement of 'evidence' is in contrast to its use in the first two chapters.

is to be made, and the assessment of this comparison has to be quantified. The totality of the evidence is denoted Ev and is such that $Ev = (M, E)$.

Consider now the odds form of Bayes' theorem (Section 1.2.3) in the forensic context of assessing the probative value of analytical findings. Replace H_1 in (1.14) by the *proposition* H_p according to which the suspect (or, the defendant, if the case has come to trial) is truly guilty. Such a choice may be appropriate at an advanced stage at trial. In the same way, H_2 is replaced by proposition H_d, which means that the suspect is truly innocent. The results of both the search for trace material and the subsequent comparative examinations, Ev, may be written as $(E, M) = (E_c, E_s, M_c, M_s)$. In agreement with a predominant part of forensic literature on probabilistic evaluation, note that, from now on, the term *proposition* is used instead of 'hypothesis'. This choice aims to emphasize the ingredients considered as relevant for the scenario under study, but note that event (or hypothesis) and proposition are taken as equivalent terms with only a different emphasis on the fact and formulation expressing it (de Finetti 1968). Also, and for the sake of simplicity, single quotation marks (' ') will be omitted when describing verbally the content of a proposition, except when a proposition is part of a sentence.

The odds form of Bayes' theorem then enables the prior odds (i.e. prior to the presentation of Ev) in favour of guilt to be updated to posterior odds given Ev, the scientific findings under consideration. This is achieved by multiplying the prior odds by the likelihood ratio, which, in this context, is the ratio of the probabilities of the findings assuming guilt and innocence of the suspect.[3] With this notation, the odds form of Bayes' theorem may be written as

$$\frac{Pr(H_p|Ev, I)}{Pr(H_d|Ev, I)} = \frac{Pr(Ev|H_p, I)}{Pr(Ev|H_d, I)} \times \frac{Pr(H_p|I)}{Pr(H_d|I)},$$

where I denotes the circumstantial information. For ease of notation, explicit mention of I is sometimes omitted, although it is always assumed as a conditional. The propositions H_p and H_d need to be mutually exclusive but not necessarily exhaustive. Thus, the use of the words 'odds' is strictly speaking anomalous. Here, the propositions are exhaustive (guilt and innocence) but in general they need not be.

Note that in the evaluation of the findings Ev, two probabilities are necessary: the probability of the finding if the suspect is guilty and the probability of this finding if the suspect is innocent. For example, it is not sufficient to consider only the probability of the outcome if the suspect is innocent and to declare that a small value of this is indicative of guilt. The probability of this outcome if the suspect is guilty also needs to be considered. Similarly, it would not be sufficient to consider only the probability of the outcome if the suspect is guilty and to declare that a high value of this is indicative of guilt. Again, the probability of this outcome if the suspect is innocent has also to be considered. A careful consideration of the

---

[3] As an aside, note that the multiplication factor that revises a prior opinion to a posterior opinion through consideration of new information is a quantity known, more properly, as the Bayes factor. It measures the change produced by the new information (or, data) in the odds when going from the prior to the posterior distribution in favour of one proposition as opposed to another. In many cases, the competing propositions have single distributions (i.e. simple vs simple hypotheses), and it can readily be shown that the Bayes factor is just the likelihood ratio of H_p to H_d, and depends only on the data at hand. There may, however, be other cases where composite hypotheses are compared and unknown parameters are involved. Here, the Bayes factor still has the form of a likelihood ratio, but it is the ratio of two marginal likelihoods obtained by integrating over the parameter space. The Bayes factor is the ratio of weighted likelihoods under the competing hypotheses, and it appears that it no longer depends only on the data. For the sake of simplicity and in accordance with forensic literature, the term *likelihood ratio* is used throughout the book. A first example of the Bayes factor is presented in (4.5).

precise meaning of these conditional probabilities is important to avoid well-known fallacious conclusions such as the fallacy of the 'transposed the conditional' or 'prosecutor's fallacy'.

Consider the likelihood ratio $Pr(Ev|H_p, I)/Pr(Ev|H_d, I)$ in some further detail, but omit explicit mention of I. In an extended form, the likelihood ratio may be rewritten as follows:

$$\frac{Pr(E|H_p, M)}{Pr(E|H_d, M)} \times \frac{Pr(M|H_p)}{Pr(M|H_d)}.$$

The second ratio in this expression, $Pr(M|H_p)/Pr(M|H_d)$, concerns the type and quantity of material found at the crime scene and on the suspect. It may also be written in an extended form as follows:

$$\frac{Pr(M_s|M_c, H_p)}{Pr(M_s|M_c, H_d)} \times \frac{Pr(M_c|H_p)}{Pr(M_c|H_d)}.$$

The value of the second ratio in this expression may be taken to be 1. The type and quantity of material at the crime scene is independent of whether the suspect is the criminal or someone else is. The value of the first ratio, which concerns the material found on the suspect given the material found at the crime scene and the guilt or otherwise of the suspect, is a matter for personal judgement and it is not proposed to consider this assignment further here. Instead, consideration will be concentrated on

$$\frac{Pr(E|H_p, M)}{Pr(E|H_d, M)}.$$

For notational convenience, M will be subsumed into I and omitted, for clarity of notation. Then,

$$\frac{Pr(M|H_p)}{Pr(M|H_d)} \times \frac{Pr(H_p)}{Pr(H_d)}$$

which equals

$$\frac{Pr(H_p|M)}{Pr(H_d|M)}$$

will be written as

$$\frac{Pr(H_p)}{Pr(H_d)}.$$

Thus,

$$\frac{Pr(H_p|Ev)}{Pr(H_d|Ev)} = \frac{Pr(H_p|E, M)}{Pr(H_d|E, M)}$$

will be written as

$$\frac{Pr(H_p|E)}{Pr(H_d|E)}$$

and

$$\frac{Pr(Ev|H_p)}{Pr(Ev|H_d)} \times \frac{Pr(H_p)}{Pr(H_d)}$$

will be written as

$$\frac{Pr(E|H_p)}{Pr(E|H_d)} \times \frac{Pr(H_p)}{Pr(H_d)}.$$

The full result is then

$$\frac{Pr(H_p|E)}{Pr(H_d|E)} = \frac{Pr(E|H_p)}{Pr(E|H_d)} \times \frac{Pr(H_p)}{Pr(H_d)},$$

or if I is included

$$\frac{Pr(H_p|E,I)}{Pr(H_d|E,I)} = \frac{Pr(E|H_p,I)}{Pr(E|H_d,I)} \times \frac{Pr(H_p|I)}{Pr(H_d|I)}.$$

The likelihood ratio is the ratio

$$\frac{Pr(H_p|E,I)/Pr(H_d|E,I)}{Pr(H_p|I)/Pr(H_d|I)}$$

of posterior odds to prior odds. The likelihood ratio converts the prior odds in favour of H_p into the posterior odds in favour of H_p. This representation also emphasizes the dependence of the prior odds on background information.

The likelihood ratio may be thought of as the *value* of the forensic or, more generally, scientific findings. Evaluating forensic scientific findings thus amounts to the assignment of a value for the likelihood ratio. This value will be denoted V.

Consider two competing propositions, H_p and H_d, and background information I. The value V of the forensic findings E is given by

$$V = \frac{Pr(E|H_p,I)}{Pr(E|H_d,I)}. \tag{3.1}$$

This expression, the likelihood ratio, converts prior odds $Pr(H_p|I)/Pr(H_d|I)$ in favour of H_p relative to H_d into posterior odds $Pr(H_p|E,I)/Pr(H_d|E,I)$ in favour of H_p relative to H_d. If logarithms are used, the relationship becomes additive:

$$\log O(H_p|E) = \log \frac{Pr(E|H_p)}{Pr(E|H_d)} + \log O(H_p),$$

where $O(H_p|E) = Pr(H_p|E)/(Pr(H_d|E)$, the odds in favour of H_p.

This has the very telling intuitive interpretation of weighing outcomes in the scale of justice, and the logarithm of the likelihood ratio is known as the *weight of evidence*, an expression widely attributed to Good (1950). Findings in support of H_p (i.e. for which the likelihood ratio is greater than 1) will provide a positive value for the logarithm of the likelihood ratio and tilt the scales in a direction towards finding for H_p. Conversely, forensic results in support of H_d (i.e. for which the likelihood ratio is less than 1) will provide a negative value for the logarithm of the likelihood ratio and tilt the scales in a direction towards finding for H_d. Outcomes for which the likelihood ratio is 1 are neutral because they lead to a zero logarithm. Such a result would leave the position of the scale unchanged. The logarithm of the likelihood ratio is $\log(Pr(E|H_p)) - \log(Pr(E|H_d))$; thus $\log(Pr(E|H_p))$ may be thought of being placed in one scale whilst $\log(Pr(E|H_d))$ is placed in the other scale, one scale representing positive values, the other negative values.

What happens when scientists are not faced with categorical data and discrete propositions, but they encounter, for example, continuous data and discrete propositions? For the sake of illustration, consider a set $\mathbf{x}$ of measurements performed on a control item, and another set $\mathbf{y}$ of measurements obtained through analysing recovered material of an unknown source. Suppose that the measurements refer to a particular analytical characteristic, such as the refractive index or the elemental composition of glass fragments. For the example here, $\mathbf{x}$ would be a set of measurements of refractive indices on fragments of a broken window at the crime scene and $\mathbf{y}$ a set of measurements of refractive indices on fragments of glass found on a suspect (e.g. the outer clothing). If the suspect was at the crime scene, then the event of transfer of fragments from the broken window might need to be considered. In turn, if the suspect was not on the crime scene, the recovered fragments may be the consequence of another mechanism of transfer (i.e. come from some other, unknown, source).

The quantitative part of the findings concerning the glass fragments in this case can be denoted by $E = (\mathbf{x}, \mathbf{y})$. The Bayes' factor is then written as follows:

$$V = \frac{f(\mathbf{x}, \mathbf{y}|H_p, I)}{f(\mathbf{x}, \mathbf{y}|H_d, I)}.$$

Bayes' theorem and the rules of conditional probability apply to probability density functions $f()$ and probabilities. The value V of the findings may be rewritten, accepting independence between $\mathbf{y}$ and $\mathbf{x}$ under H_d, as

$$V = \frac{f(\mathbf{y}|\mathbf{x}, H_p, I)}{f(\mathbf{y}|H_d, I)}.$$

See (3.2) for a similar development. In a Bayesian approach, the characteristic of interest may be able to be parameterized, for example, by the mean. Denote the parameter by θ. This parameter may vary from source (window) to source (another window). If one considers the two propositions H_p (the recovered item is from the same source as the control material) and H_d (the recovered material comes from an unknown source), then the measurements $\mathbf{x}$ are from a distribution with parameter θ_1, say, and the measurements $\mathbf{y}$ are from a distribution with parameter θ_2, say. If $\mathbf{x}$ and $\mathbf{y}$ come from the same source, then $\theta_1 = \theta_2$, otherwise $\theta_1 \neq \theta_2$. In practice, the parameter θ is not known and the analysis is based on the marginal probability densities of $\mathbf{x}$ and $\mathbf{y}$ (see also Footnote 3). The above equation for V can be rewritten as

$$V = \frac{\int f(\mathbf{y}|\theta) \int f(\mathbf{x}|\theta)\pi(\theta)d\theta}{\int f(\mathbf{x}|\theta)\pi(\theta)d\theta \int f(\mathbf{y}|\theta)\pi(\theta)d\theta},$$

where $\pi(\theta)$ represents the prior distribution of the unknown parameter θ. Therefore, the Bayes' factor does not depend only on the sample data. It is the ratio of two weighted likelihoods.

3.3 Principles of forensic evaluation and relevant propositions

It is now widely accepted that the coherent and balanced assessment of scientific findings requires forensic scientists to consider competing propositions that typically represent alternatives proposed by the prosecution and the defence to illustrate their view of the case under examination. These alternatives are formalized representations of the framework of circumstances. The forensic scientist evaluates the findings under these propositions. The formulation of the propositions represents a crucial basis for a logical approach to the evaluation of results of forensic examinations (Cook et al. 1998a). This procedure can be summarized by three key principles [e.g. Evett and Weir (1998)]. A first principle says that evaluation is only meaningful when at least one alternative proposition is considered. There may be two or more competing propositions, but there is a convention to denote them H_p and H_d in cases where there are only two. As a second principle, evaluation of scientific results (E) is based on the probability of the outcome given the propositions that are addressed, that is $Pr(E|H_p)$ and $Pr(E|H_d)$. A third precept holds that the evaluation of scientific results is to be carried out within a framework of circumstances, habitually denoted by I. The evaluation is thus conditioned not only by the competing propositions but also by the structure and content of the framework of circumstantial information I.

For these reasons, propositions play a key role in the process of evaluation. Often, propositions are considered in pairs, but there are situations where there will be three or more (see Section 7.5). This happens regularly with DNA, notably with DNA mixtures when, for example, the number of contributors to the mixture is in dispute (Biedermann et al. 2011b; Buckleton et al. 1998; Lauritzen and Mortera 2002), or when related and unrelated individuals both need to be considered as potential donors of a biological stain. A further example could be the assessment of the distance of firing based on gunshot residue particles detected on a target. In fact, there may be various competing propositions relating to the distance from which a pattern in question has been shot, depending on the position taken by, respectively, the prosecution and the defence.

It is generally possible to restrict the number of propositions to two which will be identified with the respective prosecution and defence positions. Clearly, the two propositions must be mutually exclusive. It is tempting to specify that they are exhaustive, but this is not necessary. The simplest way to achieve exhaustiveness is by adding the word 'not' into the first proposition, saying for example: 'Mr C is the man who kicked Mr Z', and 'Mr C is not the man who kicked Mr Z'. However, this gives the court no idea of the way in which the scientist has assessed the outcomes with regard to the second proposition. Mr C may not have kicked the victim, but he may have been present at the incident. Analogously, consider the proposition 'Mr B had sexual intercourse with Miss Y', and 'Mr B did not have sexual intercourse with Miss Y'. In fact, if semen has been found on the vaginal swab, then it may be inferred that someone has had sexual intercourse with Miss Y and, indeed, the typing results from the semen would be evaluated by considering their probability given that the semen came from some other man (and that Mr B has nothing to do with the case). It will help the court if this is made plain in the alternative proposition. So, the alternative could be 'Some unknown man, unrelated to Mr B, had sexual intercourse with Miss Y' (with no consideration of relatives).

In summary, the simple use of 'not' to frame the alternative proposition is unlikely to be particularly helpful to the court. In the same sense, it is useful to avoid the use of unclear words like 'contact' to describe the alleged type of action. In fact, there is a danger in using such a vague word. As noted by Evett et al. (2000c), the statement that a suspect has been in recent contact with broken glass could mean many things. Thus, there is a clear need to specify propositions accurately in a framework of circumstances. Otherwise, the scientist will not be able to address the probability of the findings in a transparent and meaningful way.

The formulation of the propositions to be considered is not an easy task for the scientist. A guiding approach to assist the scientist has been proposed by Cook et al. (1998a). It is led by the idea that propositions addressed in a judicial case depend on (a) the circumstances of the case, (b) the observations that have been made and (c) the available background information. The approach also considers a classification, called a *hierarchy*, of these propositions into three main categories or levels, that is source level, activity level and crime level.

3.3.1   Source level propositions

The evaluation of findings under propositions at source level depends on analyses and measurements (also sometimes called *intrinsic features*) made on the recovered and control items. The expression of the probative value of findings under source level propositions (e.g. 'Mr. X's pullover is the source of the recovered fibres', and 'an unknown garment is the source of the recovered fibres') does not need to take account of anything other than the analytical information obtained during examination. The probability of the findings under the first proposition (numerator) follows from a careful comparison between the two examined materials (the recovered and the control). This assessment relies on specialized knowledge about observations made on items known to come from the same source. The probability of the findings under the second proposition (denominator) is a consideration that relates to the scientist's knowledge about the occurrence of target features in the relevant population of alternative sources.

3.3.1.1   Notation

As usual, let E be the outcome, the value of which has to be assessed. Let the two propositions to be compared be denoted H_p and H_d. The likelihood ratio, V, is then $V = Pr(E|H_p)/Pr(E|H_d)$. The propositions will be stated explicitly for any particular context. For example, in cases involving DNA profiles where the alternative proposition states that a stain of body fluid did not come from a given suspect, it is common to assume that the origin of the stain is some unknown person unrelated to the suspect and not originating from the same sub-population. More generally, the conditioning on background information I also needs to be incorporated so that the likelihood ratio becomes $V = Pr(E|H_p, I)/Pr(E|H_d, I)$.

3.3.1.2   Single stain

Consider a scenario in which a blood stain has been left at the scene of the crime by the person who committed the crime. A suspect has been designated and it is desired to establish the strength of the relationship between the suspect and the crime. The forensic scientist compares the profile of the stain with the profile of control material obtained from the suspect. The two propositions of interest, under which the outcome will be evaluated, are defined as follows: 'the crime stain comes from the suspect (H_p)', and 'the crime stain comes from some (unrelated) person other than the suspect (H_d)'.

The findings, denoted by E, may be divided into two parts (E_c, E_s), where E_s refers to the DNA profile Γ of the suspect and E_c to the DNA profile Γ of the crime stain. A general formulation of the problem is given in the following. In addition, suppose that the scientist knows, from data previously collected (population studies), which profile Γ occurs in $100\gamma\%$ of some relevant population, denoted Ψ.

The value of the findings E, expressed as $V = Pr(E|H_p, I)/Pr(E|H_d, I)$, can then be developed in some further detail:

$$V = \frac{Pr(E|H_p, I)}{Pr(E|H_d, I)} = \frac{Pr(E_c, E_s|H_p, I)}{Pr(E_c, E_s|H_d, I)} = \frac{Pr(E_c|E_s, H_p, I)Pr(E_s|H_p, I)}{Pr(E_c|E_s, H_d, I)Pr(E_s|H_d, I)}.$$

Now, E_s is the analytical result that the suspect's profile is Γ. It is assumed here that a person's profile is independent of whether that person was at the scene of the crime (H_p) or not (H_d) or whether or not that person is the source of the recovered stain. Thus, $Pr(E_s|H_p, I) =$

$Pr(E_s|H_d, I)$, so that the likelihood ratio reduces to $Pr(E_c|E_s, H_p, I)$ in the numerator and $Pr(E_c|E_s, H_d, I)$ in the denominator.

If the suspect was not at the scene of the crime and therefore he is not the source of the stain (i.e. H_d is true), then the result (E_c) about the profile of the crime stain is independent of the result E_s pertaining to the profile of the suspect. Thus, $Pr(E_c|E_s, H_d, I)$ can be reduced to $Pr(E_c|H_d, I)$ and the likelihood ratio becomes

$$V = \frac{Pr(E_c|E_s, H_p, I)}{Pr(E_c|H_d, I)}. \tag{3.2}$$

The assumption that the suspect's characteristics are independent of whether he is the origin of the crime stain or not should not be made without consideration. In a context involving DNA, the question of how many other individuals in the population of possible culprits might also be expected to share this DNA profile is of great relevance. The selection of a genotype (the suspect's genotype) should modify one's expectation to select an identical one coming from the true donor of the stain (proposition H_d). This question is complicated by the phenomenon of genetic correlations because of shared ancestry (Balding and Nichols 1997). The simple calculation of profile frequencies, assuming Hardy–Weinberg equilibrium, is not sufficient for forensic evaluation when there are dependencies amongst different individuals involved in the case under examination, such as the suspect and the perpetrator (i.e. the real source of the recovered trace) featured in the proposition H_d (Buckleton and Triggs 2005).

The more common source of dependency is a result of membership in the same population and the possession of similar evolutionary histories. The mere fact that populations are finite in size means that two people taken at random from a population have a non-zero chance of having relatively recent common ancestors. Disregarding this correlation of alleles in considerations of probative value leads to an exaggeration of the strength of the outcome against the person whose genotype was found to correspond. For the sake of simplicity, in the proposed development above and in what follows, independence is assumed as it is generally not criticized in forensic branches other than DNA.

It should further be noted that, for some crime scenes, transfer of materials to an offender's clothing may very easily occur. If the characteristics of interest relate to such materials, and the offender is later identified as a suspect, the presence of such material is not independent of his presence at the crime scene. If the reasonableness of the simplifications is in doubt, then the original formulaic expression is the one which should be used.

The background information, I, may be used to assist in the determination of the relevant population from which the criminal is supposed to have come. For example, consider a case where I may include an eyewitness description of the criminal as Chinese. This is valuable because the occurrence of DNA profiles can vary between ethnic groups and, thus, affect the value of the findings.

Example 3.1 *(Prototype example for likelihood ratio assignment) Consider the assignment of values for the components of the likelihood ratio in some further detail assuming a one stain one offender case. Focus first on the numerator $Pr(E_c|E_s, H_p, I)$. This is the probability that the crime stain is of profile Γ, given that the suspect was the source of the crime stain, that the suspect has DNA profile Γ, and all other information I. This may include, for example, an eyewitness account that the criminal is Chinese. This probability is conveniently set to 1 since if the suspect was the source of the stain and has profile Γ, then the crime stain can be expected to be of profile Γ. Thus, $Pr(E_c|H_p, E_s, I) = 1$. Note that this assignment supposes* *that the laboratory's experts are in fact able to detect $E_c = E_s$ each time such a situation does exist (i.e. there are no special circumstances that could induce a difference between two entities that should be found to correspond). Further, such an extreme assignment of a value of 1 is acceptable only if categorical features are used during the process of comparison. Here, the numerator of the likelihood ratio thus represents the so-called intra-variability. The setting here, which involves a categorical description of features (i.e. allele numbers), supposes an absence of intra-variability.*

Next, consider the denominator $Pr(E_c|H_d, I)$. This term supposes that the proposition H_d is true, that is the suspect was not the source of the crime stain. The background information I is also assumed known. Together I and H_d define the relevant population. It may be that I provides no information about the criminal which will affect the scientist's assessment of the probability of the DNA profile being of a particular type. For example, I may include eyewitness testimony that the criminal was a tall, young male. However, a DNA profile (apart from the sex marker) is independent of all three of these qualities, so I bears no information for assigning the probability that the DNA profile is of a particular type.

More generally, let the relevant population be denoted Ψ, and suppose that the donor of the recovered stain is an unknown member of Ψ. The finding E_c is to the effect that the crime stain is of profile Γ. This is to say that an unknown member of Ψ is Γ. The probability of this is the probability that a person drawn at random from Ψ has profile Γ, which is γ. This is also sometimes referred to as the inter-variability. *Thus, $Pr(E_c|H_d, I) = \gamma$. The likelihood ratio V is then*

$$V = \frac{Pr(E_c|H_p, E_s, I)}{Pr(E_c|H_d, I)} = \frac{1}{\gamma}. \tag{3.3}$$

This value, $1/\gamma$, expresses the probative value of the correspondence in DNA profiles when the criminal is a member of Ψ.

As it stands, this result has some limitations. One of them relates to uncertainty with respect to the nature of the observation. Because of the sensitivity of current DNA profiling technology, it is now possible to encounter situations where it is not necessarily the case that a particular profile came from a discernible region of staining (i.e. the profile comes from material that is at the cellular level and so cannot be associated with visible staining). In such cases, it may be necessary to address what are termed propositions at 'sub-source' level. In a DNA context, common source level propositions such as 'The semen came from the suspect', and 'The semen came from some other man', might then need to be replaced by 'DNA came from the suspect', and 'DNA came from some other person' (Evett et al. 2000c). The available information and the context of the case will impact on the choice of the propositional level. However, the value V would remain $1/\gamma$. Other limitations, such as the possibility of laboratory error, are considered later in Section 7.11.

Extensions of the likelihood ratio for findings assessed under source level propositions have also been developed for more sophisticated situations involving, for example, mixtures (i.e. a stain or trace that contains a mixture of genetic material from more than one person; see also Section 7.8) and multiple traces (i.e. cases involving several stains and several offenders), notably the so-called forensic two-trace problem (see Section 6.3).

3.3.2   Activity level propositions

A further level in the hierarchy of propositions relates to an activity. This implies that the definitions of the propositions of interest have to include an action. Such propositions could be, for example, 'Mr X assaulted the victim', and 'Mr X did not assault the victim, but an unknown man committed the assault' (i.e. Mr X is not involved in the offence), or 'Mr X sat on the car driver's seat' and 'Mr X never sat on the car driver's seat'. The consequence of such an activity (the assault or the sitting on a driver's seat) is a contact (between the two people involved in the assault, or the contact between the driver and the seat of the car) and, consequently, a possible transfer of material (i.e. fibres in the example here). The scientist thus needs to consider more detailed information about the case under examination, in particular with respect to the transfer and persistence of the fibres on the receptor (e.g. the victim's pullover). Circumstances of the case such as the areas of contact between the victim and the criminal, the strength or intensity of the contact and the modus operandi are all factors that have a bearing on the answers to relevant questions like 'Is this the sort (quantity) of trace that would be seen if Mr X were the man who assaulted the victim?' or 'Is this the sort (quantity) of trace that would be seen if Mr X were not the man who assaulted the victim?' This is a principal difference with respect to the assessment of the findings under source level propositions (Section 3.3.1), which requires rather limited circumstantial information. Only I, general background information, is needed, as it could help to define the relevant population for use in the assessment of the rarity of the analytical characteristic of interest. For assessment under activity level propositions, however, evaluation cannot meaningfully be conducted without a framework of circumstances. The importance of this requirement will become clear when the scientist considers an approach based on pre-assessment (Section 3.4). This will require the expert to examine possible scenarios of the case and to assure that all relevant information for the proper assessment of the outcomes is available.

3.3.2.1   Notation and formulaic development

When activity level propositions are retained, it may typically be the case that the stain of interest cannot necessarily be regarded as coming from the criminal. This consideration is all the more important when the direction of transfer is from the scene or victim to the criminal. As an illustration, suppose that a case in which the blood of a victim has been shed. A suspect is investigated for this crime and a single blood stain of genotype Γ has been found on an item of the suspect's clothing. The suspect's genotype is different from Γ, but the victim's genotype is found to be Γ. The scientist can then consider two mutually exclusive propositions (or events) of transfer: 'the blood stain came from some background source' (T_0) and 'the blood stain was transferred during the commission of the crime' (T_n). Here, the lowercase letter n is a placeholder for the quantity or number of traces and will be developed in some further detail at a later point; here $n = 1$. As before, there are two main propositions to consider: 'the suspect assaulted the victim' (H_p) and 'the suspect did not assault the victim' (H_d). The latter propositions suppose that the suspect is not involved in any way whatsoever with the victim.

The forensic findings E to be evaluated consist of (i) a single blood stain found on the suspect's clothing and (ii) the stain's genotype determined as Γ. Here, the information that the victim's genotype is Γ is considered as part of the relevant background information I. This is a so-called scene-anchored perspective (Stoney 1991). The expression of the value of these findings, in its general form, then takes the habitual notation $V = Pr(E|H_p, I)/Pr(E|H_d, I)$.

Consider the numerator first and event T_0 initially. In this scenario, the suspect and the victim have been in contact (H_p) and no blood has been transferred to the suspect (T_0) as a result of the contact. The probability for this event is $Pr(T_0|H_p)$. Also, a stain of genotype Γ has been transferred by some other means, an event with probability $Pr(B, \Gamma)$ where B refers to the event of a transfer of a stain from a source (the background source) other than the crime scene (victim). In effect, this term expresses the scientist's probability for the findings E assuming the event of no transfer, that is $Pr(E|T_0, H_p, I)$.

Next, consider the event of transfer T_n. The probability for this event, given H_p is $Pr(T_n|H_p)$. Given T_n, H_p and knowledge about the genotype Γ of the victim, it is certain that to obtain E, the group of the transferred stain of type Γ: $Pr(E|T_n, H_p, I) = 1$. This implies also that no blood has been transferred from a background source.

To ease notation, write t_0 and t_1 for the numerical values assigned to, respectively, $Pr(T_0|H_p)$ and $Pr(T_n|H_p)$, the events of no stain and one stain being transferred during the course of a crime. In turn, let b_0 and $b_{1,1}$ denote the values assigned for the probabilities that a person from the relevant population will have, respectively, zero blood stains or one group of one blood stain on his clothing. Note that, more generally, the target on which the presence of transfer material is studied is referred to as a *receptor*. Further, let γ denote the assigned probability that a stain acquired innocently on the clothing of a person from the relevant population will be of genotype Γ. This probability assignment may be different from the population proportion Γ for members of the general population as the general population and the relevant population may be different. Then, $Pr(E|T_0, H_p, I) = Pr(B, \Gamma) = \gamma b_{1,1}$. The numerator can thus be written as follows (Evett 1984): $t_0 \gamma b_{1,1} + t_1 b_0$. The first term in this sum accounts for the event T_0 and the second for the event T_n.

Now consider the denominator. This term assumes that suspect and the victim were *not* in contact, that is the suspect has nothing to do with the incident. Arguably, the presence of the staining is the result of chance alone (i.e. legitimate transfer). The denominator thus takes the value $Pr(B, \Gamma)$, which, as defined above, equals $\gamma b_{1,1}$. The overall expression for the value of the findings thus is

$$V = \frac{t_0 \gamma b_{1,1} + t_1 b_0}{\gamma b_{1,1}} = t_0 + \frac{t_1 b_0}{\gamma b_{1,1}}. \tag{3.4}$$

Equation (3.4) requires the scientist to assign probabilities for (i) no stain being transferred during the commission of the action under investigation (t_0), (ii) one stain being transferred during the commission of such an action (t_1), (iii) no stain being transferred innocently (b_0), (iv) (one group of) one stain being transferred innocently ($b_{1,1}$) and (v) staining of body fluids on clothing being of genotype Γ (γ).

The first four of these probabilities relate to what is sometimes called *extrinsic aspects* and the fifth an *intrinsic aspect* (Kind 1994). The assignment of probabilities for extrinsic aspects is subjective, in the sense of a personal judgement formed by the forensic scientist based on his specialized knowledge. Probabilities for intrinsic aspects may primarily be informed by observation and measurement. More generally, extrinsic features can be physical attributes or descriptors (e.g. the number/quantity, position and location of stains), and intrinsic features can be descriptors of the stains (e.g. analytical characteristics such as profiles). Note that, in general, t_0 is often small in relation to $t_1 b_0/(b_{1,1}\gamma)$ and may thus be considered negligible.

Example 3.2 *(Evaluation of blood staining on a suspect under activity level propositions) Let Γ be a DNA profile (e.g. a partial profile) that the scientist expects to see with probability 0.01 in a member of a relevant population (e.g. Caucasians in England). Assume that the distribution of DNA profiles amongst stains on clothing is comparable to the distribution amongst this population. This assumption is not necessarily correct. Then $\gamma = 0.01$. Consider the results of a survey of men's clothing from which it appears reasonable that $b_0 > 0.95$ and $b_{1,1} < 0.05$.*

Suppose also that $t_1 > 0.5$. Then, leaving aside the value for t_0, the likelihood ratio becomes

$$V > \frac{0.5 \times 0.95}{0.05 \times 0.01} = 950.$$

This result indicates very strong support to the proposition that the suspect assaulted the victim. The findings are at least 950 times more likely if the suspect assaulted the victim than if he did not.

Notice that this result is considerably different from the simple expression $1/\gamma$ encountered earlier in (3.3) for evaluation under source level propositions. Here, $1/\gamma$ would lead to the value 100. This latter result would hold if $(\{t_1 b_0\}/b_{1,1})$ were approximately 1, which may mean unrealistic assumptions about the relative values of the transfer probabilities would have to be made.

These considerations emphasize that the evaluation of the analytical results should take into account the possibility of transfer of material from a source other than the suspect, which involves the assignment of probabilities for transfer and background. A main advantage of activity over source level propositions is that the evaluation under the former propositions does not strictly depend on the recovered material. In particular, it is possible to assess the fact that, for example, no blood has been recovered, and it is clearly important to assess the importance of the absence of material.

Forensic literature has reported on various further developments. Examples include cases involving material transferred from the offender to the crime scene, multiple stains and/or offenders and cross-transfer (also called *two-way transfer*) [e.g. Aitken and Taroni (2004)].

There is another complication, which is discussed briefly here and in greater detail in Chapter 5. It concerns situations in which there may be uncertainty about the connection between a person and an item belonging to that person, which is suspected to be involved in the transfer of material. Often, it may be reasonable to assume that the recovered material has a certain link with a given individual (who may be the suspect). For example, if there has been a direct and primary transfer,[4] then DNA found on a crime scene provides a direct connection between the source of that DNA and the crime scene. The reason for this is that biological material is intimately related to each individual. This may not be the case, however, with certain trace categories other than DNA, such as shoemarks or fibres. In particular, shoes or garments may − in principle − be worn by different individuals. Arguably, one should also consider the association between a given donor item (such as a garment or a pair of shoes) and a given suspect if the aim is to examine the degree of involvement of that person in the action under investigation. Chapter 5 investigates the effect that the latter issue may have on the development of a likelihood ratio under activity level propositions and illustrates this analysis through the use of Bayesian networks.

3.3.3   Crime level propositions

The so-called crime (or offence) level propositions are close to the propositions at activity level. As the term *offence* clearly points out, the propositions at this level in the hierarchy are really those of direct interest to the jury. Non-scientific information, such as whether or not a

---

[4] A primary transfer is referred here to as an *event* in which a given donor sheds material in the first place (e.g. an injured offender leaves blood stains on a broken window). Secondary transfer occurs, for example, if trace material on a suspect's clothing (e.g. traces of illegal drugs acquired as a primary transfer because of the manipulation of drugs in powder form) is transferred to another surface (e.g. a car seat).

crime occurred or whether or not an eyewitness is reliable, plays an important role in decisions made at this level.

In routine work, forensic scientists generally restrict their work to evaluation given source level propositions, notably for DNA profiling results. Evaluation using activity level propositions is more informative for legal practitioners, but it requires that an important body of circumstantial information is made available to the scientist. Unfortunately, this is often not the case because of a lack of collaboration between scientists and investigators. There should be more collaboration because of the clear limitations in the use of source level propositions, compared to the advantage of evaluation with the use of activity level propositions. The lower the level at which findings are assessed, the lower is the relevance of the results in the context of the case discussed before trial. As an illustration, recall that even if the value V of particular outcomes may be such as to add considerable support to the proposition that the stain comes from the suspect, this does not − by itself − help determine whether the stain has been transferred during the criminal action or for an innocent reason. Consequently, there is often dissatisfaction if the scientist's evaluation is restricted to source level propositions.

A formal development of the likelihood ratio using 'crime level' propositions for a situation of transfer of material from the criminal to the crime scene shows that two additional factors are of interest. The first concerns material that may be relevant (Stoney 1991, 1994). Material seized at a crime scene which came from the offender is said to be relevant in that it is helpful to the consideration of suspects as possible offenders. The second factor concerns the recognition that if the material is not relevant to the case, then it may have arrived at the scene from the suspect for innocent reasons.

3.3.3.1   Notation

A likelihood ratio has been developed by Evett (1993) for a scenario involving k offenders. It is easily considered in situations where $k = 1$, but for the generality of discussion, imagine the case of a crime committed by k offenders. A single blood stain is found at the crime scene in a position where it may have been left by one of the offenders. A suspect is found and he gives blood for comparative analysis. The suspect's blood and the crime stain are found to have the same profile Γ. This profile is shared by a proportion γ of the relevant population from which the criminals have come. As before, consider two propositions: 'the suspect is one of the k offenders' (H_p) and 'the suspect is not one of the k offenders' (H_d).

Note the difference between these propositions and those of the previous sections on source or activity level propositions. At source level, the propositions referred, respectively, to the suspect being, or not being, the donor of the blood stain found at the crime scene. Now, the propositions are stronger in an associative sense. In particular, the propositions now state that the suspect is, or is not, one of the offenders. The formal expression for the value V of the evidence is still $V = Pr(E_c|E_s, H_p)/Pr(E_c|H_d)$, where E_c stands for the profile Γ of the crime stain and E_s for the profile Γ of the suspect.

An argument needs to be constructed between the observations (referring to the stain found at the crime scene) and the target propositions according to which the suspect is or is not one of the offenders. The argument is constructed in two steps. The first step introduces a proposition that states that the crime stain came from one of the k offenders, rather than H_p according to which the suspect is one of the k offenders. The alternative proposition states that the crime stain did not come from any of the k offenders. This pair of propositions are also known as *association propositions*.

Assume that the crime stain came from one of the k offenders. The second step then involves the consideration of a proposition according to which the crime stain came from the suspect, and the alternative proposition that the crime stain did not come from the suspect. These propositions are also known as *intermediate association propositions*, but they are also readily recognized as source level propositions.

Introducing these two pairs of propositions indicates that additional factors need to be considered. These are, firstly, relevance and, secondly, innocent acquisition (for the receptor). Relevance, usually denoted r, is a probabilistic notion that expresses one's belief, given the information at hand, about the connection between the recovered stain and the the crime (i.e. whether the stain has been left by one of the offenders). Innocent acquisition, usually denoted p, is an expression of belief about whether the trace material has been acquired in a manner unrelated to the crime.

3.3.3.2 Association propositions

Consider the pair of propositions G, 'the crime stain came from one of the k offenders' and $\bar{G}$, 'the crime stain did not come from any of the k offenders'. The likelihood ratio V can now be rewritten by extending the conversation to this additional variable:

$$V = \frac{Pr(E_c|H_p, G, E_s)Pr(G|H_p, E_s) + Pr(E_c|H_p, \bar{G}, E_s)Pr(\bar{G}|H_p, E_s)}{Pr(E_c|H_d, G)Pr(G|H_d) + Pr(E_c|H_d, \bar{G})Pr(\bar{G}|H_d)}.$$

In the absence of E_c, the analytical results regarding the profile of the crime stain, knowledge of H_p and E_s, should not affect our belief in the truth or otherwise of G. Thus, $Pr(G|H_p, E_s) = Pr(G|H_p) = Pr(G)$ and $Pr(\bar{G}|H_p, E_s) = Pr(\bar{G}|H_p) = Pr(\bar{G})$.

Let $Pr(G) = r$ and $Pr(\bar{G}) = (1 - r)$, and call r the relevance term. That is, relevance is interpreted here as the probability that the stain has been left by one of the offenders. The higher the value assigned to r, the more relevant a stain is in the view of the evaluator. Thus,

$$V = \frac{Pr(E_c|H_p, G, E_s)r + Pr(E_c|H_p, G, E_s)(1 - r)}{Pr(E_c|H_d, G, E_s)r + Pr(E_c|H_d, \bar{G}, E_s)(1 - r)}. \tag{3.5}$$

3.3.3.3 Intermediate association propositions

In order to determine the component probabilities of the likelihood ratio obtained after introducing association propositions G and $\bar{G}$, it is helpful to consider further propositions, called *intermediate association propositions*, defined as follows: 'the crime stain came from the suspect' (F) and 'the crime stain did not come from the suspect' ($\bar{F}$). They allow the development of four conditional probabilities involving E_c as outlined below.

$Pr(E_c|H_p, G, E_s)$ is the probability that the crime stain would be of profile Γ if it had been left by one of the offenders (G), the suspect had committed the crime (H_p) and the suspect is of profile Γ. Introducing an extension to F and $\bar{F}$ leads to

$$Pr(E_c|H_p, G, F, E_s)Pr(F|H_p, G, E_s) + Pr(E_c|H_p, G, \bar{F}, E_s)Pr(\bar{F}|H_p, G, E_s).$$

Here, the profile of the crime stain and the suspect are both of type Γ, that is $E_c = E_s = \Gamma$. Hence, $Pr(E_c|H_p, G, F, E_s) = 1$. In the absence of E_c, F is independent of E_s so that $Pr(F|H_p, G, E_s) = Pr(F|H_p, G) = 1/k$. This is an expression of the assumption that there is nothing in the background information I to distinguish the suspect, given H_p, from the other offenders as far as blood shedding is considered. In a similar manner, $Pr(\bar{F}|H_p, G, E_s) = (k-1)/k$. Further, $Pr(E_c|H_p, G, \bar{F}, E_s) = Pr(E_c|H_p, G, \bar{F}) = \gamma$, since if $\bar{F}$ is true, E_c and E_s are independent and one of the other offenders left the stain (since G holds). In summary, thus, $Pr(E_c|H_p, G, E_s) = \{1 + (k-1)\gamma\}/k$.

Next, consider $Pr(E_c|H_p, \bar{G}, E_s)$. This is the probability that the crime stain would be of profile Γ if it had been left by an unknown person who was unconnected with the crime. Note that this is the implication of assuming $\bar{G}$ to be true. The population of people who may have left the stain is not necessarily the same as the population from which the criminals are assumed to have come. Thus, let $Pr(E_c|H_p, \bar{G}, E_s) = \gamma'$, where γ' is the probability of profile Γ amongst the people who may have left the stain. The prime $\prime$ thus indicates that it may not be the same value as γ, which relates to the population from which the criminals have come.

Consider now that the suspect is innocent and H_d is true. Then, $Pr(E_c|H_d, G, E_s)$ can be written more shortly as $Pr(E_c|H_d, G) = \gamma$, the probability that a person from the population of potential offenders has profile Γ. There is no need here to partition this probability by considering F and $\bar{F}$ because the suspect is assumed not to be one of the offenders and G is the proposition that the stain was left by one of the offenders.

The fourth probability of interest is $Pr(E_c|H_d, \bar{G}, E_s)$. Still, the suspect is assumed to be innocent (H_d), but the stain did not come from one of the offenders ($\bar{G}$). Therefore, the suspect could be the source of the crime stain so that the following extension can be made:

$$Pr(E_c|H_d, \bar{G}, F, E_s)Pr(F|H_d, \bar{G}, E_s) + Pr(E_c|H_d, \bar{G}, \bar{F}, E_s)Pr(\bar{F}|H_d, \bar{G}, E_s).$$

If F is true, $Pr(E_c|H_d, \bar{G}, F, E_s) = 1$. Also, $Pr(F|H_d, \bar{G}, E_s) = Pr(F|H_d, \bar{G})$. This probability is denoted p and refers to *innocent acquisition*. That is, the probability that the stain would have been left by the suspect even though the suspect was innocent of the offence. It is also assumed here that the propensity to leave a stain is independent of the profile of the person who left the stain. Moreover, $Pr(\bar{F}|H_d, \bar{G}, E_s) = Pr(\bar{F}|H_d, \bar{G}) = 1 - p$ and $Pr(E_c|H_d, \bar{G}, \bar{F}) = \gamma'$. Thus, in summary, $Pr(E_c|H_d, \bar{G}, E_s) = p + (1-p)\gamma'$.

Substitution of the above-mentioned four expressions for E_c in (3.5) gives the following likelihood ratio:

$$V = \frac{[r\{1 + (k-1)\gamma\}/k] + \{\gamma'(1-r)\}}{\gamma r + \{p + (1-p)\gamma'\}(1-r)} = \frac{r\{1 + (k-1)\gamma\} + k\gamma'(1-r)}{k[\gamma r + \{p + (1-p)\gamma'\}(1-r)]}. \tag{3.6}$$

Values of V for varying values of r and p are presented in Aitken and Taroni (2004). A practical example of the use of the above is given by Evett et al. (1998b) in the context of footwear marks evaluation with $k = 1$. Its formal analysis using Bayesian networks is presented in Section 6.1.2. Extensions of this particular probabilistic result have been presented for scenarios involving two stains and one offender as developed in Section 6.3. Note that the formal development pursued in this section focusses on the analysis of a scenario involving a so-called (potential) offender to scene transfer case; that is, a single stain has been found at a crime scene and there is uncertainty about the degree to which that stain is relevant. Chapter 6 develops a case involving potential transfer in the opposite direction (i.e. 'victim/scene to offender transfer').

3.4   Pre-assessment of the case

The evaluation process should start when the scientist first meets the case. It is at this stage that the scientist thinks about the questions that are to be addressed and the outcomes that may be expected. The scientist should attempt to frame propositions and think about the value of the expected findings (Evett et al. 2000c). However, there is a tendency to consider the evaluative process only as a final step of a casework examination, notably at the time of preparing the formal report. This is so even if an earlier interest in the process would enable the scientist to make better decisions about the allocation of resources. For example, consider a case of assault involving the possible cross-transfer of textile fibres between a victim and the assailant. The scientist has to decide whether to look first for potentially transferred fibres on the victim's pullover rather than for extraneous fibres on the suspect's pullover. If traces compatible with the suspect's pullover are found on the victim's pullover, then the expectation of the detection of traces from the victim's pullover on the suspect's pullover has to be reconsidered. This is based on the possibility of phenomena such as reciprocal transfer. Should the scientist have any expectations? How can they be quantified? If so, what is the interpretative consequence when those expectations are or are not met (presence or absence of extraneous material)? In such cases, issues to be considered include (i) the appropriate nature of the expectations, (ii) the quantification of the expectations and (iii) the interpretation of the presence or absence of material.

The scientist requires an adequate appreciation of the circumstances of the case so that a framework may be set up for considering the kind of examinations that may be carried out, and what may be expected from them (Cook et al. 1998b) in order for a logical decision to be made. Such a procedure of pre-assessment can be justified on a variety of grounds. An essential point is that the choice of the propositional level for the evaluation of scientific findings is carried out within a framework of circumstances, and these circumstances have to be known before any examination is made. This helps to ensure that relevant propositions are proposed (e.g. activity level propositions instead of source level propositions). This procedure may inform a discussion with the customer before any substantial decision is reached (e.g. about costs). Moreover, this process provides a basis for all scientists for consistency for proceeding. Scientists are encouraged, therefore, to consider carefully factors such as circumstantial information and the data that are to be used for the evaluation and to declare them in the final report.

The scientist should proceed by assigning probabilities for the events of finding particular configurations of trace material, given each proposition. The listing of events should cover the principal outcomes that can reasonably be expected for the kind of case under examination. Consider, for example, a case where a window is smashed and assume that the prosecution and defence propose the following propositions: 'The suspect is the man who smashed the window' (H_p) and 'The suspect was not present when the window was smashed' (H_d). The examination of a suspect's pullover will reveal a quantity Q of glass fragments, where Q can be defined, for example, as a categorical variable that covers the following states 'no', 'few' or 'many' (glass fragments). Then, the first question to be asked for the assessment of the numerator of the likelihood ratio is 'What is the probability of finding a quantity Q of matching glass fragments if the suspect is the man who smashed the window?' A second question needs to be asked for the assessment of the denominator of the likelihood ratio. It takes the following form: 'What is the probability of finding a quantity Q of matching glass fragments if the suspect was not present when the window was smashed?'

Initially, the scientist is thus asked to assess a total of six distinct probabilities: the probabilities of finding {no, few, many} matching glass fragments if the suspect is the man who smashed the window, and the probabilities of finding {no, few, many} matching glass fragments if the suspect was not present when the window was smashed. These probabilities may not be easy to assign because of the limited information available to the scientist (Cook et al. 1998a). For example, it will be very difficult to assess transfer probabilities if the scientist has no answer to questions such as 'Was the window smashed by a person or by a vehicle?' Clearly, the fact that a window was smashed by a person or a vehicle should change the amount of glass fragments the scientist expects to be transferred. It may also be relevant to know how exactly − in the sense of the modus operandi − the window was smashed. If it was indeed smashed by a person, then was that person standing close to it? Was a brick thrown through it? Information about the way in which a window was smashed is important because it represents a crucial factor for the assessment of the amount of glass potentially projected. Information on the distance between the person (if it were a person) who smashed the window and the window offers relevant information on the amount of glass fragments the scientist would expect to recover from the pullover of the person who smashed the window.

Where there is little information about the time of the alleged offence and the time at which the investigators seized the clothing, the lapse of time between the offence and the actual collection of relevant material may not be well determined. This will make it difficult to assess the probability of persistence of any transferred glass fragment. Therefore, if the scientist has only a limited amount of information about the circumstances of the case under examination, then the assessment may need to be restricted to source level propositions or to a purely descriptive reporting of the analytical results (Section 3.3.1).

Consider a further short example presented in Evett et al. (2002). A cigarette end is recovered at the scene of a burgled house. None of the family in the house smokes. A suspect is apprehended. The cigarette end is submitted for DNA profiling analyses (because the suspect denied ever being anywhere near to the house that was burgled) and the results are compared with the suspect's DNA profile. In such a case, various outcomes are possible. The outcome 'Match' is one where a single profile from the cigarette end shows the same features as that of the suspect. A 'Mixture/Match' is a mixed profile from the cigarette end which presents alleles that are also found in the suspect's profile. The outcome 'Difference' may either be a single profile with a genotype different from that of the suspect, or a two persons mixture with features that differ from the suspect's DNA profile. Finally, the outcome 'No-profile' designates a situation in which no DNA profile is detected.

If propositions at the activity level are proposed (e.g. 'the suspect is the person who smoked the cigarette' and 'some other unknown person smoked the cigarette'), then the assessment of the outcomes depends on probabilities that the scientist specifies for the possibility the DNA entered (or did not enter) the process by innocent means, the fact the person who smoked the cigarette left (or did not leave) sufficient DNA to give a profile, and the possibility that DNA of a third person entered (or did not enter) the process (third in addition to the suspect and some other person). Domain knowledge and circumstantial information about laboratory quality are necessary for such assessments. It will then be possible to propose a pre-assessment table to define a likelihood ratio for each of the four outcomes (i.e. 'match', 'mixture/match', 'difference' and 'no-profile'). It is recognized that it may be challenging to assign point values to the different factors of interest, but sensitivity analyses (see Chapter 13) can be considered as a way to explore the relative importance of the various factors.

The process of case pre-assessment can be summarized by the following steps: (i) collection of information that the scientist may need about the case; (ii) consideration of the questions that the scientist can reasonably help address, and, consequently, the level of propositions for which the scientist can reasonably choose to assess findings; (iii) definition of the relevant factors that will appear in the likelihood ratio; (iv) specification of the magnitude of the expected likelihood ratio given the background information; (v) decision on an examination strategy and analyses with which to proceed (i.e. observation of outcomes) and (vi) evaluation of the likelihood ratio and report of its value. A practical procedure using Bayesian networks is presented in Chapter 10.

Examples considering the pre-assessment for various types of trace material and related findings are presented in the literature. Cook et al. (1998b) present pre-assessment through the case of a hypothetical burglary involving potential glass fragments (an unknown quantity, Q, of recovered fragments). Stockton and Day (2001) consider an example involving signatures on questioned documents. Champod and Jackson (2000) consider a burglary case involving fibres. Booth et al. (2002) discuss a drug case. A cross-transfer (also called *two-way transfer*) case involving textiles is presented by Cook et al. (1999). This cross-transfer case shows how the pre-assessment can be updated when a staged approach is taken, which is the results of the examination of one of the garments that is used to inform a decision about whether the second garment should be examined.

3.5 Evaluation using graphical models

3.5.1 Introduction

As noted in Chapter 1, Bayesian networks have been found useful in assisting human reasoning in a variety of disciplines in which uncertainty plays an important role. In forensic science too, Bayesian networks have been proposed as a method of formal reasoning that could assist forensic scientists to understand the dependencies that may exist between different aspects of forensic findings. Historically, Aitken and Gammerman (1989) were amongst the first to suggest the use of graphical probabilistic models for the assessment of scenarios involving scientific findings. Ideas expressed there have been developed by authors such as Dawid and Evett (1997) or Garbolino and Taroni (2002). These studies provide clear examples of the relevance of Bayesian networks for assisting the evaluation of forensic findings, but explanations, if any, as to how practitioners should use the method to build their own models, are mostly brief and very general. The problem is well posed in Dawid et al. (2002), where the authors note that finding an appropriate representation of a case under examination is crucial for several reasons (viability, computational routines, etc.) and that the graphical construction is to some extent an art form, but one which can be guided by scientific and logical considerations. Whilst no explicit guidelines exist as to how one should proceed in the construction of a Bayesian network, there are some general concepts considered useful for the elicitation of sensible network structures. These are discussed below.

3.5.2 General aspects of the construction of Bayesian networks

As a preliminary note, consider the following quote from Lindley (2000, p. 303): 'A model is merely your reflection of reality and, like probability, it describes neither you nor the world, but only a relationship between you and that world. It is unsound to refer to the true model.' Underlying this quote is the idea that scientists should not expect that there is a single model to describe a given scenario of interest.[5] Depending on the level of detail, the extent

---

[5] In this context, it also appears relevant to mention the widely known quote from George Box '[e]ssentially, all models are wrong, but some are useful' (Box and Draper 1987, p. 424).

of information available, and the user's perception of reality, a variety of acceptable model constructions can be proposed. In the same sense, Schum (1994, p. 2) argued:

> 'On close examination many apparently simple inferences reveal some remarkably subtle properties that often go unrecognized. Inferences can be decomposed to various levels of "granularity". As we make finer decompositions of an inference, we expose additional and often interesting sources of uncertainty. One trouble we face is that there rarely seems to be any final or ultimate decomposition of an inference. Indeed, there may be alternative decompositions, none of which we can label as being uniquely correct.'

In what follows, this is illustrated using two models to approach the same scenario.

Example 3.3 *(Definition of the two-trace problem) Consider a two-stain scenario as described by Evett (1987). A crime has been committed by two men, each of whom left a bloodstain at the crime scene. A hypothetical characteristic Γ, which could be a conventional DNA profile, of the stains is analysed. This characteristic has some variability so that the various possible values may be denoted $\Gamma_1, \Gamma_2, \Gamma_3, ..., \Gamma_n$. Suppose that one of the two crime stains is of type Γ_1 and the other is of type Γ_2. Sometime later, a suspect is apprehended as a result of information completely unrelated to the blood characteristic considered here. Control material provided by the suspect is found to be of type Γ_1. Note that this scenario assumes that there is no information in the form of injuries and the scientific finding is confined solely to the results of the blood analysis. It is further assumed that the recovered material is relevant to the offence, that there are exactly two offenders and that there is no innocent transfer of blood to the crime scene.*

A forensic scientist may be asked to evaluate these findings with respect to the following two crime level propositions: 'the suspect was one of the two men who committed the crime' (H_p) and 'the suspect was not one of the two men who committed the crime' (H_d). It has already been noted several times in this book that the probative value of forensic findings is appropriately expressed in the form of a likelihood ratio V. In the case of the two-stain scenario, the findings E consist of two distinct elements, defined as follows: 'the bloodstains at the crime scene are of types Γ_1 and Γ_2' (E_1) and 'the suspect's blood is of type Γ_1' (E_2). A general expression of the likelihood ratio then is as follows (Evett 1987):

$$V = \frac{Pr(E|H_p)}{Pr(E|H_d)} = \frac{Pr(E_1|E_2, H_p) \times Pr(E_2|H_p)}{Pr(E_1|E_2, H_d) \times Pr(E_2|H_d)}. \tag{3.7}$$

Consider the conditioning of E_2 on H_p and on H_d on the right-hand side of (3.7). It appears to be a reasonable assumption that the probability of the suspect's blood being type Γ_1 does not depend on whether he was one of the two men who committed the crime. So, one may set $Pr(E_2|H_p) = Pr(E_2|H_d)$ and the likelihood ratio becomes

$$V = \frac{Pr(E_1|E_2, H_p)}{Pr(E_1|E_2, H_d)}. \tag{3.8}$$

3.5.3   Eliciting structural relationships

An advantage of constructing Bayesian networks on the basis of an existing probabilistic formula is that the number and definition of the nodes is already given. In the case of the two-trace scenario (Example 3.3), there are the following variables H (which can take values H_p and H_d), E_1 and E_2. These variables can be used to define nodes for a Bayesian network.

Example 3.4 *(Network structure for a two-trace scenario) Consider three nodes H, E_1 and E_2. For the time being, suppose that they have binary states; for E_1 and E_2, the two states are the values given in Example 3.3 and their negations, the bloodstains at the crime scene are not of types Γ_1 and Γ_2 and the suspect's blood is not of type Γ_1, respectively. As an aid to visualization of the potential dependencies about which one will need to decide, consider a complete undirected graph over these variables of interest [Figure 3.1(i)]. At this juncture, it is worth noting that a probabilistic formula may not only help in the definition of relevant variables. It can also contain expressions that may provide guidance in defining the number of edges and their direction. For example, the expression $Pr(E_1|E_2, H_p)$ in (3.7), to be read as 'the scientist's probability of the two blood stains at the crime scene are of type Γ_1 and Γ_2, given that the suspect's blood is of type Γ_1 and the suspect is one of the two men who committed the crime', indicates that E_1 is conditioned on both H and E_2. An appropriate graphical expression of this conditioning would be a converging connection at the node E_1 with parental variables H and E_2.*

The question remains as to whether there should be a link of some kind between H and E_2. Consider again (3.7), which can be rewritten as two separate fractions:

$$\frac{Pr(E_1|E_2, H_p)}{Pr(E_1|E_2, H_d)} \times \frac{Pr(E_2|H_p)}{Pr(E_2|H_d)}$$

The second fraction suggests that there should be an arc pointing from H to E_2. This would lead to a graphical structure as shown in Figure 3.1(ii). However, if it is understood that the probability of E_2, the suspect's blood being of type Γ_1, is the same whether he is or is not one of the two men who committed the crime, then the link $H \to E_2$ carries no inferential force. In other words, if $Pr(E_2|H_p) = Pr(E_2|H_d)$, knowledge about E_2 does not affect H along $H \to E_2$. Thus, the stated assumption serves as a formal argument for justifying the omission of the link $H \to E_2$ [Figure 3.1(iii)]. Note, however, that this does not mean that there is no influence between H and E_2. The converging connection at the node E_1 implies that H and E_2 become dependent as soon as knowledge about E_1 becomes available (Section 2.1.5).

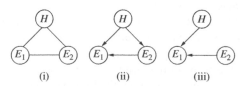

$$(i) \qquad\qquad (ii) \qquad\qquad (iii)$$

Figure 3.1   (i) Complete undirected graph over the variables E_1, E_2 and H; (ii) and (iii) possible network structures for the two-trace transfer problem.

3.5.4   Level of detail of variables and quantification of influences

An inference problem may be approached on different levels of detail and this may result in structural differences in Bayesian networks. This is an instance of a general characteristic of probabilistic modelling, as pointed out by the quotes of Lindley (2000) and Schum (1994) in Section 3.5.3. It is possible to illustrate this in case of the two-trace problem (Example 3.3) that has, so far, been studied here only on a binary level. Consider the observational variable E_1, defined as binary earlier in Example 3.3. The discovery of two crime-related traces of types Γ_1 and Γ_2 at the scene of a crime is only one of several possible outcomes. The stains may have been of some other type. For simplicity, aggregate the other possible types $\Gamma_3, \ldots, \Gamma_n$ in terms of Γ_x. The probability of Γ_i is γ_i, for $i = 1, \ldots, n$ and $\gamma_x = \gamma_3 + \cdots + \gamma_n$. The possible states (i.e. *genotypic configurations*) for the node E_1 can then be defined as follows: $\Gamma_1 - \Gamma_1, \Gamma_1 - \Gamma_2, \Gamma_1 - \Gamma_x, \Gamma_2 - \Gamma_2, \Gamma_2 - \Gamma_x$ and $\Gamma_x - \Gamma_x$. The suspect is a potential contributor to the pair of recovered stains, so it appears reasonable to assume the following possible values for the node E_2: Γ_1, Γ_2 or Γ_x. The third variable of interest, H, is concerned with the major proposition, which remains unchanged. The suspect may or may not be one of the two criminals, so H may be assumed – as before – to be a binary node.

For a numerical specification of the Bayesian network shown in Figure 3.1(iii) (graphically) unconditional probabilities are required for the nodes H and E_2, whereas conditional probabilities are required for the node E_1. The probabilities assigned to the node H are the prior beliefs held by some evaluator, such as an investigator or a court of law, about the two competing propositions. The probabilities of the node E_2 could be population proportions of the characteristics Γ_1, Γ_2 and Γ_x in a relevant database. Note that the value γ_x of Γ_x is $\sum_{i=3}^{n} \gamma_i$ and that $\sum_{i=1}^{n} \gamma_i = 1$.

More detailed study is required for the probabilities assigned to the node E_1. If H_p is true, then at least one of the two recovered stains must have the same profile as that suspect's blood. For example, if the suspect is one of the two criminals and his blood is of type Γ_1, then the only possible outcomes for the pair of recovered stains are $\Gamma_1 - \Gamma_1, \Gamma_1 - \Gamma_2$ and $\Gamma_1 - \Gamma_x$. The stain left by one of the offenders is of type Γ_1, Γ_2 or Γ_x, respectively. Assuming the characteristics of a man's blood to be independent of his tendency to criminal activity, then the probability of the second man's blood being of, for example, type Γ_2 can be assigned on the basis of the population proportion γ_2 in a relevant population. Thus, a probability such as $Pr(E_1 = \Gamma_1 - \Gamma_2 | E_2 = \Gamma_1, H_p)$ is given by $1 \times \gamma_2$. Analogous considerations apply to all the other conditional probabilities that assume H_p to be true.

If the suspect is not one of the two criminals (H_d), then the characteristics of this person's blood are – given the stated assumptions – irrelevant for determining the characteristics of the pair of recovered stains. In other words, the stains have been left by two unknown men. These two individuals may be regarded as drawn randomly from the population. If the two stains have different characteristics (i.e. $\Gamma_i - \Gamma_j$, with $i \neq j$), then the first man may have left the stain of type Γ_i, and the second man may have left the stain of type Γ_j and vice versa. Therefore, the probability of observing two stains of different type, given H_d, is $2\gamma_i\gamma_j$. On the other hand, the probability of observing two stains of the same type, such as $\Gamma_i - \Gamma_i$, is γ_i^2.

Table 3.1 summarizes the probabilities assigned to the node E_1. As may be seen, there are far more probabilities than are used in the initial likelihood ratio formula (3.8). The reason for this is that (3.8) accounts only for a particular situation, notably the two stains being of type Γ_1 and Γ_2 and the suspect's blood being of type Γ_1. The Bayesian network considered here is different in that respect because it accounts for all possible combinations of outcomes for the nodes E_1 and E_2.

Table 3.1 Probabilities assigned to the node E_1, conditional on E_2 and H for possible values of E_1, E_2 and H in Figure 3.1(iii).

H:		H_p			H_d		
E_2:		Γ_1	Γ_2	Γ_x	Γ_1	Γ_2	Γ_x
E_1:	$\Gamma_1 - \Gamma_1$	γ_1	0	0	γ_1^2	γ_1^2	γ_1^2
	$\Gamma_1 - \Gamma_2$	γ_2	γ_1	0	$2\gamma_1\gamma_2$	$2\gamma_1\gamma_2$	$2\gamma_1\gamma_2$
	$\Gamma_1 - \Gamma_x$	γ_x	0	γ_1	$2\gamma_1\gamma_x$	$2\gamma_1\gamma_x$	$2\gamma_1\gamma_x$
	$\Gamma_2 - \Gamma_2$	0	γ_2	0	γ_2^2	γ_2^2	γ_2^2
	$\Gamma_2 - \Gamma_x$	0	γ_x	γ_2	$2\gamma_2\gamma_x$	$2\gamma_2\gamma_x$	$2\gamma_2\gamma_x$
	$\Gamma_x - \Gamma_x$	0	0	γ_x	γ_x^2	γ_x^2	γ_x^2

Population proportions for Γ_1, Γ_2 and Γ_x are γ_1, γ_2 and γ_x, respectively. Propositions are H_p, the suspect was one of the two men who committed the crime, and H_d, the suspect was not one of the two men who committed the crime.

Example 3.5 *(A numerically specified Bayesian network for the two-trace problem) Imagine a scenario in which $\gamma_1 = \gamma_2 = 0.05$ and $Pr(H_p) = Pr(H_d) = 0.5$. Figure 3.2(i) shows the numerically specified Bayesian network in its initial state. Notice that the values displayed in the nodes H and E_2 are given by the unconditional probabilities assigned to these nodes. Throughout this book, probabilities displayed in the pictorial representations will take the form of numbers between 0 and 100 (i.e. probabilities are represented as percentages), as is the case with many commercially and academically available Bayesian network software systems. The values displayed for the initial state of the variable E_1 are the result of considering*

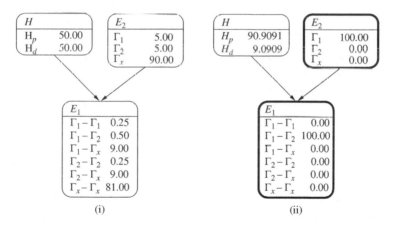

(i)   (ii)

Figure 3.2 Bayesian network for a two-trace transfer scenario. (i) Initial state of the model; probability values for E_1 are given by (3.9). The probabilities displayed in the nodes H and E_2 are the unconditional (marginal) probabilities specified for these nodes. (ii) State of the model after instantiation of E_1 and E_2. Node H has two states, H_p, the suspect was one of the two men who committed the crime, H_d, the suspect was not one of the two men who committed the crime. Node E_2 has three states, the suspect's profile is Γ_1, Γ_2 or Γ_x (something other than Γ_1 or Γ_2). Node E_1 has six states, the profiles of the bloodstains at the crime scene are one of the six possible pairs obtainable from Γ_1, Γ_2 and Γ_x.

all possible parental configurations. The initial probability of, for example, E_1 being in state $\Gamma_1 - \Gamma_2$ is thus obtained by

$$\sum_{i=1}^{x} Pr(E_1 = \Gamma_1 - \Gamma_2 | E_2 = \Gamma_i, H_p) \times Pr(E_2 = \Gamma_i) \times Pr(H_p) +$$

$$\sum_{i=1}^{x} Pr(E_1 = \Gamma_1 - \Gamma_2 | E_2 = \Gamma_i, H_d) \times Pr(E_2 = \Gamma_i) \times Pr(H_d). \tag{3.9}$$

Use of the probabilities γ_1 and γ_2 from Table 3.1, both replaced by 0.05, and $Pr(H_p) = Pr(H_d) = 0.5$, in (3.9) verifies the value of 0.50 in Figure 3.2(i) for $\Gamma_1 - \Gamma_2$.

$$Pr(E_1 = \Gamma_1 - \Gamma_2) = (1 \times \gamma_2) \times \gamma_1 \times 0.5 + (\gamma_1 \times 1) \times \gamma_2 \times 0.5$$

$$+ (0) \times \gamma_x \times 0.5 + (2 \times \gamma_1 \times \gamma_2) \times \gamma_1 \times 0.5$$

$$+ (2 \times \gamma_1 \times \gamma_2) \times \gamma_2 \times 0.5 + (2 \times \gamma_1 \times \gamma_2) \times \gamma_x \times 0.5$$

$$= (0.05 \times 0.05 \times 0.5) + (0.05 \times 0.05 \times 0.5)$$

$$+ (0.005 \times 0.05 \times 0.5) + (0.005 \times 0.05 \times 0.5)$$

$$+ (0.005 \times 0.9 \times 0.5)$$

$$= 0.005.$$

Note again that probability values in figures are expressed here on a 0–100 % scale. The value of 0.50 in Figure 3.2 corresponds to a probability for state $\Gamma_1 - \Gamma_2$ of 0.005 on a scale of 0 – 1.

The Bayesian network as described may now be used to compute the posterior probability of H given particular observations. Let us recall the scenario described at the beginning of Section 3.5.2 (Example 3.3). There are two stains of type Γ_1 and Γ_2, and the suspect's blood is of type Γ_1. These observations are communicated to the model by instantiating the nodes E_1 and E_2. This situation is shown in Figure 3.2(ii). The instantiation of the nodes E_1 and E_2 (shown with a bold border line) affects the node H, and the probability of H_p has increased. Notice that the probabilities displayed for H_p and H_d are posterior probabilities given the observed set of findings. In their extended form, H_p and H_d may be written $Pr(H_p | E_1 = \Gamma_1 - \Gamma_2, E_2 = \Gamma_1)$ and $Pr(H_d | E_1 = \Gamma_1 - \Gamma_2, E_2 = \Gamma_1)$, respectively. The value of the observed set of scientific findings may be expressed by the likelihood ratio. Based on the probabilities defined in Table 3.1, one can find that (3.8) reduces to $1/2\gamma_1$. As γ_1 was set to 0.05 a likelihood ratio of 10 is indicated. The same value can be deduced from the Bayesian network shown in Figure 3.2. Here, equal prior probabilities are assumed for the node H. In such a situation, the likelihood ratio equals the ratio of the posterior probabilities of H, namely $0.91/0.091 = 10$. Note also later that the same numerical result is obtained when using, in (6.10), the values 2, 1 and 0 for, respectively, the number of criminals, the relevance and innocent transfer terms.

3.5.5 Deriving an alternative network structure

Different Bayesian networks may be proposed for the consideration of questions surrounding the same situation. This can be illustrated on the basis of the Bayesian network discussed in

Section 3.5.2, which may be used for the evaluation of a two-stain scenario. In fact, some of the conditional probabilities of the node E_1 in the Bayesian network shown in Figure 3.2 need to be computed using profile (or genotype) probabilities, in particular when conditioning on the alternative hypothesis H_d (Table 3.1). Such computations may be time consuming and, to some degree, prone to error. An interesting question thus is whether one can propose a more convenient Bayesian network structure, so as to avoid calculations when completing probability tables.

The use of so-called compound probabilities as used, for example in the probability table for the node E_1 (see Section 3.5.4 for the definition of this node), suggests the presence of latent intermediate states that have not been represented explicitly. An essential aspect of an approach to find an alternative Bayesian network structure thus consists of examining the definitions of nodes more closely. Such an analysis may reveal hidden propositions that, if represented as distinct nodes, may avoid the use of composed probabilities.

Consider thus the case of the node E_1. In Section 3.5.4, this node was defined as the characteristics of the two recovered stains, one for each of the recovered stains. It is clear from this definition that this covers a combination of two distinct events. The node states are the following: $\Gamma_1 - \Gamma_1$, $\Gamma_1 - \Gamma_2$, $\Gamma_1 - \Gamma_x$, $\Gamma_2 - \Gamma_2$, $\Gamma_2 - \Gamma_x$ and $\Gamma_x - \Gamma_x$. One possibility for the provision of a more detailed description of the proposition E_1 consists in the adoption of distinct nodes that describe the characteristics that each of the recovered stains may assume. For example, let T_1 and T_2, each with states Γ_1, Γ_2 and Γ_x, denote the possible characteristics of, respectively, stain 1 and stain 2. It is then reasonable to postulate that both T_1 and T_2 determine the state of E_1. Consequently, a converging connection with $T_1 \rightarrow E_1 \leftarrow T_2$ is appropriate. A practical consequence of this is that the probability table associated with the node E_1 can now be completed logically using values 0 and 1 (Table 3.2).

From T_1 and T_2, the characteristics of the crime stains 1 and 2, it is then necessary to construct an argument to E_2 and H. The last two variables designate the characteristics of the suspect's blood (E_2) and the propositions according to which the suspect is or is not one of the two men who committed the crime (H). A feasible way to achieve this involves two steps.

First, consider that knowledge about the characteristics of the suspect's blood (E_2) may inform about the characteristics of the crime stains (T_1 and T_2). However, knowledge about which of the two stains, if either, has been left by the suspect is relevant here. Therefore, an additional variable F is adopted to represent the uncertainty about which stain has been left by

Table 3.2 Probabilities assigned to the node E_1, where T_1 and T_2 are the characteristics of stains 1 and 2, respectively, and E_1 is the combination of T_1 and T_2.

T_1:	Γ_1			Γ_2			Γ_x		
T_2:	Γ_1	Γ_2	Γ_x	Γ_1	Γ_2	Γ_x	Γ_1	Γ_2	Γ_x
E_1: $\Gamma_1 - \Gamma_1$	1	0	0	0	0	0	0	0	0
$\Gamma_1 - \Gamma_2$	0	1	0	1	0	0	0	0	0
$\Gamma_1 - \Gamma_x$	0	0	1	0	0	0	1	0	0
$\Gamma_2 - \Gamma_2$	0	0	0	0	1	0	0	0	0
$\Gamma_2 - \Gamma_x$	0	0	0	0	0	1	0	1	0
$\Gamma_x - \Gamma_x$	0	0	0	0	0	0	0	0	1

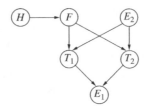

Figure 3.3 An alternative Bayesian network structure for a two-trace transfer scenario, where T_1 and T_2 are characteristics of stains 1 and 2, respectively, and F represents the uncertainty about which stain has been left by the suspect.

the suspect. Let this variable have the states 'neither' (f_0), 'stain 1' (f_1) and 'stain 2' (f_2). Both E_2 and F are chosen as parental variables for T_1 and T_2. Notice that f_0, f_1 and f_2 are mutually exclusive events. This is an implicit assumption that the suspect has left at most one trace. Second, consider that the proposition concerning which stain, if either, was left by the suspect (F) depends on whether the suspect is one of the two offenders (H). Accordingly, the node H is chosen as a parental variable for the node F. The Bayesian network shown in Figure 3.3 represents the above-mentioned two constructional considerations.

The assignment of values to the probability tables of the nodes T_1, T_2 and F is rather straightforward. Start by considering the nodes T_1 and T_2. If the suspect has left neither of the two stains (f_0), then his blood profile characteristics (E_2) are not relevant for assessing the probability of the stains being of type Γ_1, Γ_2 or Γ_x. The stains are then regarded as coming from two unknown men drawn randomly from the population. The probabilities of the stains being of type Γ_1, Γ_2 or Γ_x are given by the respective profile probabilities, that is γ_1, γ_2 and γ_x. There are further situations in which a stain is considered as originating from a man drawn randomly from the population. In particular, (i) the stain considered is T_1 and the suspect has left stain 2 (f_2), and (ii) the stain considered is T_2 and the suspect has left stain 1 (f_1). In these cases, the probability tables of the nodes T_1 and T_2 contain again the profile probability of the genetic characteristics considered, irrespective of the suspect's blood type. However, the profile of the suspect's blood becomes important whenever one considers a stain which is assumed to

Table 3.3 Probabilities assigned to the nodes T_1 and T_2, given uncertainty F about the stain left by the suspect: neither (f_0), stain 1 (f_1) or stain 2 (f_2), the profiles Γ_1, Γ_2 and Γ_x of suspect E_2 and T_1 and T_2 are the characteristics of stains 1 and 2, respectively.

F:		f_0			f_1			f_2	
E_2:	Γ_1	Γ_2	Γ_x	Γ_1	Γ_2	Γ_x	Γ_1	Γ_2	Γ_x
T_1: $\quad \Gamma_1$	γ_1	γ_1	γ_1	1	0	0	γ_1	γ_1	γ_1
Γ_2	γ_2	γ_2	γ_2	0	1	0	γ_2	γ_2	γ_2
Γ_x	γ_x	γ_x	γ_x	0	0	1	γ_x	γ_x	γ_x
T_2: $\quad \Gamma_1$	γ_1	γ_1	γ_1	γ_1	γ_1	γ_1	1	0	0
Γ_2	γ_2	γ_2	γ_2	γ_2	γ_2	γ_2	0	1	0
Γ_x	γ_x	γ_x	γ_x	γ_x	γ_x	γ_x	0	0	1

have been left by the suspect. For example, if the suspect has left stain 1 (f_1) and he is of type Γ_1, then the stain 1 is certain to be of type Γ_1, formally written $Pr(T_1 = \Gamma_1|f_1, E_2 = \Gamma_1) = 1$. Stated otherwise, a stain left by the suspect cannot have a characteristic which the suspect does not possess: $\sum_{i=1,2,x}\sum_{j=1,2}Pr(T_j = \Gamma_i|f_j, E_2 \neq \Gamma_i) = 0$ (Table 3.3).

For node F, it is relevant to consider that the scenario outlined in Section 3.5.2 (Example 3.3) assumes that there are two offenders, each of whom left exactly one stain. Notice also that both stains are regarded as actually being left by the offenders. Thus, if the suspect is one of the two offenders, then he may have left either stain 1 or stain 2 and these two possibilities are taken to be equally likely. Consequently, $Pr(f_1|H_p) = Pr(f_2|H_p) = 0.5$ and $Pr(f_0|H_p) = 0$. Given that the suspect is not one of the two offenders (H_d), neither of the two stains has been left by the suspect: $Pr(f_0|H_p) = 1$ and $\sum_{i=1,2}Pr(f_i|H_p) = 0$.

Example 3.6 *(Numerically specified, alternative Bayesian network for the two-trace problem) In order to see whether the extended Bayesian network proposed in Figure 3.3 yields the same results as those obtained with the basic network discussed in Section (3.5), consider again the numerical assignments $\gamma_1 = \gamma_2 = 0.05$ and $Pr(H_p) = Pr(H_d) = 0.5$. A Bayesian network initialized with these values is shown in Figure 3.4(i). Next, consider the scenario from Section 3.5.2 where the two recovered stains are of type Γ_1 and Γ_2, and the suspect's blood is of type Γ_1. These observations are entered in the Bayesian network by instantiating*

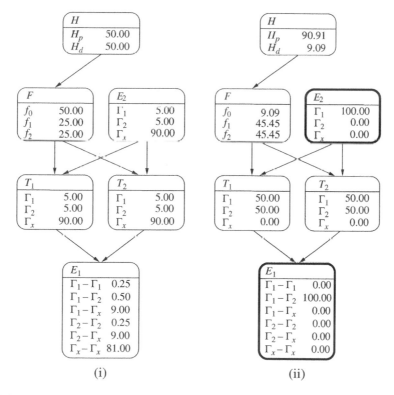

(i)   (ii)

Figure 3.4   (i) Initial state of an alternative Bayesian network for a two-trace transfer scenario and (ii) state of the Bayesian network after instantiation of the nodes E_1 and E_2. The definitions of the nodes are to be found in the text.

the nodes E_1 and E_2, respectively. These nodes are shown with a bold border in Figure 3.4(ii). This figure shows the state of the Bayesian network after entering the observations and propagating their effect. As may be seen from the values displayed in the node H, the value of the observations corresponds to a likelihood ratio of 10. This result is in agreement with the result presented in Section 3.5.4. Using the proposed Bayesian network, it is easy to verify that a likelihood ratio of $1/2\gamma_i$ is obtained in all situations in which the suspect's blood is of type Γ_i and the two recovered stains are of types $\Gamma_i - \Gamma_j$ (with $i \neq j$).

4

Evaluation given source level propositions

4.1 General considerations

Evaluation of scientific results in the light of source level propositions typically involves a rather restricted number of propositions. On the one hand, the scientist needs to express the outcomes of a comparison between material found on a crime scene or on a person of interest and control material from a potential source. Depending on the desired level of detail, this may be achieved using one or more propositions. On the other hand, the scientist needs to formulate competing propositions that state whether or not a given potential source is the true source. Throughout forensic literature, there is considerable variation in the notation used in formal analyses involving source level propositions, whereas, on a conceptual account, there is broad agreement with respect to the relevance relationships assumed between the various variables.

Bayesian networks are well suited to clarify such distinctions. Figure 4.1 exemplifies three recurrent expressions that are encountered at different junctures throughout this book. Start by considering Figure 4.1(i), which represents Bayes' theorem in one of its most general forms. Here, the variable H represents the source level propositions H_p and H_d, whereas E is an observational variable that stands for the outcomes of comparative examinations. These are typically expressed in terms of corresponding analytical characteristics. The latter variable is particular in the sense that it amounts to a summary of the entirety of the observations in terms of a single proposition. The node table of E contains the conditional probabilities $Pr(E|H_p)$ and $Pr(E|H_d)$, which make up, respectively, the numerator and denominator of the likelihood ratio (see, for example, (3.1)). Contextual information I is used when assigning these probabilities but is omitted here for ease of notation. It is also not written explicitly because I is not modelled in terms of a distinct network node. The network fragment 4.1(i) is admittedly basic but can be seen as part of more elaborate developments involving additional

Bayesian Networks for Probabilistic Inference and Decision Analysis in Forensic Science, Second Edition.
Franco Taroni, Alex Biedermann, Silvia Bozza, Paolo Garbolino and Colin Aitken.
© 2014 John Wiley & Sons, Ltd. Published 2014 by John Wiley & Sons, Ltd.

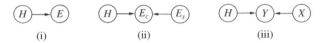

(i)   (ii)   (iii)

Figure 4.1 (i–iii) Examples of Bayesian network fragments for inference about a source level proposition H, stating that a particular potential source is or is not the source of recovered material of unknown source. The node E is binary and describes the outcomes of comparative examinations in terms of a correspondence (i.e. 'match' or 'no match' in some loose sense since a match is not strictly speaking possible). The binary nodes E_c and Y represent alternative ways to describe the features of recovered trace material, whereas E_s and X are alternative expressions for features of control material from a known source (e.g. a suspect).

variables at other propositional levels. Section 6.1.1 covers one such development where the target propositions refer to the crime level.

As noted earlier in Section 3.3.1, the findings E may also be considered in two parts (E_c, E_s), where E_s refers to the findings for the control material (i.e. the potential source), and E_c to the findings for the recovered material (i.e. trace or mark). In the particular context of DNA, this is equivalent to Evett and Weir's notation G_s and G_c for, respectively, the DNA profile (i.e. genotype) of the suspect and the recovered stain of unknown source, with the subscript c denoting 'crime related' (Evett and Weir 1998). Note further that a structurally similar network has previously been encountered in relation to the two-trace scenario (Section 3.5.3), where E_2 was used to express the suspect's genotype and E_1 the findings for the two recovered stains. In either case, the proposed converging connection reflects the general understanding according to which the probability assignment for findings relating to the recovered trace material, E_c, is conditioned upon the proposition H and knowledge about the characteristics of the potential source, expressed in terms of E_s [see also (3.2)].

It might be argued that a network construction using a minimal fragment of the kind $H \to E$ is sufficient and that it is unnecessarily elaborate to consider distinct nodes for findings regarding recovered and known materials. However, there may often be much more to say about observations than simply 'there is a correspondence in analytical features'. Moreover, it may even be that the scientist may find it difficult to actually define a so-called match. Besides obvious situations in which the compared characteristics can be acceptably summarized in terms of discrete elements (e.g. allele numbers), and thus readily compared, there may also be applications in which correspondences may only be partial or have varying degrees. This includes DNA where profiles may be incomplete. The simplistic 'match/non-match' perspective may then become difficult to apply because the scientist may need some criterion to define when a 'match' can be declared. To further complicate things, it may even be that the observations pertaining to recovered and control materials have – by definition – different information content and, thus, do not necessarily overlap. A typical example of such a situation concerns traces of DNA mixtures. Such traces contain material from multiple donors, whereas the profile of one of the potential sources, the suspect, is characterized by a single profile.

This has prompted some authors to prefer a more general notation, using Y and X for observations pertaining to recovered and known materials, respectively. For example, in the context of textile fibres (Champod and Taroni 1999), the attributes of recovered trace material may be described in terms of extrinsic features (e.g. number of fibres and position of recovery on the receptor item) and intrinsic features (e.g. physical and chemical characteristics), that is a variable Y with y denoting a particular configuration of observed attributes. The relevant forensic features of a known source are summarized in terms of the variable X with x denoting a particular set of observed attributes. The latter typically cover intrinsic features and can also include extrinsic features if results from experiments under controlled conditions are taken into account. For a similar notational convention in the context of shoe mark examinations, see Evett et al. (1998b). It may appear cumbersome to deviate to different notational choices, but preserving them at this point will help the reader make connections to discussions in existing literature. Figure 4.1 summarizes examples of Bayesian network fragments for inference about source level propositions that will be reused at various junctures in later chapters.

4.2   Standard statistical distributions

The use of the general structure of the basic network fragment $H \rightarrow E$ shown in Figure 4.1 is not limited to situations in which the nodes are taken to be binary. That is, observations E of comparative examinations need not necessarily be described in terms of a 'correspondence' (i.e. an agreement) or a difference. With a slight modification of the definition of the node states, it is also possible to approach situations in which observations are described, for instance, by numbers or ranges of numbers. Indeed, in many forensic comparison tasks, observations consist of a certain number of points of similarity or agreement, and scientists may be required to inquire about questions of the following kind: 'what is my probability of this number or quantity (e.g. none, some, few and many) of corresponding features if the first (second) proposition is true?', 'are my observations more probable under the first than under the second proposition?' That is, the scientist may need to assign probabilities for a range of possible outcomes, and these assigned probability distributions may differ depending on the conditioning hypothesis. In some situations, the scientist's probability distribution over possible outcomes may take the form of standard statistical distributions, and some Bayesian network software allows a user to take this into account when specifying node probabilities. An example for this is given in the following.

Example 4.1 *(Bayesian network for evaluating consecutive matching striations) In a paper published over a decade ago, Bunch (2000, p. 957) asked: '(...) is it possible to provide likelihood ratios in forensic firearms examination?' To approach this question, Bunch discussed a summary for observations of comparative examinations between bullets known as consecutive matching striations (CMS). To illustrate this concept, suppose that a forensic scientist is provided with a firearm seized from a suspect and a bullet recovered from a victim's body. The examiner may compare bullets test fired by the suspect's weapon with the bullet recovered from the victim. One way to quantify the extent of agreement that is noted during such a comparison process consists of counting the so-called CMS. That is, an examiner studies bullets under a macroscope and decides what is a striation and, for comparison, what striations match between the two bullets. Typically, scientists may focus on the maximum number of CMS on a bullet when compared to another bullet. It is generally thought that from experiments made under controlled conditions, that is comparisons between bullets fired by the same or different weapons, the scientist will be able to formulate expectations as to the maximum CMS count to be found for the so-called* same source *and* different source *comparisons. There has been some research intended to establish a threshold above which scientists may conclusively report a common source, but such a perspective and conclusion is not pursued or supported here.*

For a case in which a bullet is found at a crime scene and a suspect is apprehended with a gun, the following source level propositions may be of interest: 'the bullet found at the crime scene was fired from the suspect's gun' (H_p), 'the bullet found at the crime scene was fired from a gun other than the suspect's gun' (H_d). The scientific result to be considered is the observed number y of CMS. To obtain a likelihood ratio, the scientist needs to assign two probabilities, $Pr(Y = y|H_p, I)$ and $Pr(Y = y|H_d, I)$. One model for the CMS count, proposed by Bunch (2000), is a Poisson model with $Pr(Y = y|\lambda) = Pn(\lambda)$, where the parameter λ is estimated by the weighted average maximum CMS count and Y is the maximum CMS count for a particular bullet found at a crime scene when compared with a bullet fired from a gun, known here as the suspect gun. *A Bayesian analysis would use a prior distribution for λ (Taroni et al. 2010).*

Two Poisson distributions thus are required, one for pairs of bullets fired from the suspect's gun (P), $Pn(\lambda_P)$, and one for pairs of bullets fired from different guns (D), $Pn(\lambda_D)$. If the suspect's gun is the gun that fired the bullet found at the crime scene (H_p is true), then

$$Pr(Y = y|\lambda_P) = \frac{\lambda_P^y}{y!}e^{-\lambda_P} \qquad y = 0, 1, \dots$$

If the bullet found at the crime scene was fired by another gun, different from that of the suspect (H_d is true), then

$$Pr(Y = y|\lambda_D) = \frac{\lambda_D^y}{y!}e^{-\lambda_D} \qquad y = 0, 1, \dots$$

If parameters λ_P and λ_D are known, the likelihood ratio for the finding $Y = y$ can be written as follows (omitting I for shortness of notation):

$$V = \frac{Pr(Y = y|H_p)}{Pr(Y = y|H_d)} = \frac{Pr(Y = y|\lambda_P)}{Pr(Y = y|\lambda_D)} = \left(\frac{\lambda_P}{\lambda_D}\right)^y e^{\lambda_D - \lambda_P}.$$

If parameters λ_p and λ_d are not known, a prior distribution may be chosen for them and a Bayesian approach used. One such prior is the conjugate prior, the Gamma distribution (Aitken and Gold 2013), but this approach is not pursued here.

The use of CMS through a likelihood ratio V enables a summary of the findings to be in a phrase of the form 'the observations are V times more probable if H_p is true than if H_d is true'. This provides a good summary of what the statistics of CMS means in the context of determining the origin of marks present on a bullet found at a crime scene. However, one should not reduce the bullet comparison process to the sole CMS count. In reality, there are several other features that scientists may note during a comparison, including possible differences. Such additional levels of observation should receive their own assessment, but this is beyond the scope of the current example (see, for example, Section 6.2.5 for further details).

To translate the above-mentioned Poisson model into a Bayesian network, one can consider two nodes H and Y. The former is binary with states H_p and H_d, representing a pair of source level propositions. The latter node, Y, is discrete with multiple states numbered, for example 0, 1, ... , 10. Depending on the context of application, more states may be chosen, but generally the maximum CMS count tends to be no greater than 10. Following the logic outlined in Section 4.1, a network structure of the kind $H \rightarrow Y$, as shown in Figure 4.1(i), is appropriate. Help with the specification of the Poisson probabilities for the states of the node Y given H_p and H_d, respectively, is provided by Bayesian network software, such as Hugin, *that* *offer the possibility to use a special subtype of node called a numbered node. The probability table of such a node can be filled through a special functionality, called Expression builder, which enables the user to define (mathematical) expressions. There are different ways in which this functionality may be used in the current example. Using, as in Bunch (2000), the values* 3.91 *and* 1.32 *for, respectively,* λ_P *and* λ_D, *one can define two simple distinct expressions,* Poisson(3.91) *for node probabilities* $Pr(Y|H_p)$ *and* Poisson(1.32) *for node probabilities* $Pr(Y|H_d)$. *Another way to fill the probability table of the node Y is achieved with the use of a single expression, such as* Poisson(if(H=="Hp",3.91,1.32)), *for example. Note also that in this approach, the probability assigned to the state with the highest number (here* 10), *given either* H_p *or* H_d, *is not the Poisson probability for* $y = 10$, *but the cumulative probability of all Poisson probabilities greater than and equal to* 10. *This is due to the fact that the sum of the probabilities for all states of a node, for a given conditioning, must sum to* 1 *and the range of possible values for a Poisson distribution is, theoretically, infinite.*

The Bayesian network fragment described in Example 4.1 deals with a very basic extension of the network with exclusively binary variables, mentioned previously in Section 4.1 [Figure 4.1(i)]. The reduced scope of the network fragment helps to draw the focus of attention to the underlying principle. It can obviously be thought of as part to be used in a larger network. It is also worth noting that current Bayesian network software are versatile enough to specify probabilities according to other standard statistical distributions.

More generally, the Bayesian network fragment considered in Example 4.1 can also find applications in contexts other than inference of source. Evett et al. (1995), for example, used a Poisson distribution for the number of recovered glass fragments, remaining after a given time. For such an application, a network fragment of the kind $H \rightarrow Y$ could be retrained, with the states of the node H representing particular time lapses, and the numbered states of the node Y representing numbers of recovered glass fragments. The parameter of the Poisson distributions to be assigned to the states of the node Y, for each lapse of time (i.e. a conditioning defined by a state of the node H), is the expected number of glass fragments remaining after a given time lapse of interest. Again, such a network can be thought of as a local approach to deal with a particular aspect of a larger inference task. More elaborate approaches to deal with issues related to glass transfer can be found in Curran et al. (1998). See Chapter 10 for other applications of statistical distributions (e.g. Sections 10.5.2 and 10.6.2) involving binomial and beta distributions.

4.3   Two stains, no putative source

4.3.1   Likelihood ratio for source inference when no putative source is available

Questions about inference of source also arise in situations in which two or more traces are recovered, for example, at different locations where, at temporally distinct instances, crimes have been committed. At an early stage of an investigation, it may well be that no potential source is available for comparison. However, analyses may be performed on the crime stains with the aim to help address the question of whether or not the analysed traces come from the same source. Such scenarios are of interest, for example, in a context also known as *forensic intelligence* (Ribaux and Margot 2003).

To illustrate this kind of case scenario, imagine two separate offences. Naked, dead bodies have been found at two distinct locations. In both cases, trace material has been collected from the bodies. Let E_1 and E_2 denote the materials collected in the first and second case, respectively. The traces could consist of, for example, textile fibres, hairs, semen or saliva. To approach the scenario at a general level of detail, it will be assumed that the compared characteristics are discrete. In the case of textile fibres, this could be a particular combination of fibre type and fibre colour, such as red wool. For a development involving continuous data together with an extension to crime level propositions, see Taroni et al. (2006b).

Suppose that the recovered traces from the two cases are analysed and both are found to be of type A, some sort of discrete physical attribute. For the current level of discussion, it will not be necessary to specify A in further detail. For evaluation at given source level propositions, the following propositions may be defined: 'the two traces E_1 and E_2 come from the same source' (H_1) and 'E_1 and E_2 come from different sources' (H_2). Because these propositions do not directly relate to a prosecutor's or defence's standpoint, the habitually used subscripts 'p' and 'd' for the variable H have been replaced by simple numerical subscripts. Notice that other issues, such as the relevance of the recovered trace materials, are not considered. To evaluate how well the scientific findings allow one to discriminate between H_1 and H_2, a likelihood ratio may be formulated as follows:

$$V = \frac{Pr(E_1 = A, E_2 = A|H_1, I)}{Pr(E_1 = A, E_2 = A|H_2, I)} . \tag{4.1}$$

The numerator asks the question 'What is the probability that the two traces are of type A, given that they come from the same source?' Notice that this question does not only focus on whether the two traces have corresponding analytical features, given proposition H_1, which is an event with probability of 1. It also focusses on the issue of this single source being of type A. Uncertainty about the single source being of type A, or $\bar{A}$ (i.e. any feature other than A), can be taken into account by considering two propositions, abbreviated by $H_{1,A}$ (the two traces come from the same source, which is A) and $H_{1,\bar{A}}$ (the two traces come from the same source, which is a source other than A), respectively. With E denoting the conjunction of $E_1 = A$ and $E_2 = A$, the numerator can be re-written as follows:

$$Pr(E|H_1) = \underbrace{Pr(E|H_{1,A})}_{1}\underbrace{Pr(H_{1,A})}_{\gamma_A} \ \underbrace{Pr(E|H_{1,\bar{A}})}_{0}\underbrace{Pr(H_{1,\bar{A}})}_{\gamma_{\bar{A}}} = \gamma_A.$$

Here, γ_A denotes the probability by which a single source from a relevant population would be found to be of type A. Note that $Pr(E|H_{1,A}) = 1$ is an expression of the assumption that the methodology used by the scientist is capable of correctly determining the analytical features of the analysed materials (i.e. of finding and reporting type A if the common source is in fact type A).

When H_2 is true, there may be different possibilities for there being two different sources: both sources are of type A, both sources are of type $\bar{A}$, or one source is of type A and the other is of type $\bar{A}$. Write these events in terms of propositions $H_{2,AA}$, $H_{2,\bar{A}\bar{A}}$ and $H_{2,A\bar{A}}$, respectively. The denominator can be re-written as

$$Pr(E|H_2) = \sum_i Pr(E|H_{2,i})Pr(H_{2,i}), \quad \text{for } i = \{AA, A\bar{A}, \bar{A}\bar{A}\}.$$

However, the E as defined above can only be true when the distinct sources of E_1 and E_2 are both of type A, that is $H_{2,AA}$ is true. Thus, the denominator reduces to

$Pr(E|H_2) = Pr(E|H_{2,AA})Pr(H_{2,AA})$. Assuming again that analytical features are determined without error, $Pr(E|H_{2,AA})$ can be set to 1. This will leave one with the term $Pr(H_{2,AA})$, that is the probability that the two distinct sources are both of type A. Write this probability as γ_{AA}. When one assumes that the probability for being of type A is the same for the two distinct sources, and the same as for the single source considered in the numerator, then one can write $\gamma_{AA} = \gamma_A^2$. The likelihood ratio thus becomes

$$V = \frac{Pr(E_1 = A, E_2 = A|H_1)}{Pr(E_1 = A, E_2 = A|H_2)} = \frac{\gamma_A}{1 \times \gamma_A^2} = \frac{1}{\gamma_A}. \qquad (4.2)$$

This result might look trivial in the sense that it is the same as in a case in which a single trace from a single crime scene is compared to a known source (Equation (3.3)). However, this similarity only holds for the final result, after simplifying the ratio and acceptance of particular assumptions. The questions asked in the numerator and denominator of (4.1), as well as the related probability assignments, are clearly different. The term γ_A is a suitable statistic drawn from the same population, but the validity of this assumption needs to be reviewed whenever other kinds of trace material are considered. See, for example, the differences in definition of population proportions γ and γ' presented earlier in Section 3.3.3.

4.3.2 Bayesian network for a two-trace case with no putative source

There are different ways to implement the approach outlined (Section 4.3.1) in terms of a Bayesian network. One way, following a close description of the algebraic development, relies on a serial connection of the kind $H \rightarrow H' \rightarrow E$, where H' covers all of the sub-propositions $H_{1,A}, H_{1,\bar{A}}, H_{2,AA}, H_{2,A\bar{A}}$ and $H_{2,\bar{A}\bar{A}}$. Another way consists of structuring the inference processes for each trace item separately and then combine the two structures. Such a construction is explained in the following.

Start by considering one of the two cases and the trace material found in that case. The order in which the two cases and their related traces are approached is not important. Here, one of the two cases is chosen and deliberately called *first case*. Let E_1 denote the outcomes of the observations made on the item of trace material. The variable E_1 is defined in terms of the binary states A or $\bar{A}$, where $\bar{A}$ covers all potential outcomes other than A. Information on the observed attributes of this first item is used to draw an inference about the characteristics of the source from which it comes. Let S_{E_1} denote the characteristics of the source of the trace material in the first case. This variable is also binary and has the states A and $\bar{A}$. Logically, the observation of the characteristics of the trace material is directly dependant on the attributes of the source. Thus, graphically, a direct edge may be drawn from S_{E_1} to E_1.

The trace material recovered in the second case may be evaluated in the same way. A variable E_2 is thus used to describe the observations made on the second trace, and a variable S_{E_2} is used to describe the characteristics of the source. These two variables are connected so that the following network fragment is obtained: $S_{E_2} \rightarrow E_2$.

Next, a variable H is defined to represent the proposition according to which E_1 and E_2 come from the same source (H_1) or come from different sources (H_2). The node H needs to be combined with the two network fragments $S_{E_1} \rightarrow E_1$ and $S_{E_2} \rightarrow E_2$ in some meaningful way. One possibility is to condition S_{E_2} on both H and S_{E_1}, as shown in Figure 4.2, and to assign

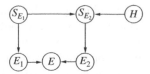

Figure 4.2 A Bayesian network for a scenario in which two stains are found at two different scenes. Nodes E_1 and E_2 denote the outcomes of the observations made on the trace material at the first and second crime scenes, respectively, with two states A and $\bar{A}$ for each node. Node E is Boolean and in state 'true' if and only if both nodes E_1 and E_2 are in state A. Nodes S_{E_1} and S_{E_2} have two states, the source of the first or second item of evidence is of type A or type $\bar{A}$. Node H represents the proposition according to which E_1 and E_2 come from the same source (H_1) or come from different sources (H_2).

the following conditional probabilities:

$$Pr(S_{E_2} = k | S_{E_1} = l, H_m) = \begin{cases} 1, & \text{for } k, l = \{A, \bar{A}\}, \ m = 1, \ k = l, \\ 0, & \text{for } k, l = \{A, \bar{A}\}, \ m = 1, \ k \neq l, \\ \gamma_A, & \text{for } k = A, \ m = 2, \\ 1 - \gamma_A, & \text{for } k = \bar{A}, \ m = 2. \end{cases} \qquad (4.3)$$

This assignment clarifies that if H_2 holds, the probability of the source of the second item being of type A $(\bar{A})$ does not depend on the features of the source of the first trace (i.e. node S_{E_1}).

The conditional probabilities associated to the observational nodes E_1 and E_2 reflect the degree to which the characteristics observed by the scientist are indicative of the characteristics of the sources from which the materials originate. Generally, scientists tend to prefer methods for which $Pr(E_i = A | S_{E_i} = A) \to 1$ and $Pr(E_i = A | S_{E_i} = \bar{A}) \to 0$, where $i = \{1, 2\}$ refers to the two traces (or cases).

The Bayesian network proposed in this section is somewhat more detailed than (4.2) developed previously. The major difference is that a distinction is introduced between the observations made on the trace materials (nodes E_1 and E_2) and the actual characteristics of their sources (nodes S_{E_1} and S_{E_2}). The latter are unobserved variables not explicitly represented in the derivation of the likelihood ratio formula. Moreover, two distinct variables are used to represent the analytical features of the recovered items of trace material. These two nodes can be summarized in terms of a single Boolean node E, which is in state 'true' if and only if both nodes E_1 and E_2 are in state A.

The general structure of the model may look laborious at first view, but should convey greater flexibility in specifying conditional probabilities than the algebraic development, in which several simplifying assumptions were made. For example, if one wishes to assume that, given H_2, different population proportions γ_A apply for the two sources, then this may directly be specified in the probability tables of the nodes representing the features of the source of each trace item; that is, the value for γ_A used for $Pr(S_{E_2} = A | H_2)$ in (4.3) may be different from the value γ_A assigned to $Pr(S_{E_1})$.

4.3.3   An alternative network structure for a two trace no putative source case

Figure 4.3 shows an alternative network for a case in which two traces from distinct crime scenes are compared, but no potential source is available. Unlike the network discussed in Section 4.3.2, the model in Figure 4.3 represents the observed analytical characteristics (i.e. fibre type and colour) of each item of trace material in terms of distinct nodes $E_{i,\{t,c\}}$. In this notation, i refers to the trace number (i.e. 1 or 2), t to fibre type and c to fibre colour. The features of a potential common source for the two items of trace material are modelled by the nodes CS_t and CS_c, where the subscripts t and c refer again to fibre type and colour, respectively. The term *potential common source* is used here to express the idea that uncertainties about the current state of the variables CS_t and CS_c will be allowed to influence the nodes $S_{i,\{t,c\}}$ (for $i = \{1, 2\}$) only if the two items of trace material do in fact come from a common source (i.e. H_1 is true). Note that the nodes referring to fibre colour are conditioned on the fibre type because one may wish to specify the population proportion for fibre colour given fibre type. For the purpose of the current example, only binary nodes are used. The states for nodes representing fibre colour (i.e. $E_{i,c}$ and $S_{i,c}$) are 'red' and 'other', and the states for nodes representing fibre type (i.e. $E_{i,t}$ and $S_{i,t}$) are 'wool' and 'other'.

The values specified in the probability tables associated with the observational nodes $E_{i,\{t,c\}}$ can account for the relative performance of the examiner. For example, when it is thought that the examiner will always recognize the type of a fibre type as wool when it is indeed wool, and never declare a fibre type as wool when in fact it is not, then the following assessments are appropriate: $Pr(E_{i,t} = \text{wool}|S_{i,t} = \text{wool}) = 1$ and $Pr(E_{i,t} = \text{wool}|S_{i,t} = \text{other}) = 0$, for $i = \{1, 2\}$. The argument applies analogously to the nodes $E_{i,c}$ as far as the observation of the fibre's colour is concerned.

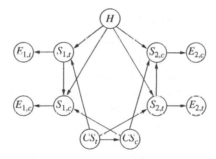

Figure 4.3   An alternative network structure for a scenario in which two stains are found on two different scenes. Node H stands for the proposition 'The two traces come from the same source' (with states 'H_1: yes', 'H_2: no'). Node $E_{i,t}$ stands for the observed fibre type of trace i, node $E_{i,c}$ for the observed colour of trace i, node $S_{i,t}$ for fibre type of the source of trace i, node $S_{i,c}$ for fibre colour of the source of trace i, node CS_t for fibre type of a potential common source and node CS_c for fibre colour of a potential common source. Note that $i = 1$ and $i = 2$ denote the first and the second trace, respectively. Nodes representing a fibre type have the two states 'wool' and 'other', and nodes representing a fibre colour have states 'red' and 'other'.

The probability tables associated with the source nodes $S_{i,\{t,c\}}$ contain either logical probabilities or population proportions. Consider, for example, the probabilities assigned to nodes $S_{i,t}$ given H_1. If the two traces come from a common source (i.e. H_1 holds), and the common source consists of wool, then, logically, $Pr(S_{i,t} = \text{wool}|H_1, CS_t) = 1$ with $i = \{1, 2\}$ denoting trace 1 and 2, respectively. Analogously, one can specify $Pr(S_{i,t} = \text{other}|H_1, CS_t = \text{other}) = 1$. However, if the two traces do not have a common source (H_2 being true), then the current state of CS_t is not relevant for the current state of the nodes $S_{i,t}$. In the latter case, the probability of the source of a trace being wool is just given by the appropriate population proportion γ_w of wool, without regard to the current state of CS_t. Thus, $Pr(S_{i,t} = \text{wool}|H_2) = \gamma_w$ and $Pr(S_{i,t} = \text{other}|H_2) = 1 - \gamma_w$.

The probability tables of the nodes $S_{i,c}$ are completed in a similar way. When H_1 is true, that is the two traces come from a common source, then the probability of $S_{i,c}$ being red, for example, depends on the probability of the common source being red: thus $Pr(S_{i,c} = red|H_1, CS_c = red) = 1$ and $Pr(S_{i,c} = red|H_1, CS_c = \text{other}) = 0$. Note that when H_1 is true, $S_{i,c}$ is independent of $S_{i,t}$ and thus is not written in the conditional. The situation is different when the two traces do not come from a common source (i.e. H_2 is true). Then, the probability of the source of a given trace being red, for example, is dependent solely on the fibre type. The variable γ_r denotes the relevant population proportion for the colour red among woollen sources of fibres. The term γ_r' denotes the appropriate population proportion for red colour among sources of fibres other than wool. This distinction can be made because of the directed edge between the nodes $S_{i,t}$ and $S_{i,c}$. If the values γ_r and γ_r' would be judged the same, then this edge could be omitted so as to retain only a single assignment γ_r.

4.4   Multiple propositions

4.4.1   Form of the likelihood ratio

The discussion in Section 4.2 pointed out that the general model for inference of source, shown in Figure 4.1, is not restricted to cases in which the node for the observations, E, is binary. Multiple states can also be considered for the main proposition H. As an example, consider a case involving results of comparative handwriting examinations (for a case involving DNA, see Sections 7.5 and 7.6). When observations relate to a questioned signature, competing propositions of interest may be formulated as follows: 'The suspect wrote the questioned signature' (H_p) and 'The questioned signature was not written by the suspect, but by another person' (H_d). Although this pair of propositions clearly suggests an activity (i.e. writing), it is not misleading to say that, as a consequence of writing, a given person can be regarded as the 'source' of the handwriting of interest. As is widely recognized, uncertainty about the writing conditions described as, for example, 'ordinary' or 'unusual' [e.g. Köller et al. (2004)] may present a complicating aspect in handwriting cases. For example, an 'unusual' situation may account for the true writer's intent to disguise, neutralize, or imitate. In an example described by Köller et al. (2004), two categories of circumstances of this kind are proposed to extend the above target propositions as follows: 'The suspect wrote the questioned signature under ordinary conditions' (H_{p1}), 'The suspect wrote the questioned signature under unusual conditions' (H_{p2}) and 'The questioned signature was not written by the suspect, but by another person, under normal conditions' (H_{d1}). 'The questioned signature was not written by the suspect, but by another person, under unusual conditions' (H_{d2}).

Suppose next that the results of the comparative handwriting examinations are available. They consist of closely corresponding handwriting features. That is, the graphic features of the questioned writing are well covered by the range of variation observed in comparison writings known to come from the suspect. Moreover, no unexplainable differences are observed. Let these findings be summarized by a binary variable E. For such a single item of information, and multiple competing propositions, Bayes' theorem can be formulated as follows:

$$Pr(H_{p1}|E,I) = \frac{Pr(E|H_{p1},I)Pr(H_{p1}|I)}{\sum_i Pr(E|H_i,I)Pr(H_i|I)}, \quad \text{for } i = \{p1,p2,d1,d2\} . \tag{4.4}$$

This expression is sometimes mentioned because the formulation of a likelihood ratio in the light of multiple propositions involves a less familiar form. One aspect is that, for example, the prior probabilities $Pr(H_i|I)$, for $i = \{p1,p2,d1,d2\}$, intervene. That is, for instance, if one seeks to consider the value of E for comparing H_{p1} with $(H_{p2}, H_{d1}, H_{d2}) = \bar{H}_{p1}$, the likelihood ratio takes the following form (Aitken and Taroni 2004):

$$\frac{Pr(E|H_{p1},I)}{Pr(E|\bar{H}_{p1},I)} = \frac{Pr(E|H_{p1},I)\left[\sum_i Pr(H_i|I)\right]}{\sum_i Pr(E|H_i,I)Pr(H_i|I)}, \quad \text{for } i = \{p2,d1,d2\} . \tag{4.5}$$

Note that such an expression should be named Bayes factor rather than likelihood ratio because the ratio involves uncertainties about hypotheses. The forthcoming sections analyse and track these results in different ways through Bayesian networks.

It is acknowledged at this point that the description and definition of the findings E as well as the set of competing target propositions H may be framed in a different way, according to the scientist's preferences and needs. In Stockton and Day (2001), for instance, more detailed categories termed *some, few and many similarities* are used as descriptors of the outcomes of comparative handwriting examinations. Also, it may be questioned if an extension to multiple propositions will still make an evaluation focussed enough to be useful for court. For the discussion here, the extension is accepted and used only for exploratory purposes and to illustrate the use of Bayesian networks as a means to clarify the inferential framework. The resulting Bayesian networks can serve as a starting point for studying further extensions and refinements of modelling assumptions that readers may wish to change.

4.4.2   Bayesian networks for evaluation given multiple propositions

4.4.2.1   Model 1

An immediate and rather obvious way to cope with multiple propositions is to take the network structure shown in Figure 4.1(i) with a variable E for the observations and a variable H for the target propositions. However, instead of a node H with only two propositions H_p and H_d, a total of four states H_{p1}, H_{p2}, H_{d1} and H_{d2} are defined. This will allow one to consider posterior probabilities for each of these multiple propositions, given the findings E, as defined by (4.4). It is not possible, however, to use the Bayesian network $H \rightarrow E$ with multiple states for node H to find the components of a likelihood ratio of the kind $Pr(E|H_p)/Pr(E|H_d)$. The reason for this is that, habitually, one is able to instantiate to 'known' only a single state of a given node, and this will result in setting the probability of the remaining states to zero. For the case considered here, however, there are two propositions for the numerator, H_{p1} and H_{p2}, and

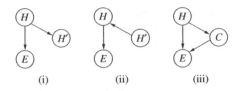

(i)   (ii)   (iii)

Figure 4.4 Bayesian networks for evaluating scientific findings E (with possible states E and $\bar{E}$) that are relevant for inference about the set of multiple target propositions $\{H_{p1}, H_{p2}, H_{d1}, H_{d2}\}$, represented in terms of the node H, as used in models (i) and (ii) (Examples 4.2 and 4.3). The node H' is binary and regroups the pair of propositions H_{p1} and H_{p2} in a state H_p and the pair H_{d1} and H_{d2} in a state H_d. In model (iii), states H_p and H_d are associated with node H, whereas node C, with states 'normal' and 'unusual', accounts for the assumed two conditions under which the questioned handwriting was produced.

these cannot be set to 'true' simultaneously. One way to avoid this complication is to regroup the propositions H_{p1} and H_{p2} in a single state H_p of a new node H' (Taroni and Biedermann 2005). Analogously, one may regroup propositions H_{d1} and H_{d2} in a single state H_d. That is, one can consider H_p to be established whenever either H_{p1} or H_{p2} holds and H_d to be established whenever either H_{d1} or H_{d2} is the case. Here, H_p and H_d refer to the authorship by the suspect and an unknown person, respectively, under any writing circumstance (i.e. normal or unusual). The binary node H' thus is adopted as a direct descendant of the node H, as shown in Figure 4.4(i). The probability table for the node H' is completed using assignments of zero and one as follows:

$$Pr(H' = H_p | H_i) = \begin{cases} 1, & \text{for } i = \{p1, p2\}, \\ 0, & \text{for } i = \{d1, d2\}. \end{cases} \tag{4.6}$$

Further, note that $Pr(H' = H_d | H_i) = 1 - Pr(H' = H_p | H_i)$ for all $i = \{p1, p2, d1, d2\}$.

Example 4.2 *(Likelihood ratio with multiple propositions for evaluation of results of comparative handwriting examinations) Consider the Bayesian network shown in Figure 4.4(i) and the following conditional probabilities for the findings E, given the various competing propositions H_i, as proposed by Köller et al. (2004):*

$$Pr(E | H_i) = \{0.95, 0.06, 0.01, 0.95\}, \quad \text{for } i = \{p1, p2, d1, d2\}.$$

Since the node E is binary, the conditional probabilities $Pr(\bar{E}|H)$ are given by $1 - Pr(E|H)$. For the node H, still following Köller et al. (2004), the probabilities

$$Pr(H_i) = \{0.45, 0.05, 0.02, 0.48\}, \quad \text{for } i = \{p1, p2, d1, d2\},$$

are specified. The initial state of the Bayesian network is shown in Figure 4.5(i). In order to obtain a value for the numerator of the likelihood ratio, that is $Pr(E|H_p)$, one needs to set the probability of the state H_p of the node H' to one. The effect of this is twofold.

Firstly, one can observe that the node H displays posterior probabilities 0.9 and 0.1 for, respectively, states H_{p1} and H_{p2}. The probabilities of the states H_{d1} and H_{d2} are zero. This is a direct consequence of the numerical specification defined by (4.6), which states that H_p is *associated with H_{p1} and H_{p2}. In an application of Bayes' theorem for inference about H, on the basis of H', this may lead to zero posterior probabilities. For example, $Pr(H_{d1}|H' = H_p) = 0$ because the fraction*

$$[Pr(H' = H_p|H_{d1})Pr(H_{d1})]/Pr(H' = H_p)$$

contains a zero likelihood in the numerator. When the probabilities are reduced in the two states H_{d1} and H_{d2}, then it must be redistributed over the remaining two states H_{p1} and H_{p2}. This redistribution is made proportionally, that is the probabilities for H_{p1} and H_{p2} keep their relative magnitudes. Initially, they were 0.45 and 0.05, respectively, as shown in Figure 4.5(ii). Given H_p, they became 0.9 and 0.1. This is a process of normalization obtained by dividing the probabilities H_{p1} and H_{p2} by their sum:

$$Pr(H = H_{p_i}|H' = H_p) = \frac{Pr(H_{p_i})}{\sum_i Pr(H_{p_i})}, \quad \text{for } i = \{p1, p2\}. \tag{4.7}$$

The two values defined by (4.7) can be interpreted as probabilities for, respectively, normal and unusual writing conditions in the event of the suspect being the writer of the questioned signature. Notice that this argument applies analogously to the states H_{d1} and H_{d2} for a situation in which H_d is assumed to hold. As shown by Figure 4.5(iii), it is then the case that $Pr(H_{d1}|H' = H_d) = 0.04$ and $Pr(H_{d2}|H' = H_d) = 0.96$.

The node E displays the probability of E given H_p. This probability is obtained by the rule of the extension of the conversation. Applied to the setting here, this is

$$Pr(E|H' = H_p) = Pr(E|H_{p1})Pr(H_{p1}|H_p) + Pr(E|H_{p2})Pr(H_{p2}|H_p) . \tag{4.8}$$

The extension to H_{d1} and H_{d2} is omitted from notation here because, given H_p, the probabilities of these two states are zero. As can be seen, for obtaining a value for the numerator of the likelihood ratio, one needs to consider an extension to H_{p1} and H_{p2}, but it would not

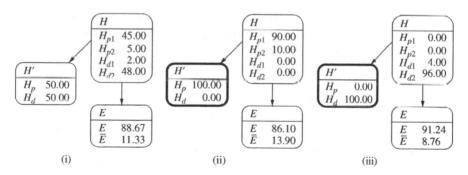

Figure 4.5 Expanded representation of the Bayesian network defined in Figure 4.4(i) for evaluating scientific findings E (with possible states E and $\bar{E}$) that are relevant for inference about the set of multiple target propositions $\{H_{p1}, H_{p2}, H_{d1}, H_{d2}\}$, represented in terms of the node H. The node H' is binary and regroups the pair of propositions H_{p1} and H_{p2} in a state H_p and the pair H_{d1} and H_{d2} in a state H_d. (i) shows the network in its initial state, whereas (ii) and (iii) show situations in which the node H is set to, respectively, H_p and H_d (instantiated nodes are indicated by a thicker border).

be correct to weight the conditional probabilities $Pr(E|H_{p1})$ and $Pr(E|H_{p2})$ only by $Pr(H_{p1})$ and $Pr(H_{p2})$ because the latter two probabilities do not sum to one. Instead, one needs to work with $Pr(H_{p1}|H' = H_p)$ and $Pr(H_{p2}|H' = H_p)$ as defined by (4.7). On the basis of these considerations, the likelihood ratio is as follows:

$$V = \frac{Pr(E|H_p)}{Pr(E|H_d)}$$

$$= \frac{Pr(E|H_{p1})Pr(H_{p1}|H_p) + Pr(E|H_{p2})Pr(H_{p2}|H_p)}{Pr(E|H_{d1})Pr(H_{d1}|H_d) + Pr(E|H_{d2})Pr(H_{d2}|H_d)}$$

$$= \frac{0.95 \times 0.9 + 0.06 \times 0.1}{0.01 \times 0.04 + 0.95 \times 0.96} = \frac{0.861}{0.9124} \approx 0.94 \ .$$

The reader can verify that all of these numerical results entirely agree with the algebraic approach described by Köller et al. (2004).

4.4.2.2   Model 2

An alternative approach to dealing with multiple propositions has been described by Buckleton et al. (2006). These authors represent the principal pair of competing propositions, H_p and H_d, in terms of a root node. Applied to the case studied here, this leads to a graphical representation as shown in Figure 4.4(ii). The main difference to the Bayesian network discussed in Section 4.3.2 is the reversed arc between the two nodes H' and H. While the states of these two nodes remain the same, their probability tables change. Here, the node H' does not receive entering arcs from other nodes. Its node table thus contains unconditional probabilities $Pr(H_p)$ and $Pr(H_d)$. For purely technical reasons, a probability of 0.5 is assigned here to each of the latter two states. The reader may choose other values, but in essence these remain irrelevant as soon as one focusses on the evaluation of the probability of the finding E, given each of the competing target propositions. That is, instantiating H' to one of its states will render the probabilities initially assigned to this node ineffective.

More attention needs to be given to the probability table of the node H, which now contains conditional probabilities. Here, one is directed to enquire, given H_p, about the probability of the suspect writing under normal (H_{p1}) and unusual (H_{p2}) conditions. These probabilities are given by (4.7). The probabilities assigned to H_{d1} and H_{d2} are zero because they do not apply to situations in which H_p holds. Given H_d, the opposite holds and one specifies probabilities for H_{d1} and H_{d2}, whereas zero probabilities are assigned to H_{p1} and H_{p2}. Here, too, (4.7) can be used, replacing $\{p1, p2\}$ by $\{d1, d2\}$.

Example 4.3 *(Likelihood ratio evaluation with alternative Bayesian network structure for multiple propositions) Consider again the numerical assignments from the questioned handwriting case described by Köller et al. (2004), outlined in Example 4.2. Using the network structure shown in Figure 4.4(ii) and following the argument outlined above, application of (4.7) leads to the following assignments of conditional probabilities for the node H:*

$$Pr(H_{p1}|H_p) = 0.9, \quad Pr(H_{p2}|H_p) = 0.1 \ \text{and} \ Pr(H_i|H_p) = 0 \ \text{for} \ i = \{d1, d2\}$$

when H_p is assumed to hold, and

$$Pr(H_{d1}|H_d) = 0.04, \quad Pr(H_{d2}|H_d) = 0.96 \ \text{and} \ Pr(H_i|H_d) = 0 \ \text{for} \ i = \{p1, p2\},$$

when the alternative proposition H_d is assumed to be true.

The assignment of these probabilities can also be understood as a transcription of the prior probabilities initially defined in Example 4.2, that is

$$Pr(H_{p1}|I) = 0.45, \quad Pr(H_{p2}|I) = 0.05, \quad Pr(H_{d1}|I) = 0.02, \quad Pr(H_{d2}|I) = 0.48,$$

in terms of two nodes H and H', rather than one. Although $Pr(H_{p1}|H_p) = 0.9$ and $Pr(H_{p2}|H_p) = 0.1$, the node H' will display $Pr(H_{p1}) = 0.45$ and $Pr(H_{p2}) = 0.05$ when the node H is not instantiated, that is when H_p holds with a probability of only 0.5. An analogous argument applies for the states H_{d1} and H_{d2} given H_d.

The structural relationship between the nodes E and H is the same for the models shown in Figures 4.4(i) and (ii). The probability assignments for the node table for E thus remain unchanged as well. The overall numerical output at the node E, given instantiations made at the node H, will also be in entire agreement with the findings in Example 4.2. In particular, the probability at the observational node E is still obtained according to the considerations defined by (4.8). More generally, the Bayesian network described in this example would display exactly the same numerical output as the one shown in Figure 4.5 (Example 4.2) for all nodes and all situations (i.e. initial state and propagated state).

4.4.2.3   Model 3

There is a further way to cope with the handwriting scenario outlined in Section 4.4.1, but it diverges slightly with respect to the Models 1 and 2 discussed above. As a distinctive feature, this third model relies on a decomposition of the main set of multiple propositions H_{p1}, H_{p2}, H_{d1} and H_{d2} into two separate pairs of propositions (Biedermann et al. 2012d). A first pair of propositions is represented by a node H, with two states H_p and H_d as defined initially in Section 4.4.1, and used also in the previous two models shown in Figure 4.4(i) and (ii). Note solely that, in the previous two models, the states H_p and H_d were associated with a node H', rather than H as done now for Model 3. A second pair of propositions focusses exclusively on the writing circumstances. This pair of propositions is represented by a node C with states referring to, respectively, normal (C) and unusual $(\bar{C})$ writing circumstances.

This separation offers a crucial insight for the currently studied scenario by clarifying that the initial set of propositions H_i (for $i = \{p1, p2, d1, d2\}$) is a combination of propositions relating to the writer of the questioned signature and propositions relating to writing circumstances. The variable relating to writing circumstances appears twice: a normal setting appears in both H_{p1} and H_{d1}, whereas an unusual setting is associated with H_{p2} and H_{d2}. It is generally advisable to avoid compound or otherwise unnecessarily intertwined propositions. They should be defined as simply as possible within the context of circumstances as it appears to the scientist.

The nodes H and C can be associated with the observational variable E in terms of a Bayesian network structure as shown in Figure 4.4(iii). The probabilities for the two states H_p and H_d are as chosen in the previous section. Again, their choice is not crucial because likelihood ratio evaluation focusses on the probability of the findings E given instantiations of the node H. The probability table of the node C contains values for normal (C) and unusual $(\bar{C})$ writing conditions, given that the suspect is the author of the signature (H_p) and given that a person other than the suspect is the writer (H_d). As may be seen, these probability assignments correspond to those made previously for the states H_{p1}, H_{p2}, H_{d1} and H_{d2} of node H in Model 2 [Figure 4.4(ii)]:

$$Pr(C|H_p) = Pr(H_{p1}|H_p) = 0.9 \quad \text{and} \quad Pr(C|H_d) = Pr(H_{d1}|H_d) = 0.04$$

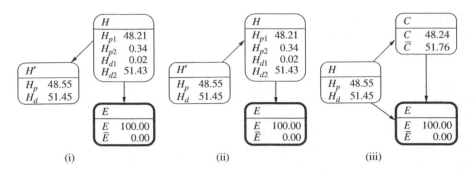

Figure 4.6 (i–iii) Expanded representation of the Bayesian networks defined in Figure 4.4. The observational nodes E are instantiated (indicated by a thicker node border). The three networks show an equivalent output at the node with the main pair of propositions H_p and H_d.

for normal writing conditions and

$$Pr(\bar{C}|H_p) = Pr(H_{p2}|H_p) = 0.1 \quad \text{and} \quad Pr(\bar{C}|H_d) = Pr(H_{d2}|H_d) = 0.96$$

for unusual writing conditions. The probability table associated with the node E reflects the conditioning by the two nodes H and C. The various combinations of the states of these two nodes correspond to the initial set of propositions $\{H_{p1}, H_{p2}, H_{d1}, H_{d2}\}$. For this reason, the probabilities assigned previously in Example (4.2) for $Pr(E|H_i)$, where $i = \{p1, p2, d1, d2\}$ can be reused here as follows:

$$Pr(E|C, H_p) = Pr(E|H_{p1}) = 0.95, \quad \text{and} \quad Pr(E|\bar{C}, H_p) = Pr(E|H_{p2}) = 0.06$$

for situations in which H_p holds and

$$Pr(E|C, H_d) = Pr(E|H_{d1}) = 0.01, \quad \text{and} \quad Pr(E|\bar{C}, H_d) = Pr(E|H_{d2}) = 0.95$$

for situations in which H_d holds. Notice that, for given conditions, the probability $Pr(\bar{E}|C, H) = 1 - Pr(E|C, H)$.

For the Bayesian network considered in Figure 4.4(iii), the likelihood ratio slightly diverges from that in the previous models but only at a notational level. For example, in order to obtain the numerator of the likelihood ratio, the variable E must be conditioned on H_p, while allowing for uncertainty about C. This leads to the following:

$$Pr(E|H_p) = Pr(E|C, H_p)Pr(C|H_p) + Pr(E|\bar{C}, H_p)Pr(\bar{C}|H_p) . \qquad (4.9)$$

This expression is equivalent to that defined in (4.8) because of the way in which the various component probabilities have been interpreted. In particular, there is an equivalence of the conditional probabilities for C given H in (4.9) and the conditional probabilities for H given H' in (4.8). A development analogous to (4.9) applies for the denominator of the likelihood ratio.

Figure 4.6 illustrates a situation in which one considers the posterior probabilities of the target propositions H_p and H_d, in the light of the observation E. It shows that the quantitative output of all Bayesian network structures described throughout this section lead to equivalent results.

5

Evaluation given activity level propositions

This chapter focusses on general methodological aspects of Bayesian network construction for forensic inference based on examination results regarding the so-called *transferable materials* such as DNA, glass, fibres, paint or gunshot residues. The starting points are general likelihood ratio formulae proposed in scientific literature to deal with the assessment of such findings. The main features of the various Bayesian networks will be illustrated through examples where the propositions of interest refer to an activity (i.e. so-called activity level propositions). The findings will be either expressed in terms of reported corresponding features or described in terms of *intrinsic* and *extrinsic* features. The latter level of description will present a further instance of the use of the notation y and x introduced earlier in Section 4.1.

In many examples, it is reasonable to assume that particular material has a direct relationship with a given individual (which may be the suspect). For example, assume a direct primary transfer.[1] For example, DNA found at a crime scene provides a direct connection between the crime scene and the source of the DNA − whoever that source is. The reason for this is that biological material is intimately related to each individual. This may not be the case, however, with certain categories of traces or marks other than DNA, such as shoemarks or fibres. Typically, shoes and garments may have been worn by different individuals. One thus also needs to establish a connection between a given donor item (such as a garment or a pair of shoes) and a given suspect in order to reason about that person's alleged activities. This chapter investigates both aspects, that is direct and indirect connections, in terms of likelihood ratio developments given activity level propositions. These arguments will be illustrated through the use of Bayesian networks. A particular issue, cross-transfer of material, is also developed and discussed.

---

[1] A primary transfer is the terminology given to an event in which a given donor sheds material in the first place (e.g. an injured offender leaves blood stains on a broken window). Secondary transfer (not considered here) occurs, for example, if trace material on a person's clothing (e.g. traces of illegal drugs acquired as a primary transfer due to manipulating drugs in powder form) is transferred to another surface (e.g. a car seat).

Bayesian Networks for Probabilistic Inference and Decision Analysis in Forensic Science, Second Edition.
Franco Taroni, Alex Biedermann, Silvia Bozza, Paolo Garbolino and Colin Aitken.
© 2014 John Wiley & Sons, Ltd. Published 2014 by John Wiley & Sons, Ltd.

Transfer traces represent an interesting and challenging area of research in forensic science essentially because the subtle phenomena of transfer, persistence and recovery of traces play an important role. However, care should be taken in order to account correctly for such factors. The aim here is to show that incorporating these factors in the evaluation can be clarified substantially through the use of Bayesian networks.

It is useful to emphasize again that when attempting to model large and complex domains, it may be preferable to limit the number of variables in the model, at least at the beginning, in order to keep the structure and the resulting probabilistic inferential implications tractable. Throughout the remainder of this chapter, therefore, there will be limited number of variables. Notwithstanding this, some target nodes, for example the node representing the event of transfer, will be studied in more detail and possible extensions will be proposed. Generally, the analyses will be directed to select the most important variables, described by nodes, in order to simplify the description of the situation and facilitate the elicitation of probabilities. Proceeding in this way, the work will concentrate on a focussed number nodes, fewer arcs and smaller state spaces, while keeping in mind that the model should faithfully reflect the reality of the scenario investigated.

5.1   Evaluation of transfer material given activity level propositions assuming a direct source relationship

5.1.1   Preliminaries

As noted earlier in Chapter 3, relevant propositions at the activity level include the description of an explicit action. Such an approach attempts to reply closely to judicial needs expressed by parties at trial (Taroni et al. 2013). Activity level propositions could be, for example, 'Mr X assaulted the victim' and 'Mr X did not assault the victim', meaning that some other man assaulted the victim and that Mr X was not involved in the action. The alternative proposition may also be formulated as 'Mr X has nothing to do with the incident' so as to further clarify a view that separates Mr X from the case. Another pair of activity level propositions could be 'Mr X sat on the car driver's seat' and 'Mr X never sat on the car driver's seat'. An important aspect or consequence of an activity like assaulting a victim or sitting on a driver's seat is the occurrence of a direct contact, such as between two people involved in an assault or between the driver and the seat of a car. A further consequence that needs to be considered in case of contact is the occurrence of the phenomenon of transfer of material (e.g. fibres).

According to arguments outlined in the following sections, it is of important to acknowledge the origin of transfer. It is of importance to enquire about whether transfer refers to an alternative source potentially involved in the action under investigation (e.g. another person being the aggressor under the alternative proposition) or whether transfer refers to a legitimate event according to which the recovered material is part of background (e.g. an alternative activity by the suspect).

Notice that the use of the term *contact*, as previously mentioned, is considered in the context of a logical consequence of an action. Although this rather vague definition does not tell much about the kind of contact, it can be useful to incorporate this event structurally in a Bayesian network in terms of a distinct node, essentially because it allows the scientist to take into account the possibility of a legitimate contact between the persons involved in the scenario under investigation. Notice that this can be meaningful even though the suspect may claim not be involved in the alleged offence.

Figure 5.1 Generic Bayesian network structure describing a scenario involving the transfer of material. Each node has two states. Node H, the main proposition, with states H_p, 'The suspect assaulted the victim', and H_d, 'Some person other than the suspect assaulted the victim'. Node T, with states T, 'There was a transfer from the victim', and $\bar{T}$, 'There was not a transfer from the victim'. Node E, with states E, 'The trace found on the suspect is reported to be similar to the control coming from the victim', and $\bar{E}$, 'The trace found on the suspect is not similar to the control coming from the victim'.

5.1.2 Derivation of a basic structure for a Bayesian network

Figure 5.1 shows a standard Bayesian network that may be used to describe the potential transfer of material from the crime scene (or the victim) to the offender. In this kind of scenario, the material found on the person of interest comes from the victim or scene or it is present on the suspect by chance alone because, given the alternative proposition, the suspect is not involved in the action under investigation.

Recall notation introduced earlier in Section 3.3.2, where T was used to denote the event of transfer, along with the conventional variables H and E denoting, respectively, the main propositions of interest (at activity level) and the scientific findings regarding the recovered material. The aim of this section is to show how previously discussed formulaic development can be translated into a simple graphical model. Note that such a model can be further developed to deal with more complex cases and outcome specifications, in particular descriptions of intrinsic and extrinsic features to specify components of node E. This will be pursued at a later point in this chapter.

For the time being, consider the three variables E, T and H. The first of these variables, E, denotes the reported correspondence between characteristics observed on the recovered material and the control material. This variable is conditioned on the event T, which covers the event of transfer, persistence and recovery of material. However, E is also conditioned on the main proposition H, in the same way as T is. In turn, the variable H is not conditioned on any other event. These relevance relationships have previously been discussed in terms of conditional probabilities such as $Pr(E|T, H)$ and $Pr(T|H)$ (Section 3.3.2). On the basis of these considerations, one can propose the network structure shown in Figure 5.1.

Example 5.1 *(Fibre transfer to the offender) Imagine a scenario involving transfer material, say fibres, shed by the victim's garment. This can lead to transfer to the offender, typically in a case of a physical assault. Suppose further that, as a result of the investigation, a suspect is arrested. The outer clothing of the suspect is taken for scientific examination. Extraneous (foreign) fibres are found on the clothing. The chemical and physical characteristics of the recovered foreign fibres are the same as the victim's jumper. The model shown in Figure 5.1 can be used to represent this scenario by choosing the following binary node definitions. The node H covers the activity level propositions H_p, 'The suspect is the person who committed the aggression against the victim', and H_d, 'The suspect is not the person who committed the aggression against the victim (i.e. the suspect is not involved in the incident)'. In turn, the* *node T accounts for the event of transfer (and related phenomena), with states T, 'There has been transfer (persistence or recovery) of material during the commission of the crime', and $\bar{T}$, 'There was no transfer (persistence or recovery) of material during the commission of the crime'. The outcome node E refers to the results of the expert's analysis. The two states of this node are E, 'The characteristics of the recovered and control materials are reported to match', and $\bar{E}$ 'The scientist reports different characteristics for recovered and control materials'.*

Recall from (3.4) that several distinct probabilities need to be assigned. Here they are used to complete the probability tables of the various nodes. In particular, it is necessary to specify the probability of transfer during the commission of the crime (t), the probability of no background (b_0) of extraneous traces, that is no innocent transfer of material, the probability of innocently transferred traces so that there is a presence of background material (b), and, finally, the probability of encountering the reported analytical characteristics among fibres on clothings (γ). The last probability is an expression of the rarity of the observed analytical features. Because node E depends on both variables H and T, a total of eight values need to be assigned to the probability table of E (Table 5.1).

Note that if H_d is true, that is, the suspect is not the person who committed the aggression against the victim, then the event of transfer is not considered; no extension is made to incorporate the event T in the denominator of the likelihood ratio (3.4). As pointed out by the assignments in Table 5.1, whatever the value of the variable T given H_d, the value for E is $b \times \gamma$, the probability of background fibres with matching characteristics.

The conditioning of the node T on H implies two probabilities in the event that H_p is true, that is the probability of transfer, $Pr(T|H_p) = t$, and the probability of no transfer, $Pr(\bar{T}|H_p) = 1 - t$. If the suspect is not the person who assaulted the victim (H_d), then the event of transfer is not considered and the assignments $Pr(T|H_d) = 0$ and $Pr(\bar{T}|H_d) = 1$ apply. However, a closer look at the currently discussed scenario shows that, actually, a number n of fibres have been recovered on the suspect's clothing. To obtain a more explicit representation, one could thus rename the two states of the variable T as $T_{i=n}$ and $T_{i=0}$, or T_n and T_0, for short. However, this would lead to a slight complication with respect to the node probabilities, which need to sum to one. Here, in a strict sense, it would not be correct to say that t_0 and $1 - t_n$ sum to one because, besides the number of recovered fibres n, there may be various other quantities of transferred fibres, and these events require their own probability assignment. To avoid this breach in the laws of probability, an additional state, such as T_{other}, could be defined.

Table 5.1 Conditional probability table associated with the node E, where E has two states, E, the presence of fibres on the receptor whose characteristics are reported to match those of the control fibres from the victim, $\bar{E}$, there is a difference between the observed characteristics on materials.

H :		H_p		H_d	
T :	T	$\bar{T}$	T	$\bar{T}$
E : E	b_0	$b \times \gamma$	$b \times \gamma$	$b \times \gamma$
$\bar{E}$	$1 - b_0$	$1 - (b \times \gamma)$	$1 - (b \times \gamma)$	$1 - (b \times \gamma)$

Variable T has two states, T, there was a transfer from the victim to the criminal, $\bar{T}$, there was not a transfer from the victim to the criminal. Variable H has two states, H_p, the suspect is the person who committed the aggression against the victim, and H_d, the suspect is not the person who committed the aggression against the victim.

Such a state would regroup all possibilities for transfer other than zero and the actual number n of recovered fibres. This would allow the sum of all specified probabilities, t_0, t_n and t_{other}, to sum up to 1. The next example illustrates this idea in further detail.

Example 5.2 *(Transfer of glass fragments) Consider a new scenario involving glass fragments (Curran et al. 2000, p. 155). A suspect is apprehended 30 minutes after a breakage. The breakage was believed to have been performed by hitting the window with a club. Subsequently, the offender entered the premises through the broken window. The offender is believed to have left through a door at the rear of the premises. Items taken or ransacked were not in the broken glass zone. Only one perpetrator was suspected. One group of four fragments of glass is found as a result of a full search of the clothing (T-shirt and the surfaces and pockets of jeans) of the apprehended suspect. The laboratory reports the event that the recovered fragments formed a group whose refractive index agrees with the mean of the measurements on the control fragments. Denote this latter event by E.*

Choosing target propositions of the kind H_p, 'The suspect is the man who smashed the window', and H_d, 'The suspect is not the man who smashed the window', the likelihood ratio for such a 'one group-one control' scenario is

$$V = \frac{Pr(E|H_p, I)}{Pr(E|H_d, I)} = \frac{t_4 b_0 + t_0 b_{1,4}\gamma}{b_{1,4}\gamma} = t_0 + \frac{t_4 b_0}{b_{1,4}\gamma}. \tag{5.1}$$

Note that there is no standard notation in forensic literature. Thus, what previously was called b_0 is sometimes expressed by p_0, where p stands for 'presence'. Consequently, $b_{g,n}$ becomes $p_g \times s_n$, where s stands for 'size'. The final likelihood ratio thus becomes:

$$V = t_0 + \frac{t_4 p_0}{p_1 s_4 \gamma}.$$

Keep this last notation in mind and refer to the network shown in Figure 5.1. The conditional probability table for node E takes into account two states for variable H and three states for variable T, covering an exhaustive set of transfer events T_0, T_4 and T_{other}.

Given H_p, the following probabilities are assigned:

$$Pr(E|T_i, H_p) = \begin{cases} p_1 s_4 \gamma, & \text{for } i = 0, \\ p_0, & \text{for } i = 4, \\ 0, & \text{for } i = other, \end{cases}$$

It is worth noting that in the event of transfer of another number of glass fragments ($n \neq 4$), the model here considers it impossible that one group of four fragments would be found, hence $Pr(E|T_{other}, H_p) = 0$. This reflects the assumption that in the event of transfer, there is no background material. Stated otherwise, it is considered impossible to have a joint occurrence of background and transfer that would occur in the finding of one group of four fragments.

Given the alternative proposition, the assignment $Pr(E|T_i, H_d) = p_1 s_4 \gamma$ is retained ($\forall$ i). This assignment reflects the view that the event of transfer is not considered, given H_d.

Node T depends on node H that implies, given H_p, a probability of no transfer, $Pr(T_0|H_p) = t_0$, a probability of transferring four glass fragments, $Pr(T_4|H_p) = t_4$, and a summary probability for all other transfer events, defined as $Pr(T_{other}|H_p) = 1 - t_0 - t_4$. Given H_d, the event of transfer is not developed, so that the assignment $Pr(T_0|H_d) = 1$ is retained.

5.1.3   Modifying the basic network

Recall the fibres scenario discussed in Example 5.1. Let the node H stand for the main propo-sition 'The suspect assaulted the victim', T for the event 'There was a transfer from the victim' and E for the finding 'The trace found on the suspect is similar to the control coming from the victim'. As mentioned at the beginning of this chapter, the transfer of material represents the consequence of a contact. It could thus be of interest to incorporate this consideration explicitly in a network in terms of a distinct proposition C, defined as 'The victim has been in contact with the suspect'.

Logically, if the suspect assaulted the victim, he had a physical contact with the victim. Such a contact may involve a transfer of evidential material, such as fibres from the victim's garment. It is for this reason that node H may be directly linked to nodes C and T. Moreover, a converging connection at the node T may be adopted so that one obtains $H \rightarrow T \leftarrow C$. This will allow one to specify distinct conditional probabilities for T, given different outcomes for the variables H and C, as discussed in further detail below.

Note that the structural relations among the nodes H, C and T as shown in Figure 5.2 provide a modelling example for an indirect relationship. It is an expression of the view that the event of assaulting a victim influences the investigator's assessment of the probability of transfer both directly (through $H \rightarrow T$) and indirectly (through $H \rightarrow C \rightarrow T$). Such relation-ship assumptions need to be carefully reviewed in order to see whether they correctly represent one's perception of the problem. In the case at hand, the variables H, C and T, are mutually dependent and any two of these will remain so upon observation of a third.

Descriptions of case scenarios may provide useful assistance in examining whether particular dependence and independence statements are acceptable. For example, it may be that the probability of transfer given contact is assumed to be different given H_p and H_d. In turn, in a special case in which one assumes $Pr(T|C, H_p) = Pr(T|C, H_d)$ and $Pr(T|\bar{C}, H_p) = Pr(T|\bar{C}, H_d)$, the directed edge between the nodes H and T would become superfluous and may be removed from the network structure, leaving only the serial connection $H \rightarrow C \rightarrow T$.

More generally, the conditional probabilities associated with the node C leave room for considerable interpretation. Consider $Pr(C|H_p)$ first and write it c for short. This is the probability of contact given the appropriate proposition, for example, the proposition that the suspect assaulted the victim. This probability assignment depends largely on the circum-stances of the case, in particular the information on how the assault was committed by the

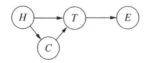

Figure 5.2   Bayesian network for transfer from the victim or scene to the criminal. Each node has two states. Node H, the main proposition, with states H_p, the suspect assaulted the victim, and H_d, some person other than the suspect assaulted the victim. Node C, with states C, the victim has been in contact with the suspect, and $\bar{C}$, the victim has not been in contact with the suspect. Node T, with states T, there was a transfer from the victim, and $\bar{T}$, there was not a transfer from the victim. Node E, with states E, the trace found on the suspect is similar to the control coming from the victim or scene, and $\bar{E}$, the trace found on the suspect is not similar to the control coming from the victim or scene.

offender. Next, consider the probability of contact given the negation of H (i.e. the proposition that the suspect did not assault the victim). This probability can be different from zero, depending on the framework of circumstances. For example, the suspect could have been in contact with the victim for reasons other than the assault. Denote this probability $Pr(C|H_d) = d$.

The probability that a transfer from the victim occurred depends on both propositions, H and C. Information concerning the nature of the crime and the position where the trace was found may lead to different probability assignments. Denote these $Pr(T|C, H_p) = t$ and $Pr(T|C, H_d) = s$. It is obvious that there is no possibility for transfer when there was no contact. Therefore, one should specify $Pr(T|\bar{C}, H_p) = 0$ and $Pr(T|\bar{C}, H_d) = 0$. Notwithstanding, one should consider that for some types of trace material, physical contact may not be a necessary requirement for transfer to occur (e.g. blood splatter).

On a structural account, it is worth noting that the node for the event of transfer, T, screens off the node for the event of contact, C, from the observational node E. That is, given knowledge of the occurrence of the event of transfer, the probability of the outcome of the comparative examinations (node E) will no longer be affected by knowledge about the node C. It seems natural to postulate that, if a transfer occurred, $Pr(E|T) = 1 \times b_0$, where b_0 represents the probability of no group of extraneous material being present (i.e. an absence of trace material). If there was no transfer, and corresponding fibres are observed, these fibres should be there for another reason than transfer linked to a contact (innocent or not) with the victim. It can thus be assumed that $Pr(E|\bar{T}) = b_{1,n}\gamma$, that is the probability of transfer other than from the victim. This term is the product of two probabilities. Suppose that one group of n fibres is on the suspect's clothes before the assault for innocent reasons. The probability of this event is denoted $b_{1,n}$, where the first subscript, 1, denotes the number of groups and the second subscript, n, denotes the size of the single group. The factor γ denotes the probability that the characteristics of this group of fibres are the same as those of the control fibres. For ease of notation, the probability $b_{1,n}\gamma$ of this joint event is denoted γ_b, that is $\gamma_b = b_{1,n}\gamma$.

The likelihoods, calculated from these assumptions and the probabilistic dependence assumptions represented in Figure 5.2, are as follows:

$$Pr(E|H_p) = b_0 ct + [c(1 - t) + (1 - c)]\gamma_b \tag{5.2}$$

and

$$Pr(E|H_d) = b_0 ds + [d(1 - s) + (1 - d)]\gamma_b . \tag{5.3}$$

The numerator, (5.2), is obtained by 'extending the conversation' to the variable T, that is $Pr(E|H_p) = Pr(E|T)Pr(T|H_p) + Pr(E|\bar{T})Pr(\bar{T}|H_p)$, where $Pr(E|T) = b_0$. The conditional probability of T given H_p, that is $Pr(T|H_p)$, needs to take into account uncertainty about C, leading to

$$Pr(T|C, H_p)Pr(C|H_p) + Pr(T|\bar{C}, H_p)Pr(\bar{C}|H_p) = tc.$$

Finally, $Pr(E|\bar{T}) = \gamma_b$ and $Pr(\bar{T}|H_p)$ is obtained in an extension to C, that is by

$$Pr(\bar{T}|C, H_p)Pr(C|H_p) + Pr(\bar{T}|\bar{C}, H_p)Pr(\bar{C}|H_p) = (1 - t)c + (1 - c).$$

The denominator, (5.3), is also obtained by taking into account uncertainty about T, that is by writing $Pr(E|H_d)$ as $Pr(E|T)Pr(T|H_d) + Pr(E|\bar{T})Pr(\bar{T}|H_d)$, where $Pr(E|T) = b_0$. The conditional probability $Pr(T|H_d)$ is obtained by

$$Pr(T|C, H_d)Pr(C|H_d) + Pr(T|\bar{C}, H_d)Pr(\bar{C}|H_d) = sd,$$

and $Pr(\bar{T}|H_d)$ by

$$Pr(\bar{T}|C, H_d)Pr(C|H_d) + Pr(\bar{T}|\bar{C}, H_d)Pr(\bar{C}|H_d) = (1 - s)d + (1 - d).$$

Moreover, $Pr(E|\bar{T}) = \gamma_b$.

From this development, one can find that, if $c = 1$ and $d = 0$ (or $s = 0$), then the likelihood ratio becomes

$$V = \frac{b_0 t + (1 - t)\gamma_b}{\gamma_b} = \frac{b_0 t_n + t_0 \gamma_b}{\gamma_b}$$

as presented in (5.1). Therefore, the numerical results obtained using the networks shown in Figures 5.1 and 5.2 and in a simple serial connection $H \rightarrow T \rightarrow E$ are identical.

However, one of the conditional probabilities, γ_b, associated with node E is the product of two other probabilities, $b_{1,n}$ and γ. This is not completely satisfactory since the specification of probabilities with several components may be time-consuming and, to some extent, also prone to error. It will be shown later that the Bayesian network can be modified in order to avoid the development of probability tables with compound probabilities (Section 5.1.4).

Example 5.3 (*Evaluation of transfer material under varying case circumstances*) *Imagine a scenario in which a group of fibres has been found on the jacket of Mr X, who has been arrested by the police because it is suspected that he physically assaulted Miss Y. The characteristics of these fibres are different from the fibres of the suspect's own jacket but indistinguishable from those of clothing of the victim. Reconsider the Bayesian network shown in Figure 5.2. The two states of the node H are H_p, 'The suspect assaulted the victim', and H_d, 'The suspect did not assault the victim'. Probabilities t and s relate to transfer (node T) under H_p and H_d, respectively, and are conditioned on background information concerning the circumstances of the case under examination. In fact, if the group of fibres has been found on the jacket, it may be reasonable to assume that they result from the assault, so that $t > s$ (i.e. $Pr(T|C, H_p) > Pr(T|C, H_d)$). On the other hand, if fibres are found on the lower part of the trousers, then the assessment might change and the assumption $t = s$ could be made. This reflects the view that the occurrence of a transfer for such a group of fibres would be the same, irrespective of the truth or otherwise of H. Notice also that, given the circumstances of the case, it could be reasonable to assume $t = 1$.*

Another possibility may be to consider $c = 1$ and $t = 1$ (i.e. no uncertainty about the occurrence of contact and transfer given H_p) and to suppose that the defence strategy is to assume that the suspect was at the scene of the crime for reasons unconnected to the crime and that he had contact with the victim (e.g. he helped the victim after the assault while waiting for the police to arrive). The latter assumption leads to $Pr(C|H_d) = d = 1$. The likelihood ratio then becomes

$$V = \frac{b_0}{b_0 s + (1 - s)\gamma_b}.$$

When, in addition, one assumes $s = 0$, that is no transfer from the victim occurred under H_d, the likelihood ratio reduces to

$$V = \frac{b_0}{\gamma_b}.$$

Example 5.3 illustrates that the circumstances of a case are a fundamental element of the analysis. The epistemic probabilities c, d, t and s are crucial considerations that have a bearing on the probative value. In addition, the probability of a group of fibres being present on the receptor beforehand, denoted $b_{1,n}$, represents a further relevant consideration.

Example 5.4 *(Fibres found on a car seat) Consider a scenario involving a car that belongs to a man who is suspected of abducting a woman and attempting to rape her. There is a single group of foreign red woollen fibres that have been collected from the passenger seat of the car. The victim was wearing a red woollen pullover. According to the suspect, no-one other than his wife ever sits on the passenger seat. In addition, the car seats have been vacuumed recently. The suspect denies that the victim has ever been in contact with the car. In such a case, the main issue of concern, H, is H_p, 'The victim sat on the passenger seat of the suspect's car', and H_d, 'The victim has never sat on the passenger seat of the suspect's car'. The node C in Figure 5.2 refers to the proposition 'The victim has been in contact with the seat'. In the case here, it appears reasonable to assume that $Pr(C|H_p) = c = 1$ and $Pr(C|H_d) = d = 0$. The likelihood (5.2) is then $b_0 t + (1 - t)\gamma_b$. Given H_d, the transfer probability $Pr(T|C, H_d) = s = 0$, so the likelihood (5.3) is then γ_b. Thus, the likelihood ratio is*

$$\frac{b_0 t + (1 - t)\gamma_b}{\gamma_b} .$$

Consider $t = 1$ and the fact that, as mentioned above, no-one other than the wife ever sits on the passenger seat and also that the car seats have been vacuumed recently. The probability b_0 should therefore be considered as being close to 1. Then, the likelihood ratio becomes approximately $1/\gamma_b$.

5.1.4   Further considerations about background presence

Consider again the scenario of Example 5.4 involving fibres found on a car seat. The initial Bayesian network with the variables H, C, T and E is as shown in Figure 5.2. In addition, let there be a further node, denoted B, to represent an event known as *background*. This event is thought of as a 'group of fibres with corresponding analytical features' that may be present on the passenger seat by chance alone. Let the node for such an additional variable have the two states B, 'Presence of a group of corresponding background fibres', and $\bar{B}$, 'There is no background of corresponding fibres, or there is no background at all (i.e. there are no fibres at all on the receptor)'. Here, the term *correspondence* refers to a non-differentiation with respect to the control material from the victim.

As a consequence of introducing such a node B, it becomes necessary to revise the definition of the states of the outcome node E. In particular, two different situations are now taken into account in the event $\bar{E}$, that is (i) the presence of fibres on the receptor that do not correspond to the control from the victim and (ii) no fibres at all on the receptor. The probabilities for the table of the node E can now be specified in terms of the following logical values: $Pr(E|\bar{T}, \bar{B}) = 0$, and 1 for all other combinations of conditioning state values of the nodes T and B (i.e. $\{T \cap B\}, \{T \cap \bar{B}\}, \{\bar{T} \cap B\}$). Moreover, for a given conditioning, the probability assigned to the state $\bar{E}$ is one minus that assigned to E given the same conditioning.

Notice that the proposed modification using a distinct node B for modelling background presence does not avoid the complication of having compound probabilities as described in Section 5.1.3. It merely shifts the problem from node E to node B because it is the table of this latter node that contains a compound probability. In fact, the probability of the event of a background group of corresponding fibres is a combination of the probability $b_{1,n}$ (the presence

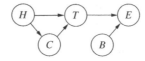

Figure 5.3 Extension of the Bayesian network for transfer from the victim or scene to the criminal defined earlier in Figure 5.2. Each node has two states. Node H, the main proposition, with states H_p, 'The suspect assaulted the victim', and H_d, 'Some person other than the suspect assaulted the victim'. Node C, with states C, 'The victim has been in contact with the suspect', and $\bar{C}$, 'The victim has not been in contact with the suspect'. Node T, with states T, 'There was a transfer from the victim', and $\bar{T}$, 'There was not a transfer from the victim'. Node E, with states E, 'The trace found on the suspect is reported to be similar to the control coming from the victim', and $\bar{E}$, 'The trace found on the suspect is not similar to the control coming from the victim, or there is no trace material at all on the receptor'. Node B is the newly added variable with states B, defined as 'There is a matching group of background fibres', and $\bar{B}$, 'There is no background of matching fibres or there is no background at all'. Note that the variable E could also refer to, for example, fibres found on a car seat.

of a single group of fibres of size n), and γ, which is an expression of the occurrence of the fibres' characteristics in a relevant population.

Figure 5.4 represents a comparison between the Bayesian network described in this section (Figure 5.3) and the one described in Section 5.1.3 (Figure 5.2). For both networks shown in Figure 5.4, probabilities have been assessed so that $c = s = t = 1$ and $d = 0$. The only differences are as follows. In the network shown in Figure 5.4(i), the conditional probability $Pr(E|\bar{T})$ is given by γ_b, for which the value 0.05 has been assumed. Recall that γ_b is an abbreviation for the product of the probabilities $b_{1,n}$ and γ. In the network shown in Figure 5.4(ii), the probability table of the node E takes the logical values of 0 and 1 as noted earlier in this section. In turn, the unconditional probabilities specified for the states B and $\bar{B}$ of the node B are, respectively, $b_{1,n}\gamma$ (i.e. γ_b) and $1 - b_{1,n}\gamma$ (i.e. $1 - \gamma_b$). As may be read from the expanded nodes, both evaluations provide the same result, as far as the support for the proposition H is concerned. This example thus illustrates that structurally different Bayesian networks may lead to the same results. However, this depends largely on the way in which the various probabilities are interpreted and defined.

The fact that the currently discussed network, Figure 5.4(ii), still contains compound probabilities, notably those associated with the states of the node B, may be an inconvenience, but it is one that may be overcome by an extension illustrated in Figure 5.5. In this network, the node B is still present, but for convenience, it is named P now, short for 'presence' (of corresponding fibres). This node has a slightly different definition, due to the fact that it is no longer a root node but conditioned on two other nodes, N and Ty. The former node, N, accounts for the number of compatible fibre groups recovered on the receptor. In order to keep the argument simple, only two states are assumed here, namely 0 groups (n_0) and 1 group (n_1). The latter node, Ty, represents the analytical features of a background fibre group (i.e. the fibre type). Here, the type is generically assumed to be either 'corresponding' (Ty) or 'non-corresponding' ($\bar{T}y$). It is now straightforward to see that the probabilities associated with the states n_1 and n_0 are, respectively, $b_{1,n}$ and $1 - b_{1,n}$. For the states Ty and $\bar{T}y$ of the node Ty, the probabilities are, respectively, γ and $1 - \gamma$. In turn, the conditional probability table of the node P can now be completed with logical values, that is 1 for $Pr(P|n_1, Ty)$ and 0 for all other combinations

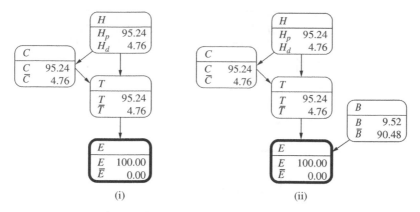

Figure 5.4 (i) and (ii) Comparison of the Bayesian networks shown in Figures 5.2 and 5.3 with the node E instantiated. Each node has two states. Node H, the main proposition, with states H_p, 'The suspect assaulted the victim', and H_d, 'Some person other than the suspect assaulted the victim'. Node C, with states C, 'The victim has been in contact with the suspect', and $\bar{C}$, 'The victim has not been in contact with the suspect'. Node T, with states T, 'There was a transfer from the victim', and $\bar{T}$, 'There was not a transfer from the victim'. Node E, with states E, 'The trace found on the suspect is reported to be similar to the control coming from the victim', and $\bar{E}$, 'The trace found on the suspect is not similar to the control coming from the victim'. Node B has states B, defined as 'There is a matching group of background fibres', and $\bar{B}$, 'There is no background of matching fibres or there is no background at all'.

of conditioning state values of the nodes N and Ty (i.e. $\{n_1 \cap \bar{T}y\}$, $\{n_0 \cap Ty\}$ and $\{n_0 \cap \bar{T}y\}$). Again, as for the node E, for a given conditioning, the probability assigned to the state $\bar{P}$ is one minus that assigned to P given the same conditioning.

Figure 5.5 provides a numerical example. Assume, as before, $c = s = t = 1$ and $d = 0$. Further, let the probabilities $b_{1,n}$ and γ take the values 0.5 and 0.01, respectively. This particular choice of the values for the latter factors is made in order to reproduce the initial values specified for B and $\bar{B}$ in the network shown in Figure 5.4(ii). The evaluation of the scientific findings using the network displayed in Figure 5.5 (i.e. by instantiating the node E) will thus yield the same result as each of the networks evaluated in Figure 5.4.

5.1.5 Background from different sources

In Section 5.1.3, an extended structure has been proposed for evaluating scenarios involving the one-way transfer of material. Here, the converging connection at the node T, that is $H \rightarrow T \leftarrow C$, is considered in more detail. Some of the various conditional probabilities for C and T given different outcomes for the variable H are also be discussed.

Consider a case of suspected trafficking of illicit drugs. A suspect is apprehended and his clothing is examined for the presence of traces of drugs, such as cocaine, heroine or the like. The presence (typically in trace quantities) of such substances may be used to infer something about recent activities of the individual of interest in relation to drugs. This could involve a direct contact with a primary source (e.g. during preparation and packaging) or exposure to a contaminated environment. This belief is based in part on the generally accepted assumption that trace quantities of drugs are less commonly found on persons unrelated to drug offences.

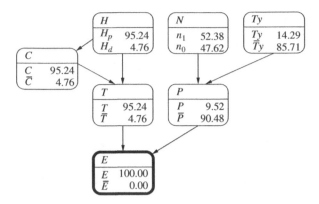

Figure 5.5 Extension of the Bayesian network shown in Figure 5.3 with the node E instantiated. Each node has two states. Node H, the main proposition, with states H_p, 'The suspect assaulted the victim' and H_d, 'Some person other than the suspect assaulted the victim'. Node C, with states C, 'The victim has been in contact with the suspect', and $\bar{C}$, 'The victim has not been in contact with the suspect'. Node T, with states T, 'There was a transfer from the victim', and $\bar{T}$, 'There was not a transfer from the victim'. Node E, with states E, 'The trace found on the suspect is reported to be similar to the control coming from the victim', and $\bar{E}$, 'The trace found on the suspect is not similar to the control coming from the victim'. Node P has states P, 'There is a presence of corresponding fibres', and $\bar{P}$, 'There is not a presence of matching fibres'. Node N has states n_0, 'No compatible groups recovered on the receptor', and n_1, 'One compatible group recovered on the receptor'. Node Ty has states Ty, 'The characteristics of the background fibres correspond to those of the control', and $\bar{T}y$, 'The characteristics of the background fibres do not correspond to those of the control.'

A suspect's clothing can be vacuumed and material thus collected analysed by a detection apparatus. A portable ion mobility spectrometer (IMS) may be appropriate for such purposes. Imagine, for example, that a positive result has been obtained for some target substance, such as cocaine. In practice, such a result will need to be confirmed by a second, independent method, such as gas chromatography/mass spectrometry (GC/MS). Notice, however, that the probabilistic analysis pursued hereafter will solely be concerned with the results obtained by the IMS.

Consider an assessment of the analytical results given the following activity level propositions: 'The suspect is involved in illegal drug trafficking, reselling and activities related thereto' (H_p), and 'The suspect is not involved in such activities (drug offences)' (H_d). The definitions of the nodes C and T will be taken to be slightly different from those in Section 5.1.3. Consider the variable C first. Here, C represents the proposition that the suspect has been in direct contact with illicit drugs. This may happen, for example, when packing or otherwise manipulating illicit drugs. These substances may be in different conditions, such as powder or compressed into blocks. The transfer that occurs during a direct contact with such items will be referred to as a *primary transfer*, whereas the item from which the transfer originates will be denoted as a *primary source*. Notice, however, that transfer may also be secondary or tertiary. These types of transfer may be illustrated as follows. Imagine a person whose clothing has been contaminated with cocaine through a primary transfer. That person takes a seat in a car that may lead to a transfer of minute quantities of cocaine from the person's clothing to the car seat, for example. This sort of transfer would be termed *secondary transfer*. Sometime later, another person sits on the same car seat. This is when a 'tertiary transfer' may occur, that is transfer from the car seat to the clothing of the second person. In view of these phenomena, a rather general definition is retained for the node T, that is the event 'Transfer of a quantity Q of illicit drugs'.

For the scenario introduced above, the conditional probabilities of the proposed network need to be interpreted in the light of case circumstances. Associated with the node C, for example, is the probability of contact given the proposition that the suspect engages in drug trafficking, $Pr(C|H_p) = c$. This probability can reasonably be assigned as close to 1. In turn, the probability of a direct contact given the proposition that the suspect has no relation at all with trafficking of illegal drugs, $Pr(C|H_d) = d$, can taken to be close to 0.

The assignment of transfer probabilities is made in view of the conditioning on both propositions, H and C. The assessments here depend crucially on the nature of the crime and the strategy chosen by the defence. Start by considering $Pr(T|C, H_p) = t$, the probability of transfer given the suspect is involved in drug trafficking and given that he has been in contact with a primary source. Generally, understanding of the phenomena involved in such situations suggests that t should be set close to or equal to 1. Related to this assignment is $Pr(T|C, H_d) = s$, the probability of transfer given the suspect has been in contact with a primary source but is not involved in drug trafficking. This probability can also be set equal to 1. In fact, the probability of transfer given a contact with a primary source can be considered as being independent of whether or not the individual is involved in illegal activities.

In Section 5.1.3, it has been assumed that there is no possibility for fibre transfer when there was no contact between the victim and the offender. This was a reasonable assumption because the trace type was textile fibres. Consequently, both $Pr(T|\bar{C}, H_p)$ and $Pr(T|\bar{C}, H_d)$ were set to 0. For the type of trace material in the current scenario, this assumption may not necessarily hold because there may be a possibility of secondary or tertiary transfer. Let $Pr(T|\bar{C}, H_p)$ be denoted by t', the probability of transfer given no direct contact with a primary source despite involvement in drug trafficking. It is through consideration of t' in terms of a non-zero probability that one may allow for the event of a secondary or tertiary transfer. Such transfer may occur, for example, through exposure to a contaminated environment (e.g. a room or a vehicle). Similarly, denote $Pr(T|\bar{C}, H_d)$ more shortly by s', the probability of a secondary or tertiary transfer, given that the suspect is not involved in drug affairs and given that there was no contact with a primary source. Probability s' may also be assigned a value different from zero.

Note that the assignments t' and s' relate to two different settings that can be thought of in terms of populations. The probability t' can be informed by studying contaminations on persons known to be involved in drugs affairs. The probability s' may be assessed in reference to a population of innocent individuals. One also needs to be aware of alternative defence strategies. Consider, for example, the defence proposition 'The suspect is not involved in drug affairs but is on good terms with a third individual who is known to be involved in drug trafficking'. Such a relationship could also be the result of the suspect's professional activities (e.g. if he is a social worker and in regular contact with people having drug problems). In such a situation, $Pr(T|\bar{C}, H_d)$ can be assessed with respect to a population composed of individuals who have such (non-criminal) relations with other individuals linked to drug affairs. This may lead to an increased value assigned to s'.

In scenarios in which $Pr(T|\bar{C}, H_p)$ and $Pr(T|\bar{C}, H_d)$ are not set equal to 0, a node B, defined in Section 5.1.4 as a node representing a possible matching background, is not needed. Similar situations involving second or tertiary transfer are regularly encountered in connection with other kinds of trace material, such as fibres or gunshot residues.

It may appear difficult to assign values for all of the various factors mentioned above, rendering the evaluation of the probative value of the recovered trace material a complicated task. Indeed, scenarios involving trace quantities of illicit drugs often relate to a highly particular set of circumstantial information that may not have previously been studied through surveys or simulations. Subjective expert judgements will therefore remain an integral part of evaluations. Methods exist that allow the treatment of such assessments in a probabilistically coherent way while the explicit specification of numerical probabilities is not a necessary requirement. Further details on this are given in Chapter 13. Section 13.1.2 provides a detailed discussion of an example involving trace quantities of drugs.

5.1.6   An alternative description of the findings

In Section 5.1.2, a fibre scenario has been developed where the fibres recovered on a person of interest either may be associated with the offence or have been present for reasons unconnected to the offence. In what follows, potential transfer in the opposite direction is considered, that is a situation in which the recovered fibres may be those left by the offender. The construction of a Bayesian network for this kind of case will show that (3.4) is a special case of a more general formula. The particular point of such a scenario is that transfer is also considered given the alternative proposition (i.e. not only presence by chance alone as is the case for a setting in which material is recovered on a person of interest).

Imagine the case of a stolen car used in a robbery on the day of its theft. One hour later, the car is abandoned. During the night, the stolen vehicle is found by the police. On the polyester seats (lower and upper back), a group of n extraneous textile fibres is collected. The day following the robbery, a suspect is apprehended. His red woollen pullover is seized and submitted to the laboratory. So far in this chapter, the scientific findings have been captured in terms of a single outcome node E. This node may be developed in further detail, similarly to what has been done previously for the node representing background (Section 5.1.4). In particular, instead of considering the presence (or not) of a correspondence in analytical features between the recovered material and the control material from a potential source, the scientist can consider characteristics of the recovered material and the control material, separately.

Using the notation introduced in Section 3.2, the material of interest M consists of two distinct parts. On the one hand, there is the material recovered on the car seat, denoted M_c, where c indicates the assumed relation to the criminal incident. On the other hand, there is material extracted from the suspect's pullover, denoted M_s, where s indicates the suspect as the source of this known material that is being used for comparative analyses. Node E consists of E_c and E_s, denoting the measurements taken on M_c and M_s respectively. Thus, as mentioned above, considering observations on known and questioned material separately implies that the term *correspondence* is no longer required to describe the findings. The notation $Ev = (M, E)$ is used to summarize the overall available information.

For convenience, let y denotes the measurements E_c taken on the recovered material and x the measurements E_s taken on the control material. Thus, in the context here, the forensic findings are a consideration of the extrinsic (physical attributes such as quantity and position) and intrinsic features (chemical or physical descriptors such as analytical results), characterizing the group of n recovered red woollen fibres, summarized by y, and the extrinsic (physical attributes such as the sheddability) and intrinsic features of known red woollen fibres generated by the suspect's pullover, summarized by x. The likelihood ratio is then $V = Pr(y, x|H_p, I)/Pr(y, x|H_d, I)$, where H_p is the proposition 'The suspect sat on the driver's seat of the stolen car' and H_d the proposition 'The suspect has never sat on the driver's seat of the stolen car'. Note that H_d implies that another person sat on the driver's seat of the stolen car. This point is important in the assessment of the transfer probabilities as will be seen later. Using the third law of probability, the likelihood ratio can be written as follows:

$$V = \frac{Pr(y, x|H_p, I)}{Pr(y, x|H_d, I)} = \frac{Pr(y|x, H_p, I)}{Pr(y|x, H_d, I)} \times \frac{Pr(x|H_p, I)}{Pr(x|H_d, I)}. \tag{5.4}$$

It is reasonable to assume that the probability of the characteristics of the suspect's pullover, x, does not depend on whether or not the suspect sat on the driver's seat of the stolen car. So, the second ratio on the right-hand side of (5.4) equals 1 and the likelihood ratio is reduced to

$$V = \frac{Pr(y|x, H_p, I)}{Pr(y|x, H_d, I)} .$$

A further common assumption is that, given H_d, x is not relevant when assessing the probability of y. The denominator of V may then be written $Pr(y|H_d, I)$.

Next, a variable T with states T_n and T_0 is defined. The variable T_n denotes the event of a transfer of a group of n fibres to the driver's seat in the course of the crime, including persistence and successful recovery of these fibres. This implies that the group consisting of n fibres has not been there before. The variable T_0 denotes the event of no group of n fibres being transferred in the course of the crime. Note that, generally, the variable T accounts not only for the phenomenon of transfer but also for aspects such as persistence and recovery. Considering these association propositions, and omitting the notation I for background information, the likelihood ratio V extends to

$$\frac{Pr(y|x, H_p, T_n)Pr(T_n|x, H_p) + Pr(y|x, H_p, T_0)Pr(T_0|x, H_p)}{Pr(y|H_d, T_n)Pr(T_n|H_d) + Pr(y|H_d, T_0)Pr(T_0|H_d)} . \tag{5.5}$$

The eight conditional probabilities in this formula should be assessed in the light of the case circumstances.

The term $Pr(y|x, H_p, T_n)$ represents the probability of observing a group of n red woollen fibres on the car seat given that the suspect wore a red woollen pullover, he sat on the driver's seat of the stolen car and a group of fibres was transferred during the activity, had persisted and was recovered successfully. This probability can be assigned as $1 \times b_0$. This is an expression of the belief that y is the joint occurrence of a crime-related transfer (factor 1) and the probability of 0 groups being on the driver's seat beforehand (factor b_0).

The term $Pr(T_n|x, H_p)$ represents the probability that a group of n red woollen fibres was transferred, had persisted and was recovered successfully from the driver's seat, given that the suspect sat on the driver's seat of the stolen car. This probability is written more shortly as t_n. It depends on physical characteristics (e.g. sheddability) of the suspect's pullover. It is assumed that the characteristics are those of the control material because the scientist assesses the probability under H_p. Note that the phenomenon of transfer is not only characterized by material which is transferred but also by the phenomena of persistence and recovery.

The term $Pr(y|x, H_p, T_0)$ is the probability that a group of n red woollen fibres is observed on the driver's seat, given that the suspect wore a red woollen pullover, he sat on the driver's seat of the stolen car and there was no transfer of a group of fibres during this activity. If there has been no transfer, this means that the group of fibres was present on the seat before the activity. Let this be represented by a term $b_{1,n} \times \gamma$, where $b_{1,n}$ represents the probability of a single group of n (i.e. a comparable number) fibres being present on the driver's seat by chance alone, and γ is an expression of the rarity of the compared characteristics on y in extraneous groups of similar sizes of fibres found on car seats.

The last term in the numerator is $Pr(T_0|x, H_p)$. It represents the probability that no group of fibres was transferred from the suspect's pullover to the driver's seat, had persisted and was recovered from there. This probability, t_0, is assigned, given that the suspect sat on the driver's seat, H_p. In summary, the numerator of the likelihood ratio is therefore $b_0 t_n + b_{1,n} \gamma t_0$.

Note that in Example 5.4, the denominator did not require a detailed development as it was assumed that the group of fibres, had it not been transferred by the victim, was present on the receptor for reasons unconnected to the crime. A different approach will be followed here. Notably, the group of recovered fibres will be considered to have been left potentially by the (true) offender, which, under H_d, is someone other than the suspect. The terms comprising the denominator of (5.5) are outlined below.

The first term in the denominator, $Pr(y|H_d, T_n)$, represents the probability of observing a group of n red woollen fibres, given that the suspect never sat on the driver's seat of the stolen car, that there was a transfer of a group of fibres during the activity and that this group had persisted and was successfully recovered. If the suspect never sat on the driver's seat, but transfer did occur, this means the driver's seat did not have this group of fibres before the commission of the crime. The event of the shared characteristics thus is one of chance. The probability of interest is therefore $b_0 \times \gamma$.

The term $Pr(T_n|H_d)$ represents the probability that a group of n fibres was transferred, had persisted and was successfully recovered from the driver's seat, given that the suspect never sat on the driver's seat of the stolen car. This probability, call it t'_n, has to be assigned remembering that transfer occurred from the (true) offender's garment, which is an item different from the suspect's garment. This probability thus depends on the physical characteristics of an unknown garment.

The term $Pr(y|H_d, T_0)$ is the probability that a group of n red woollen fibres is observed on the driver's seat, given that the suspect never sat on the driver's seat and that no transfer of fibres occurred during the activity. If there was no transfer, fibres were present on the seat before the commission of the crime. The probability of the joint occurrence of a group of foreign fibres on the driver's seat and a set of characteristics described by y then is $b_{1,n} \times \gamma$.

The last term in the denominator is $Pr(T_0|H_d)$ and represents the probability that no group of fibres was transferred from the offender's garments to the driver's seat. This probability, denoted t'_0, assumes that the suspect never sat on the driver's seat and, thus, another individual sat in the stolen car.

In summary, the denominator of the likelihood ratio is thus $b_0 \gamma t'_n + b_{1,n} \gamma t'_0$ and the likelihood ratio in (5.5) can be written as

$$V = \frac{b_0 t_n + b_{1,n} \gamma t_0}{b_0 \gamma t'_n + b_{1,n} \gamma t'_0}. \tag{5.6}$$

This expression illustrates that, theoretically, $Pr(T_n|x, H_p) \neq Pr(T_n|H_d)$ and $Pr(T_0|x, H_p) \neq Pr(T_0|H_d)$. Practically, as the probabilities are assigned through controlled experiments using the garments involved under propositions H_p and H_d, it is reasonable that they are different.

5.1.7   Bayesian network for an alternative description of findings

The probabilistic analysis proposed in the previous section involves, in the main, four variables, covering various propositions and types of findings, notably H, T, X and Y. These variables can be used to define nodes of a Bayesian network. The main proposition, a node H, will have two states defined as H_p, 'The suspect sat on the driver's seat of the stolen car', and H_d, 'The suspect never sat on the driver's seat of the stolen car (i.e. another person sat on that seat)'. A node T can be defined for the event of transfer, persistence and recovery of a group of fibres. Such a variable has three states, T_0, T_n and T_{other}, denoting the transfer of a group of 0 fibres, a group of n fibres and a group of fibres of size other than 0 and n, respectively. Note that the component T_{other} does not appear in the likelihood ratio formula (5.6). It is needed, however, as a state of a network node T because the sum of the probabilities, for a given conditioning, must add to 1. Recall that if such a state were not used, this would unrealistically imply that $Pr(T_0) = 1 - Pr(T_n)$. The variable representing the intrinsic features of the known suspect's pullover is represented by a node X with two states, x and $\bar{x}$. For example, x may stand for red woollen fibres. For the sake of simplicity, let $\bar{x}$ denotes the union of all sets of possible features other than those described as x. As far as Y is concerned, it is tempting to restrict the definition to the intrinsic and extrinsic descriptors only. However, further considerations will be necessary in order to obtain a sensible definition of the states of Y. The way in which the states of Y are defined is partially determined by the further variables on which Y depends and how these further variables are defined. For the time being, let Y be simply denoted with the term *findings*, whereas y designates a particular outcome, described in terms of intrinsic and extrinsic features.

Based on the eight conditional probabilities previously developed (Section 5.1.6), several structural dependencies may be defined. First, consider the relevance of variables H, T and X for Y, in terms of the conditioning. This means that the four variables can be related structurally as shown in Figure 5.6. In turn, the variable T, denoting 'transfer', depends on both X and H. This dependency leads to $H \to T \leftarrow X$, a structural aspect that is also shown in Figure 5.6. More generally, this structure highlights the importance of the intrinsic features of X. The physical attributes, such as the sheddability of the control material, determine its capacity to transfer fibres. In turn, note that the network's node for the proposition of transfer takes into account the three phenomena transfer, persistence and recovery. A more detailed development of the 'transfer' node T is given in Section 5.1.8.

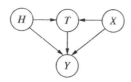

Figure 5.6   Bayesian network for a case involving transfer material. Node H has two states H_p and H_d denoting two competitive activity level propositions. Variable T has three states, T_0, T_n and T_{other}, denoting the transfer of a group of 0, a group of n fibres and a group of fibres other than 0 and n, respectively. Variable Y denotes the findings with state y designating a particular outcome, described in terms of intrinsic and extrinsic features, for a single group of recovered fibres. The second state, $\bar{y}$, represents all combinations of features other than those described by y. Variable X is similarly defined but for the control material (i.e. the suspect's garment).

Table 5.2 Conditional probabilities assigned to the node T.

H:	H_p		H_d	
X:	x	$\bar{x}$	x	$\bar{x}$
T : T_0	t_0	$1/3$	t_0'	t_0'
T_n	t_n	$1/3$	t_n'	t_n'
T_{other}	$1 - t_0 - t_n$	$1/3$	$1 - t_0' - t_n'$	$1 - t_0' - t_n'$

Variable T has three states, T_0, T_n and T_{other}, denoting the transfer of a group of 0, a group of n fibres and a group of fibres other than 0 and n, respectively. Node H has two states, H_p, the suspect sat on the driver's seat of the stolen car, H_d, the suspect has never sat on the driver's seat of the stolen car, but another person sat on that seat. Probability t_i is the probability that a group of $i = \{0, n\}$ fibres has been transferred, persisted and recovered from the suspect's pullover. Probability t_i' is the probability that a group of $i = \{0, n\}$ fibres has been transferred, persisted and recovered from an alternative source other than the suspect's pullover.

Because the nodes H and X will be instantiated when using the Bayesian network, the initial probabilities assigned in the table associated with these nodes become irrelevant for the likelihood ratio of interest (5.5). The only requirement is that the probability assignments, for a given conditioning, sum up to 1. For purely technical reasons, one can thus specify $Pr(H_p) = Pr(H_d) = 0.5$ and $Pr(x) = Pr(\bar{x}) = 0.5$.

The probability table for node T is more elaborate. This node has entering arcs from X and H, which implies the structure given in Table 5.2. Note that inspection of the likelihood ratio formula discussed in Section 5.1.6 can provide relevant indications on how to complete the probability table of the node T. First, in the event of H_p, the suspect's garment was in contact with the car seat. This is why only probabilities for transfer assuming x, the observed features of the suspect's garment, are actually considered. Probabilities for transfer given $\bar{x}$ (and H_p) do not enter the calculations in any way. Thus, they do not require further discussion and can be specified by default as $1/3$. Note that this is a purely technical assignment and any other set of values summing to one may be used. When H_d is true, then knowledge of x about the features of the suspect's garment does not affect the probability of transfer, so values t_0' and t_n' are assigned. These refer to the probability of transfer from the true offender's garment, which is different from that of the suspect, described as x.

Following Figure 5.6, the node Y has three parent nodes H, T and X. This implies a structure for the node probability table as given in Table 5.3 for the conditioning H_p. Given H_d, the variable X is not taken into account and the probabilities for Y can be summarized more shortly as follows:

$$Pr(Y = y | T_i, H_d) = \begin{cases} \gamma_b, & \text{for } i = 0, \\ b_0 \gamma, & \text{for } i = n, \\ 0, & \text{for } i = other, \end{cases} \tag{5.7}$$

and $Pr(Y = \bar{y} | T_i, H_d) = 1 - Pr(Y = y | T_i, H_d)$. The factors γ_b and γ denote the relevant population proportions of the compared characteristics from y among, respectively, alternative sources (garments from potential offenders) and similar sized extraneous groups of fibres found as part of the background on car seats. Table 5.3 also contains some default values of 0.5, which, as explained above for Table 5.2, are purely technical assignments because they do not enter the expression of the likelihood ratios considered in this application. Note further

Table 5.3 Conditional probabilities assigned to the node Y under the main proposition H_p.

H:			H_p			
T:	T_0		T_n		T_{other}	
X:	x	$\bar{x}$	x	$\bar{x}$	x	$\bar{x}$
Y : y	γ_b	0.5	b_0	0.5	0	0
$\bar{y}$	$1 - \gamma_b$	0.5	$1 - b_0$	0.5	1	1

Variable Y denotes the findings with state y designating a particular outcome, described in terms of intrinsic and extrinsic features, for a single group of recovered fibres. The second state, $\bar{y}$, represents all combinations of features other than those described by y. Variable X is similarly defined but for the control material (i.e. the suspect's garment). Variable T has three states, T_0, T_n and T_{other}, denoting the transfer of a group of 0, a group of n fibres and a group of fibres other than 0 and n, respectively. The term γ_b is the probability that the characteristics of a group of fibres correspond by chance alone to the characteristics of the control fibres.

that the model constructed here does not consider situations in which y could be observed in the event of a transfer of a number of fibres other than n, which is also reflected in the reference likelihood ratio formula (5.5). The Bayesian network shown in Figure 5.6 and specified with probability tables as noted here above leads to results that agree with (5.1) and (5.6). The only difference between the latter two equations is that in the former no transfer is admitted under hypotheses H_d, which means that $Pr(T_n|x, H_d) = 0$.

5.1.8   Increasing the level of detail of selected propositions

The development of more elaborate Bayesian networks is of interest for two main reasons. A first reason is to enable the solution of an intrinsic complication related to entries of conditional probability tables, notably the so-called composite probabilities (e.g. $b_1\gamma$ or γ_b). Second, one may wish to expand a particular node in order to take into account additional information related to the node of interest.

To deal with the particular issue of composite probabilities at the node Y, for example, an additional variable may be introduced in order to account for the fact that the scientist may observe y even if no transfer occurred, that is T_0 was true. One thus needs to inquire about circumstances under which this may be the case. One obvious possibility is that a group of fibres described as y was present on the receptor before the event of interest (proposition H_p) occurred. Call this event B, short for 'background'. It can be introduced in a network in terms of a root node B with a direct influence on Y. The states of this variable are $\bar{B}$ and B with (unconditional) probabilities b_0 and $b_{1,n}$, representing, respectively, the probabilities of no foreign group being present and of the joint occurrence of a single group of n foreign fibres on the driver's seat and a set of characteristics described by y.

A consequence of this change in the network structure is that, given the definitions of the nodes X, T and B, the states of the node Y will need to be refined. That is, so far, the node Y is characterized by the state y, representing the observed features (e.g. red woollen fibres with specified extrinsic characteristics), and state $\bar{y}$, representing all combinations of features other than those described by y. However, two further situations now need to be considered, due to the extended set of parental variables of Y. The first situation is concerned with the joint occurrence of T_n and B. Under these circumstances, the scientist is faced with two groups of fibres. The state of Y corresponding to this situation could thus be denoted 'Two groups'. The second situation is one in which both T_0 and $\bar{B}$ are true, that is no fibres at all are present.

In summary, this leads to four different states for the node Y, notably y, $\bar{y}$, *Two groups* and *No group of fibres*. Notice that some of these states may be specified in more detail, as required. For example, the state *Two groups* may be divided into a state representing two groups of fibres being present, with each group having the features y, a state representing two groups of fibres, with each group having features different from y, and so on. The consequence of this would be that the size of the conditional probability table increases. However, the Bayesian network described at this point is designed to represent a given scenario only, that is one involving one group of n foreign fibres. No further states will thus be adopted here for the node Y.

A further illustration of a more elaborate representation of a selected proposition can be considered in Bayesian networks that contain a node T as developed so far in this chapter. Recall that such a node T stands for the events of transfer, persistence and recovery of material on some kind of receptor (i.e. a person or a scene). In situations where more detailed information is available on these three distinct phenomena, the node T may be extended to a more elaborated Bayesian network. In Example 5.5, some aspects of fibre scenarios that have been previously discussed are used to approach a new scenario involving glass fragments, another frequently encountered kind of trace material.

Example 5.5 *(A more detailed approach to the event of transfer in a case involving broken glass) Imagine a case in which a burglar smashed the window of a house. A suspect is arrested and a quantity, say Q_r, of glass fragments is recovered from the surface of his clothing. These fragments correspond, in some sense, with the type of glass of the house window. Instead of using a single transfer node T, as depicted, for example, in Figure 5.2, it is possible to extend the network to a level of greater detail.*

Start by considering that the quantity of glass fragments recovered from the suspect's pullover (Q_r) depends on both the unknown quantity of glass fragments that have persisted (Q_p) and the performance of the search technique. The latter factor may be assessed by the proportion of glass fragments lifted from the pullover, denoted by a variable P_l. For such a variable, a variety of different states may be adopted as required. A very general approach could consist in adopting, for example, a binary node with states 'good' and 'poor', thus expressing the performance of the search procedure as evaluated in previous experiments under controlled conditions.

The quantity of glass fragments that have persisted on the suspect's clothing, Q_p, depends on the quantity Q_t of glass fragments transferred and on the proportion P_s of glass fragments that were shed between the time of transfer and the time of the examination of the pullover. The quantity of transferred glass fragments (Q_t) depends on the appropriate ultimate probandum (H). The states of the variables can be 'none', 'few', 'many', or 'none', 'small' and 'large'.

Further distinctions may also be made at the node Q_t, which represents the quantity of transferred glass fragments. As noted above, Q_t is assumed to depend on the variable H that covers propositions for the action committed by the true offender. Assignments for Q_t will naturally vary for different assumptions. These may cover, for example, the assumptions that the suspect broke the window, he only stood nearby when someone else broke it or the suspect had nothing to do with the event under investigation.

A variety of other variables can be proposed, but it should be emphasized that results from simulations under controlled conditions should be considered in order to support the argument that a given variable affects or does not affect another variable. For the assessment

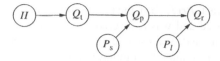

Figure 5.7   A Bayesian network fragment for evaluating sources of uncertainty relating to a scenario involving a transfer of glass fragments. Node Q_r represents the quantity of glass fragments recovered from the suspect's pullover. Node Q_p represents the quantity of glass fragments that have persisted. Node Q_t represents the quantity of glass fragments transferred. Node P_l represents the proportion of glass fragments lifted from the pullover, that is the performance of the search procedure. It can be a binary node with states 'good' and 'poor'. Node P_s represents the proportion of glass fragments that were shed between the time of transfer and the time of the examination of the pullover. Node H represents the ultimate probandum, which is the action committed by the offender. It has two states, H_p, the suspect broke the window, and H_d, some other person broke the window.

of uncertainties in relation to the phenomenon of transfer of glass fragments, factors such as the type of window under investigation, the size of the window and the distance between the window and the receptor (note that this information is closely related to the way the window has been broken) may be considered (Curran et al. 2000).

Figure 5.7, inspired by Halliwell et al. (2003), illustrates a Bayesian network for some of the variables mentioned in this example. Notice that, on the one hand, developments of the kind considered here tend to have fewer analogies with the traditional likelihood ratio formulae proposed in forensic literature. The reason for this is that more variables with varying definitions and relationships are considered. In part, this is a direct consequence of the graphical environment, which allows one to consider more variables (i.e. sources of uncertainty) with less effort. On the other hand, the graphical structures become more tentative, as there may be fewer reference points, for example in the form of likelihood ratio formulae, against which a model may be compared.

5.1.9   Evaluation of the proposed model

In situations where the presence of the material given the alternative proposition is considered as one of chance alone because the alleged activity is assumed not to have happened (see also the scenario developed in Section 5.1.2, concerning the potential transfer of fibres from the victim to the offender), the numerator of the likelihood ratio is as in (5.6), but the denominator changes. In fact, if it is assumed that the alleged activity did not occur, then there is no need to develop $Pr(y|H_d)$ using the association propositions T_n and T_0. For such a scenario, the likelihood ratio is

$$V = \frac{b_0 t_n + b_{1,n}\gamma t_0}{b_{1,n}\gamma}. \tag{5.8}$$

The aim is to try to incorporate into a single Bayesian network the scenario of the stolen car (5.6) and to be flexible enough to consider other scenarios, such as (5.8).

A Bayesian network as shown in Figure 5.6 allows the scientist to cope with both situations. A change from the situation involving an alternative donor of the recovered material to the situation involving chance alone is operated through changes in the probability table associated to node T. Notably, when the alternative proposition H_d is that the suspect never sat on the car seat, thus, the group of n foreign fibres is present by chance alone, then the value of t'_n will be set to zero. In fact, $Pr(T_n|H_d)$, the probability that a group of n red woollen fibres was transferred, had persisted and was recovered successfully from the driver's seat, given that the suspect never sat on the driver's seat of the stolen car, represents an impossible event because the alleged activity is assumed not to have happened.

In order to use the described network to obtain the components of the likelihood ratio, different instantiations need to be made. For evaluating the numerator, the nodes H and X need to be set to, respectively, the states H_p and x. For evaluation of the denominator, it is sufficient to instantiate the node H to H_d. Notice here that any instantiations made at the node X would not affect the probabilities associated with Y. This is a consequence of the probability assignment defined in (5.7). Note further that the Bayesian network shown in Figure 5.6 can be extended to consider more complex situations, such as cross-transfer (Section 5.2), or cases involving more than one relevant group of material (Gittelson et al. 2013a).

5.2   Cross- or two-way transfer of trace material

The Bayesian network described previously in Section 5.1.2 (Figure 5.1) can be used to approach another category of forensically relevant situations, that is cases in which a direct contact between two persons or objects, or an object and a person, may have occurred. In such cases, a cross- or two-way transfer of trace material may take place. Consider the following example, simplified here in order to ease the discussion of the general idea. A vehicle is used in a robbery. An hour later it is abandoned, as reported by an eyewitness, and then seized by the police a few hours later still. On the polyester seats, which were recently cleaned with a car vacuum cleaner, extraneous textile fibres are collected. The car owner lives alone and has never lent his vehicle to anyone. It is then assumed that the vehicle was stolen. The owner wears nothing but cotton. The day following the robbery, a suspect is apprehended, his red woollen pullover and denim jeans are confiscated. On the driver's seat, one group of relevant foreign fibres is collected. It consists of a large number of, say, n red woollen fibres. This finding, denoted E_1, is a combination of the form $\{y_1, x_1\}$, where y_1 refers to the recovered fibres on the car seat and x_1 refers to known (control) material from the suspect's red woollen pullover. The evaluation will assume that the group of fibres on the driver's seat could be transferred from the offender's clothing. That is, it is a group of fibres that appears to be relevant in the context of the case. Further, it is assumed that the suspect was wearing the pullover of interest here at the time of the offence. That is, the association between the pullover and the suspect is not questioned as in scenarios developed later in Section 5.3.

A particular term invoked in this example is that of a foreign fibre group (FFG). An FFG consists of fibres that can be distinguished from fibres from a known source (associated with either the suspect or an object such as a car seat). On both the suspect's pullover and denim jeans, there is an FFG. It consists of 20 extraneous black fibres. These fibres correspond, in some sense, to the fibres of which the driver's seat is composed. Note that this only means that the fibres cannot be discriminated from the fibres composing the seat using the comparative techniques used by the scientist (i.e. it is not meant that the seat is the source of the extraneous fibres). This finding, denoted E_2, is a combination of the form $\{y_2, x_2\}$, where y_2 refers to the 20 recovered fibres on the suspect's clothes and x_2 refers to known material from the driver's seat.

Let the competing propositions for this case refer to an activity, such as 'The suspect sat on the driver's seat of the stolen car (H_p)' and 'The suspect never sat on the driver's seat of the stolen car (H_d)'. Because of the potentially reciprocal nature of transfer, the two sets of recovered traces should be considered as related, depending on the main proposition H. In fact, given H_p, if material is found that characterizes a potential transfer in one direction, then the expert might expect to find trace material characterizing transfer in the other direction. Stated otherwise, the presence of material transferred in one direction provides information about the potential presence of material transferred in the other direction. On the other hand, if H_d holds (i.e. the suspect has nothing to do with the case), then knowledge about material found on the car seat should not affect one's expectations to find material on the suspect's pullover. A formal analysis will clarify the relevance of this aspect.

From Chapter 3, with I omitted for ease of notation, Bayes' formula for the combined set of scientific findings is

$$\frac{Pr(H_p|E_1, E_2)}{Pr(H_d|E_1, E_2)} = \frac{Pr(E_2|E_1, H_p)}{Pr(E_2|E_1, H_d)} \times \frac{Pr(E_1|H_p)}{Pr(E_1|H_d)} \times \frac{Pr(H_p)}{Pr(H_d)}.$$

The value of the scientific results is then

$$V = \frac{Pr(E_2|E_1, H_p)}{Pr(E_2|E_1, H_d)} \times \frac{Pr(E_1|H_p)}{Pr(E_1|H_d)}. \tag{5.9}$$

Following previous discussion (Section 5.1.6), the second ratio in (5.9) is developed as follows:

$$\frac{Pr(E_1|H_p)}{Pr(E_1|H_d)} = \frac{b_0 t_n + b_{1,n} \gamma_1 t_0}{b_0 \gamma_1 t'_n + b_{1,n} \gamma_1 t'_0}, \tag{5.10}$$

where γ_1 is the population proportion for the characteristics from y_1 in extraneous groups of fibres of similar size found on seats of stolen cars, determined with reference to some background database of fibres. Equation (5.10) reduces to $1/\gamma_1$, if two assumptions can be accepted. First, given the case circumstances, one may consider that background material on the car's seat would be absent. This leads to the assignment $b_{1,n} = 0$ and $b_0 = 1$. Second, if the suspect has never sat on the driver's seat of the stolen car (H_d), another individual (apart from the owner) sat on it, so the transfer characteristics of the unknown garment of that individual, the true offender, are of importance. The transfer probabilities t and t' refer to, respectively, the probabilities of transfer from the suspect's and the true offender's garment. Treating these probabilities as equal leads to $1/\gamma$ for (5.10). Note that these assumptions, and others hereafter, are made for the sole reason of keeping the complexity of the analytical form of the resulting likelihood ratio at a reduced level. The reader is not required to share the assumptions and it goes without saying the Bayesian network can cope with a complete analysis.

The first ratio on the right-hand side in (5.9) accounts for a group of 20 fibres (y_2) present on the suspect's clothing. In the numerator, it is assumed that this group of fibres is, potentially, the result of transfer while the suspect sat on the car's seat. In the denominator, the presence of y_2 is considered as being part of background presence. If the suspect did not sit on the car's seat, the fibres found on his pullover are there by chance alone as previously supposed in scenarios involving glass fragments (e.g. Example 5.2). Thus, this denominator can be written as $b^*_{1,20} \gamma_2$. Note that the '$*$' in this notation is used to distinguish the assignment of a probability for the occurrence of fibres on the pullover from some source other than the car seat from that used for the background presence of fibres on the car seat.

The numerator $Pr(E_2|E_1, H_p)$ needs a more detailed analysis. More formally, it may be written as follows:

$$Pr(E_2|E_1, T_2, H_p)Pr(T_2|E_1, H_p) + Pr(E_2|E_1, \bar{T}_2, H_p)Pr(\bar{T}_2|E_1, H_p). \tag{5.11}$$

Given the event of transfer T_2 from the car seat to the suspect's pullover, the observation E_1 of the fibres on the car seat (corresponding to the suspect's pullover) does not influence the conditional probability of E_2. Thus, T_2 makes E_2 and E_1 independent, and one can write $Pr(E_2|E_1, T_2, H_p) = Pr(E_2|T_2, H_p)$ and $Pr(E_2|E_1, \bar{T}_2, H_p) = Pr(E_2|\bar{T}_2, H_p)$. Equation (5.11) thus becomes

$$\underbrace{Pr(E_2|T_2, H_p)Pr(T_2|E_1, H_p)}_{b_0^*} + \underbrace{Pr(E_2|\bar{T}_2, H_p)Pr(\bar{T}_2|E_1, H_p)}_{b_{1,20}^* \gamma_2}, \qquad (5.12)$$

with b_0^* and $b_{1,20}^* \gamma_2$ following habitual notation for, respectively, zero background and background of one group of comparable size and with compatible analytical features. The probability of the event of transfer T_2, conditional on H_p and E_1, that is $Pr(T_2|E_1, H_p)$, can be extended by considering the event of transfer T_1, conditional on the result E_1. Assuming T_2 and E_1 are conditionally independent given T_1, the term $Pr(T_2|E_1, H_p)$ becomes

$$\overbrace{Pr(T_2|T_1, H_p)}^{u_{20|T_1}} Pr(T_1|E_1, H_p) + \overbrace{Pr(T_2|\bar{T}_1, H_p)}^{u_{20|\bar{T}_1}} Pr(\bar{T}_1|E_1, H_p), \qquad (5.13)$$

with $u_{n|\cdot}$ (here $n = 20$) representing a conditional transfer probability. This assignment is conditional on the state of the variable T_1, that is the event of transfer of a group of foreign fibres to the car seat. This conditional transfer probability is highly case dependent as it is strongly influenced by the kind of textile materials involved, in particular their properties (e.g. sheddability). The conditional transfer probability thus requires a case-tailored assessment.

The probability $Pr(T_1|E_1, H_p)$ is obtained using Bayes' theorem. Using habitual notation and assignments, the following transformation can be applied:

$$Pr(T_1|E_1, H_p) = \frac{Pr(E_1|T_1, H_p)Pr(T_1|H_p)}{Pr(E_1|T_1, H_p)Pr(T_1|H_p) + Pr(E_1|\bar{T}_1, H_p)Pr(\bar{T}_1|H_p)}$$

$$= \frac{b_0 t_n}{b_0 t_n + b_1 \gamma_1 t_0}. \qquad (5.14)$$

For the sake of simplicity, consider again the previously supposed extreme situation with $b_0 = 1$ and $b_{1,n} = 0$, concerning the background on the driver's seat. In such a situation, $Pr(T_1|E_1, H_p) = 1$ and, thus, $Pr(\bar{T}_1|E_1, H_p) = 1 - Pr(T_1|E_1, H_p) = 0$. This expresses the view that if there was no background on the driver's seat, but corresponding fibres are found, then it is the event of transfer that led to this finding. Consequently, (5.13) becomes $Pr(T_2|E_1, H_p) = u_{n|T_1}$ and hence $Pr(\bar{T}_2|E_1, H_p) = (1 - u_{n|T_1})$. Thus, in summary, the numerator as specified by (5.12) becomes $b_0^* u_{20|T_1} + b_{1,20}^* \gamma_2 (1 - u_{20|T_1})$.

Note that in such a situation, the probability $u_{n|\bar{T}_1}$ does not appear in the final expression. However, if the assumption of no background (i.e. $b_0 = 1$ and $b_{1,20} = 0$) for the car seat is relaxed, then $u_{n|\bar{T}_1}$ needs to be assigned. This latter term refers to the event of transfer from the car seat to the criminal, given that no transfer occurred in the opposite direction.

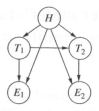

Figure 5.8 Bayesian network structure describing a scenario involving a potential cross-transfer of trace material. Each node has two states. Node H, the main proposition, with states H_p, 'The suspect sat on the car seat', and H_d, 'The suspect never sat on the car seat'. Node T_1, with states T_1, 'There was a transfer from the offender to the car seat', and $\bar{T}_1$, 'There was not a transfer to the car seat'. Node E_1, with states E_1, 'The fibres found on the car seat are found to correspond to those of the suspect's pullover', and $\bar{E}_1$, 'The fibres found on the car seat are not found to correspond to those of the suspect's pullover'. Nodes T_2 and E_2 similarly denote, respectively, transfer to the suspect's pullover and the finding of corresponding fibres on the suspect's pullover.

Combining the various developments made above, one can re-write (5.9) as follows:

$$V = \frac{b_0^* u_{20|T_1} + b_{1,20}^* \gamma_2 (1 - u_{20|T_1})}{b_{1,20}^* \gamma_2} \times \frac{1}{\gamma_1}. \tag{5.15}$$

Figure 5.8 illustrates a possible Bayesian network for this scenario. It consists of a combination of two networks of the kind shown in Section 5.1.2 but retaining only a single node for the main proposition H. The assumed relevance relationship between the two sets of findings is expressed in terms of a connection between the two network fragments. The main consideration of dependency is given by the arrow between the nodes T_1 and T_2, representing distinct events of transfer. This dependency expresses the view that the occurrence of an event of transfer in one direction can affect one's assessment of the occurrence of an event of transfer in the opposite direction. The extent of this influence depends, however, on the way in which the node tables are specified.

Equation (5.13) clarifies that, potentially, the probability for the event of transfer from the seat to the suspect (under H_p) can vary according to the truth or otherwise of T_1. That is, $u_{n|T_1}$ can be different from $u_{n|\bar{T}_1}$. If, however, one judges these probabilities to be the same, then the link between T_1 and T_2 would not be needed because one would suppose that knowledge of the state of the variable T_1 would not affect one's assessment of the probability of the event T_2. Besides, note that given H_d (i.e. the suspect did not sit on the car seat), a zero probability is assigned to the event T_2 because the event is supposed not to have taken place.

The conditional transfer probability $u_{n|\bar{T}_1}$ might appear difficult to conceptualize, but note that with particular assumptions (i.e. $b_0 = 1$ and $b_{1,n} = 0$) that reduce (5.14) to 1, only a single conditional transfer probability, $u_{n|T_1}$, needs to be assigned. It is often useful to examine the behaviour of the likelihood ratio, (5.15), for different values of its components. Recovered trace material characterizing a possible cross-transfer does not necessarily increase the probative value. This point is discussed, for example, by Champod and Taroni (1999) and Aitken and Taroni (2004).

5.3   Evaluation of transfer material given activity level propositions with uncertainty about the true source

5.3.1   Network structure

The general approach for developing a likelihood ratio and a Bayesian network structure for evaluation given activity level propositions can be applied to various categories of findings other than DNA, for example fibres and shoemarks. Such categories differ from DNA in the sense that they are not 'intrinsic' to a given individual; that is to say, leaving aside biological anomalies and other special cases, a given individual has, as far as the current level of analytical detail is concerned, one and only one DNA profile. It cannot be deliberately changed. Most people, however, almost certainly have more than one pullover or more than one pair of shoes. Thus, with such items, it is necessary to relate a particular pullover or a given pair of shoes to a particular suspect. It is not sufficient to focus solely only on a possible relationship between fibres (or shoemarks) and a crime scene. In the examples presented earlier in this chapter, it was tacitly assumed that there was no uncertainty at all about the known source. In what follows, this assumption is relaxed.

In cases involving textile fibres, a potential relationship pertains, in the first place, to the suspect's pullover and fibres recovered on a scene of a crime. This does not necessarily cover a relationship between this suspect and the recovered fibres, and an extended likelihood ratio development is needed in order to account for the possibility of the suspect being a wearer of the pullover (Taroni et al. 2012). The problem of interest thus is that of uncertainty about the characteristics of the item actually worn by the suspect in the event that he committed the action of interest. It may not be known if the item worn by the suspect during the alleged facts is in fact the item available (and analysed) as a known source. Define this event in terms of a proposition, say A. The node X previously used in Figure 5.6 is renamed here, for the sake of clarity, as K, short for 'known source', with two states x and $\bar{x}$. Together, nodes A and K condition a newly introduced variable, called S, short for 'suspect's source'. Note that this is a variable that is not directly observed. The probability of whether this node is in the state x or $\bar{x}$ depends on both propositions A and K. This is shown in Figure 5.9 in terms of a converging connection. The node S, in turn, is linked to the nodes Y and T.

In summary, the main difference between Figures 5.6 and 5.9 is that there is no longer a variable X, which refers to the characteristics of the item available as a control and which is assumed to be that which was worn by the suspect if he truly is the criminal. A distinction is now made between the item available as a known source and the 'true' (yet unobserved) source worn by the suspect (under H_p).

5.3.2   Evaluation of the network

The numerator of the likelihood ratio, $Pr(Y|H_p)$, specifies that the probability of the features of the recovered fibres (node Y) depends, given H_p, on whether a transfer occurred (T) and on the characteristics of the true source (S), that is the actual donor item. One can thus write:

$$Pr(Y|H_p) = Pr(Y|T, S, H_p) \times Pr(T, S|H_p). \tag{5.16}$$

Note that, for shortness of the algebraic development, the node T is confined to two states only, T and $\bar{T}$ (or T_n and T_0). For the likelihood ratio, the outcome $Y = y$ is of interest. In order to keep the same compact notation as in previous chapters, $Pr(y)$ is shorthand for $Pr(Y = y)$.

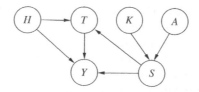

Figure 5.9 Bayesian network for evaluating transfer material when there is uncertainty about whether or not the potential donor item available as a known source is the item actually worn by the suspect. Node H has two states, H_p, the suspect sat in the driver's seat of the stolen car, and H_d, the suspect has never sat on the driver's seat of the stolen car, another person sat on that seat. Node T has two states, T_n and T_0, denoting the transfer of a group of n and of 0 fibres, respectively. Node S represents the intrinsic features of the known suspect's pullover. It has two states, x and $\bar{x}$. For example, x may be red woollen fibres having a specified set of intrinsic features and $\bar{x}$ the union of all sets of possible features other than x. The node Y stands for the recovered fibres and has two states. State y designates a particular outcome, described in terms of intrinsic and extrinsic features, for a single group of fibres. State $\bar{y}$ represents all combinations of features other than those described by y. Node K has two states, x and $\bar{x}$, and node A has also two states, A and $\bar{A}$, where A represents the event that the suspect has worn the item referred to as K, the known source.

Similarly, consider $Pr(S)$, $Pr(\bar{S})$, $Pr(K)$ and $Pr(\bar{K})$. $Pr(T)$ and $Pr(\bar{T})$ denote $Pr(T = T_n)$ and $Pr(T = T_0)$, respectively.

The knowledge about the characteristics of the known source is part of the conditioning. Therefore, from the dependencies specified in the Bayesian network shown in Figure 5.9, expression (5.16) can be further developed by considering the known source K:

$$Pr(y|K, H_p) = Pr(y|T, S, H_p) \times Pr(T|S, H_p) \times Pr(S|K)$$
$$+ Pr(y|T, \bar{S}, H_p) \times Pr(T|\bar{S}, H_p) \times Pr(\bar{S}|K)$$
$$+ Pr(y|\bar{T}, S, H_p) \times Pr(\bar{T}|S, H_p) \times Pr(S|K)$$
$$+ Pr(y|\bar{T}, \bar{S}, H_p) \times Pr(\bar{T}|\bar{S}, H_p) \times Pr(\bar{S}|K).$$

The probability that the suspect's source (S), that is the actual source under H_p, has the characteristics described as x, depends on the characteristics of the known source (K) and on the probability that the suspect wore the known source, represented by the node A. One can therefore write:

$$Pr(S|K) = Pr(S|K, A)Pr(A) + Pr(S|K, \bar{A})Pr(\bar{A}).$$

Notice that given $\bar{A}$, that the suspect did not wear the item available as a known source, K is irrelevant for the assessment of the probability of S:

$$Pr(S|K) = Pr(S|K, A)Pr(A) + Pr(S|\bar{A})Pr(\bar{A}).$$

Logically, if A is true, then the characteristics of the known source correspond to that of the suspect's source: $Pr(S|K, A) = 1$. If $\bar{A}$ is true, then the probability that the suspect's source is still of type x is given by the probability that the item effectively worn by the suspect, other than the known source, would be of type x. Let this probability be represented by the term γ'. The letter term is marked with a prime because it is potentially different from the general population proportion of the fibre characteristic described as x. The reason for this is that there may be information in the framework of circumstances that indicates that the collection of textile garments in possession of the suspect is not representative of the general population proportions of the various fibre types. Next, consider $Pr(A)$, the expression of uncertainty about whether or not the suspect wore the known source. Let this be abbreviated by the term w. In summary, thus, the probability of the suspect's source being of type x is given by $Pr(S|K) = w + \gamma'(1 - w)$. Pursuing the analysis of the numerator, one can incorporate the latter expression along with assignments defined earlier in Section 5.1.2 and write

$$
Pr(y|K, H_p) = \underbrace{Pr(y|T, S, H_p)}_{b_0} \times \underbrace{Pr(T|S, H_p)}_{t_n} \times \underbrace{Pr(S|K)}_{w+\gamma'(1-w)}
$$

$$
+ \underbrace{Pr(y|T, \bar{S}, H_p)}_{0} \times Pr(T|\bar{S}, H_p) \times Pr(\bar{S}|K)
$$

$$
+ \underbrace{Pr(y|\bar{T}, S, H_p)}_{b_{1,n} \times \gamma} \times \underbrace{Pr(\bar{T}|S, H_p)}_{t_0} \times \underbrace{Pr(S|K)}_{w+\gamma'(1-w)}
$$

$$
+ \underbrace{Pr(y|\bar{T}, \bar{S}, H_p)}_{b_{1,n} \times \gamma} \times \underbrace{Pr(\bar{T}|\bar{S}, H_p)}_{t_0''} \times \underbrace{Pr(\bar{S}|K)}_{1-[w+\gamma'(1-w)]}
$$

$$
= b_0 t_n [w + \gamma'(1 - w)]
$$
$$
+ b_{1,n} \gamma t_0 [w + \gamma'(1 - w)]
$$
$$
+ b_{1,n} \gamma t_0'' [1 - (w + \gamma'(1 - w))] .
$$

Writing δ short for $w + \gamma'(1 - w)$, the numerator becomes

$$
Pr(y|K, H_p) = b_0 t_n \delta + b_{1,n} \gamma t_0 \delta + b_{1,n} \gamma t_0''(1 - \delta)
$$
$$
= b_0 t_n \delta + b_{1,n} \gamma [t_0 \delta + t_0''(1 - \delta)] . \tag{5.17}
$$

Note that t_0'' refers to the probability of no transfer of the true source, given that this source has characteristics different from those described as x (i.e. those seen on the known source K). It can reasonably be conceptualized as an average probability of transfer from all potential sources described as different from x.

This term is taken into account here because, given H_p, it is no longer certain that the item actually worn by the suspect is of type x. In fact, given $\bar{A}$, it may be that he has worn an item with characteristics other than x. However, in order for the recovered fibres to be of type $y = x$ in such a case, no transfer must have occurred from the item actually worn by the suspect. The probability t_0'' is distinguished from t_0 because, potentially, the probability of no transfer may be different for the item available as a known source. It is also distinguished from the probability t_0', which is reserved for no transfer from the item worn by the true offender under H_d.

It is worth emphasizing that the above-mentioned development has a potential effect only for cases in which there is uncertainty about whether or not the suspect wore the item available as a known source. It is readily seen that if there no such uncertainty, that is $Pr(A) = w = 1$, then δ becomes 1 and the numerator reduces to

$$b_0 t_n + b_{1,n} \gamma t_0 .$$

This result is in agreement with the numerator given earlier in (3.4) and (5.1). A further aspect of the development considered up to this point is that a single probability of transfer t_n is assigned to the term $Pr(T|S, H_p)$. This is an expression of the assumption that the probability for an event of transfer from an item other than K, but still with characteristics described as x, would be the same as that retained for K.

In the denominator of the likelihood ratio, $Pr(y|H_d)$, it is assumed that a person other than the suspect is the offender. Thus, the clothing available from the suspect as a known source cannot be the donor of the evidential fibres, and none of the variables S, K and A needs to be taken into account. Only the occurrence of transfer from the item worn by the true offender needs to be considered. Therefore, the denominator may be written:

$$Pr(y|H_d) = \underbrace{Pr(y|H_d, T)}_{b_0\gamma} \ \underbrace{Pr(T|H_d)}_{t'_n} + \underbrace{Pr(y|H_d, \bar{T})}_{b_{1,n}\gamma} \ \underbrace{Pr(\bar{T}|H_d)}_{t'_0}$$

$$= b_0\gamma t'_n + b_{1,n}\gamma t'_0 . \tag{5.18}$$

Combining (5.17) and (5.18) leads to the following likelihood ratio:

$$V = \frac{b_0 t_n \delta + b_{1,n}\gamma[t_0\delta + t''_0(1 - \delta)]}{b_0\gamma t'_n + b_{1,n}\gamma t'_0} . \tag{5.19}$$

5.3.3 Effect of varying assumptions about key factors

With respect to (5.1), (5.19) contains some additional variables. It is thus useful to examine the implications of (5.19) in the context of selected scenarios. A first example for this was considered in Section 5.3.2, when the probability that the suspect wore the item referred to as the *known source* was set to 1. It was found that this reduced the numerator to a result found earlier in this chapter.

Another categorical setting is one in which it is assumed that the suspect did not wear the item available as a known source, that is $Pr(A) = w = 0$. Consequently, the term $\delta = w + \gamma'(1 - w)$ reduces to $\delta = \gamma'$, that is the probability that the item actually worn by the offender would have analytical characteristics described as x, in the same way as the properties of the known source. One can take this a step further and assume, for example, that the suspect could not have worn such an item or could not have had access to such an item, so that one would need to set γ' to zero. Then, logically, δ would become zero, too. In such a situation, there was no item with characteristics described as x. Equation (5.19) expresses this by setting the first two terms of the sum in the numerator to zero, that is the events for transfer and no transfer from a source of type x. The only possibility that then remains for there being target fibres is dependent on there being no transfer from the actual item worn by the suspect (with characteristics different from x) and one group of background fibres with corresponding characteristics:

$$b_0 t_n \delta + b_{1,n}\gamma t_0\delta + b_{1,n}\gamma t''_0(1 - \delta) = b_{1,n}\gamma t''_0 \quad \text{(for } \delta = 0\text{)}.$$

On the other hand, if it is known that the actual item worn by the suspect (different from the known source) is of type x, that is $\gamma' = 1$, then $\delta = 1$ and the numerator becomes

$$b_0 t_n \delta + b_{1,n} \gamma t_0 \delta + b_{1,n} \gamma t_0''(1 - \delta) = b_0 t_n + b_{1,n} \gamma t_0 \quad \text{(for } \delta = 1).$$

This numerator accounts for any recovered fibres with corresponding features that have arisen by a transfer from the item of type x that was actually worn (but different from the known source) by the suspect or that there has been no transfer from that item but there is a group of background fibres with the corresponding features. As may be seen, the latter situation of background fibres represents yet another setting in which the form of the generic numerator of (3.4) is obtained. This is a consequence of the assumption that the probabilities of transfer and the complementary event 'no transfer' from the actual source, which is assumed different from the known source but with the same characteristics x, are the same as those of the known source.

Notice further that if γ' is set to one, then the numerator will remain constant for whatever value is chosen for w, the probability that the suspect wore the known source. This stems from the fact that for $\gamma' = 1$, the term δ becomes one:

$$\delta = w + \gamma'(1 - w) = 1, \quad \text{for } \gamma' = 1.$$

6

Evaluation given crime level propositions

Chapter 3 introduced general elements concerning the construction and the evaluation of Bayesian networks for forensic inference problems. As in previous chapters, these ideas are extended here to a more systematic study of the evaluation of results of forensic examinations in various scientific disciplines. Particular attention is drawn to so-called transfer material (e.g. fibres and glass fragments), marks (e.g. shoemarks and firearms/toolmarks) and various types of transferable biological trace matter containing DNA. This chapter mainly focuses on the analysis and discussion of likelihood ratios that involve propositions at a hierarchical level known in the context as *offence (or crime) level*. Literature on the topic (e.g. Evett 1993; Stoney 1994) has considered such formulaic developments for cases in which material is found on a crime scene (or victim), sometimes referred to as *offender to scene (or victim) transfer* cases. These settings involve a single stain found on a crime scene, but with possible uncertainty about the degree to which that stain is relevant (i.e. uncertainty about whether or not the stain has been left by the offender). Extensions to multiple stains or multiple offenders have also been described (Garbolino and Taroni 2002). Yet, another variation of evaluative settings with propositions at crime level deals with potential transfer in the opposite direction, that is when material is collected on a person of interest or their belongings (e.g. the clothing of a suspect). This latter category of situations can also be approached through likelihood ratio analyses, implemented in terms of Bayesian networks.

6.1 Material found on a crime scene: A general approach

6.1.1 Generic network construction for single offender

A very general abstraction of forensic case scenarios involving results of scientific examinations considers a single stain found at the crime scene. Suppose, for example, that the stain

Bayesian Networks for Probabilistic Inference and Decision Analysis in Forensic Science, Second Edition.
Franco Taroni, Alex Biedermann, Silvia Bozza, Paolo Garbolino and Colin Aitken.
© 2014 John Wiley & Sons, Ltd. Published 2014 by John Wiley & Sons, Ltd.

is blood, and its DNA profile is different from the victim's profile. The stain was found at a position where it may have been left by the offender. As part of the background knowledge, it is assumed, for simplicity, that there was only one offender. A development without this latter restriction is presented later (Section 6.1.4). For the case introduced here, suppose further that the crime stain and blood provided by a suspect for comparative analyses share the same DNA profile.

For evaluation of such a result given crime level propositions (H), the following competing positions may be formulated: 'the suspect is the offender (H_p)' and 'some unknown person is the offender (H_d)'. As mentioned in Section 3.3.3, an argumentative connection is needed between these propositions and the observations, that is the analytical result for the stain found on the crime scene. This connection is operated in two steps. The first step extends the argument to an additional proposition that states that the crime stain came from the offender, and an alternative proposition according to which the crime stain did not come from the offender.

Start by assuming that the crime stain came from the offender. A second step then involves the consideration of a proposition that the crime stain came from the suspect and the related alternative proposition that the crime stain did not come from the suspect. Let H be the network node representing the ultimate probandum (i.e. the propositions 'the suspect is the offender' vs 'some other person is the offender'). Given the additional considerations mentioned earlier, two further nodes are also defined: 'the crime stain came from the offender' (node G) and 'the crime stain came from the suspect' (node F). In order to account for the forensic findings, a node E is defined for the proposition 'the suspect's blood and the blood stain found at the scene of the crime have the same DNA profile'. These nodes are all binary.

An appropriate graphical representation of the logical relationships amongst these propositions is shown in Figure 6.1, discussed earlier in Section 2.1.7. In addition to the probabilistic dependencies shown in Figure 6.1, the following five assumptions about particular values in the conditional probability tables of the network are reasonable.

(i) Concerning the node E, one should consider that if the crime stain came from the suspect, then the DNA profile of suspect's blood certainly corresponds to that of the crime stain: $Pr(E|F) = 1$. Note that no distinction is made here between a true correspondence and a *reported* correspondence as proposed, for example, by Thompson et al. (2003) or Aitken et al. (2003). This issue is discussed later in Section 7.11. (ii) If the crime stain did not come from the suspect, then the probability that the DNA profile of the crime stain corresponds to the DNA profile of the suspect's blood is given by the genotype proportion γ in the relevant population: $Pr(E|\bar{F}) = \gamma$ (Balding and Nichols 1994). (iii) For node F, one can consider that

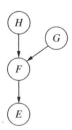

Figure 6.1   A Bayesian network describing a one trace one offender scenario. Each node has two states. H, the suspect is (H_p) or is not (H_d) the offender, and G, the crime stain did or did not come from the offender. F, the crime stain did or did not come from the suspect. E, the suspect's blood and the blood stain found at the crime scene share the same DNA profile.

if the suspect is the offender and the crime stain came from the offender, then certainly the stain came from the suspect: $Pr(F|G, H_p) = 1$. Notice, however, that for other types of marks or traces, such as shoemarks, values different from 1 may be appropriate (see, for example, Section 6.2). (iv) If the suspect is the offender and the stain did not come from the offender, then the crime stain did not come from the suspect: $Pr(F|\bar{G}, H_p) = 0$. (v) If the suspect is not the offender and the stain came from the offender, then certainly the crime stain did not come from the suspect: $Pr(F|G, H_d) = 0$.

As one may see, the directed graph of Figure 6.1 is, in a sense, a 'degenerate Bayesian network'. This is primarily due to the fact that the network contains conditional probabilities equal to zero and one, which itself is a direct consequence of the particular choice of the variables. The choice fits the purpose of minimizing the number of probabilities that are needed and whose values are different from certainty (1) and impossibility (0). One such probability is the genotype proportion γ that may be assigned on the basis of population data. Further probabilities different from zero and one are $Pr(H_p)$, $Pr(G)$ and $Pr(F|\bar{G}, H_d)$. The probability $Pr(H_p)$ can be thought of as the evaluator's initial belief that the suspect is the offender given the non-scientific information pertaining to the case. This idea has been mentioned in the example presented in Section 3.5.2, where the values assigned to the node H represented the (prior) beliefs held by some evaluator, such as an investigator or court of law, before consideration of the scientific information. The term $Pr(G)$ expresses the relevance of the trace material of interest, that is the probability that the blood stain is relevant for the case. This assignment, too, is based on the information available. In turn, $Pr(F|\bar{G}, H_d)$ is the probability that the stain originates from the suspect, given he is not the offender and the stain is not relevant to the case. In other words, this is the probability that the stain would have been left by the suspect even though he was innocent of the offence.

6.1.2   Evaluation of the network

A Bayesian network with a qualitative network specification, together with an assignment of values to probability tables, allows one to evaluate target probabilities. Such target probabilities may be those that define a likelihood ratio, which is of primary importance to forensic scientists. Formally, the numerator and denominator are given by, respectively, $Pr(E|H_p)$ and $Pr(E|H_d)$ (see also Section 3.2). These terms can be derived from the proposed Bayesian network (Figure 6.1) as outlined below.

Start by considering the numerator of the likelihood ratio, $Pr(E|H_p)$. For assessing this term, uncertainty in relation to F is relevant, so that it is necessary to write

$$Pr(E|H_p) = Pr(E|F, H_p)Pr(F|H_p) + Pr(E|\bar{F}, H_p)Pr(\bar{F}|H_p). \tag{6.1}$$

Assumptions (1) and (2) from Section 6.1, and the result that F 'screens off' E from H, that is $Pr(E|F, H_p) = Pr(E|F)$, reduce (6.1) to

$$Pr(E|H_p) = Pr(F|H_p) + \gamma Pr(\bar{F}|H_p). \tag{6.2}$$

This equation states that the likelihood of the hypothesis given the corresponding profiles is equal to the sum of the likelihood of the hypothesis, given that the stain came from the suspect, and the likelihood of the hypothesis, given that the stain did not come from the suspect multiplied by the genotype proportion γ.

Next, the likelihood of H_p given F may be calculated by 'extending the conversation' (Lindley 1991) to G, that is the proposition that the stain is relevant and it did not come from an extraneous source:

$$Pr(F|H_p) = Pr(F|G, H_p)Pr(G|H_p) + Pr(F|\bar{G}, H_p)Pr(\bar{G}|H_p). \tag{6.3}$$

As $Pr(F|\bar{G}, H_p) = 0$, that is assumption (iv) introduced in Section 6.1.1, (6.3) reduces to

$$Pr(F|H_p) = Pr(F|G, H_p)Pr(G|H_p). \tag{6.4}$$

According to assumption (iii), F is certain if both H and G are true. In addition, H and G are probabilistically independent.[1] Thus, (6.4) can be further reduced to

$$Pr(F|H_p) = Pr(G). \tag{6.5}$$

As may now be seen, the first term in (6.2) is simply the 'relevance' probability. A second major term in (6.2) is the likelihood of H_p given $\bar{F}$. Considering again assumptions (iii) and (iv) from Section 6.1, it can be shown that $Pr(\bar{F}|H_p)$ is $Pr(\bar{G})$,[2]

$$Pr(\bar{F}|H_p) = Pr(\bar{F}|G, H_p)Pr(G|H_p)$$
$$+Pr(\bar{F}|\bar{G}, H_p)Pr(\bar{G}|H_p) = Pr(\bar{G}). \tag{6.6}$$

Using the above considerations, notably the results of (6.5) and (6.6), the numerator of the likelihood ratio (6.2) can be written as follows:

$$Pr(E|H_p) = Pr(G) + Pr(\bar{G})\gamma.$$

Consider now the denominator of the likelihood ratio, $Pr(E|H_d)$. In analogy to (6.1), uncertainty in relation to F enters the consideration, so one can write

$$Pr(E|H_d) = Pr(E|F, H_d)Pr(F|H_d) + Pr(E|\bar{F}, H_d)Pr(\bar{F}|H_d).$$

Given that F 'screens off' E from H, that is $Pr(E|F, H_d) = Pr(E|F)$, and invoking assumptions (i) and (ii), the formula for the denominator can be written as

$$Pr(E|H_d) = Pr(F|H_d) + \gamma Pr(\bar{F}|H_d).$$

This sum has the same general structure as (6.2) with the difference that the alternative proposition H_d is used, rather than H_p.

The likelihood of H_d given F, and given $\bar{F}$, can be calculated as before by extending the conversation to give $Pr(F|H_d) = Pr(F|\bar{G}, H_d)Pr(\bar{G}|H_d)$ and $Pr(\bar{F}|H_d) = Pr(G|H_d) + Pr(\bar{F}|\bar{G}, H_d)Pr(\bar{G}|H_d)$ since $Pr(\bar{F}|G, H_d) = 1$.

---

[1] For a graphical illustration of this, consider Figure 6.1 where no direct link exists between H and G.
[2] Remember that $Pr(\bar{F}|G, H_p) = 0$ if $Pr(F|G, H_p) = 1$ and $Pr(\bar{F}|\bar{G}, H_p) = 1$ if $Pr(F|\bar{G}, H_p) = 0$.

Therefore, considering, as above, that G and H are dependent only conditionally on F (i.e. in the absence of knowledge about F, nodes G and H are independent; see also Footnote 1), the likelihood ratio V is given by

$$V = \frac{Pr(E|H_p)}{Pr(E|H_d)}$$

$$= \frac{Pr(G) + Pr(\bar{G})\gamma}{Pr(F|\bar{G}, H_d)Pr(\bar{G}) + [Pr(G) + Pr(\bar{F}|\bar{G}, H_d)Pr(\bar{G})]\gamma}.$$

Next, assume that $Pr(G) = r$ and $Pr(F|\bar{G}, H_d) = p$. The former is the 'relevance' probability, whereas the latter is the probability that the crime stain was left innocently by someone who is now a suspect. It is assumed here that the propensity to leave a stain is independent of the DNA profile of the person leaving the stain (Aitken 1995). This leads to the following (simplified) version of Evett's formula (Evett 1993):

$$V = \frac{r + (1-r)\gamma}{p(1-r) + r\gamma + (1-p)(1-r)\gamma}$$

$$= \frac{r + (1-r)\gamma}{r\gamma + (1-r)[p + (1-p)\gamma]}. \tag{6.7}$$

Note that to assess the relevance term, the scientist has to answer the following question: 'if we find a trace, how sure can one be that it is relevant to the case of interest?' If the personal assessment of relevance is maximal, $r = 1$, then V reduces to its simplest form, $1/\gamma$. Note that the conditional probabilities associated with the links in Figure 6.1 are fixed and independent of the critical epistemic probabilities r and p. The Bayesian network can thus enjoy a fairly wide inter-subjective acceptance, leaving room for disagreement in the assignment of values for r and p.

The Bayesian network shown in Figure 6.1 can be helpful in justifying (6.7) by providing an intuitive model for it. It shows pictorially the dependence, and independence assumptions made by the scientist, and the probability assessments needed to evaluate Evett's formula. Moreover, it is a standard model, although the simplest, which can be used in any 'one-trace' case. It could relate to traces such as fibres or fingermarks, for example.

6.1.3   Extending the single-offender scenario

Consider the following scenario proposed by Champod and Taroni (1999). An offender entered the rear of a house through a hole which he cut in a metal grille. Here, the offender attempted to force entry but failed, and a security alarm went off. He left the scene. About 10 min after the offence, a suspect wearing a red pullover is apprehended in the vicinity of the house following information from an eyewitness who testified that they saw a man wearing a red pullover running away from the scene. At the scene, a tuft of red fibres was found on the jagged end of one of the cut edges of the grille. Notice that in such a scenario, the eyewitness testimony justifies the assumption that the offender wore a red pullover. This is one of the aspects that will be introduced in the Bayesian network proposed for this scenario.

Fibres from the suspect's pullover are found to correspond, in some sense, to the red fibres from the tuft recovered on the scene. Evaluating this finding means, in the context here, the assessment of a correspondence in, for example, fibre type and colour of the fibres given a pair of competing propositions H_p and H_d. When considering the case at crime level propositions, the following definition may be chosen: 'the suspect is the offender (H_p)' and 'an unknown person is the offender (H_d)'.

The scenario considered here closely resembles the one-stain scenario considered in Section 6.1.1. The Bayesian network for the evaluation of one-stain scenarios, shown in Figure 6.1, may thus serve as a starting point for approaching the fibre case outlined above. As suggested by Garbolino and Taroni (2002), for example, a new proposition can be considered in order to account for the eyewitness testimony. Here, this additional binary proposition is denoted R and states that 'The suspect wore a red pullover at the time the crime was committed'. If it is admitted that this proposition R is independent of the crime level proposition H, 'The suspect is the offender', and proposition G, 'The red fibres have been left by the offender', then R can be added as a new node with only one outgoing arc. This arc is pointing to the node for the source level proposition F, 'The red fibres came from the suspect'. Event E is now 'The red fibres found at the grille correspond to the fibres of a pullover belonging to the suspect'. Figure 6.2 shows a graphical representation of these assumptions.

The probability assignment $Pr(R)$ for the table of the new root node can be based on the fact that the suspect was wearing a red pullover 10 min after the attempted crime and on the basis of other information. In the case considered here, it is reasonable to admit that this probability is very close to one. Notwithstanding, one may also assign a lower probability in case of particular circumstances. For example, it may be that the suspect was apprehended only several hours after the crime. This illustrates the subjective (epistemic) nature of this probability assignment.

From the probabilistic dependencies shown in Figure 6.2, it follows that conditional probabilities need to be assessed for the variable E. If F is true (i.e. the fibres come from the suspect's pullover), then the probability of a correspondence between the suspect's pullover and fibres from the crime scene can be considered maximal: $Pr(E|F) = 1$. If F is false (i.e. the fibres do not come from the suspect's pullover), then the probability of E is given by the

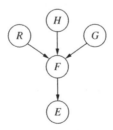

Figure 6.2 Bayesian network for evaluating a single-offender scenario involving fibres recovered on the scene of a crime. Each node has two states. For H, these are H_p, the suspect is the offender, and H_d, the suspect is not the offender. For R, these are R, the suspect wore a red pullover at the time the crime was committed, and $\bar{R}$, the suspect was not wearing a red pullover at the time the crime was committed. For G, these are G, the red fibres have been left by the offender, and $\bar{G}$, the red fibres were not left by the offender. For F, these are F, the red fibres came from the suspect, and $\bar{F}$, the red fibres did not come from the suspect. For E, these are E, the red fibres found at the grille show the same analytical characteristics as the fibres of a pullover belonging to the suspect, and $\bar{E}$, the red fibres found at the grille do not correspond to the fibres of a pullover belonging to the suspect.

chance occurrence of corresponding fibres. This value may be interpreted as the proportion γ of the relevant population that shows the fibre characteristics of interest: $Pr(E|\bar{F}) = \gamma$.

For the variable F, eight conditional probabilities need to be assigned. They correspond to the eight possible combinations of the outcomes of events H, G and R. Start by considering the situation in which the suspect is the offender (H_p), he was wearing a red pullover (R) and the fibres came from the offender (G). Then, certainly, the fibres came from the suspect: $Pr(F|R, G, H_p) = 1$. In turn, if the suspect is not the offender (H_d), he was wearing a red pullover (R) and the fibres did not come from the offender ($\bar{G}$), then the probability that the fibres came from the suspect (F) is the same as the probability that he passed through the hole in the grille for innocent reasons. Indeed, this is a very highly improbable setting, but account is taken of it in order to consider the most general case. Thus, write $Pr(F|R, \bar{G}, H_d) = p$ for this situation, in analogy to the notation used in Section 6.1.2.

The remaining conditional probabilities are clearly equal to zero. In fact, if the suspect is the offender (H_p), the fibres come from the offender (G) and he was not wearing a red pullover at the time the crime was committed ($\bar{R}$), then it is impossible that the red fibres found at the grille come from a pullover belonging to the suspect. This is so, similarly, (i) if the suspect is the offender (H_p), he was wearing a red pullover at the time the crime was committed (R), but the corresponding fibres did not come from the offender ($\bar{G}$), and (ii) if the suspect is the offender (H_p), he was not wearing a red pullover at the time the crime was committed ($\bar{R}$), and the corresponding fibres did not come from the offender ($\bar{G}$). Under proposition H_d (the suspect is not the offender), the probability the red fibres came from the suspect are clearly an impossible event. The sole exception is described in the situation expressed in terms of the probability p.

Therefore, writing $Pr(G) = r$ for the probability of relevance and $Pr(R) = q$ for the probability that the suspect wore a red pullover at the time the crime was committed, and using the laws of probability, the following likelihood ratio is obtained:

$$V = \frac{Pr(E|H_p, I)}{Pr(E|H_d, I)} = \frac{rq + (1 - rq)\gamma}{r\gamma + (1 - r)[pq + (1 - pq)\gamma]}. \qquad (6.8)$$

The likelihood of H_p given E, $Pr(E|H_p, I)$, is now the sum of the probability of the scenario where the trace is relevant and the suspect was wearing a red pullover when the crime was committed (rq) and of the probability of the negation of that scenario ($1 - rq$) multiplied by the relevant fibre type proportion (γ).

The likelihood of H_d given E, $Pr(E|H_d, I)$, is the sum of the probability of the scenario where the trace is relevant ($r\gamma$) and of the probability of the scenario where the trace is not relevant ($1 - r$) and it has been left either by the innocent suspect wearing the red pullover (pq) or by another innocent person (($1 - pq)\gamma$).

Comparing (6.8) with (6.7), one may see that if $q = 1$, then (6.8) equals the likelihood ratio described by (6.7). Notice also that if $r = 1$ (i.e. the fibres found at the crime scene came from the offender), then the likelihood ratio reduces to

$$V = \frac{Pr(E|H_p, I)}{Pr(E|H_d, I)} = \frac{q + (1 - q)\gamma}{\gamma}. \qquad (6.9)$$

Example 6.1 (*Single-offender scenario with case-relevant trace material*) *Consider a single-offender case with q and γ taking the values 0.5 and 0.01, respectively. If the recovered*

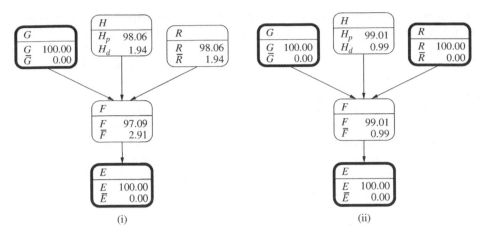

Figure 6.3 Bayesian network for the evaluation of a single-offender fibre scenario. Each node has two states. For H, these are H_p, the suspect is the offender, and H_d, the suspect is not the offender. For R, these are R, the suspect wore a red pullover at the time the crime was committed, and $\bar{R}$, the suspect was not wearing a red pullover at the time the crime was committed. For G, these are G, the red fibres have been left by the offender, and $\bar{G}$, the red fibres were not left by the offender. For F, these are F, the red fibres came from the suspect, and $\bar{F}$, the red fibres did not come from the suspect. For E, these are E, the red fibres found at the grille show the same analytical characteristics as the fibres of a pullover belonging to the suspect, and $\bar{E}$, the red fibres found at the grille are different from the fibres of a pullover belonging to the suspect. Evaluation of the posterior probability of the proposition H when the red fibres found at the grille are found to be indistinguishable from the fibres of a pullover belonging to the suspect (i.e. node E is instantiated to E). Assumptions are (i) relevance, $Pr(G) = r = 1$, the probability the suspect wore a red pullover at the time the crime was committed, $Pr(R) = q = 0.5$ and the relevant fibre type proportion, $\gamma = 0.01$, and (ii) $r = 1$, $q = 1$ and $\gamma = 0.01$. Instantiated nodes are shown with a bold border.

material is considered relevant (i.e. $Pr(G) = 1$), then the likelihood ratio is about 50. This can be deduced from the posterior probability $Pr(H_p|E, I) = 0.9806$ shown in Figure 6.3(i) and may also be verified using (6.9). Notice that equal prior probabilities have been specified for the states of the node H. If both r and q are one, then the likelihood ratio reduces to $1/\gamma$ and the posterior probability of $Pr(H_p|E, I)$ is equal to 99.01. This situation is shown in Figure 6.3(ii).

6.1.4   Multiple offenders

Consider a case in which there is an eyewitness report not only of one offender but also of several offenders, say k, entering through the hole which they cut in the metal grille. One of them was seen to wear a red pullover. A formal mathematical development of this scenario has previously been discussed in Section 3.3.3. The graphical model proposed in Section 6.1 (Figure 6.1) allows one to approach the modified scenario considered here, involving k offenders, by a modification of the probability table of the node F. In fact, under assumption (iii) (Section 6.1.1), if the suspect is the offender and the crime stain came from the offender, then certainly the stain came from the suspect: $Pr(F|G, H_p) = 1$. However, in scenarios involving more than one offender, one may consider $Pr(F|G, H_p) = 1/k$. In fact, if the suspect is one of the k offenders and if the crime stains come from one of the k offenders, the probability that the crime stain comes from the suspect is $1/k$ under the assumption of equal probabilities amongst the offenders. Such an assessment depends largely on the way in which the crime was committed and other background information from the case.

The network proposed in Figure 6.2 represents an alternative model to deal with the k-offenders scenario. It is presented in what follows. The two alternative propositions that define the states of node H are now H_p, 'The suspect is one of the k offenders', and H_d, 'The suspect is not one of the k offenders'. Proposition G becomes the binary proposition 'The red fibres came from one of the k offenders'.

Referring to a graph as shown in Figure 6.2, it may be accepted that the assumptions regarding the probability table of the node F, mentioned in Section 6.1.3, hold. Then, (6.8) is still applicable in this case and, if it is assumed that the fibres certainly come from one of the offenders (i.e. $r = 1$), (6.9) may again be obtained. The value q for the probability that the suspect was wearing the pullover (node R) depends again on the particular information available in the case. If the suspect has been apprehended shortly after the attempted robbery wearing a red pullover, a value close to 1 seems appropriate.

A different scenario would be one in which the suspect has been apprehended the following day or later. Here, a value for q of $= 1/k$ has been proposed, for example in Evett (1993). This leads to the following likelihood ratio:

$$V = \frac{1/k + (1 - 1/k)\gamma}{\gamma}.$$

As with any assumption with respect to particular factors, it is important to consider them in the light of any relevant additional information that one may have to assist in the identification of the person wearing the red pullover. When there is no eyewitness statement that one of the offenders was wearing a red pullover, the assumption that $q = 1/k$ may not be warranted. Moreover, the assumption concerning $Pr(F|R, G, H_p)$ may need to be changed as well. Here, it can be interpreted as an additional epistemic probability, denoted $Pr(F|R, G, H_p) = f$, for which a value less than 1 may be found to be appropriate. Indeed, even if the suspect is one of the offenders (H_p), he wore a red pullover (R) and the fibres came from one of the offenders (G), then it is still possible that another offender could have worn similar clothing and left the trace. The likelihood ratio, V, if $r = 1$ is assumed, is then

$$V = \frac{fq + [(1 - f)q]\gamma}{\gamma}.$$

The above-mentioned developments are mathematical and graphical representations of evaluation given crime level propositions such as 'the suspect is the offender' or 'the suspect is one of the k offenders'. Elicitation of probability plays an important role in this context. It should be given careful attention, taking account of information available from the case under investigation. The major advantage of the use of Bayesian networks here is to help forensic scientists focus on the relationships amongst the propositions involved and the assessment of the various target probabilities, whilst calculations can largely be confined to appropriate software tools.

6.1.5   The role of the relevant population

Recall the scenario introduced in Section 3.3.3 where there is a single crime stain but multiple offenders, k. The likelihood ratio under propositions at the crime level has the following form:

$$V = \frac{r\{1 + (k-1)\gamma\} + k\gamma'(1-r)}{k[\gamma r + \{p + (1-p)\gamma'\}(1-r)]}. \tag{6.10}$$

Note that (6.10) contains two assignments for the population proportion. These values refer to the relevant populations. In Evett's example on DNA profiling results (Evett 1993), the values γ and γ' are taken to be equal. This expresses the idea that there is no reason to suppose that the occurrence of a given genetic trait depends on the 'criminal origin' of the donor (i.e. the donor is or is not one of the member of the group of k criminals). The structure of the Bayesian network shown in Figure 6.2 reflects this assumption.

More generally, the relevant question is 'Should the scientist specify a direct link between node G and the observational node (e.g. E)?' The answer to this question depends on the scientist's view of the problem for a situation in which the state $\bar{F}$ of node F holds (i.e. the suspect is not the source of the recovered trace material). In such a situation, it may be of importance to know whether or not the trace comes from one of the offenders. The reason for this is that one may have different population proportions (relating to the analytical characteristic) for people with and without criminal background. That is, strictly speaking, if the suspect is not the source of the crime stain and the crime stain comes from one of the k offenders, then the probability of the observation of a DNA profile correspondence between the suspect and the crime stain depends on the population proportion γ of the genetic profile in a population of people with a criminal background. Note that the suspect's (genetic) characteristics are considered irrelevant in such a case. Under the assumption that the crime stain comes neither from the suspect nor from one of the k offenders, the probability of observing the results (i.e. profile or other analytical characteristic of the recovered stain) depends on the population proportion γ' in a population of innocent people. Again, the suspect's characteristics are irrelevant for this consideration. The assumption $\gamma = \gamma'$ should thus be considered carefully.

6.2   Findings with more than one component: The example of marks

6.2.1   General considerations

In the scenario considered in Section 6.1, the results of the comparative forensic examinations were thought of in terms of a 'match'. It is worth to emphasize again that this view does not distinguish between a true and a reported match. Although it is widely considered as a viable strategy in many situations, the so-called match-approach involves a considerable simplification of reality. For example, when comparing a crime mark and an impression (or print) made under controlled laboratory conditions, scientists will only rarely note a 'perfect' correspondence — if ever. In practice, there will usually be similarities as well as differences so that the very concept of a 'match' is merely a more or less appropriate expression to use (Friedman 1996), depending on the case at hand.

Another way to look at observations and findings is to consider them in terms of distinct groups. For example, one may distinguish between observations relating to the mark from the crime scene and observations relating to the control material (or impression). Each of these sets may themselves consist of multiple sets of observations. When comparing footwear marks, for example, a distinction may be made between, on the one hand, traits relating to a shoe's manufacturing characteristics, such as size or sole pattern, and, on the other hand, acquired features, such as wear pattern. Such distinct partitions of observations have also been called *components* (Evett et al. 1998b).

If one decides to consider findings in terms of distinct components, it becomes essential to address the question of how the probative value of each component is to be assessed. Further, one needs to consider the question of how the probative values of distinct components are logically to be combined. Evett et al. (1998b) proposed a probabilistic approach for this in the context of footwear marks. The aim of the presentation here is the use of Bayesian networks as a means not only to deepen the understanding of this approach but also to clarify the underlying assumptions.

Start by considering the problem given source level propositions. To represent the positions of the prosecution and defence, the following definition may be chosen: 'shoe X is the source of the mark (F)' and 'the mark was left by some unknown shoe $(\bar{F})$'. Shoe X is a shoe of a person of interest. This shoe, X, is used to make prints under controlled laboratory conditions. Observations made on them are denoted x. Likewise, let y denote the observations made on the crime mark. For now, focus solely on the sets of observations x and y without distinguishing between marks due to features of manufacture and acquired features. The likelihood ratio is then

$$V = \frac{Pr(x, y|F)}{Pr(x, y|\bar{F})}.$$

Applying the third law of probability, V becomes

$$V = \frac{Pr(y|x, F)}{Pr(y|x, \bar{F})} \frac{Pr(x|F)}{Pr(x|\bar{F})}. \tag{6.11}$$

It appears reasonable to assume that, in the absence of knowledge of y, uncertainty about x is not affected by the truth or otherwise of F, that is $Pr(x|F) = Pr(x|\bar{F})$. In addition, if the crime mark was not made by shoe X, then knowledge about x can be considered as irrelevant[3] for assessing the probability of y. This allows one to write $Pr(y|x, \bar{F}) = Pr(y|\bar{F})$. The likelihood ratio (6.11) then reduces to

$$V = \frac{Pr(y|x, F)}{Pr(y|\bar{F})}. \tag{6.12}$$

The conditioning of y on x and F is explicit in (6.12). A Bayesian network over the three variables X, Y and F would thus involve a converging connection as shown in Figure 6.4(i). Note that the lower case letters x and y are used to denote particular observations for the variables X and Y.

6.2.2 Adding further propositions

In Section 6.2.1, the results of comparative examinations of footwear marks were considered with respect to source level propositions. In order to operate an extension to crime level propositions, the following definition may be chosen: 'the suspect is the offender (H_p)' and

---

[3] Notice that such an assumption is not valid in the context of DNA (Balding and Nichols 1994). The assumption has also been questioned for other kinds of forensic traces (Aitken and Taroni 1997).

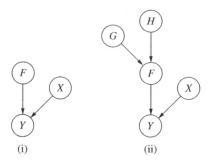

(i)   (ii)

Figure 6.4 Bayesian networks for evaluating footwear marks given (i) source level and (ii) crime level propositions. Node F has two states, the mark was left by shoe X and the mark was left by some unknown shoe. Node X is concerned with observations made on reference prints obtained from shoe X. Node Y is associated with the observations made on the crime mark. Node G has two states, the footwear mark was left by the offender and the footwear mark was left by someone other than the offender. Node H has two states, the suspect is the offender and some unknown person is the offender.

'some unknown person is the offender (H_d)'. As noted in Section 6.1, evaluation given crime level propositions requires one to take into account relevance as a further factor. For the purpose of the current discussion, consider thus the following two propositions: 'the footwear mark was left by the offender (G)' and 'the footwear mark was left by someone other than the offender $(\bar{G})$'.

Figure 6.4(ii) shows an extended model of the network introduced at the end of the previous section [Figure 6.4(i)]. The variables G and H appear as two distinct parental nodes for the variable F. Note that the presence of an arc is just as informative as the absence of an arc. By adding the nodes G and H, the network is augmented by further dependence and independence properties. These may be justified by the two following assumptions. First, in the absence of knowledge of F, the probability that the footwear mark was left by the offender is not dependent on whether or not the suspect is the offender: $Pr(G|H_p) = Pr(G|H_d) = Pr(G)$. Here, in analogy to the network discussed in Section 6.1, $Pr(G)$ denotes the relevance term. Second, besides X, only F is directly relevant for assessing the probability of Y. Notably, given F or $\bar{F}$, the probability of Y is independent of G and H. In other words, F 'screens off' Y from G and H. Note that discussion from Section 6.1.5, concerning the potential dependency between the node G and the observational node (here, Y), should be kept in mind in the current analysis. In what follows, however, independence is assumed.

6.2.3   Derivation of the likelihood ratio

The numerator of the likelihood ratio is the probability of the observations on the crime mark, given the observations on the test impression, under the assumption that the suspect is the offender: $Pr(y|x, H_p)$. Considering uncertainty in relation to F, the numerator of V, $Pr(y|x, H_p)$ in the context of this discussion, becomes

$$Pr(y|x, H_p) = Pr(y|x, F, H_p)Pr(F|x, H_p)$$

$$+ Pr(y|x, \bar{F}, H_p)Pr(\bar{F}|x, H_p). \qquad (6.13)$$

According to the second assumption introduced at the end of Section 6.2.2, knowledge about H would not affect the belief in y if F were already known (i.e. that the mark was left by shoe X). So, $Pr(y|x, F, H_p)$ can be reduced to $Pr(y|x, F)$. In the absence of knowledge about y, uncertainty about F is not affected by knowledge about x, so $Pr(F|x, H_p)$ can be written as $Pr(F|H_p)$. (6.13) is now

$$Pr(y|x, H_p) = Pr(y|x, F)Pr(F|H_p) + Pr(y|x, \bar{F})Pr(\bar{F}|H_p). \tag{6.14}$$

Next the likelihood of H_p given F, $Pr(F|H_p)$, may be derived. As in Section 6.1.2, 'extend the conversation' to G and obtain

$$Pr(F|H_p) = Pr(F|G, H_p)Pr(G|H_p) + Pr(F|\bar{G}, H_p)Pr(\bar{G}|H_p). \tag{6.15}$$

If the suspect is the offender and the crime mark did not come from the offender, then the crime stain was not made by shoe X (assuming no other person than the suspect wore shoe X) so $Pr(F|\bar{G}, H_p) = 0$. From this assumption and the probabilistic independence between H and G, (6.15) reduces to

$$Pr(F|H_p) = Pr(F|G, H_p)Pr(G). \tag{6.16}$$

Similar arguments apply to the likelihood of H_p given $\bar{F}$, which in its extended form is

$$Pr(\bar{F}|H_p) = Pr(\bar{F}|G, H_p)Pr(G|H_p) + Pr(\bar{F}|\bar{G}, H_p)Pr(\bar{G}|H_p). \tag{6.17}$$

From the assumption $Pr(F|\bar{G}, H_p) - 0$ made above, it follows that $Pr(\bar{F}|\bar{G}, H_p) = 1$. Again, G and H are probabilistically independent, so (6.17) becomes

$$Pr(\bar{F}|H_p) = Pr(\bar{F}|G, H_p)Pr(G) + Pr(\bar{G}). \tag{6.18}$$

Application of (6.16) and (6.18) to (6.14) gives

$$Pr(y|x, H_p) = Pr(y|x, F)Pr(F|G, H_p)Pr(G)$$
$$+Pr(y|x, \bar{F})[Pr(\bar{F}|G, H_p)Pr(G) + Pr(\bar{G})]. \tag{6.19}$$

The denominator is concerned with the probability of y given H_d and x. Considering again that F 'screens off' y from H, the denominator can be written as

$$Pr(y|x, H_d) - Pr(y|x, F)Pr(F|H_d) + Pr(y|x, \bar{F})Pr(\bar{F}|H_d).$$

Evaluation of $Pr(F|H_d)$ and $Pr(\bar{F}|H_d)$ separately gives

$$Pr(F|H_d) = Pr(F|G, H_d)Pr(G|H_p) + Pr(F|\bar{G}, H_d)Pr(\bar{G}|H_d) \tag{6.20}$$

and

$$Pr(\bar{F}|H_d) = Pr(\bar{F}|G, H_d)Pr(G|H_p) + Pr(\bar{F}|\bar{G}, H_d)Pr(\bar{G}|H_d). \tag{6.21}$$

Consider two assumptions. First, if the suspect is not the offender and the crime mark was left by the offender, then the probability of the crime mark being made by shoe X is zero: $Pr(F|G, H_d) = 0$. It is therefore assumed that the true offender, if he is not the same person as the suspect, could not have worn shoe X. Second, if the suspect is not the offender and the crime mark was not made by the offender, then the probability of the crime mark being made by shoe X is zero also: $Pr(F|\bar{G}, H_d) = 0$. This is an expression of the belief that the suspect could not have left the crime mark for innocent reasons. This probability was denoted p in Section 6.1.

These assumptions imply that both $Pr(\bar{F}|G, H_d)$ and $Pr(\bar{F}|\bar{G}, H_d)$ are 1. Application of these assumptions to (6.20) and (6.21), and observing again the independence between G and H, gives

$$Pr(F|H_d) = 0 \quad \text{and} \quad Pr(\bar{F}|H_d) = Pr(G) + Pr(\bar{G}) = 1.$$

This simplifies the denominator considerably and it can now be written as

$$Pr(y|x, H_d) = Pr(y|x, \bar{F}). \tag{6.22}$$

The combination of (6.19) and (6.22) gives the overall likelihood ratio

$$V = \frac{\left(\begin{array}{c} Pr(y|x, F)Pr(F|G, H_p)Pr(G) \\ +Pr(y|x, \bar{F})\{Pr(\bar{F}|G, H_p)Pr(G) + Pr(\bar{G})\} \end{array} \right)}{Pr(y|x, \bar{F})}. \tag{6.23}$$

This result can be simplified if the following assumptions are made. First, let $Pr(G)$, the relevance term, be abbreviated by r. Arguably, $Pr(\bar{G}) = 1 - r$. Then, second, consider $Pr(F|G, H_p)$, the probability that the suspect was wearing shoe X, given that he was the offender and that he let the footwear mark. This factor is abbreviated by w. Its complement, $Pr(\bar{F}|G, H_p)$, is therefore $1 - w$. Third, define $Pr(y|x, \bar{F}) = Pr(y|\bar{F}) = \gamma$ as the relevant population proportion. Application of these assumptions to (6.23) yields

$$V = \frac{Pr(y|x, F)rw + \gamma[(1 - w)r + (1 - r)]}{\gamma}.$$

This can further be simplified to

$$V = rw\frac{Pr(y|x, F)}{\gamma} + (1 - rw). \tag{6.24}$$

V, deduced from the Bayesian network shown in Figure 6.4(ii), is thus in agreement with the formula proposed by Evett et al. (1998b). As a point of discussion, compare the likelihood ratio formulae (6.7) in the one-trace scenario with Equation (6.24) for the footwear mark scenario. These two formulae might appear to have very little in common if judged only by their general appearance. However, comparison of the corresponding Bayesian networks, that is Figure 6.1 and 6.4(ii), illustrates a strong resemblance. This may be taken as an indication of the coherence of the proposed probabilistic solutions for these two scenarios.

6.2.4   Consideration of distinct components

So far, the analysis in this chapter has considered y and x as global assignments for the observations made on the crime mark and control prints respectively. Figure 6.4(ii) shows a graphical summary of this. As noted at the beginning of Section 6.2, observations made on a footwear mark, y, may be partitioned into distinct components. For example, a component y_m may be defined with the aim of describing traits that originate from the features of manufacture of the shoe that left the mark y. In the same way, y_a may be considered for aspects of a footwear mark that are thought to originate from the acquired features of the shoe that produced the mark y. Then, with $x = (x_m, x_a)$, an analogous notation can be used to describe the distinct levels of observation made on prints obtained from the suspect's shoe under controlled conditions. In summary, thus, the terms (y_m, y_a) and (x_m, x_a) will express a mark's or print's visible traits that originate from, respectively, the manufacturing and acquired features of the source of the mark or print.

Graphically, a logical extension from nodes Y and X to nodes (Y_m, Y_a) and (X_m, X_a) can be obtained by creating copies of the nodes Y and X. Such a Bayesian network is shown in Figure 6.5(i). Notice, however, that the network shown in Figure 6.5(i) is, to some degree, a minimal representation, providing room for possible extensions. For example, one may consider that the nodes with sub-scripts m and a denote the true presence of manufactured and acquired features. Thus, no distinction is made between the observation of what are taken to be features in a mark or print and the actual presence of characteristics in a given source (i.e. a shoe). Distinct observational nodes for each component may thus be adopted, allowing the scientist to consider the potential of observational error. Such a network is shown in Figure 6.5(ii) where observational nodes labelled with a ''' are present. The absence of an arc between Y_m and Y_a (or, X_m and X_a) assumes independence between features of manufacture and characteristics acquired through wear. As noted by Evett et al. (1998b), the validity of such an assumption depends on the type of wear and the way in which it has been described. The formula for V that corresponds to the Bayesian network shown in Figure 6.5 is an extension of (6.24):

$$V = rwV_m V_a + (1 - rw). \tag{6.25}$$

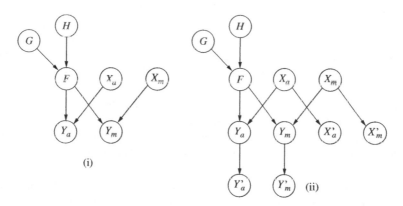

(i)

(ii)

Figure 6.5 Bayesian networks for evaluating footwear marks, partitioning observations into components. Node F has two states, the mark was left by shoe X and the mark was left by some unknown shoe. Nodes X_m and X_a refer to, respectively, manufacturing and acquired characteristics of the shoe X (i.e. the control material or prints). Nodes Y_m and Y_a denote features on the crime mark deemed to be associated with, respectively the manufacturing and acquired features of the shoe at the origin of the crime mark. The nodes labelled with a ''' represent observations relating to the features represented by the respective parental node. Node G has two states, the footwear mark was left by the offender and the footwear mark was left by someone other than the offender. Node H has two states, the suspect is the offender and some unknown person is the offender.

The factors V_m and V_a represent to following ratios:

$$V_m = \frac{Pr(y_m|x_m, F)}{Pr(y_m|\bar{F})}, \quad V_a = \frac{Pr(y_a|x_a, F)}{Pr(y_a|\bar{F})}.$$

Example 6.2 *(Burglary case with footwear marks) Suppose a burglary was committed. Entry was gained through a forced back door. The forensic scientist examining the scene recovered a right footwear mark on the linoleum floor near the point of entry. The mark was lifted and submitted for laboratory examination. The crime mark showed general pattern markings typically found on a given series of running shoes, say B, produced by manufacturer A. Several weeks later, burglary was attempted at a local supermarket. As the alarm went off, the offender left the scene. A nearby police patrol arrested a man fleeing the scene. He was found to wear a pair of running shoes of type B from manufacturer A. Laboratory examination revealed that prints made with the suspect's right shoe (under controlled laboratory conditions) closely corresponded to the crime mark taken from the linoleum floor of the premises burglared a few weeks earlier. The size corresponded and agreements were found in general pattern and small cut marks. However, the suspect's shoe was slightly more worn and bore some cut marks, which were not present in the mark recovered on the scene. How can these findings be evaluated using a Bayesian network?*

Assume that Figure 6.5 provides an appropriate representation of the various issues related to the evaluation of the scientist's observations. The target propositions are at the crime level and defined as follows: 'the suspect is the offender (H_p)' and 'some unknown person is the offender (H_d)'. The proposed model allows for uncertainty concerning the relevance of the crime mark with respect to the offence under consideration. Uncertainty is also taken into account with respect to the question of whether the suspect, if he is the offender and the crime mark is relevant, wore the shoe at time of the offence. The findings are described in terms of features relating to the manufacture and to the use of the shoe(s) at the origin of the compared elements. The quantitative assessment of these factors, that is conditional and (graphically) unconditional probabilities, is discussed below in more detail.

The value to be assigned to the state H_p of the node H is the prior probability of guilt. As was mentioned in the previous examples, this probability lies outside the competence of the forensic expert. Imagine that other available (scientific) information that could tend to associate the suspect to the first burglary is only weak, so that an evaluator (investigator or court of law) would hold a low prior belief, such as $Pr(H_p) = 0.01$.

For the node G, one needs to assign the relevance term $Pr(G)$, previously denoted r. Although the fact that the crime mark was found near the point of entry suggests that the mark is relevant, a rather conservative value is assumed here. Consider a probability of 0.5, for example.

The node F contains one probability, w, which requires particular attention. It is the probability that the crime mark was made by the suspect's shoe, given that the mark is relevant and the suspect is the offender. Clearly, this probability depends on the number of pairs of shoes in the possession of the suspect and how frequently he wears them. In the case at hand, the suspect was wearing the pair of running shoes at the time of arrest. As this pair of shoes bore marks of wear, it is reasonable to assume that there is some chance that the suspect would have worn the pair of shoes if he were the offender. Thus, consider a probability of 0.5 for

$Pr(F|G, H_p)$. For all configurations of the two parental nodes G and H other than $\{G, H_p\}$, the value 0 is assigned to the state F of the probability table of the node F.

For the node Y_m, two probabilities are of primary interest. One is $Pr(y_m|x_m, F)$, that is the probability of marks of manufacturing features observed on the crime mark, given that it has been made by the suspect's shoe and assuming knowledge about the manufacturing features of the suspect's shoe. A probability of 1 or close to 1 may be found appropriate here. A second probability is $Pr(y_m|x_m, \bar{F})$, that is the probability of the observed marks of manufactured features, given that the mark had been left by some other running shoe. This probability may be assigned on the basis of information on sales and distribution. One may also refer to data collected within a forensic laboratory. For the discussion here, assume the assignment of a probability of 0.01. Here, knowledge about x_m is clearly irrelevant, so it is permissible to set $Pr(y_m|x_m, \bar{F}) = Pr(y_m|\bar{F})$. If the suspect's shoe has left the crime mark but the shoe's features of manufacture $\bar{x}_m$ are different from those described as x_m, then the probability of the mark bearing traits of manufacturing features described as y_m can be considered impossible: $Pr(y_m|\bar{x}_m, F) = 0$.

For the node Y_a, there are also two main probabilities to consider: $Pr(y_a|x_a, F)$ and $Pr(y_a|x_a, \bar{F})$. For assessment of the former, it is necessary to answer a question of the kind 'what is the probability to observe the marks of acquired features in the crime mark if this mark has been left by the suspect's shoe?' Recall that the suspect's shoes exhibited slightly more wear and acquired features than the crime marks, so a value less than one is indicated, say 0.2 for the purpose of illustration. For assessment of the second probability mentioned above, it is necessary to answer a question of the kind 'what is the probability to observe the marks of acquired features in the mark if this mark had been left by some other shoe?' Assume that the expert considers factors such as degree of wear as well as shape and the relative position of cut marks, on the basis of which an assignment of, say, $1/1000$ is made. Further discussion of issues that pertain to the choice of relevant databases for assigning values to the components on the likelihood ratio at source level can also be found in Biedermann et al. (2012c). In analogy to the node Y_m, node X_a is irrelevant for Y_a in cases where $\bar{F}$ is true: $Pr(y_a|x_a, \bar{F}) = Pr(y_a|\bar{x}_a, \bar{F}) = Pr(y_a|\bar{F})$. The remaining probability, $Pr(y_a|\bar{x}_a, F)$, can be taken to be 0, as was done for $Pr(y_m|\bar{x}_m, F)$.

The tables of the nodes X_m and X_a contain the probabilities $Pr(x_m)$ and $Pr(x_a)$ for the suspect's shoe's manufacturing and acquired features being of the described kind. As both nodes X_m and X_a will usually be instantiated whilst using the Bayesian network for inference, the probabilities assigned to these nodes are not crucial. Assume, for illustration, $Pr(x_m) = 0.01$ and $Pr(x_a) = 0.001$.

The remaining probabilities needed to complete the Bayesian network either are the logical complements of the probabilities discussed above or have already been defined through various assumptions made in Section 6.2.3. A Bayesian network initialized with these values is shown in Figure 6.6(i). To evaluate the results of comparative examinations between crime mark and control prints, one would typically instantiate the nodes Y_m, Y_a, X_m and X_a. This situation is shown in Figure 6.6(ii). This figure shows that the posterior probabilities for H_p have changed from 0.01 to 0.9806 (rounded to four decimals). This corresponds to a likelihood ratio of approximately 5000. This result is in agreement with the one obtained in (6.25). Using the various probabilities specified in the previous paragraphs, this formula leads to

$$V = 0.5 \times 0.5 \times (1/0.01) \times (0.2/0.001) + (1 - 0.5 \times 0.5) \simeq 5000.$$

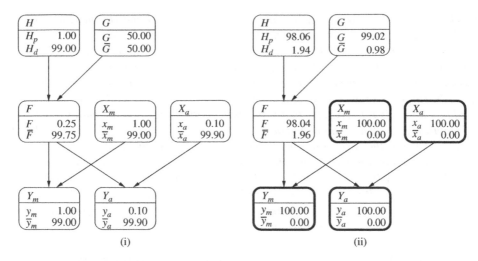

Figure 6.6 (i) Initialized Bayesian network for evaluating results of comparative footwear mark examinations and (ii) state of the Bayesian network after instantiating the nodes Y_m, Y_a, X_m and X_a (shown with a bold border). Node F has two states, F, the mark was left by shoe X, and $\bar{F}$, the mark was left by some unknown shoe. Nodes X_m and X_a are associated with the marks left by, respectively, the manufacturing and acquired features of the shoe X. Nodes Y_m and Y_a relate to observations described as marks of, respectively, manufacturing features and acquired characteristics of the shoe that left the crime mark. Node G has two states, G, the footwear mark was left by the offender, and $\bar{G}$, the footwear mark was left by someone other than the offender. Node H has two states, H_p, the suspect is the offender, and H_d, some unknown person is the offender.

Some readers may be skceptical about the various epistemic probabilities used throughout Example 6.2. However, these numerical values should not be seen as rigid assignments. Although exact values may be hard to specify in practice, one should also keep in mind the importance of considering the relative magnitude of the different factors, notably those used in (6.25). Given that the factors r and w will usually have values less than 1, it is clear that to obtain a likelihood ratio greater than 1 the values of the components V_m and V_a play an essential role. Consider V_m for example. One question relates to the values to be specified for the component probabilities of V_m, that is $Pr(y_m|x_m, F)$ and $Pr(y_m|\bar{F})$, but it is equally important to understand that the more the values assigned to $Pr(y_m|x_m, F)$ and $Pr(y_m|\bar{F})$ diverge, the more the likelihood ratio V will increase. In a more formal notation, one can note that component probabilities satisfying $Pr(y_m|x_m, F) > Pr(y_m|\bar{F})$ and $Pr(y_a|x_a, F) > Pr(y_a|\bar{F})$, that is $V_m > 1$ and $V_a > 1$, lead to an increase in the value of the likelihood ratio. Such purely qualitative considerations are particularly well suited for addressing situations involving unique sets of circumstances. In forensic casework, qualitative reasoning can usefully supplement quantitative Bayesian networks. Chapter 13 presents a more detailed discussion of this topic.

6.2.5 An extension to firearm examinations

The two-component approach to footwear marks as discussed in Section 6.2.4 may similarly be applied to other kinds of traces where observations can be divided into distinct groups of observations. Marks present on fired bullets provide a good example of this.

When it is believed that a particular bullet found on the scene of a crime has been fired by a particular weapon found in the possession of a suspect, a firearms examiner may be asked to perform comparative examinations between the bullet from the scene and test-fired bullets from the suspect's weapon (under controlled laboratory conditions). The results of such comparative examinations may be used to support an evaluative opinion with regard to the assistance the observations might provide in addressing matters that the court is deliberating, for example whether or not the bullet was fired from the barrel of the suspect's weapon. Expressed in very general terms, a firearm examiner's work will cover two main stages. Usually, features of manufacture will be examined first. These include descriptors such as a bullet's calibre as well as the number, width, angle and twist of land and groove impressions. When these points of comparison correspond to the observable features of the test-fired bullets, further examinations may be undertaken. In such a second step, the examiner will compare microscopic marks originating from marks on the inside of the barrel. The inner surface of a barrel holds a distinct set of imperfections and irregularities originating from, for example, its manufacture and can also arise due to corrosion, use and cleaning/maintenance.

Although close correspondences in microscopic surface features may be observed between bullets that have been fired through the same barrel, there are a number of factors that may seriously compromise the conclusions that can be drawn from such comparative examinations. The condition of the recovered bullet represents one such limiting factor. As a result of striking surfaces of different kinds, bullets recovered on a scene of a crime may be damaged or partially missing. Another factor is the condition of the suspect's weapon. Often, considerable time may have elapsed between the crime and the examination of a suspect's firearm. In the meantime, the suspect's weapon may have been fired and cleaned a number of times. In addition, the weapon may have been stored in unfavourable conditions (e.g. humidity). Consequently, differences may be observed between a questioned bullet and bullets fired under controlled laboratory conditions even though they were fired through the same barrel.

Although this is only a very general summary of the many subtleties that comparative examinations of marks left by firearms may entail, it appears sufficient to illustrate that practising forensic scientists face a difficult evaluative task. They are required to account for different sets of observations, each of which being described by similarities and differences. In addition, scientists may be asked to consider the findings in the light of a distinctive configuration of circumstantial information I. The standpoint towards evaluating marks left by firearms as outlined hereafter (Example 6.3) is based on a general framework with the aim of incorporating elements of a weapon's manufacture and acquired features.

Example 6.3 (*Bayesian network for results of comparative mark examinations in a shooting case*) *Consider a hypothetical shooting incident during which some private property has been severely damaged. On the scene, a bullet was extracted from the wooden frame of a window. No other elements, such as cartridges or other bullets, were found on the scene. There is no* *immediate suspect, but a few months later, a man is apprehended on the basis of information completely unrelated to the recovered materials. This suspect is found to be in possession of a firearm. Laboratory examinations revealed that the manufacturing features of this weapon leave characteristic marks that 'correspond' to those present on the bullet recovered on the scene. However, only a few correspondences were found in acquired features. A question of interest thus is how a Bayesian network can be used to guide scientists in the evaluation of such findings.*

In analogy to Sections 6.2.1–6.2.4, consider crime level propositions, defined here as follows: 'the suspect fired the recovered bullet (H_p)' and 'some unknown person fired the recovered bullet (H_d)'. Notice that in the current scenario, there is no uncertainty about the relevance of the recovered bullet. The bullet is necessarily relevant for the damage it caused. Thus, no relevance node, earlier named G, is defined. However, it is necessary to consider a node F for source level propositions. It acts as a child variable for H and is defined as follows: 'the bullet was fired by the suspect's weapon (F)' and 'the bullet was fired by some unknown weapon ($\bar{F}$)'. In addition, nodes Y_m and Y_a stand for those observable marks on a bullet that are thought to originated from, respectively, manufacturing and acquired features of the barrel through which it was fired. The nodes X_m and X_a relate to the same features as observed on bullets fired with the suspect's weapon under controlled laboratory conditions. The dependencies amongst these variables can be expressed in terms of the network structure as shown in Figure 6.5. The only difference is that no node G is retained. The various node probabilities may be assigned as outlined below.

The table of the node H contains the probability of the suspect being the shooter prior to the consideration of the forensic results. This probability may be arrived at by considering the extent to which there is other available information that tends to associate the suspect with the shooting incident. As was mentioned in earlier chapters, the assessment of prior probabilities for such propositions lies outside the scientist's area of competence. In addition, an assessment of the likelihood ratio does not require such priors to be specified. The reason for specifying these probabilities here is of purely technical nature because they are needed to run the model. For the purpose of illustration, let $Pr(H_p) = Pr(H_d) = 0.5$.

For the node F, two conditional probabilities need to be assigned. One is $Pr(F|H_p)$, abbreviated by w, and denotes the probability that the suspect's weapon has been used to fire the recovered bullet, given that the suspect is the offender. This probability depends, in part, on the number of weapons in possession of the suspect or the number of weapons to which the suspect had access at the time the crime happened. But, it also depends on the suspect's eventual preferences in using one or another of these weapons. Imagine that no other weapons have been found in possession of the suspect and that there is information that the suspect did not have access to a weapon other than the one being seized. The factor w could then be taken to be 0.99, for example. The second target probability for the node F is $Pr(F|H_d)$. It relates to the event that the suspect's weapon was used for firing the recovered bullet, given that the suspect is not the offender. The assignment of a probability for this event depends, in part, on the number of persons that potentially had access to the suspect's weapon during the time the crime happened. For the scenario considered here, assume that it does not appear conceivable that someone other than the suspect could have used the suspect's weapon, so that $Pr(F|H_d) = 0$.

For the observational node Y_m, several assignments are necessary. Given the characteristics of manufacture of the suspect's firearm and given that it has fired the bullet found on the scene, one will almost certainly have corresponding characteristics of manufacture on *this bullet. For convenience, assume a value of 1 for $Pr(y_m|x_m, F)$. If the bullet has not been fired by the suspect's weapon, then its characteristics (node X_m) are irrelevant for assigning a probability for Y_m. Arguably, it is permissible to set $Pr(y_m|, x_m, \bar{F}) = Pr(y_m|\bar{x}_m, \bar{F}) = Pr(y_m|\bar{F})$. In the event that the bullet has been fired by another weapon, the probability of y_m depends on the proportion in the relevant population of potential sources that have this configuration of manufacturing features. Assume that there is a laboratory database available allowing the assignment of a value of 0.05 for $Pr(y_m|\bar{F})$. If the bullet were fired by the suspect's weapon whose manufacturing characteristics were described by $\bar{x}_m$, one may consider it impossible to have correspondences with the manufacturing features of the bullet, described by y_m. Arguably, $Pr(y_m|\bar{x}_m, F)$ is set to 0.*

Given that the suspect's firearm was seized a few months after the shooting incident, the firearm may have been exposed to a number of constraints, such as shooting, cleaning and storage conditions of different kinds. Consequently, there is some chance that the firearm will leave a pattern of marks that may considerably differ from the kind of marks the firearm produced a few months ago. Therefore, one may assume a moderate probability for observing a few correspondences in acquired features if the bullet were shot by the suspect's weapon, comparable to what one may expect to see if the bullet were shot by some other firearm. For example, let $Pr(y_a|x_a, F) = Pr(y_a|\bar{F}) = 0.1$. Again, consideration of x_a can be omitted if $\bar{F}$ is true. The remaining factor, $Pr(y_a|\bar{x}_a, F)$, can, as was done for the node x_m, be considered impossible.

The tables of the nodes X_m and X_a contain the probabilities for encountering, respectively, the manufacturing and acquired features on the control material. These probabilities can be assigned, in the most general setting, according to feature proportions in the relevant population. Again, different sources of information may be used to assign these probabilities, such as databases or expert knowledge. For the purpose of illustration, let $Pr(x_m)$ and $Pr(x_a)$ be 0.01 and 0.001 respectively. Notice, however, that the evaluation of the likelihood ratio is not necessarily influenced by the values that have actually been chosen: when evaluating the numerator of the likelihood ratio, the nodes X_m and X_a are set to 'known' (that is assumed to be certain), and, when evaluating the denominator, knowledge about the actual states of X_m and X_a may be irrelevant. The latter is notably the case when H_d implies $\bar{F}$ with certainty and $Pr(y|x, \bar{F}) = Pr(y|\bar{x}, \bar{F})$ holds.

In the case at hand, the findings consist of similarities and differences in marks present on both the recovered bullet and the bullets fired with the suspect's weapon. The value of these findings can be expressed by a likelihood ratio V. Under the stated assumptions, (6.25) reduces to $V = wV_mV_a + (1 - w)$. Using the probabilities defined above, one can thus find:

$$V = 0.99(1/0.05)(0.01/0.01) + (1 - 0.99) = 19.81. \qquad (6.26)$$

A likelihood ratio of the same magnitude can be obtained with the Bayesian network. When instantiating the relevant observational nodes in Figure 6.7, that is Y_m, Y_a, X_m and X_a, the Bayesian network will display the posterior probabilities 0.95195 and 0.04805 for, respectively, states H_p and H_d (Figure 6.7). Their ratio is $0.95195/0.04805 \simeq 19.81$. Note that the high number of decimals is solely retained here to illustrate the agreement between this result and (6.26). It is not suggested that reports should amount to such precision.

Note that assigning the same probability to the event of observing the acquired features y_a being present on the crime bullet, given that it were fired by the suspect's weapon (F), characterized by the acquired features described as x_a, and given that if it were fired by some

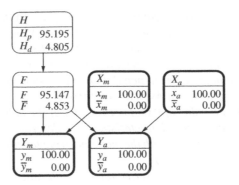

Figure 6.7 Bayesian network for evaluating results of comparative examinations of fired bullets. The observational variables Y_m, Y_a, X_m and X_a are instantiated (shown with a bold border). Node F has two states, F, the suspect's weapon has been used to fire the recovered bullet, and $\bar{F}$, the recovered bullet has been fired from some other weapon. Nodes X_m and X_a are associated with the observable bullet marks that originated from, respectively, manufacturing and acquired features of the barrel of the suspect's weapon. Nodes Y_m and Y_a are associated with the observable marks on the bullet from the crime scene that originated from, respectively, manufacturing and acquired features of the barrel through which this bullet was fired. Node H has two states, H_p, the suspect fired the incriminated bullet, and H_d, some unknown person fired the incriminated bullet.

other weapon ($\bar{F}$), that is $Pr(y_a|x_a, F) = Pr(y_a|\bar{F})$, implies $V_a = 1$. The likelihood ratio thus is confined to the factors w and V_m, so that $V = wV_m + (1 - w)$.

As noted in Section 6.2.4, the various numerical assignments required for the probability tables can be a point of discussion. An agreement on particular numerical values is not a necessary requirement for using the described Bayesian network. Valuable insight can often be gained if the user studies different values for some of the key assignments, in order to see how they affect the outcomes. Such analyses, also known as *sensitivity analyses*, will be discussed in more detail in Chapter 13.

The Bayesian network discussed in this section allows the joint evaluation of results at different levels of detail, in particular manufacturing and acquired features, together with uncertainty about the event that the suspect's weapon was used, given that the suspect is the shooter, expressed in terms of the probability w. Moreover, a meaningful evaluation can be achieved even though part of the findings may be neutral. In Example 6.3, a neutral finding was encountered in connection with y_a, the marks from acquired features present on the recovered bullet. The overall support for the node H remained positive, however, because of the relative rarity of the manufacturing features, expressed by the probability $Pr(y_m|\bar{F})$. This probability determines, together with $Pr(y_m|x_m, F)$, the value of V_m. In particular, as $Pr(y_m|x_m, F) > Pr(y_m|\bar{F})$, the factor V_m is greater than one and this tends to increase the likelihood ratio V (Equation (6.26)). In summary, thus, the proposed Bayesian network approach provides a pictorial representation of how to reason from results of comparative forensic examinations to selected propositions of interest. Through the use of a Bayesian network, a deeper understanding may be gained of how, and at which level, selected sources of uncertainty intervene.

6.2.6   A note on the likelihood ratio

In Examples 6.2 and 6.3, a likelihood ratio for the probative value of results of comparative examinations of footwear marks and marks present on fired bullets was obtained from the ratio of the posterior probabilities of the proposition H. For example, posterior probabilities at the node H were obtained by instantiating the observational nodes as shown in Figures 6.6 and 6.7. Notice, however, that the ratio of the posterior probabilities of the node H can be directly interpreted as the likelihood ratio *only* if the prior probabilities assumed for H are equal. Moreover, it is only the *ratio* that is equivalent and not the individual terms that make up the numerator and the denominator.

Forensic scientists will usually avoid working with posterior probabilities of the major proposition of interest. Instead, they focus on the probability of the findings given the competing propositions of interest. Thus, scientists may wish to find the component values of the likelihood ratio by making different instantiations at the node H. That is, instantiating either H_p or H_d and then observing the probabilities for the findings. This is a readily feasible strategy whenever one has only one observational node as, for example, in the Bayesian network discussed in Section 6.1. However, the application of this procedure may not readily be clear if there is more than one observational node. As an example, consider the Bayesian networks for evaluating results of mark examinations (Section 6.2.4 and 6.2.5). When instantiating the node H, changes will occur in the values of the two observational nodes, that is Y_m and Y_a. A question thus is how the overall likelihood ratio may be found in such a case.

A possible approach could be one that relies on an additional node in order to aggregate the distinct groups of findings. Let this summary node be denoted Y and have entering arcs from both Y_m and Y_a. The node Y is binary with states y and $\bar{y}$. The former state stands for a combination of characteristics described as y_m and y_a on the recovered trace or mark. The latter state, $\bar{y}$, stands for a combination of characteristics different from y_m and y_a. The definition of the node Y thus is a logical combination of the two nodes Y_m and Y_a, and the probability table associated with this node contains values equating 0 and 1. That is, if both y_m and y_a hold, then the state y of the node Y takes probability 1. In all other cases, it takes the value 0.

Example 6.4 *(Burglary case with footwear marks − continued) As an example of the use of a Bayesian network that contains an extension to a summary node for distinct groups of findings, consider again the scenario involving footwear marks discussed in Section 6.2.4 The general structure of an extended Bayesian network is as shown in Figure 6.8. The numerical specification is the same as that outlined in Example 6.2. The only difference is that a node Y has been added according to the definition given above (Section 6.2.6).*

Figure 6.8(i) represents the evaluation of the prosecution's case. Here, the probability of the findings, y_m and y_a, or y for short, is evaluated given H_p, and x_m and x_a, the features of the control object. The nodes H, X_m and X_a, are instantiated (shown with a bold border) and a probability of about 0.05 is obtained for y. An evaluation of the defence case is shown in Figure 6.8(ii). Here, the node H is instantiated to H_d. Notice that no instantiations are made in the nodes X_m and X_a. Features present on the control object do not affect consideration of the observations on the recovered mark, findings y, when assuming the alternative proposition to be true. In the case evaluated here, this leads to a probability 0.00001 for y given H_d.

These evaluations represent a so-called top-down analysis. The numerator and denominator of the likelihood ratio have been found separately according to the following formula:

$$V = \frac{Pr(y|x_m, x_a, H_p)}{Pr(y|H_d)}.$$

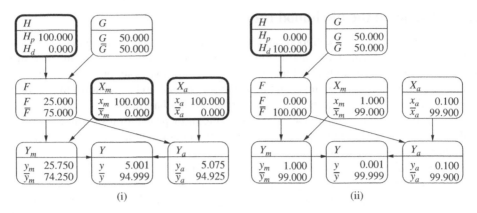

Figure 6.8 Extended Bayesian networks for evaluating results of comparative examinations of footwear marks: (i) evaluation of the probability of the findings given the proposition H_p ('the suspect is the offender') and knowledge about the features of the control object (nodes H, X_m and X_a, are instantiated and shown with a bold border), (ii) evaluation of the probability of the findings given the alternative proposition H_d ('some unknown person is the offender', node H is instantiated and shown with a bold border). Nodes X_m and X_a are associated with the observable marks of manufacturing (x_m) and acquired (x_a) features of control material (the suspect's shoe). States $\bar{x}_m$ and $\bar{x}_a$ denote manufacture and acquired marks, respectively, different from x_m and x_a. Node Y has two states, y, a combination of characteristics described as y_m and y_a is present on recovered mark, and $\bar{y}$, a combination of characteristics different from y_m and y_a is present on recovered mark. In nodes Y_m and Y_a, states $\bar{y}_m$ and $\bar{y}_a$ denote marks originating from manufacturing and acquired features, respectively, different from y_m and y_a. Node F has two states, the crime mark has been made by the suspect's shoe X and has not been made by the suspect's shoe. Node G has two states, the mark was left by the offender and the mark is associated with someone other than the offender.

In the case analysed here, this ratio of two conditional probabilities leads to a value of 5000, which is in agreement with the value obtained in Section 6.2.4. Note that in the latter section, the likelihood ratio was found in a 'bottom-up' strategy, that is by means of the posterior probabilities of the main propositions (node H).

6.3   Scenarios with more than one trace: 'Two stain-one offender' cases

The Bayesian networks discussed so far in this chapter concentrate on the evaluation of single traces or stains given crime level propositions. This reserved comment for further concepts, such as relevance and innocent acquisition (Section 6.1). In addition, attention has been drawn to different levels of detail, or components, that may be considered when describing forensic results (Section 6.2). The discussion of such cases, involving a single offender, can be extended to scenarios that involve more than one stain. Imagine, for example, the scenario described by Stoney (1994), where two bloodstains, with profiles Γ_1 and of Γ_2, are recovered at the scene of a crime. It is not known which, if either, of the two stains are relevant. A suspect is found who is of profile Γ_1. The result of interest thus is, here, a correspondence between the suspect's characteristics and those of one of the two stains.

A formal approach to this scenario is based on the definition of the following probabilities:

$$Pr(\text{The stain of profile } \Gamma_1 \text{ is from the offender}) = r_1,$$

$$Pr(\text{The stain of profile } \Gamma_2 \text{ is from the offender}) = r_2,$$

$$Pr(\text{Neither stain is from the offender}) = 1 - r_1 - r_2.$$

If H_p is true (i.e. the suspect is the offender), there are three components to the probability of the observed correspondence. First, when the stain of profile Γ_1 is from the offender, then it is sure that the profile of the suspect will correspond. The probability for this event is thus $1 \times r_1$. Second, if the stain of profile Γ_2 is from the offender, then it is impossible to find a correspondence with the profile of the suspect. This event has probability zero (i.e. $0 \times r_2$) since the suspect is assumed to be the offender and only one offender is assumed. Third, when neither stain is from the offender, an event with probability $(1 - r_1 - r_2)$, there is a probability γ_1 of a correspondence between the suspect's profile (Γ_1) and the crime stain of the same profile. The probability of the combination of these events is $(1 - r_1 - r_2)\gamma_1$.

The above-mentioned three components are mutually exclusive and the probability in the numerator of the likelihood ratio is the sum of these three probabilities, namely $r_1 + (1 - r_1 - r_2)\gamma_1$. If H_d is true (i.e. the suspect is not the offender), the probability of a correspondence is as before, namely γ_1. The likelihood ratio is then

$$V = \frac{Pr(\Gamma_1, \Gamma_2|H_p)}{Pr(\Gamma_1, \Gamma_2|H_d)} = \frac{r_1 + (1 - r_1 - r_2)\gamma_1}{\gamma_1}. \tag{6.27}$$

A graphical structure for the 'two stain-one offender' scenario may be obtained in two steps. Start by considering the case at the source level, where two propositions are of interest. One is represented by the node F and deals with the question of which of the two stains, if either, comes from the suspect. The possible states of this node are f_0, neither of the two stains comes from the suspect, f_1, the stain of profile Γ_1 comes from the suspect, and f_2, the stain of profile Γ_2 comes from the suspect. The other proposition relates to the findings and is represented by the node E. It is defined as 'a correspondence in Γ_1 between the profile of the suspect and the profile of one of the crime stains'. This variable may be either true (state e_1) or false (state e_2).

The node F relates to an unobservable proposition and its nature is such as to affect the probability of the outcomes represented by the variable E. The node F is thus chosen as a graphical parent. A second step in deriving a network structure then consists of constructing an argument from the source level propositions concerning the origin of the stain, that is node F, to the ultimate crime level propositions, H_p and H_d. This passage may be operated with use of two nodes, H and G. Node H represents the crime level propositions 'the suspect is the offender (H_p)' and 'the suspect is not the offender (H_d)'. In turn, the node G, expresses the relevance of the recovered stains. For this node, three states may be defined: g_0, neither of the two stains comes from the offender, g_1, the stain of profile Γ_1 comes from the offender, g_2, and the stain of profile Γ_2 comes from the offender. Notice that g_0, g_1 and g_2 are mutually exclusive states. This implies that no more than one, if either, of the two stains may be relevant.

The proposition as to which, if either, of the two stains has been left by the suspect (node F) depends on whether the suspect is the offender (node H) and which of the two stains has been left by the offender (node G). These three nodes thus combine in a converging connection such that $G \to F \leftarrow H$. The considerations mentioned above thus lead to an overall network structure, which is analogous to the network discussed in Section 6.1 (Figure 6.1), applicable to one-trace transfer cases assuming one offender.

The quantitative part of the proposed Bayesian network amounts to a series of assignments. The node H has the smallest probability table, containing only a pair of (prior) probabilities for the guilt (H_p) and innocence (H_d) of the suspect. Three probability assignments are needed for the node G. They express relevance considerations for the pair of recovered stains. In agreement with notation from previous discussion of the concept of relevance (e.g. Section 6.1), the probability of stain 1 being left by the offender is denoted r_1 and the probability of stain 2 being left by the offender is denoted r_2. Consequently, there is a probability of $1 - r_1 - r_2$ (assuming $r_1 + r_2$ less than 1) that neither of the stains came from the offender. These three probabilities are assigned to the states g_1, g_2 and g_0, respectively.

The table of the node F contains conditional probabilities. If the suspect is the offender (H_p) and a crime stain is relevant (i.e. it has been left by the offender), then it has been left by the suspect since the suspect and the offender are one. Formally, $Pr(f_i|g_j, H_p) = 1$, for $i = j = 0, 1, 2$. Conversely, for all $i \neq j$ $(i, j = 0, 1, 2)$ $Pr(f_i|g_j, H_p) = 0$. If the suspect is not the offender (H_d), then it may be assumed that none of the stains, irrespective of their relevance, comes from the suspect: $Pr(f_0|g_j, H_d) = 1$ for all j. Conditional probabilities are also required for the node E. If the suspect has left the stain of profile Γ_1, then certainly one will find a correspondence in Γ_1, thus $Pr(e_1|f_1) = 1$. There cannot be a match in Γ_1 if the stain of type Γ_2 comes from the suspect, thus $Pr(e_1|f_2) = 0$. The probability of a correspondence in Γ_1, given the suspect left neither of the two stains, equals the population proportion of profile Γ_1, written $Pr(e_1|f_0) = \gamma_1$.

Example 6.5 *(Two equally relevant traces, one offender) Consider a hypothetical case in which the assumption is made that either of two stains is as likely to be left by the offender as neither of the two stains so that $Pr(r_0) = Pr(r_1) = Pr(r_2) = 1/3$. Take the Γ_1 population proportion γ_1 to be 0.05. Figure 6.9(i) shows a fully specified Bayesian network for such a case, with equal prior probabilities for the ultimate probandum H. Figure 6.9 (ii) and (iii) illustrates the evaluation of the probability of the findings (at the node E) given the prosecution's and defence's cases, respectively. First, consider the the numerator of the likelihood ratio, that is $Pr(e_1|H_p)$. From (6.27) and the values defined above, one obtains the following: $r_1 + (1 - r_1 - r_2)\gamma_1 = 1/3 + 1/3 \times 0.05 = 0.35$. This is also the value obtained in node E in Figure 6.9(ii).*

Recall that, given H_p, there are three components to the probability of the findings. When using the described Bayesian network in an appropriate computerized environment, such component probabilities may be readily found. One needs to set the node H to H_p and then make various instantiations at the node G. That is, if the state g_1 of the node G is instantiated, then the node E would display the value 1 for the state e_1. This correctly reflects the understanding that a correspondence in Γ_1 will be certain. If G is set to the state g_2, the node E would show the value 0 for the state e_1, that is an impossible event. Finally, if the node G is set to g_0, then the node E would indicate the probability $\gamma_1 = 0.05$ for the state e_1. Notice that the states of the Bayesian network given these operations are not shown in Figure 6.9. Readers are invited to evaluate these themselves.

Next, consider the value of the denominator of the likelihood ratio. If H_d is true, then the probability of a correspondence is γ_1 (see also (6.27)). This situation is shown in Figure 6.9(iii). The node E displays 0.05 for the state e_1, a value earlier defined for γ_1. Note that under H_d, the probability of the findings is given by γ_1, irrespective of instantiations made at the node G.

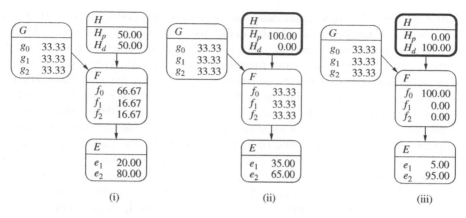

Figure 6.9   Evaluation of the Bayesian network describing the 'two stain-one offender' sce-nario: (i) initial state, (ii) the prosecution's case (H_p is true) and (iii) the defence case (H_d is true). Node E is stands for the proposition 'a correspondence in Γ_1 is found'. This obser-vational variable may either be true (state e_1) or false (state e_2). Node F has three states, f_0, neither of the two stains comes from the suspect; f_1, the stain of profile Γ_1 comes from the suspect and f_2, the stain of profile Γ_2 comes from the suspect. Node G has three states: neither stain is from the offender (g_0), the stain of profile Γ_1 is from the offender (g_1), the stain of profile Γ_2 is from the offender (g_2). Node H has two states, H_p, the suspect is the offender, and H_d, the suspect is not the offender. Instantiated nodes are shown with a bold border.

In summary, separate evaluations of $Pr(e_1|H_p)$ and $Pr(e_1|H_d)$ have led to a likelihood ratio of $0.35/0.05 = 7$. The same value may be obtained when the Bayesian network is used to compute the posterior probabilities of the target propositions H (a situation not shown in Figure 6.9). Instantiating E to e_1 will change the prior probabilities assigned to H_p and H_d to 0.875 and 0.125, respectively, the ratio of which is 7.

It is worth noting that some special cases can be distinguished. As r_1 and r_2 tend to zero, indicating that neither stain is relevant, then the likelihood ratio tends to 1. The proposed Bayesian network offers a clear illustration of this. When the node G is instantiated to g_0, knowing the variable E to be in state e_1 or e_2 would no longer affect the probability of H. A likelihood ratio of 1 provides no support for either proposition, a result in this case which is entirely in agreement with the information that neither stain is relevant. Similarly, one may verify that when assuming $r_1 = r_2 = 0.5$, $V - 1/2\gamma_1$. For $r_1 = 1$, $V = 1/\gamma_1$. As $r_2 \to 1$, then $r_1 \to 0$ and $V \to 0$. All of these are perfectly reasonable results.

6.4   Material found on a person of interest

6.4.1   General form

Previous sections in this chapter focussed on likelihood ratio development and on graphical modelling for cases in which material is found on a crime scene or a victim, potentially left or transferred by the offender. Another category of situations, studied below, focusses on trace material found on a person of interest (e.g. a suspect). Such situations are also sometimes referred to as *scene to offender transfer* cases, but this expression is avoided here because it implies an assumption of relevance of the recovered material.

Earlier in Chapter 3, following the definition given by Stoney (1991, 1994), the term *relevance* has been associated with the offender. In fact, the scenario of interest was one in which the stain is found at the crime scene and so that it is intuitively reasonable to consider whether or not the recovered stain actually came from the offender (or not). With respect to material found on a person of interest, that is in a potential case of transfer in the opposite direction, the classical definition of relevance cannot be applied. The reason for this is that, for a stain found on a person of interest, it is not helpful to ask if the stain came from the offender. Instead, it is of interest to know whether or not the stain originates from the scene (or, in some context, from a given victim). This suggests that a different definition of the so-called relevance term is needed. It may be tempting to define relevance in a case in which a stain is found on a person of interest in terms of 'the stain comes from the victim', so as to assure a symmetry to the definition in scenarios developed in Section 6.1. This does not appear feasible, however, because such a definition corresponds to the existing category of source level propositions. For the kind of potential transfer settings discussed here, it appears more reasonable to assume that the relevance of a stain found on a person of interest depends on whether or not the particular category of trace material to which the recovered stain belongs (e.g. blood) was actually produced during the course of the offence, depending on what is known through the framework of circumstances. Relevance in this context could thus be interpreted as a property of the type (or category) of trace, such as blood, rather than the stain itself.

Imagine a case in which a victim has been stabbed. A suspect is found on the basis of investigative information unrelated to any of the scientific findings. The clothing of this individual is submitted for examination and found to have foreign staining (i.e. cannot be attributed to the person of interest because of different analytical characteristics) in abundant quantities and of fresh appearance on it. In addition, the staining is in a position that is compatible with the dynamics of the incident under investigation (i.e. an attack on the victim).

Following discussion earlier in this book, a common notational convention in such contexts uses letter y for characterizing the observations made on the stain found on the suspect and letter x for the observation made on the control material from the victim. In the case here, suppose that the same genetic characteristic is observed for y and x. Let Γ_1 be the particular genotype observed and then $y = x = \Gamma_1$. Considering a single-offender scenario, the main propositions of interest at the crime level are 'the suspect is the offender (H_p)' and 'some other person is the offender (H_d)'. In order to discriminate between these two propositions on the basis of observations y and x, a likelihood ratio of the form $V = Pr(y|x, H_p)/Pr(y|x, H_d)$ can be considered. This initial form can be further developed following the procedure outlined in Section 3.3.3, by introductory association and intermediate association hypotheses.

Association hypotheses refer to what has previously been called the *relevance* term. Here, they are interpreted as a statement about whether or not the actual category of trace material is one that is compatible with the offence, given the known circumstances. The following hypotheses thus are defined: 'the victim shed blood during the offence (B)' and 'the victim did not shed blood during the offence $(\bar{B})$'. The initial form of the likelihood ratio can thus be extended as follows:

$$V = \frac{Pr(y|x, B, H_p)Pr(B|x, H_p) + Pr(y|x, \bar{B}, H_p)Pr(\bar{B}|x, H_p)}{Pr(y|x, B, H_d)Pr(B|x, H_d) + Pr(y|x, \bar{B}, H_d)Pr(\bar{B}|x, H_d)}.$$

In the absence of knowledge about y, the probability of the victim shedding blood (B) can be assumed to be independent of whether or not the suspect is the offender as well as of the analytical characteristics of the victim's blood. In fact, the analytical characteristics of the victim's blood (x) have no effect on sheddability. Therefore,

$$Pr(B|x, H_p) = Pr(B|x, H_d) = Pr(B|x) = Pr(B), \qquad \text{and} \qquad (6.28)$$

$$Pr(\bar{B}|x, H_p) = Pr(\bar{B}|x, H_d) = Pr(\bar{B}|x) = Pr(\bar{B}). \qquad (6.29)$$

Considering the shedding of blood as a property of relevance for the crime of interest, denote $Pr(B) = r$ and $Pr(\bar{B}) = 1 - r$. This simplifies the likelihood ratio to

$$V = \frac{Pr(y|x, B, H_p)r + Pr(y|x, \bar{B}, H_p)(1 - r)}{Pr(y|x, B, H_d)r + Pr(y|x, \bar{B}, H_d)(1 - r)}. \qquad (6.30)$$

It is then necessary to consider intermediate association hypotheses on whether or not the particular foreign stain found on the suspect actually comes from the victim. These propositions can be defined as 'the foreign stain found on the suspect comes from the victim (A)' and 'the foreign stain found on the suspect comes from someone other than the victim ($\bar{A}$)'. In order to incorporate this proposition, one needs to apply the law of total probability to the various conditional probabilities of y contained in the numerator and the denominator of (6.30).

6.4.2   Extending the numerator

Start by considering the first term, $Pr(y|x, B, H_p)$. By taking account of proposition A, the conditional probability may be extended to

$$Pr(y|x, A, B, H_p)Pr(A|x, B, H_p) + Pr(y|x, \bar{A}, B, H_p)Pr(\bar{A}|x, B, H_p).$$

If the stain truly comes from the victim (A), then, assuming error-free analyses, it will be certain to find a correspondence with the victim's blood. This is not affected by knowledge of blood shedding by the victim (B) or of the suspect being the offender (H_p). Therefore, $Pr(y|x, A, B, H_p) = Pr(y|x, A) = 1$.

$Pr(A|x, B, H_p)$ is the probability that the stain found on the suspect comes from the victim, given that the victim shed blood and that the suspect is the offender. This term is given by the probability (b_0) that the suspect did not have any foreign blood on him (e.g. his clothing, or hands) beforehand multiplied by the probability (t_c) that this blood actually was transferred during the crime, persisted and was successfully recovered: $Pr(A|x, B, H_p) = Pr(A|B, H_p) = b_0 t_c$. In the absence of knowledge of y, knowledge about x has no bearing on this assessment and can be omitted from the conditioning. Notice that t_c is a highly subtle assignment that requires a very careful inspection of the framework of circumstances. Most importantly, one should consider information about the position in which the bloodstain was found, its quantity (or abundance) as well as its freshness (with regard to the time elapsed between the examination of the suspect and the occurrence of the crime). It is also of importance to consider whether transfer could actually have occurred. This may not necessarily be the case, even though the victim shed blood. For example, the probability of transfer during a crime may be low if the kind of aggression did not involve an interaction between victim and offender that was sufficiently close to allow a transfer to have occurred (e.g. strangling a victim is something quite different from beating a victim with a long metal bar). Note further that, from the above, one also obtains $Pr(\bar{A}|x, B, H_p)$ as $(1 - b_0 t_c)$.

Next, consider the term $Pr(y|x, \bar{A}, B, H_p)$. If the stain does not come from the victim ($\bar{A}$), the probability that the characteristics of the true source of the stain would match those of the victim's blood is given by the proportion γ of the relevant population that has the characteristic of interest. Note, however, that the probability of interest should refer to the occurrence of the characteristics amongst foreign bloodstains on offenders of this particular category of crime. For the purpose of the current discussion, it is assumed that this shift from individuals of the population of potential sources to stains on offenders does not affect the assignment of the value γ. Observe also that, given $\bar{A}$, that is that the victim is not the source, information on bleeding of the victim (B) and the victim's characteristics x do not affect y. Thus, $Pr(y|x, \bar{A}, B, H_p) = Pr(y|\bar{A}) = \gamma$.

Given the above,

$$Pr(y|x, B, H_p) = Pr(y|x, A, B, H_p)Pr(A|x, B, H_p)$$

$$+Pr(y|x, \bar{A}, B, H_p)Pr(\bar{A}|x, B, H_p)$$

$$= 1 \times b_0 t_c + \gamma(1 - b_0 t_c)$$

$$= b_0 t_c + \gamma(1 - b_0 t_c).$$

One can proceed analogously with the second conditional probability of y in the numerator of (6.30). Extending $Pr(y|x, \bar{B}, H_p)$ to the proposition A leads to

$$Pr(y|x, A, \bar{B}, H_p)Pr(A|x, \bar{B}, H_p) + Pr(y|x, \bar{A}, \bar{B}, H_p)Pr(\bar{A}|x, \bar{B}, H_p).$$

Assuming, as before, error-free analyses, a correspondence between y and x can be taken as certain if the victim truly is the source of the stain (A), independently of knowledge about bleeding by the victim (proposition B) and the truth or otherwise of the proposition H (i.e. the suspect being the offender): $Pr(y|x, A, \bar{B}, H_p) = Pr(y|x, A) = 1$.

If the suspect is the offender, H_p, but there was no bleeding of the victim during the crime (i.e. the presence or absence of blood is not relevant in this scenario, $\bar{B}$), then the crime stain must be present for reasons unconnected with the crime at hand. Let a denote this probability of fortuitous presence: $Pr(A|x, \bar{B}, H_p) = a$. This is a further assignment that crucially depends on the framework of circumstances. For example, one's assessment should be obviously different in a case in which the victim and the suspect are completely unknown to each other (in the extreme, they never met before), compared to a case in which the victim and the suspect know each other (e.g. are a couple and live together). In the former case, a value close to zero would seem appropriate, whereas in the latter case, a value close to one may be meaningful. Related to this assignment is $Pr(\bar{A}|x, \bar{B}, H_p)$, which is given by $(1 - a)$.

The probability of observing the characteristics of the crime stain, y, if it does not come from the victim, can be conceptualized, as mentioned above, in terms of γ, the proportion of foreign bloodstains on offenders that have the target characteristic: $Pr(y|x, \bar{A}, \bar{B}, H_p) = \gamma$.

The above-mentioned assignments combine as follows:

$$Pr(y|x, \bar{B}, H_p) = Pr(y|x, A, \bar{B}, H_p)Pr(A|x, \bar{B}, H_p)$$

$$+Pr(y|x, \bar{A}, \bar{B}, H_p)Pr(\bar{A}|x, \bar{B}, H_p)$$

$$= a + \gamma(1 - a).$$

In summary, the numerator of the likelihood ratio becomes

$$Pr(y|x, H_p) = \underbrace{Pr(y|x, B, H_p)r}_{b_0 t_c + \gamma(1 - b_0 t_c)} + \underbrace{Pr(y|x, \bar{B}, H_p)(1 - r)}_{a + \gamma(1 - a)}$$

$$= [b_0 t_c + \gamma(1 - b_0 t_c)]r + [a + \gamma(1 - a)](1 - r). \qquad (6.31)$$

6.4.3 Extending the denominator

Allowing for uncertainty about the source level proposition A modifies the first conditional probability of y in the denominator of (6.30), that is the term $Pr(y|x, B, H_d)$, as follows:

$$Pr(y|x, A, B, H_d)Pr(A|x, B, H_d) + Pr(y|x, \bar{A}, B, H_d)Pr(\bar{A}|x, B, H_d).$$

One can proceed analogously for $Pr(y|x, \bar{B}, H_d)$. This conditional probability in the denominator of (6.30) extends to

$$Pr(y|x, A, \bar{B}, H_d)Pr(A|x, \bar{B}, H_d) + Pr(y|x, \bar{A}, \bar{B}, H_d)Pr(\bar{A}|x, \bar{B}, H_d).$$

As argued in Section 6.4.2, if the victim truly is the source of the crime stain, then, assuming error-free analyses, y and x will surely correspond. Therefore, one can write

$$Pr(y|x, A, B, H_d) = Pr(y|x, A, \bar{B}, H_d) = 1.$$

Conversely, if the victim is truly not the source of the crime stain, $\bar{A}$, and the suspect is not the offender, H_d, then the probability of observing the analytical characteristic at hand depends on the proportion of foreign bloodstains on innocent people that have this characteristic. Let this probability be denoted γ'. As previously discussed in Section 6.1.5, this probability is, at least conceptually, different from γ because (by assuming H_d to be true) it relates to foreign blood staining on innocent people. This distinction may be relaxed in the case of analytical characteristics of bloodstains if one assumes that the occurrence of these characteristics is not affected by whether or not a given stain is present on an offender or an innocent person. Such an assumption may need to be re-examined for other kinds of trace material, such as extraneous textile fibres, for example. For generality of the development pursued at this point, the distinction will be maintained so that

$$Pr(y|x, \bar{A}, B, H_d) = Pr(y|x, A, B, H_d) = \gamma'.$$

If the suspect is not the offender, H_d, the probability that the stain found on him comes from the victim depends, in part, on whether the victim's blood could have been transferred for innocent reasons. This assessment is conditioned on the event of bleeding during the offence (i.e. B is true) so that it becomes relevant to consider if the suspect could have acquired this bloodstain even though he is innocent (e.g. as a bystander or someone who tried to help the victim). This relates to a particular complication that has already been discussed elsewhere in literature on evaluation given activity level propositions. In particular, Cook et al. (1998a) noted that a suspect might not have kicked a victim, but he might have been present at the time the incident happened. Denote the probability for such an event of transfer t_i, as opposed to t_c (i.e. a crime-related transfer). Further, it is required that no stain was present beforehand, so that a probability b_0 is also needed. One may thus assign $b_0 t_i$ to the term $Pr(A|x, B, H_d)$ and the complement, $(1 - b_0 t_i)$, to $Pr(\bar{A}|x, B, H_d)$.

Finally, if the suspect is not the offender, H_d, and there has been no bleeding of the victim during the offence, $\bar{B}$, the probability of the suspect having a bloodstain of the victim is given by the probability of accidental presence, written a, as defined previously:

$$Pr(A|x, \bar{B}, H_d) = a \text{ and } Pr(\bar{A}|x, \bar{B}, H_d) = 1 - a.$$

In summary,

$$\underbrace{Pr(y|x, A, B, H_d)}_{1}\underbrace{Pr(A|x, B, H_d)}_{b_0 t_i} + \underbrace{Pr(y|x, \bar{A}, B, H_d)}_{\gamma'}\underbrace{Pr(\bar{A}|x, B, H_d)}_{1-b_0 t_i}.$$

and

$$\underbrace{Pr(y|x, A, \bar{B}, H_d)}_{1}\underbrace{Pr(A|x, \bar{B}, H_d)}_{a} + \underbrace{Pr(y|x, \bar{A}, \bar{B}, H_d)}_{\gamma'}\underbrace{Pr(\bar{A}|x, \bar{B}, H_d)}_{1-a}.$$

Incorporating these assignments in the denominator of Equation (6.30) gives

$$Pr(y|x, H_d) = Pr(y|x, B, H_d)r + Pr(y|x, \bar{B}, H_d)(1 - r)$$

$$= [b_0 t_i + \gamma'(1 - b_0 t_i)]r + [a + \gamma'(1 - a)](1 - r). \qquad (6.32)$$

6.4.4   Extended form of the likelihood ratio

Combining (6.31) and (6.32) leads to a likelihood ratio of the following form:

$$V = \frac{Pr(y|x, H_p)}{Pr(y|x, H_d)} = \frac{[b_0 t_c + \gamma(1 - b_0 t_c)]r + [a + \gamma(1 - a)](1 - r)}{[b_0 t_i + \gamma'(1 - b_0 t_i)]r + [a + \gamma'(1 - a)](1 - r)}. \qquad (6.33)$$

This expression involves several factors, such as the probability of background staining of comparable nature (b_0), the rarity of analytical characteristics (γ), the probability of transfer (t), relevance (r) and the probability of presence of staining by chance (a). This result may be simplified if the framework of circumstances is such that some assumptions can reasonably be introduced. Different scenarios are discussed in Biedermann and Taroni (2011).

6.4.5   Network construction and examples

As a starting point, consider the basic network illustrated in Figure 6.10(i) with node H for the crime level propositions and nodes Y and X for, respectively, the observations made on the crime stain and on control material from the victim. This converging connection, $H \rightarrow Y \leftarrow X$, reflects the conditioning defined in the general form of the likelihood ratio $V = Pr(y|x, H_p)/Pr(y|x, H_d)$. Each of the three nodes has two states: H_p and H_d for node H, y and $\bar{y}$ for node Y and x and $\bar{x}$ for node X. Here, x and y stand for the observation of a given characteristic (i.e. analytical feature) denoted Γ_1. Arguably, $\bar{x}$ and $\bar{y}$ denote all characteristics other than Γ_1.

In an argument to establish the relevance of the kind of trace material found in the case at hand, a so-called association proposition B was introduced. This proposition is added as a conditional for the proposition that accounts for the observations made on the stain found on the

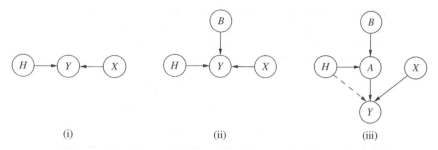

$$(i) \qquad\qquad\qquad (ii) \qquad\qquad\qquad (iii)$$

Figure 6.10   (i–iii) Bayesian network structures for modelling inference based on results of the analysis of transfer material found on a person of interest. Variable H represents the crime level propositions (e.g. assault) put forward by, respectively, the prosecution (H_p) and the defence (H_d). Node A has states 'the stain on the suspect comes from the victim (A)' and 'the stain on the suspect comes from some other person ($\bar{A}$)'. Node B stands for 'the victim was bleeding during the offence' (with $\bar{B}$ denoting the negation of B). Nodes Y and X stand for the analytical (e.g. genetic) characteristics of, respectively, the stain found on the suspect and control material from the victim. For explanation of the dotted edge, see Section 6.4.5.

suspect (y). Through Assumptions (6.28) and (6.29), the proposition B (i.e. bleeding of the victim during the offence) is considered independent of the characteristics of the victim's blood x and the principal proposition H (i.e. whether or not the suspect is the offender). An appropriate graphical representation of these relationships is shown by the model in Figure 6.10(ii).

The values assigned to the probability tables of nodes H and X are not crucial for the current analysis. Again the reason for this is that these nodes will be instantiated during the use of the Bayesian network. This is different for node B. The probabilities assigned to this node are relevance probabilities $Pr(B) = r$ and $Pr(\bar{B}) = 1 - r$.

The network constructed so far does not readily allow one to assign probabilities to the node Y in the same way as outlined in Section 6.4 because the intermediate association proposition A (i.e. the proposition according to which the crime stain comes from the victim) is not yet specified graphically. This latter proposition, represented in terms of a node A, acts as a direct conditioning for Y. In turn, A is conditioned on both B (i.e. the relevance of blood) and the main proposition H (i.e. the suspect is the offender). It is not conditioned, however, on the node representing the victim's characteristics, X. Knowledge of the victim's genotype should not affect our belief about whether or not the victim is the source of the stain found on the suspect, as long as nothing is known about the characteristics of that stain. Arguably, as shown in Figure 6.10(iii), no direct connection is specified between nodes X and A.

The probability table for node A, shown in Table 6.1, is completed according to the various assignments discussed in Section 6.4. In particular, if bleeding of the victim did not occur during the offence ($\bar{B}$ is true), the probability that the stain found on the suspect comes from the victim depends on the presence by chance, denoted a, which is independent of whether or not the suspect is in fact the offender (i.e. H_p and H_d). This assignment can be seen in columns three and five of Table 6.1. If the victim shed blood during the offence (B is true), the probability of there being a bloodstain from the victim on the suspect depends on the probability of no background (b_0) multiplied by the probability of transfer. For the latter factor, the probability t_c applies if the suspect is the offender (H_p), whereas t_i (short for 'innocent

Table 6.1 Probability table for the node A with states 'the stain on the suspect comes from the victim (A)' and 'the stain on the suspect comes from some other person $(\bar{A})$'.

$H:$	H_p		H_d	
$B:$	B	$\bar{B}$	B	$\bar{B}$
A	$b_0 t_c$	a	$b_0 t_i$	a
$\bar{A}$	$(1 - b_0 t_c)$	$(1 - a)$	$(1 - b_0 t_i)$	$(1 - a)$

The proposition B stands for 'the victim was bleeding during the offence' (with $\bar{B}$ denoting the negation of B). Variable H represents the crime level propositions of, respectively, the prosecution (H_p) and the defence (H_d). Factor a represents the probability of adventitious presence, factor b_0 the probability of zero background staining and factors t_c and t_i probabilities for crime-related and innocent transfer, respectively.

transfer') is used in the event that the suspect is not the offender (H_d). These probability assignments are given in columns two and four of Table 6.1.

It is worth noting that by introducing proposition A, node B no longer acts as a parent for node Y. This expresses the view that the assessment of the probability of observing the characteristics of the recovered stain (Y) would not be influenced by knowledge about the victim's shedding of blood during the offence (node B) if it were also known that the victim is the source of the stain (node A). Knowledge of 'source' thus is sufficient for assigning the probability of observing analytical characteristics for a given stain. It does not depend on the way in which the person of interest acquired the staining that is found to correspond.

This perception may not hold analogously for node H. If the victim is not the source of the stain found on the suspect $(\bar{A})$, the probability of observing the characteristic of interest depends on the population proportion. As noted above, this value may be assigned differently according to whether one is looking at foreign stains on criminals or one is looking at foreign stains on innocent people. This distinction is acknowledged by distinguishing between the population proportion γ (for the occurrence of the characteristic of interest amongst stains found on criminals) and γ' (for the occurrence of the characteristic of interest amongst stains found on innocent individuals). If this distinction is to be reflected by the Bayesian network, then the graph structure that is needed is the one with the dotted edge pointing from node H to node Y as shown in Figure 6.10(iii). If this distinction is not required, that is $\gamma = \gamma'$, then the direct connection between H and Y can be omitted. This is further clarified in Table 6.2 that contains the probabilities for node Y in a graph structure with a direct link between node H and node Y. Given the assumption $\gamma = \gamma'$, the table would contain the same set of values for the conditionals H_p and H_d. This indicates that the arc between H and Y entails no inferential force and could then be omitted.

Example 6.6 *(Relevant material found on a person of interest) Consider, as discussed throughout this section, a case of assault in which staining (e.g. biological) is found on a person of interest. In a first approach, assume that the rarity of the corresponding analytical characteristic (e.g. a DNA profile in case of blood) for stains found adventitiously on offenders, expressed in terms of the factor γ, does not differ from that for stains present on innocent individuals (factor γ'). Thus, $\gamma = \gamma'$. In addition, assume that blood is a relevant type of transfer material. This may be a robust assumption in a case in which it is not contested that the victim*

Table 6.2 Probability table for node Y, the analytical (e.g. genetic) characteristic of the stain found on the suspect.

$H:$	H_p				H_d			
$A:$	A		$\bar{A}$		A		$\bar{A}$	
$X:$	x	$\bar{x}$	x	$\bar{x}$	x	$\bar{x}$	x	$\bar{x}$
$Y:$ y	1	0	γ	γ	1	0	γ'	γ'
$\bar{y}$	0	1	$(1-\gamma)$	$(1-\gamma)$	0	1	$(1-\gamma')$	$(1-\gamma')$

Node X stands for the characteristic of control material from the victim. Lower case letters y and x denote a particular characteristic, such as Γ_1, and $\bar{y}$ and $\bar{x}$ denote characteristics other than Γ_1. Variable H represents the crime level propositions (e.g. assault) put forward by, respectively, the prosecution (H_p) and the defence (H_d). Node A has states 'the stain on the suspect comes from the victim (A)' and 'the stain on the suspect comes from some other person ($\bar{A}$)'.

was injured and bleeding. Using agreed notation, relevance is defined here in terms of the proposition B (i.e. 'the victim shed blood during the offence') and assigned the probability $Pr(B) = r = 1$. Note, however, that for other types of trace material, such as saliva, one may need to allow for uncertainty about relevance (e.g. when it may be difficult to judge whether the victim left saliva on the offender). The case with blood staining not being relevant, that is when $Pr(B) = r = 0$, is not discussed in further detail at this point because, from (6.33), it can be seen directly that the likelihood ratio reduces to 1. This result is reasonable since it says that, given that the victim did not bleed during the offence, blood of the same type as that of the victim found on the suspect would not help one to discriminate between the principal propositions (i.e. whether or not the suspect is the offender).

The above-mentioned assumptions about γ, γ' and B allow one to reduce (6.33) to the following equation:

$$V = \frac{b_0 t_c + \gamma(1 - b_0 t_c)}{b_0 t_i + \gamma(1 - b_0 t_i)}. \tag{6.34}$$

In this result, the probability for the adventitious presence, $Pr(A|\bar{B}) = a$, of a stain from the victim on the suspect is no longer present. This is a consequence of the fact that the event of interest has been defined as the presence of the victim's blood on the suspect for situations in which bleeding by the victim did not occur during the crime (i.e. blood is not a relevant category of trace material in the case). Thus, by assuming that blood is relevant in the case ($r = 1$), then adventitious presence is no longer an issue [the terms on the right in both the numerator and the denominator of (6.34) cancel out].

Next, assume also that the circumstances are such that it is impossible for there to be no background of blood staining on the suspect: $b_0 = 0$. Such an assumption may appear unrealistic. However, it is made here for exploratory reasons because extreme assumptions often provide special cases that represent good means for examining logical implications. So, assuming $b_0 = 0$ means that some blood must necessarily be present as background. However, since one is actually observing a bloodstain, this stain must be the one that is the background. Arguably, the stain at hand should not allow one to say anything about whether or not the suspect is the offender. Looking at the likelihood ratio, this conclusion is actually confirmed, *since by assuming $b_0 = 0$, (6.34) reduces to*

$$V = \frac{b_0 t_c + \gamma(1 - b_0 t_c)}{b_0 t_i + \gamma(1 - b_0 t_i)} = \frac{\gamma}{\gamma} = 1.$$

On the other hand, if one assumes it impossible to have background blood staining on the suspect, that is $b_0 = 1$, then the likelihood ratio becomes

$$V = \frac{t_c + \gamma(1 - t_c)}{t_i + \gamma(1 - t_i)} = \frac{t_c(1 - \gamma) + \gamma}{t_i(1 - \gamma) + \gamma}. \tag{6.35}$$

The only remaining factors in this result are the population proportion γ and the transfer probabilities t_c and t_i, that is the probabilities of transfer of blood, given that the suspect is, or is not, the offender.

From this, one can consider two further possibilities. On the one hand, assume an extreme value of $\gamma = 1$; the characteristic of interest is found in all potential sources. Obviously, such a characteristic has no interest from a forensic point of view because it does not allow one to discriminate between potential sources. The likelihood ratio reflects this appropriately in that it reduces to $\gamma/\gamma = 1$. On the other hand, one can also assume a value for γ that is close to zero. This may be of interest to biological trace material, where γ may often attain particularly low values. The likelihood ratio then approximates to

$$V = \frac{t_c(1 - \gamma) + \gamma}{t_i(1 - \gamma) + \gamma} \approx \frac{t_c}{t_i}.$$

In summary, in cases in which the victim shed blood during the crime (i.e. blood as a category of trace material is relevant to the case), and (i) the suspect can reasonably be assumed to have no blood staining on him as a background (i.e. $b_0 = 1$), and (ii) the population proportion of the observed corresponding characteristic tends to be particularly low, the likelihood ratio becomes dominated by the probabilities of transfer of blood under the assumptions that the suspect is, or is not, the offender. This is expressed by the probabilities of 'criminal' (t_c) and 'innocent' (t_i) transfer (including persistence and recovery).

This conclusion is overridden in one particular case. If 'innocent' transfer (t_i) is believed to be impossible and 'criminal' transfer (t_c) is believed to be necessarily true, then the population proportion becomes again a dominant factor. In fact, for $t_c = 1$ and $t_i = 0$, the likelihood ratio in (6.35) reduces to

$$V = \frac{1}{\gamma}.$$

This is a well-known result in forensic literature on evaluation of discrete data for source level propositions. In particular, this result agrees with Evett's (Evett 1993) offence level likelihood ratio for offender to scene transfer in a setting in which there is a single offender and a single crime stain with a relevance probability that is maximal (i.e. $r = 1$).

Example 6.7 *(Relaxing assumptions on relevance) When there is uncertainty about the relevance of the kind of trace material found on a person of interest, then there are two products to consider in both the numerator and the denominator of the likelihood ratio (6.33). This makes it less obvious to explore the likelihood ratio in its general form.*

Two assumptions may be made. Firstly, as was done in Example 6.6, let the population proportions for stains found on suspects and innocent people be equal: $\gamma = \gamma'$. Secondly, *consider that, for settings in which blood shedding by the victim did not occur (i.e. blood is not relevant), there is no possibility for a bloodstain from the victim to be present on the suspect by chance alone (proposition A): $Pr(A|\bar{B}, H_p) = Pr(A|\bar{B}, H_d) = a = 0$. Such an assumption may be acceptable in situations in which, according to the framework of circumstances, the suspect and victim are not known to each other and there is no possibility that a stain from the victim could have arrived on the suspect's clothing.*

According to this last assumption, any stain present on the suspect under the assumption of the victim not having shed blood during the offence must come from some person other than the victim. The probability that this individual will have the observed analytical characteristic is γ. The products on the right-hand-side in both the numerator and the denominator of (6.33) thus reduce to (where the former assumption of equal population proportions is also made): $\gamma(1 - r)$. Equation (6.33) thus becomes

$$V = \frac{[b_0 t_c + \gamma(1 - b_0 t_c)]r + \gamma(1 - r)}{[b_0 t_i + \gamma(1 - b_0 t_i)]r + \gamma(1 - r)},$$

which further reduces to

$$V = \frac{t_c b_0 r(1 - \gamma) + \gamma}{t_i b_0 r(1 - \gamma) + \gamma}.$$

For given values of the population proportion γ and the probabilities of transfer t_c and t_i, the value of the likelihood ratio will have an upper limit for $r = b_0 = 1$. As soon as uncertainty about the relevance (factor r), the absence of background staining (factor b_0), or both is introduced, the likelihood ratio will be reduced still further. However, this effect will not be so pronounced as the population proportion γ decreases.

There is an analogy between this result and considerations in the previous example in that as the population proportion decreases so the contribution of the transfer probabilities t_c and t_i to the value of the likelihood ratio increases. For γ close to zero, and acknowledging that this weakens the contribution of the relevance factor (r) and the probability of no background staining (b_0), the likelihood ratio approximates to

$$V \approx \frac{t_c}{t_i}.$$

7

Evaluation of DNA profiling results

Interest in the probabilistic evaluation of DNA profiling results has grown considerably during the past 20 years. Topics such as the assessment of transferred DNA traces in the form of mixtures, the consideration of error probabilities and the effects of database searches have been responsible for the increase in interest in forensic statistics amongst forensic scientists. There are other complications in DNA analysis such as the possibility of the occurrence of a mutation, which have to be considered in certain applications, typically kinship analysis. There are many aspects to the probabilistic evaluation of DNA profiling results, which is reflected in the size of the published literature on the use of Bayesian networks, which consider these evaluative topics. DNA profiling analyses are, arguably, the most reported applications of Bayesian networks. As will become apparent throughout this chapter, an important aspect of Bayesian network modelling for DNA profiles relies on the proper representation of basic entities, such as genes and genotypes, which may be made at different levels of detail. Such principal network fragments may be repeatedly invoked, given recognized and extensive biological theory (e.g. Mendelian inheritance), to represent the genetic constitution of various observed and unobserved individuals. This explains in part why rigorous Bayesian network models may be constructed even for scenarios that are large and complex at the outset.

7.1 DNA likelihood ratio

Imagine a crime has been committed. Examination of the scene has revealed the presence of a single blood stain. From the position where the stain was found, as well as its condition (e.g. freshness), investigators believe the stain to be relevant to the case. The crime stain is submitted to a forensic laboratory where some kind of DNA typing technique is applied. Imagine further that, on the basis of information completely unrelated to the recovered blood, a suspect

Bayesian Networks for Probabilistic Inference and Decision Analysis in Forensic Science, Second Edition.
Franco Taroni, Alex Biedermann, Silvia Bozza, Paolo Garbolino and Colin Aitken.
© 2014 John Wiley & Sons, Ltd. Published 2014 by John Wiley & Sons, Ltd.

is apprehended. Blood provided by this individual for comparison purposes is subjected to the same DNA profiling analyses. For a situation in which the suspect is found through a search in a database, refer to Section 7.10.

If the profile of the blood of the suspect and the crime stain were found to be of the same type, Γ_1 for instance, then a question of interest to recipients of expert information is 'What evidential value is to be assigned to this correspondence?' As argued throughout this book, forensic scientist reply to such a question most appropriately if their reasoning and reporting is guided by a likelihood ratio. In order to consider this in some further detail here, let E denote the correspondence between the profile of the blood provided by the suspect and the profile of the crime stain. The term *correspondence* will not be considered further at present. It is possible to introduce a distinction between a 'true' correspondence and an 'observed' correspondence; issues, needs and implications related to this additional level of distinction are considered in further detail in Section 7.11.

Imagine further that the assignment of the likelihood ratio, with I representing the available non-scientific background information, is based on a pair of propositions H_p and H_d at the source level, defined as follows: 'The suspect is the source of the stain (H_p)' and 'An unknown person, unrelated to the suspect, is the source (H_d)'. The likelihood ratio considers the probability of the profile correspondence given each of these propositions and is written follows:

$$V = \frac{Pr(E|H_p, I)}{Pr(E|H_d, I)}. \tag{7.1}$$

The variable E may be considered as a combination of two distinct components, E_s and E_c, denoting the DNA profile of the suspect and the crime stain, respectively. Following the arguments set out in Section 3.3.1, the likelihood ratio can then be expressed as

$$V = \frac{Pr(E_c|E_s, H_p, I)}{Pr(E_c|E_s, H_d, I)}. \tag{7.2}$$

Assume that the DNA typing system is sufficiently reliable that distinct items of biological material, but from a common source, will be found to correspond (in the analysed genetic traits) and that there are no false negatives. Take the case in which both the profile of the suspect and the crime stain are of type Γ_1, say. Then, assuming that the suspect is the source of the crime stain and knowing that he is of type Γ_1, the probability of the recovered stain being of type Γ_1 is 1: $Pr(E_c = \Gamma_1|E_s = \Gamma_1, H_p, I) = 1$.

When the alternative proposition is assumed to be true (H_d), the suspect and the donor of the stain are two different persons. Assuming that the DNA profiles of the suspect and the true donor of the stain are independent, one is allowed to write then $Pr(E_c = \Gamma_1|E_s = \Gamma_1, H_d, I)$ as $Pr(E_c = \Gamma_1|H_d, I)$. This term is generally thought of as *profile probability*, that is the probability with which an unknown person in a relevant population is believed to have a Γ_1 profile.

Usually, a DNA profile consists of pairs of alleles at several loci. An individual is said to be *homozygous* at a particular locus if the genotype consists of two indistinguishable alleles, say A_iA_i, for example. If a genotype consists of two different alleles (e.g. $A_iA_j, i \neq j$) the individual is considered as *heterozygous* at the locus of interest. Individual allele probabilities may be used to calculate genotype probabilities, denoted P_{ij}. Assuming Hardy–Weinberg equilibrium,

$$P_{ij} = 2\gamma_i\gamma_j \ (i \neq j),$$

$$= \gamma_i^2 \ (i = j). \tag{7.3}$$

Notice that these considerations involve a widely accepted simplification. In reality, the probative value of a correspondence between the profile of a crime stain and that of a suspect needs to take into account the fact that there is a person (the suspect) who has already been seen to have that profile (of type Γ_1). The term of interest is a conditional probability $Pr(E_c = \Gamma_1 | E_s = \Gamma_1, H_d, I)$, which can be quite different from $Pr(E_c = \Gamma_1 | H_d, I)$ (Weir 2000). Stated otherwise, knowledge of the suspect's genotype should affect one's uncertainty about the offender's genotype in cases where, for example, the offender is a close relative to the suspect or both the suspect and the offender share ancestry because of membership in the same population.

In fact, observation of a gene in a sub-population increases the chance of observing another of the same type. Hence, within a sub-population, DNA profiles with corresponding allele types are more common than suggested by the independence assumption, even when two individuals are not directly related. The conditional probability of interest, in this context also termed *random match probability* or, more properly, *conditional genotype probability*, incorporates the effect of population structure or other dependencies between individuals, such as that imposed by family relationships (Balding and Nichols 1994). Systematic incorporation of such elements of dependency in Bayesian network models is challenging and leads to more complex models (Green and Mortera 2009; Hepler and Weir 2007).

The form of the likelihood ratio presented in (7.2) is a general one. More technical presentations can be found, for instance, in Dawid and Mortera (1996). Generally, more detailed developments may be highly case dependent, in particular when account has to be taken of additional genotypic information on target individuals, such as close relatives. A major aim of this chapter is to clarify that evaluative subtleties in probabilistic procedures for DNA profiling results can be captured and tracked both efficiently and elegantly through Bayesian networks.

7.2   Network approaches to the DNA likelihood ratio

7.2.1   The 'match' approach

Consider the variable E as a global assignment for the event that the profile of the crime stain and the suspect correspond. In Chapter 4, the discussion focussed on a basic two-node network fragment allowing knowledge about E to be used for drawing an inference to propositions at the source level. If the latter propositions are represented by a binary node H, then an appropriate Bayesian network fragment is of the form $H \rightarrow E$. Besides the prior probabilities that need to be specified for the node H, two further probabilities are essential for the quantitative definition of this network fragment: $Pr(E|H_p, I)$ and $Pr(E|H_d, I)$. It has been assumed that if the suspect were the source of the crime stain, then the profile of suspect's blood would certainly be found to correspond: $Pr(E|H_p, I) = 1$. If the crime stain came from a person other than the suspect, the probability of the DNA profiling results may be interpreted as the profile probability: $Pr(E|H_d, I) = \gamma$. As noted above, these two assignments represent the numerator and denominator, respectively, of the likelihood ratio formulated by (7.1).

7.2.2   Representation of individual alleles

The representation of outcomes of DNA profiling analyses in terms of a single node E, representing a correspondence of two profiles (Section 7.2.1), is sometimes used in more general discussions of patterns of reasoning based on scientific findings (e.g. Garbolino and

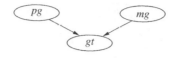

Figure 7.1   Representation of a genotype, node *gt*, with *pg* and *mg* denoting the alleles inherited from the paternal (*p*) and maternal (*m*) sides, respectively.

Taroni 2002). Often, however, it is useful, or even necessary, to consider a less coarse level of detail when evaluating DNA profiling results. An important step ahead in that respect was achieved by Dawid et al. (1999, 2002), who proposed more fine-grained network fragments focusing on individual genes and genotypes. These authors have shown how appropriate graphical structures for Bayesian networks can be derived from initial pedigree representations of forensic identification problems. Parts of their approach are basic representational schemes, usable as repeatable modules in analogous situations.

An example of this is shown in Figure 7.1. Here, the node *gt*, representing a genotype, is modelled as a logical combination of the alleles inherited from the mother and father, respectively. These parentally inherited genes are represented by the nodes *mg* and *pg*, to be read 'maternal gene' and 'paternal gene', respectively. A gene, denoted A here for example, can take one of several different forms, also known as *alleles*. Suppose that there are n alleles at gene A. These alleles may be denoted $A_1, A_2, \ldots, A_n$. In the Bayesian network shown in Figure 7.1, the states A_1, A_2 and A_x are assumed for the nodes *mg* and *pg* where the third state, A_x, is an aggregation of all unobserved alleles $A_3, \ldots, A_n$. This is an expression of the view according to which a gene is represented by a random variable and a realized value of such a variable is an allele. On the basis of the set of alleles assigned to the nodes *pg* and *mg*, it is then possible to define the states $A_1 A_1, A_1 A_2, A_1 A_x, A_2 A_2, A_2 A_x$ and $A_x A_x$ for the genotype node *gt*.

Using these notions, consider now the construction of a Bayesian network for evaluating the DNA likelihood ratio. Instead of reducing the findings to E, as was assumed by the so-called match approach outlined earlier in Section 7.2.1, present the typing results for the suspect and the crime stain as distinct components: $E = (E_s, E_c)$. Recall from Section 7.1 that the variables E_s and E_c were used, rather generally, to denote the *profile* of the suspect's blood and the crime stain, respectively. For ease of argument, the definition of the findings is now restricted to the typing results of a single locus or marker. Thus, let *sgt* and *tgt*, notation for 'suspect genotype' and 'trace genotype' respectively, denote the allelic configuration at a certain locus, where the term *trace* is simply used as an alternative expression for 'crime stain'.

Modelling the genotype of the suspect and crime stain separately suggests the need of two network fragments of the kind shown in Figure 7.1. One such network fragment consists of the nodes 'suspect genotype' (*sgt*), 'suspect paternal gene' (*spg*) and 'suspect maternal gene' (*smg*). The other contains the nodes 'trace genotype' (*tgt*), 'trace paternal gene' (*tpg*) and 'trace maternal gene' (*tmg*). Next, a logical combination is required between these two network fragments and the target propositions at the source level. This may be achieved by conditioning the paternal genes of the crime stain by both the paternal genes of the suspect and the propositions at the source level. This is an expression of the idea that if the suspect were known to be the source of the crime stain, then his allelic configuration provides relevant information for the allelic configuration of the crime stain. Graphically, the following connections are adopted: $H \to tpg$, $H \to tmg$, $spg \to tpg$ and $smg \to tmg$. The resulting Bayesian network is shown in Figure 7.2. If the states of the variables are defined as mentioned earlier, then the numerical specification outlined in the following may be adopted.

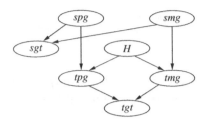

Figure 7.2 Bayesian network for evaluating DNA typing results obtained from a crime stain and blood provided by a suspect. H, the suspect is the source of the crime stain; sgt and tgt, genotype of the suspect and the crime stain, respectively; spg and smg, suspect paternal and suspect maternal genes and tpg and tmg, trace paternal and trace maternal genes.

The probability tables associated with the nodes spg and smg contain unconditional probabilities that represent the rarity of these alleles in a relevant population. The probabilities of the alleles A_1 and A_2 are denoted γ_1 and γ_2. The probability of an allele being different from A_1 and A_2 (i.e. one of $A_3, \ldots, A_n$ or A_x) is then $1 - \gamma_1 - \gamma_2 = \gamma_x$. The assumption made here is that these probabilities are fixed. More generally, notice however that the actual values of these probabilities are not crucial for the intended inference task. The reason for this is that, given H_p, the suspect's known genotype will determine the actual probabilities of the respective parental gene nodes, whereas under H_d, the probability distributions of the nodes tpg and tmg (and, thus, tgt) are taken to be independent of the nodes associated with the suspect's genotype.

For nodes sgt and tgt, the probability tables can logically be completed using values equating 0 and 1. For example, if $spg = A_1$ and $smg = A_2$, then $sgt = A_1A_2$. Stated otherwise, all the states of a genotype node not strictly containing the alleles that represent the actual realizations of the respective parental gene nodes must be false.

For nodes tpg and tmg, if the crime stain comes from the suspect (i.e. $Pr(H = H_p) = 1$), then the state of a trace parental gene must equate the state of the respective suspect parental gene. That is, $Pr(tpg = A_i | spg = A_i, H_p) = 1$ and $Pr(tmg = A_i | smg = A_i, H_p) = 1$ for $i = 1, 2, x$. If the crime stain does not come from the suspect, then the probability of a crime stain parental gene being A_1, A_2 or A_x is given by the allele proportions in the population of interest. For example, we can write $Pr(tpg = A_i | spg = A_i, H_d)$ more shortly as $Pr(tpg = A_i | H_d) = \gamma_i$, for $i = 1, 2, x$. Again, the model here assumes that these probabilities are fixed.

A summary of these assignments can be found in Tables 7.1–7.3.

Table 7.1 Unconditional probabilities applicable to the nodes modelling the suspect's parental genes, that is spg and smg.

spg, smg:		
A_1		γ_1
A_2		γ_2
A_x		$\gamma_x = 1 - \gamma_1 - \gamma_2$

Possible node states are A_1, A_2 and A_x, with $A_x = A_3, \ldots, A_n$.

Table 7.2 Conditional probabilities assigned to the nodes 'trace paternal gene' (*tpg*) and 'trace maternal gene' (*tmg*).

H:	H_p			H_d		
spg(smg):	A_1	A_2	A_x	A_1	A_2	A_x
tpg(tmg): A_1	1	0	0	γ_1	γ_1	γ_1
A_2	0	1	0	γ_2	γ_2	γ_2
A_x	0	0	1	γ_x	γ_x	γ_x

Factor H has two states, H_p, the suspect is the source of the crime stain, and H_d, the suspect is not the source of the crime stain. The nodes, suspect paternal gene, *spg*, and suspect maternal gene, *smg*, have three states, alleles A_1, A_2 and A_x, with $A_x = A_3, \ldots, A_n$.

Table 7.3 Conditional probabilities applicable for a genotype node, that is, *sgt* or *tgt*.

spg(tpg):	A_1			A_2			A_x		
smg(tmg):	A_1	A_2	A_x	A_1	A_2	A_x	A_1	A_2	A_x
sgt(tgt): A_1A_1	1	0	0	0	0	0	0	0	0
A_1A_2	0	1	0	1	0	0	0	0	0
A_1A_x	0	0	1	0	0	0	1	0	0
A_2A_2	0	0	0	0	1	0	0	0	0
A_2A_x	0	0	0	0	0	1	0	1	0
A_xA_x	0	0	0	0	0	0	0	0	1

The nodes, suspect (trace) paternal gene, *spg* (*tpg*), and suspect (trace) maternal gene, *smg* (*tmg*), have three states, alleles A_1, A_2 and A_x ($A_x = A_3, \ldots, A_n$).

Example 7.1 *(Bayesian network for a one stain one-offender scenario) Imagine DNA profiling results for a single marker. The suspect's blood and the crime stain are both found to be of type A_1A_2. Consider a Bayesian network as shown in Figure 7.2 and let the states of the parental gene nodes consist of, respectively, the states A_1 A_2 and A_x. Let γ_1 and γ_2 be the population proportions for alleles A_1 and A_2, respectively. A value of 0.05 is assumed for both γ_1 and γ_2.*

For evaluation of the numerator of the likelihood ratio, one needs to consider the probability of the crime stain being of type A_1A_2, given that the prosecution's case is true and given that the suspect's blood is of type A_1A_2. To this end, the node H is set to H_p and the node sgt is set to A_1A_2. In the node tgt, one can now read, as shown in Figure 7.3(i), the probability of the trace being of type A_1A_2 given the stated conditions: $\Pr(tgt = A_1A_2|sgt = A_1A_2, H_p) = 1$.

The denominator of the likelihood ratio is a consideration of the probability that the stain is of type A_1A_2 assuming that the defence case is true. An evaluation of this scenario is shown in Figure 7.3(ii). The node H is instantiated to H_d and the effect of this is propagated to the node representing the findings for the trace, tgt, which displays $\Pr(tgt = A_1A_2|H_d) = 0.005$ (0.50%). Notice that when assuming H_d to be true, any information regarding the suspect's genotype has no effect on the probability of the crime stain's genotype.

As a result of this analysis, one obtains a likelihood ratio of $1/0.005 = 200$. This result is in agreement with what may be obtained using the so-called match approach: here the Bayesian network consists of $H \to E$ and the likelihood ratio is given by $1/2\gamma_1\gamma_2$. Notice further that

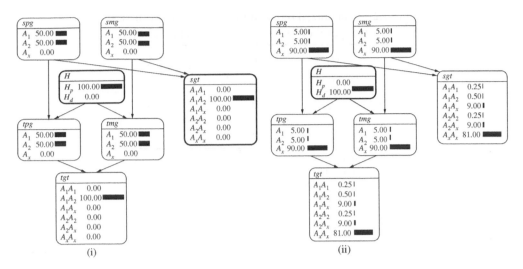

Figure 7.3   Evaluation of the likelihood ratio in a case where the findings consist of a crime stain and a suspect with same genotype, that is A_1A_2: (i) the prosecution's case and (ii) the defence case. Node H has two states, H_p, the suspect is the source of the crime stain, and H_d, the suspect is not the source of the crime stain. Nodes spg, smg, tpg and tmg, the paternal and maternal genes for the suspect and the trace have three states A_1, A_2 and A_x. Nodes sgt and tgt, the suspect and trace genotypes, have six states, $A_1A_1, A_1A_2, A_1A_x, A_2A_2, A_2A_x$ and A_xA_x.

a 'top-down' analysis has been performed here. That is, the node H has been alternatively instantiated to H_p and H_d. By doing so, it is not necessary to care about prior probabilities for the node H, since they do not enter the probability calculations. This view conforms to the requirement according to which forensic scientists should solely focus on the ratio of the probabilities of the findings given the pair of competing propositions considered, that is the likelihood ratio.

These results may appear elementary and, thus, an approach using Bayesian networks unnecessarily technical. At this point, however, the main purpose was only that of illustrating the general principle. The modelling approach will demonstrate better its added value when more advanced situations are considered in later parts of this chapter (e.g. with missing individuals or particular alternative propositions).

7.2.3   Alternative representation of a genotype

The representation of an individual's genotype in terms of a single node gt, as shown in Figure 7.1, is a general one. It has the primary merit of illustrating the logic of specifying Bayesian network structures for inheritable traits. The generic approach using allele designations such as A_1, A_2 and A_x may however be viewed as a limitation in practical usage because an adjustment of allele probabilities may be necessary in order to account for the allelic configuration of the items examined in a given case. One could avoid this by specifying, for each parental gene nodes pg and mg, the full repertory of alleles at a given locus, but this leads to a rapid increase in the size of the probability table for the node gt. For the purpose of illustration, consider the locus TH01 for which one may observe, depending on the characteristics of a

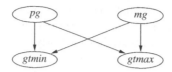

Figure 7.4   Representation of an individual's genotype as the minimum (*gtmin*) and maximum (*gtmax*) of the two parental gene nodes *pg* and *mg*, inherited via the paternal and maternal lines, respectively.

chosen item, up to eight distinct alleles. With eight alleles, there are $[8(8+1)]/2 = 36$ possible genotypes and these would make up the states of the node *gt*. Along with parental gene nodes *pg* and *mg* with eight states each, this would lead to a node table for the node *gt* of size $8 \times 8 \times 36 = 2304$. Although current Bayesian network software offers ways to specify large node tables automatically using expressions, such a modelling approach may become increasingly impractical for larger networks.

One way to avoid this consists in separating the node *gt* into two distinct nodes *gtmin* and *gtmax*. These latter two nodes represent, respectively, the minimum and the maximum of the two parental gene nodes *pg* and *mg*. That is, for a given pair of alleles observed for a given individual, one represents the smaller and the larger (in terms of the allele number) of these separately. By linking both nodes *gtmin* and *gtmax* to both *pg* and *mg*, one accounts for the reality that one can observe the allele number, but not tell whether an allele is actually inherited paternally or maternally. The node tables of *gtmin* and *gtmax* are completed logically using zeros and ones. For example, $Pr(gtmin = i|pg = i, mg = j) = 1$ for $i \leq j$ and $Pr(gtmin = j|pg = i, mg = j) = 1$ for $i \geq j$, where i and j cover the set of eight alleles for the locus TH01. Although there are still $8 \times 8 \times 8 = 512$ probabilities to assign to each of these nodes, this can readily be handled in current Bayesian network software through the specification of expressions. In Hugin language, for example, it is possible to specify the probability tables for the nodes *gtmin* and *gtmax* through, respectively, the expressions `min(pg,mg)` and `max(pg,mg)`.

Genotypic findings are entered in a network fragment of the kind shown in Figure 7.4 by specifying separately each of the two observed alleles. This is the main difference with respect to the network shown earlier in Figure 7.1 where an observed genotype is entered as a single instantiation of the node *gt* to a state representing a pair of alleles. The constructional approach implied by Figure 7.4 is discussed in further detail in later sections that focus on the topic of object-oriented Bayesian networks. In order to keep the representational complexity at a tractable level, the remainder of this chapter relies on the generic genotype representation introduced in Section 7.2.2. It is assumed that the networks presented in forthcoming sections can also be realized with the more detailed genotype modelling approach outlined in this section.

7.3   Missing suspect

One may ask whether there is a need for such sophisticated models, such as Figure 7.2, when the same result can also be obtained by basic algebraic calculus and arithmetic, such as (7.3). Although this is a legitimate question to ask, Dawid et al. (2002) point out that the very aim of using a probabilistic network approach is to extend the considerations to scenarios where one needs to account for genetic information of further individuals. This may be necessary if, for example, biological material from one or more target individuals cannot be obtained but that from one or more close relatives is available. Clearly, a purely arithmetic solution to such problems can become increasingly difficult.

In order to illustrate this aspect, consider again a blood stain of some relevance found on the scene of a crime. The principal suspect is unavailable for profiling. However, the suspect is known to have a brother and this individual is willing to provide reference material (i.e. blood) for analysis. It is of interest to ask how one is to use knowledge of the brother's genotype for inferring something about the suspect's genotype. In addition, it will also be of interest to draw an inference to the propositions at source or crime level. These are typical examples of questions for which meaningful answers may be found through the use of Bayesian networks.

As a starting point, consider again the Bayesian network shown in Figure 7.2. This model accounts for genotypic information available from the suspect and the crime stain. Now, if one were to extend the considerations to the genotype of the suspect's brother, a further network fragment as shown in Figure 7.1, representing the genotype of an individual (i.e. the brother), must be incorporated in the existing network. Let the brother's genotype be modelled by three nodes arranged in a converging connection: $bpg \rightarrow bgt \leftarrow bmg$. The nodes are defined as follows: brother genotype (bgt), brother paternal gene (bpg) and brother maternal gene (bmg). It is such a network fragment that needs to be logically combined with the network relating the suspect's genotype to that of the crime stain.

The suspect and the brother will be assumed here to be full brothers, that is they share the same mother and father. For such a setting, it is the gene configurations of the parents of the mother and the father that are relevant for determining the configuration of the suspect's and the brother's parental genes. Let the parental genes of the mother and the father be defined as follows: mother paternal gene (mpg), mother maternal gene (mmg), father paternal gene (fpg) and father maternal gene (fmg). This enables one to adopt two sets of connections. On the one hand, the suspect's and the brother's paternal gene nodes (spg and bpg) are both conditioned on their father's parental gene nodes (fpg and fmg). This results in the following set of arcs: $fpg \rightarrow spg$, $fmg \rightarrow spg$, $fpg \rightarrow bpg$ and $fmg \rightarrow bpg$. On the other hand, the suspect's and the brother's maternal gene nodes (smg and bmg) are both conditioned on their mother's parental gene nodes (mpg and mmg). The set of adopted arcs is then as follows: $mpg \rightarrow smg$, $mmg \rightarrow smg$, $mpg \rightarrow bmg$ and $mmg \rightarrow bmg$.

These structural assumptions are shown graphically in Figure 7.5. Note that the node representing the suspect's genotype, sgt, has been maintained although the aim is to evaluate a scenario in which the suspect is missing. The reason for this is that useful insight might be gained in how knowledge about the brother's genotype effects the probability distribution of the node modelling the suspect's genotype.

The Bayesian network discussed here contains further differences with respect to the one discussed in Section 7.2. Notably, an extension is made from the source node, here denoted F, to the propositions at the crime level, denoted H. This transition is made analogously to the procedure described in Section 6.1. In particular, a node G, defined as 'the stain comes from the offender', is adopted in order to account for the uncertainty in relation to the relevance of the recovered stain.

The node probability tables are essentially the same as discussed before, except for those of the nodes representing the suspect's and the brother's parental genes. These nodes do not contain unconditional probabilities as was the case for the network shown in Figure 7.2. In the Bayesian network discussed here, the parental gene nodes of the suspect and the brother

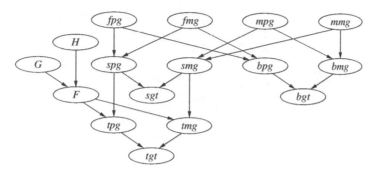

Figure 7.5 Bayesian network for evaluating DNA typing results when genotypic information from the suspect's brother is available. Node H is the crime level node with two states, H_p, the suspect is the offender, and H_d, the suspect is not the offender. Node F is the source level node, with two states, F, the suspect is the source of the crime stain, and $\bar{F}$, the suspect is not the source of the crime stain. Node G is the relevance node, with two states, G, the crime stain was left by the offender, and $\bar{G}$, the crime stain was not left by the offender. Nodes $fpg, fmg, mpg, mmg, spg, smg, bpg, bmg, tpg$ and tmg are gene nodes, with three states, A_1, A_2 and A_x. Nodes sgt, bgt and tgt are genotypic nodes, with six states, $A_1A_1, A_1A_2, A_1A_x, A_2A_2, A_2A_x$ and A_xA_x. Note that f denotes father, p paternal, m mother (if in first place) or maternal (if in second place), s suspect, b brother and t trace (if in first place).

Table 7.4 Conditional probabilities assigned to the node spg, the suspect's paternal gene.

fpg:		A_1			A_2			A_x	
fmg:	A_1	A_2	A_x	A_1	A_2	A_x	A_1	A_2	A_x
spg: A_1	1	0.5	0.5	0.5	0	0	0.5	0	0
A_2	0	0.5	0	0.5	1	0.5	0	0.5	0
A_x	0	0	0.5	0	0	0.5	0.5	0.5	1

Nodes fpg and fmg are the paternal and maternal genes of the suspect's father. In each case, the gene can take one of three forms, A_1, A_2 and A_x.

are conditioned on the parental genes of the mother or the father. For illustration, consider the probability table for the node spg, where the possible alleles are, as before, A_1, A_2 and A_x. For $i = 1, 2, x$, the probability table contains $Pr(spg = A_i | fpg = fmg = A_i) = 1$, whereas for $i, j = 1, 2, x$ and $i \neq j$, it contains

$$Pr(spg = A_i | fpg = A_i, fmg = A_j) = Pr(spg = A_j | fpg = A_i, fmg = A_j) = 0.5.$$

A summary of the probability table of the node spg is given in Table 7.4. This probability table is applicable analogously for the nodes smg, bpg and bmg.

Note that the current Bayesian network can readily be modified for analysing other scenarios where a probabilistic study would generally be very tedious if approached in a purely arithmetic way. For example, when biological material is available from the suspect's parents (mother, father or both), a Bayesian network as shown in Figure 7.6 could be used, although,

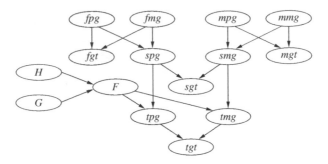

Figure 7.6 Bayesian network for evaluating DNA typing results when genotypic information from the suspect's parents is available. Node H is the crime level node with two states, H_p, the suspect is the offender, and H_d, the suspect is not the offender. Node F is the source node, with two states, F, the suspect is the source of the crime stain, and $\bar{F}$, the suspect is not the source of the crime stain. Node G is the relevance node, with two states, G, the crime stain was left by the offender, and $\bar{G}$, the crime stain was not left by the offender. Nodes $fpg, fmg, mpg, mmg, spg, smg, tpg$ and tmg are gene nodes, with three states, A_1, A_2 and A_x. Nodes fgt, mgt, sgt and tgt are genotypic nodes, with six states, $A_1A_1, A_1A_2, A_1A_x, A_2A_2, A_2A_x$ and A_xA_x. Note that f denotes father, p paternal, m mother (if in first place) or maternal (if in second place), s suspect and t trace (if in first place).

at this point, numerical examples are not considered in further detail. Instead, Section 7.4 continues the discussion with the study of further modifications.

7.4   Analysis when the alternative proposition is that a brother of the suspect left the crime stain

7.4.1   Revision of probabilities and networks

In Section 7.2, Bayesian networks were discussed for the evaluation of the so-called one-trace one-offender scenarios in which the suspect's blood was found to be of the same type as that of a crime stain. Under the general assumptions stated there, the value of the likelihood ratio reduced to the inverse of the profile probability of the observed combination of characteristics. For example, the likelihood ratio for a correspondence in the genotype A_1A_2 with the propositions chosen at the source level was given as $1/2\gamma_1\gamma_2$, where γ_1 and γ_2 denote, respectively, the individual probabilities of the alleles A_1 and A_2.

A modification to this scenario was discussed in Section 7.3 where it was assumed that the individual suspected of leaving the stain was unavailable. For such a scenario, a Bayesian network was constructed and discussed (Figure 7.5), allowing an inference of the suspect's genotype to be made when genotypic information of a sibling of the suspect, that is the suspect's brother, is available.

In this section, yet another scenario will be considered. Imagine a case in which the alternative proposition forwarded by the defence is not, as was assumed in the previous scenarios, that another (unrelated) person left the trace but that a sibling of the suspect, such as a brother, is the source of the stain. Whilst unrelated individuals are rather unlikely to share the same alleles at a certain locus, brothers will share zero, one or two alleles. However, it may well be that a brother is not available for DNA typing because he may be missing or refuse to co-operate. In such a case, the inferential problem consists of inferring the genotypic probabilities of the brother, given knowledge of the suspect's genotype.

An approach for evaluating such scenarios has been described by Evett (1992). This approach considers a pair of propositions at the source level, denoted F here: 'The suspect left the crime stain (F)' and 'A brother of the suspect left the crime stain $(\bar{F})$'. There is a crime stain and a suspect, and both were found to be of genotype A_1A_2. The numerator of the likelihood ratio considers the probability of this correspondence, given that the suspect is the source of the stain. Following the development discussed in Section 7.1, this probability can be set to 1. It is less obvious how the denominator ought to be evaluated, but one may gain a reasonable idea of the order of magnitude of the denominator if one imagines a case in which the alleles A_1 and A_2 are both very rare. Then, it is very likely that one of the suspect's parents has a genotype of the kind A_1A_x and the other parent has a genotype of the kind A_2A_x. Again, A_x is notation for the set of alleles $A_3, \dots, A_n$. Next, assume that the alleles A_1 and A_2 are so rare that parental configurations such as A_1A_1/A_2A_x and A_1A_2/A_2A_x may be ignored. If the parental genotype configuration is A_1A_x/A_2A_x, then the probability that a brother of the suspect would be of type A_1A_2 is $1/4$. Consequently, one obtains a likelihood ratio of 4. Notice, however, that the likelihood ratio is in fact slightly less than 4, since the probability of parental genotype configuration being A_1A_x/A_2A_x is not exactly 1.

Two principal issues must be considered in the construction of a Bayesian network for this scenario. First, an argument must be constructed that allows for the revision of the probability of the brother's genotype, given knowledge about the suspect's genotype. Basically, this is equivalent to the problem discussed in Section 7.3, where knowledge about a brother's genotype was used to infer something about the suspect's genotype. The corresponding graphical approach, shown in Figure 7.5, may thus serve as a starting point here.

Secondly, an appropriate graphical structure should allow for an evaluation of the following two probabilities: first, $Pr(tgt|sgt, F)$, that is the probability of the crime stain's genotype (tgt), given knowledge about the suspect's genotype (sgt) and given the proposition that the suspect is the source of the crime stain (F), and second, $Pr(tgt|sgt, \bar{F})$, that is the probability of the crime stain's genotype (tgt), given knowledge about the suspect's genotype (sgt) and given the proposition that a brother of the suspect is the source of the crime stain $(\bar{F})$.

A graphical structure that satisfies these requirements is shown in Figure 7.7. This structure shows several differences with respect to the network and the model previously discussed (Figure 7.5). In particular, the parental gene nodes associated with the crime stain (nodes tpg and tmg) are explicitly conditioned on the parental gene nodes of the brother (nodes bpg and bmg). This conditioning is needed for evaluating the probability of the genotype of the crime stain given the alternative proposition. This latter proposition states that a brother of the suspect is the source of the stain. Further, the propositions addressed are at the source level, represented by the node F. No extension is made to the crime level. However, a node B is adopted as a parental node for F. The node B is binary with states 'yes/no' and is used to represent uncertainty that may exist in cases where it may not be known whether the suspect has in fact a brother. Another difference is that the genotype nodes of the parents are explicitly represented: mgt and fgt denote the mother's and father's genotypes, respectively. Although the scenario under consideration assumes the parents to be unavailable, these nodes will later be used here to evaluate particular cases, notably the above-mentioned development by Evett (1992).

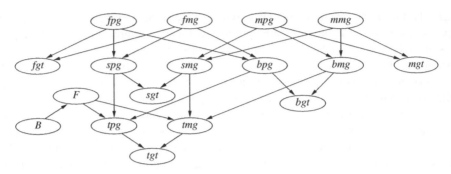

Figure 7.7 Bayesian network for evaluating DNA typing results in a one-stain scenario when the alternative proposition at the source level is that the stain was left by the suspect's brother. Node F is the source node, with two states, F, the suspect is the source of the crime stain, and $\bar{F}$, a brother of the suspect is the source of the crime stain. A node B acts as a parental node for F. Node B has two states: B, the suspect has a brother, and $\bar{B}$, the suspect does not have a brother. Nodes $fpg, fmg, mpg, mmg, spg, smg, bpg, bmg, tpg$ and tmg are gene nodes, with three states, A_1, A_2 and A_x. Nodes fgt, mgt, sgt, bgt and tgt are genotypic nodes, with six states, $A_1A_1, A_1A_2, A_1A_x, A_2A_2, A_2A_x$ and A_xA_x. Note that f denotes father, p paternal, m mother (if in first place) or maternal (if in second place), s suspect, b brother and t trace (if in first place).

For these structural relationships, node probabilities are adopted as follows. Concerning nodes tpg and tmg, the configuration equals that of the suspect for situations in which the suspect is the source of the crime stain. That is, for $i = 1, 2, x$, one has $Pr(tpg = A_i | spg = A_i, F) = 1$ and $Pr(tmg = A_i | smg = A_i, F) = 1$. Note that the brother's actual parental gene configuration is irrelevant under F, the suspect being the source of the crime stain. Table 7.5 provides a summary of these assignments. Notice that this table applies analogously to the node tmg. If a brother of the suspect is the source of the crime stain, then it is the parental gene configuration of this individual that determines the actual state of the parental gene configuration of the crime stain: for $i = 1, 2, x$, $Pr(tpg = A_i | bpg = A_i, \bar{F}) = 1$ and $Pr(tmg = A_i | bmg = A_i, \bar{F}) = 1$. Thus, $\bar{F}$ being true, it is the suspect's parental gene configuration that is irrelevant here. Table 7.6

Table 7.5 Conditional probabilities assigned to the node tpg, the paternal gene of the crime stain, assuming F to be true, that is the suspect is the source of the crime stain.

F :					F				
spg:		A_1			A_2			A_x	
bpg:	A_1	A_2	A_x	A_1	A_2	A_x	A_1	A_2	A_x
tpg: A_1	1	1	1	0	0	0	0	0	0
A_2	0	0	0	1	1	1	0	0	0
A_x	0	0	0	0	0	0	1	1	1

Nodes spg and bpg are paternal gene nodes for the suspect and the brother, with three states, A_1, A_2 and A_x. The relationship is obtained from Figure 7.7.

Table 7.6 Conditional probabilities assigned to the node tpg, the paternal gene of the crime stain, assuming $\bar{F}$ to be true, that is a brother of the suspect is the source of the crime stain.

F :				$\bar{F}$					
spg:	A_1			A_2			A_x		
bpg:	A_1	A_2	A_x	A_1	A_2	A_x	A_1	A_2	A_x
tpg: A_1	1	0	0	1	0	0	1	0	0
A_2	0	1	0	0	1	0	0	1	0
A_x	0	0	1	0	0	1	0	0	1

Nodes spg and bpg are paternal gene nodes for the suspect and the brother, with three states, A_1, A_2 and A_x. The relationship is obtained from Figure 7.7.

provides a summary of the conditional probabilities applicable for the nodes tpg and tmg, given that the brother is the source of the crime stain.

The table of node F contains the probability that the suspect or a brother left the crime stain, and this depends on whether or not the suspect has in fact a brother. So, if the suspect has in fact a brother, that is B is true, default probabilities of 0.5 are assigned here for $Pr(F|B)$ and $Pr(\bar{F}|B)$. Notice that, in practice, these values will be assessed in the light of the circumstantial information I. If the suspect has no brother, that is $\bar{B}$ is true, then certainly the suspect must have left the stain, so $Pr(F|\bar{B}) = 1$ and $Pr(\bar{F}|\bar{B}) = 0$.

Nodes fgt and mgt are genotype nodes according to the definition provided in Section 7.2. Thus, conditional probabilities as defined in Table 7.3 apply to these nodes.

Example 7.2 *(Bayesian network for case where the alternative proposition specifies a brother) For illustration, consider again a case in which both a crime stain and a suspect were found to be of type A_1A_2. As in the examples discussed in the previous paragraphs, the allele probabilities γ_1 and γ_2 are assumed to be 0.05. Figure 7.8 represents this scenario. As the suspect is assumed to have a brother (node B is instantiated), there are equal prior probabilities for the two possible outcomes of the variable F, that is the propositions at the source level. Arguably, the ratio of the posterior probabilities displayed in the node F, given the instantiation $tgt = A_1A_2$ and $sgt = A_1A_2$, will equal the likelihood ratio. This situation is shown in Figure 7.8. From the node F, it may be found that $78.35/21.65 = 3.62$. A likelihood ratio less than 4 has thus been obtained, in agreement with the result described by Evett (1992).*

The Bayesian network in Figure 7.8 may be used to examine further implications of the approach of Evett (1992), notably the upper and lower limits of the likelihood ratio. Consider the upper limit first. The rarer the alleles A_1 and A_2, the more likely it becomes that one of the two parents is of type A_1A_x and the other parent is of type A_2A_x. Assuming the parental genotype configuration to be in fact A_1A_x/A_2A_x, then, as mentioned earlier in this section, a likelihood ratio of 4 is obtained. This result may be obtained by use of the Bayesian network shown in Figure 7.8. If, in addition to the instantiations already made, one sets the node fgt to A_1A_x and the node mgt to A_2A_x (or vice versa), then the node F would display the probabilities 0.8 and 0.2, respectively. This corresponds to a likelihood ratio of 4. Notice also that for a parental genotype configuration A_1A_x/A_2A_x, the Bayesian network correctly displays a probability of $1/4$ for the brother's genotype being of type A_1A_2.

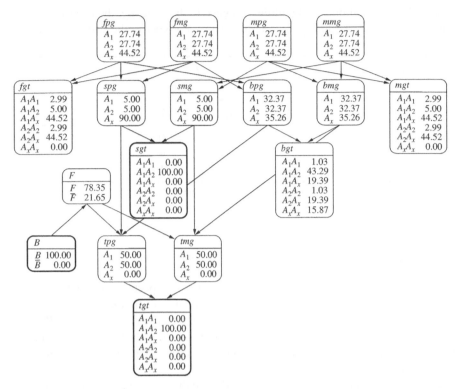

Figure 7.8 Bayesian network for evaluating DNA typing results in a one-stain scenario when the alternative proposition at the source level is that the stain was left by the suspect's brother. The nodes B, tgt and sgt are instantiated, meaning that the suspect is assumed to have a brother and that information is available about the suspect's and the crime stain's genotype. Node F is the source node, with two states, F, the suspect is the source of the crime stain, and $\bar{F}$, a brother of the suspect is the source of the crime stain. Node B has two states: B, the suspect has a brother, and $\bar{B}$, the suspect does not have a brother. Nodes $fpg, fmg, mpg, mmg, spg, smg, bpg, bmg, tpg$ and tmg are gene nodes, with three states, A_1, A_2 and A_x. Nodes fgt, mgt, sgt, bgt and tgt are genotypic nodes, with six states, $A_1A_1, A_1A_2, A_1A_x, A_2A_2, A_2A_x$ and A_xA_x. Note that f denotes father, p paternal, m mother (if in first place) or maternal (if in second place), s suspect, b brother and t trace (if in first place).

Consider next the lower limit of the likelihood ratio. Following Evett (1992), the probability of a full sibling's genotype (fsgt) being, for example, A_1A_2 may be obtained via using the following formula:

$$Pr(fsgt = A_1A_2 | sgt = A_1A_2) = \frac{1}{Pr(A_1A_2)} \sum_i Pr^2(fsgt = A_1A_2 | \phi_i)Pr(\phi_i). \quad (7.4)$$

The variable ϕ_i denotes the genotypic configuration of the parents. Notice that with the two alleles A_1 and A_2 being very common, the genotypic configuration of the parents is less likely to contain the allele A_x. For illustration, imagine an extreme case in which the probability for both A_1 and A_2 is taken to be 0.5. The possible parental configurations can then be reduced to A_1A_1/A_2A_2, A_1A_1/A_1A_2, A_1A_2/A_2A_2 and A_1A_2/A_1A_2. *Evaluation of (7.4) according to these parameters yields 5/8, the inverse of which, 1.6, represents the lower limit of the likelihood ratio.*

In order to track this result with the Bayesian network described so far, the unconditional probabilities of the nodes fpg, fmg, mpg and mmg need to be changed so that $Pr(A_1) = Pr(A_2) = 0.5$ *and* $Pr(A_x) = 0$. *Then, three queries may be addressed. Only the last of these is represented graphically because of limitations of space. These queries allow illustrate the flexibility of Bayesian networks and their wide range of possibilities for the study of alternative scenarios.*

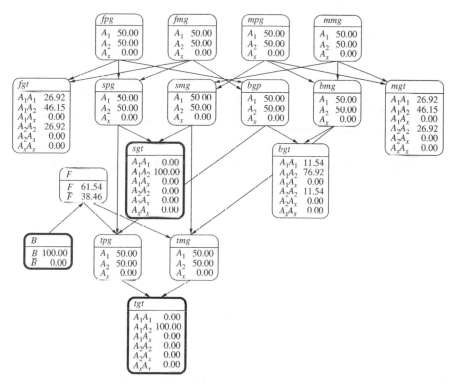

Figure 7.9   Bayesian network for evaluating the lower limit of the likelihood ratio in a one-stain scenario involving DNA typing results. The alternative proposition at the source level is that the stain was left by a brother of the suspect. The allele probabilities for A_1 and A_2 have been set to 0.5 (50%). The nodes B, tgt and sgt are instantiated, meaning that the suspect is assumed to have a brother and that information is available about the suspect's and the crime stain's genotype. Node F is the source node, with two states, F, the suspect is the source of the crime stain, and $\bar{F}$, a brother of the suspect is the source of the crime stain. Node B has two states: B, the suspect has a brother, and $\bar{B}$, the suspect does not have a brother. Nodes $fpg, fmg, mpg, mmg, spg, smg, bpg, bmg, tpg$ and tmg are gene nodes, with three states, A_1, A_2 and A_x. Nodes fgt, mgt, sgt, bgt and tgt are genotypic nodes, with six states, $A_1A_1, A_1A_2, A_1A_x, A_2A_2, A_2A_x$ and A_xA_x. Note that f denotes father, p paternal, m mother (if in first place) or maternal (if in second place), s suspect, b brother and t trace (if in first place).

Start by considering an evaluation of the probability that a brother of the suspect is of the same type as the suspect, that is A_1A_2. This is operated by instantiating the node sgt (suspect genotype) to A_1A_2. The Bayesian network would propagate this information and update the probability distribution of the brother's genotype node (bgt). In the case at hand, the probability of node bgt being in state A_1A_2 would increase from 0.5 to 0.625. This is in agreement with what is obtained by (7.4), that is 5/8.

Second, an evaluation of the denominator of the likelihood ratio may be obtained if, in addition to the node sgt, the node representing the proposition at the source level, F, is instantiated to $\bar{F}$. Then, the Bayesian network will display the value 0.625 for the trace genotype node (tgt) being in state A_1A_2. The inverse of this value, 1.6, represents the likelihood ratio.

An alternative way to obtain the likelihood ratio is in terms of the posterior probabilities of the propositions at the source level (node F). After initializing the Bayesian network, the genotype nodes of the suspect and the crime stain (nodes sgt and tgt) are both instantiated to A_1A_2. In addition, the node B is set to 'true'. Then, the node F displays the posterior probabilities of the propositions at the source level, given the information that the suspect and the crime stain are of the same genotype, that is A_1A_2. The values of these two probabilities are, respectively, 0.6154 and 0.3846. As may be seen, their ratio is 1.6. Figure 7.9 provides a graphical representation of this evaluation.

7.4.2   Further considerations on conditional genotype probabilities

As outlined in Section 7.4.1, it may be of interest to assign a probability for an untyped person having a particular genotype, given typing information for a relative (i.e. a brother). More generally, situations involving relatives are quite common in routine laboratory work. In many civil cases, for example, people seek to inquire about questions of relatedness, such as 'Are we brothers? Are we half-brothers or cousins?' This implies an interest in conditional genotype probabilities. Following general notational convention used throughout this chapter, a conditional genotype probability can be written as $Pr(Agt|Bgt)$, where Agt and Bgt stand for genotypes gt of persons A and B, respectively. Balding and Nichols (1994) proposed a methodology that refers to the notion of 'allele identical by descent', denoted ibd for short [for further technical details, see, for example, Weir (1996)] because relatedness is a consequence of sharing common ancestral alleles. The expression 'identical by descent' is thus used to denote alleles that have descended from a single ancestral allele.

Two individuals, such as brothers as in the example above, possess four alleles at a given locus. This allows one to specify three probabilities: a probability, say $Pr(Z_0)$, for observing zero allele that are identical by descent between the two individuals, a probability $Pr(Z_1)$ of observing one allele that is identical by descent and a third probability $Pr(Z_2)$ of observing two alleles that are identical by descent. A basic example that provides a simple explanation of how the values of these three probabilities are assigned relates to the association between a child and a father. Given the principle of Mendelian inheritance, a child should possess an allele from their biological father. Arguably, one of the child's alleles is identical by descent to a father's allele. Therefore, if the relation of interest is that of father and child, then the values of interest are $Pr(Z_0) = 0, Pr(Z_1) = 1$ and $Pr(Z_2) = 0$; the probabilities that zero or two alleles are ibd under the hypothesis of a father–child relationship are equal to 0. On the other hand, the probability that one allele is ibd equals 1. Tables offering the values of $Pr(Z_i)$ (for $i = 0, 1, 2$) for different relationships are available in the literature (Balding and Nichols 1994).

Example 7.3 *(Conditional genotype probabilities for siblings) To further illustrate the notion of ibd, consider again the case of a scientist who is interested in the probability that an untyped person A has the profile A_1A_2 given that a typed person B, his brother, has also the profile A_1A_2. Using H to denote the proposition that the two persons are brothers, the probability of interest can be written in terms of the extension of conversation:*

$$Pr(Agt|Bgt, H) = \sum_{i=0}^{2} Pr(Agt|Bgt, H, Z_i)Pr(Z_i),$$

where $Agt = Bgt = A_1A_2$ and $i = 0, 1, 2$. Writing this in further detail lead to

$$Pr(Agt = A_1A_2|Bgt = A_1A_2, H) = \overbrace{Pr(Agt = A_1A_2|Bgt = A_1A_2, H, Z_0)0.25}^{2\gamma_1\gamma_2}$$

$$+ \overbrace{Pr(Agt = A_1A_2|Bgt = A_1A_2, H, Z_1)0.5}^{\frac{1}{2}\gamma_1 + \frac{1}{2}\gamma_2}$$

$$+ \overbrace{Pr(Agt = A_1A_2|Bgt = A_1A_2, H, Z_2)0.25}^{1}$$

$$= \frac{1}{2}\gamma_1\gamma_2 + \frac{1}{4}(\gamma_1 + \gamma_2 + 1).$$

Note that the component probability $\frac{1}{2}\gamma_1 + \frac{1}{2}\gamma_2$ expresses the view that, with probability 0.5, person A shares allele 1 ibd with person B and possesses allele 2 with allele proportion γ_2, and, with probability 0.5, shares allele 2 ibd with person B and possesses allele 1 with allele proportion γ_1.

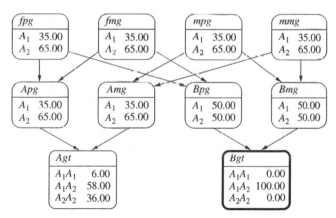

Figure 7.10 Bayesian network for finding the probability of the genotype *gt* of a person, for example *A*, given knowledge of the genotype *gt* of another person, for example *B*. Nodes *fpg, fmg, mpg, mmg, Apg, Amg, Bpg* and *Bmg* are gene nodes (of a hypothetical diallelic locus), with two states A_1 and A_2. Nodes *Agt* and *Bgt* are nodes representing genotypes, with three states, A_1A_1, A_1A_2 and A_2A_2. Note that *f* denotes father, *p* paternal, *m* mother (if in first place) or maternal (if in second place). The bold outer border of node *Bgt* indicates that it is instantiated.

Using part of the Bayesian network shown in earlier in Figure 7.5, it is possible to find conditional probabilities for any combination of genotypes that siblings may have. As an example, consider Figure 7.10 for a hypothetical diallelic locus with possible alleles A_1 and A_2. This network allows one to address a question of the following kind: What is the probability that two persons, full brothers, will have the same genotype (e.g. A_1A_2)? Using as illustrative allelic proportions $\gamma_1 = 0.2$ and $\gamma_2 = 0.8$ for, respectively, alleles A_1 and A_2, one can find that the probability of interest is 0.58. Figure 7.10 shows this result for the state A_1A_2 of the node Agt, which stands for the genotype of person A. The same network can readily be used to find other conditional probabilities. For example, imagine that person B has genotype A_1A_1 and it is of interest to find the probability for the brother's genotype (i.e. person A) being, for instance, A_1A_2. The proposed Bayesian network allows one to find the result 0.48 (details not shown here).

7.5   Interpretation with more than two propositions

Consider again a situation in which the finding E consists of a DNA profile of a stain of body fluid found at a crime scene and of the DNA profile of a suspect that corresponds, in some sense, to that of the crime stain. Until now, such scenarios have been evaluated with respect to pairs of propositions H, introduced in Section 7.1. At source level, these are, typically, the suspect is the source of the crime stain (H_p) and another person, unrelated to the suspect, left the crime stain (H_d).

However, it may well be that the relevant population contains close relatives of the suspect. It is important to account for this possibility, as argued by Buckleton and Triggs (2005), since close relatives are far more likely to have corresponding genetic traits than other members of the population. The defence may thus require forensic scientists to incorporate in their analysis more than one proposition. An approach for such a situation, using posterior probabilities, has been discussed by Evett (1992). It considers the following three propositions: 'The suspect left the crime stain (H_p)', 'An unknown member of the population left the crime stain (H_{d1})' and 'A brother of the suspect left the crime stain (H_{d2})'.

Let θ_0, θ_1 and θ_2 denote the prior probabilities for H_p, H_{d1} and H_{d2}, respectively, so that $\theta_0 + \theta_1 + \theta_2 = 1$. Assume that $Pr(E|H_p) = 1$. Denote $Pr(E|H_{d1})$ by ϕ_1 and $Pr(E|H_{d2})$ by ϕ_2. Notice however that here ϕ has a different definition than the one used previously in Section 7.4.1. It is further assumed that H_d, the complement of H_p, is the conjunction of H_{d1} and H_{d2}. The posterior probability of H_p given E can now be written as follows:

$$Pr(H_p|E) = \frac{Pr(E|H_p)\theta_0}{Pr(E|H_p)\theta_0 + Pr(E|H_{d1})\theta_1 + Pr(E|H_{d2})\theta_2}$$

$$= \frac{\theta_0}{\theta_0 + \phi_1\theta_1 + \phi_2\theta_2}. \tag{7.5}$$

Similarly, the posterior probability of H_d given E is given by

$$Pr(H_d|E) = \frac{\phi_1\theta_1 + \phi_2\theta_2}{\theta_0 + \phi_1\theta_1 + \phi_2\theta_2}.$$

Hence, the posterior odds in favour of H_p are

$$\frac{\theta_0}{\phi_1\theta_1 + \phi_2\theta_2}. \tag{7.6}$$

It is worth mentioning at this point that an assessment of the findings through (7.6) requires prior probabilities to be specified for the propositions put forward by the prosecution and the defence. One way to specify prior probabilities for Equation (7.6) is to consider what may be called a *pool* of possible suspects. For the purpose of illustration, suppose that the size of such a suspect population pool, including the defendant, covers N individuals. In addition, let n denote the number of brothers of the suspect, which are also included in the suspect population. Thus, there are $N - n$ unrelated individuals amongst the N members of the suspect population (ignoring cousins). If the other information in the case (i.e. other than DNA) does not distinguish between these N individuals, then it can be argued that the prior probability for each and every member of the suspect population is $1/N$. Consequently, $\theta_0 = 1/N$, $\theta_1 = (N - n)/N$ and $\theta_2 = (n - 1)/N$.

Using these figures and assuming $N >> n$, (7.6) leads to posterior odds for H_p approximately equal to

$$\frac{1}{\phi_1 N + \phi_2(n - 1)}.$$

From a practicing scientist's point of view, it may be of interest to study and handle this formula by means of a Bayesian network. This could also allow one to point out any structural relationships with respect to the inference problems discussed elsewhere in this chapter. This question can be approached by reconsidering the Bayesian network described in Section 7.4 (Figure 7.7). This network may be modified in order to account for the distinctive properties of the problem considered here.

One of the modifications that is necessary concerns the definition of the source node F. Here, F will be assigned the same states as those for variable H stated immediately above in this section. The change from notation H to F is to ensure consistency between Figure 7.9, which has binary states, and Figure 7.11, for which the following three states are necessary: 'The suspect is the source of the crime stain (F_p)', 'A member of the population of potential donors, different from the suspect and his brother, is the source of the crime stain (F_{d1})' and 'A brother of the suspect left the crime stain (F_{d2})'.

As a consequence of this, more probabilities must be specified for the probability tables of the trace parental gene nodes (*tpg* and *tmg*). In particular, it is necessary to specify probabilities for the tables of each trace parental gene node, given that a member of the relevant population, different from the suspect and the brother, left the crime stain. In addition to the Tables 7.5 and 7.6 already defined in Section 7.4, probabilities must thus be added that comply with the following:

$$Pr(tpg = A_i | F_{d1}) = \gamma_i, \quad i = 1, 2, x. \tag{7.7}$$

Analogous probabilities apply for the node *tmg*. Notice that (7.7) holds irrespective of the actual state of the parental gene nodes of the suspect and the brother. At this point, notice that one could avoid the specification of probabilities γ_i by representing explicit parental gene nodes of an alternative source (not shown in Figure 7.11), different from the suspect and a brother. In such a case, the trace parental gene nodes would copy the actual state of the parental gene nodes of that unknown source. This is achieved by specifying, as was done for the conditioning on the suspect's and the brother's parental gene nodes, values of 0 and 1 in the probability tables of the trace parental gene nodes.

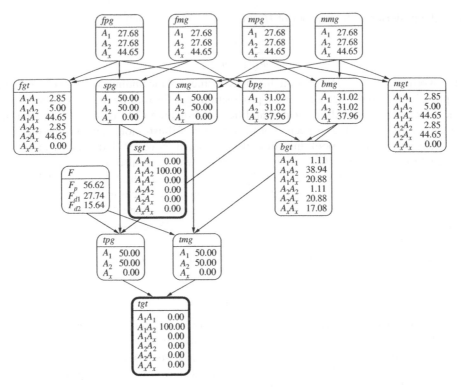

Figure 7.11 Bayesian network for evaluating DNA typing results with respect to more than one alternative proposition. The allele proportions for A_1 and A_2 have been set to 0.05. Node F is the source node, with three states, F_p, the suspect is the source of the crime stain, F_{d1}, a random member of the population is the source of the crime stain, and F_{d2}, a brother of the suspect left the crime stain. Nodes $fpg, fmg, mpg, mmg, spg, smg, bpg, bmg, tpg$ and tmg are gene nodes, with three states, A_1, A_2 and A_x. Nodes fgt, mgt, sgt, bgt and tgt are genotypic nodes, with six states, $A_1A_1, A_1A_2, A_1A_x, A_2A_2, A_2A_x$ and A_xA_x. Note that f denotes father, p paternal, m mother (if in first place) or maternal (if in second place), s suspect, b brother and t trace (if in first place).

A second modification of Figure 7.7 concerns the node B, which is omitted here for the time being. Consequently, there will be unconditional probabilities that need to be specified for the node F. One possibility for assigning these probabilities, by making assumptions about the size of a suspect population, has been outlined above.

Example 7.4 *(Bayesian network for multiple source level propositions) Consider a numerical example for the Bayesian network discussed so far in this section. In order to facilitate comparisons with results obtained in earlier sections, allele proportions of 0.05 (more generally, γ_1 and γ_2) for both A_1 and A_2, are assumed. The pool of potential suspects is set to 100 individuals, including the suspect and one brother. Then, following Evett (1992), the initial probabilities – denoted θ in (7.5) – obtained for the node F are $Pr(F_p) = 0.01$, $Pr(F_{d1}) = 0.98$ and $Pr(F_{d2}) = 0.01$. The remaining probabilities are assigned as outlined above.*

The findings consist of profiling results for the crime stain and the control material from the suspect, both of which are found to be of type A_1A_2. These observations are communicated to the model by instantiating the nodes tgt and sgt. The corresponding state of the Bayesian *network is shown in Figure 7.11. The allele proportions for the traits A_1 and A_2 have been set equal to 0.05, as mentioned above, but these values are not shown in Figure 7.11 because it displays a propagated network and not an initialized Bayesian network. A Bayesian network in a propagated state no longer shows the initial probabilities for the parental gene nodes. Node F indicates the posterior probabilities of the various competing hypotheses at source level. For example, the probability that the suspect is the source of the crime stain, given the observations, is 0.5662.*

The same result can be obtained from (7.5):

$$Pr(H_d|E) = \frac{\theta_0}{\theta_0 + \phi_1\theta_1 + \phi_2\theta_2}$$

$$= \frac{0.01}{0.01 + 0.98 \times 0.005 + 0.01 \times 0.2763} = 0.5662.$$

The probability of encountering the genotype A_1A_2 for a member of the target population, ϕ_1, is 0.005, given generally by $2\gamma_1\gamma_2$. Notice that it may also be read off the initialized Bayesian network. Before any observations are entered, all genotype nodes (sgt, bgt, tgt, fgt and mgt) display 0.005 for the state A_1A_2.

The probability of the findings given a brother of the suspect is the source of the crime stain, written ϕ_2, is another parameter in the denominator that may require further explanation. This parameter depends directly on the probability of a brother sharing the same characteristics as the suspect. In order to express the idea that a brother is more likely to have a corresponding profile than an unrelated member of the suspect population, scenarios proposed in the literature have, for example, assumed a value of 10^{-2} for a brother versus 10^{-6} for each of the (unrelated) individuals of the suspect population (Balding 2000).

When working with a Bayesian network, a value for ϕ_2 can be inferred from available knowledge about the suspect's genotype. Notably, the probability of the crime stain being of type A_1A_2, given that a brother of the suspect is the source of the stain (F instantiated to F_{d2}) and given that the suspect is of type A_1A_2, may be processed as a query. The value obtained is 0.2763. This result can be found to agree with formulae published in literature. Following Buckleton et al. (2005), for example the conditional probability for an untyped brother being of type A_1A_2, $Pr(bgt = A_1A_2|sgt = A_1A_2)$, is given by (leaving aside a sub-population correction)

$$\frac{(1 + (\gamma_1 + \gamma_2) + 2\gamma_1\gamma_2)}{4}.$$

For $\gamma_1 = \gamma_2 = 0.05$, this leads to $(1 + (0.05 + 0.05) + 2 \times 0.05 \times 0.05)/4 = 0.2763$.

7.6   Evaluation with more than two propositions

In Section 7.5, the assessment of DNA profiling results has been extended to more than two propositions. This has led to consideration of posterior odds in favour of one of the specified propositions. However, when assessing scientific results, working with likelihood ratios is preferable. As likelihood ratios are used to compare propositions in pairs, some meaningful procedure is required to combine propositions, when more than two of them need to be considered.

An approach for comparing more than two propositions has been described, for example, by Aitken and Taroni (2004). Consider a number n of competing exclusive propositions

$H_1, \ldots, H_n$. Let E denote the observations to be evaluated under each of the n propositions and consider $Pr(E|H_i)$, with $i = 1, \ldots, n$. If the prior probabilities of each of the propositions H_i are available, the ratio of the probability of E given each of the pair of competing propositions H_1 and $\bar{H}_1 = (H_2, \ldots, H_n)$ can be evaluated as follows:

$$\frac{Pr(E|H_1)}{Pr(E|\bar{H}_1)} = \frac{Pr(E|H_1)\{\sum_{i=2}^{n} Pr(H_i)\}}{\sum_{i=2}^{n} Pr(E|H_i)Pr(H_i)}.$$

Applying this approach to the scenario discussed in Section 7.5, the source level propositions F_p ('The suspect is the source of the crime stain'), F_{d1} ('A member of the population of potential donors, different from the suspect and his brother, is the source of the crime stain') and F_{d2} ('A brother of the suspect left the crime stain') could be combined to H_p ('The suspect is the source of the crime stain') and H_d ('The suspect is not the source of the crime stain, that is a member of the population of potential donors, different from the suspect, or a brother of the suspect, is the source of the crime stain').

With E denoting a global expression of the results of the DNA typing analyses, the following form of the likelihood ratio thus is obtained:

$$\frac{Pr(E|H_p)}{Pr(E|H_d)} = \frac{Pr(E|F_p)\{Pr(F_{d1}) + Pr(F_{d2})\}}{Pr(E|F_{d1})Pr(F_{d1}) + Pr(E|F_{d2})Pr(F_{d2})}. \tag{7.8}$$

Using the values defined in Section 7.5, (7.8) provides the following result:

$$\frac{Pr(E|H_p)}{Pr(E|H_d)} = \frac{1 \times \{0.98 + 0.01\}}{0.005 \times 0.98 + 0.2763 \times 0.01} = 129. \tag{7.9}$$

The approach considered here can also be tracked within a Bayesian network. For this purpose, it is necessary to slightly modify the Bayesian network described previously in Section 7.5 (Figure 7.11). In particular, an additional binary node, termed H, that regroups the various competing propositions, needs to be introduced. Because this node H will be used to evaluate a Bayes factor, it will be assigned the two states H_p and H_d as described above. Considering H_p to be true whenever the suspect is in fact the source of the crime stain (F_p), and considering H_d to be true only if either a member of the suspect population (F_{d1}) or a brother of the suspect (F_{d2}) is the source of the crime stain, the conditioning $F \rightarrow H$ can be adopted. The node probabilities assigned to H are described in Table 7.7. It follows from this that the prior probability of H_p equals the probability assigned to F_p, whereas the prior probability of H_d is the sum of the probabilities assigned to F_{d1} and F_{d2}.

Example 7.5 (*Bayes factor evaluation with multiple propositions*) *Figure 7.12(i) shows an evaluation of the numerator of the Bayes factor for the setting considered so far in this section. The suspect is known to be of type A_1A_2. The node sgt is thus instantiated to A_1A_2. Under the assumption that the suspect is the source of the crime stain, H_p, the probability of the crime stain being A_1A_2 can be found to be 1, as shown in node tgt. An evaluation of the denominator of the Bayes factor is provided by Figure 7.12(ii). Here, the node H is instantiated to H_d and the node tgt indicates the probability of the crime stain being of type A_1A_2, that is 0.0077. The likelihood ratio is 129, given by $1/0.00774$ (using the additional accuracy provided by the Bayesian network software), in agreement with the result obtained using (7.9).*

Notice that the Bayesian network described in this section is usable for addressing a wide range of different scenarios simply by changing the prior probabilities assigned to the node F.

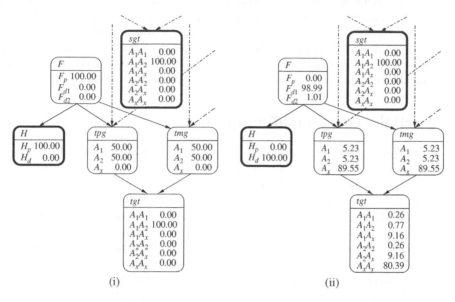

(i)   (ii)

Figure 7.12 Partial representations of the Bayesian network shown in Figure 7.11, including an extension to a node H, regrouping the various source level propositions of the node F such that $H_p = F_p$, the suspect is the source of the crime stain, and H_d is the union of F_{d1} and F_{d2}, either a member of the suspect population or a brother of the suspect is the source of the crime stain. Evaluation of (i) the numerator and (ii) the denominator of the likelihood ratio follows from instantiation of (i) H_p and (ii) H_d. Instantiation has also been made at *sgt*. The allele population proportions for A_1 and A_2 have been set to 0.05. Node F is the source node, with three states, F_p, the suspect is the source of the crime stain, F_{d1}, a member of the suspect population is the source of the crime stain, and F_{d2}, a brother of the suspect left the crime stain. Nodes *tpg* and *tmg* are gene nodes, with three states, A_1, A_2 and A_x. Nodes *sgt* and *tgt* are genotypic nodes, with six states, $A_1A_1, A_1A_2, A_1A_x, A_2A_2, A_2A_x$ and A_xA_x. Note that p denotes paternal, m denotes maternal, s denotes suspect and t denotes trace (if in first place).

Table 7.7 Conditional probabilities assigned to the node H in Figure 7.12 such that $H_p = F_p$, the suspect is the source of the crime stain, and H_d is the union of F_{d1} and F_{d2}, either a member of the suspect population or a brother of the suspect is the source of the crime stain.

F :	F_p	F_{d1}	F_{d2}
H : H_p	1	0	0
H_d	0	1	1

For example, if it is not an issue that a brother of the suspect may be a potential source of the crime stain, then the initial probability assigned to F_{d2} is 0 and the Bayesian network yields a likelihood ratio equal to $1/2\gamma_1\gamma_2$, where γ_1 and γ_2 are the population proportions of the alleles A_1 and A_2, respectively. On the other hand, if the suspect and a brother are the only possible sources of the crime stain, then F_{d1} would be set to 0 and the Bayesian network would yield the same results as described in Section 7.4: the likelihood ratio is slightly less than 4.

7.7   Partially corresponding profiles

So far in this chapter, Bayesian networks have been constructed and discussed for situations in which there is a full correspondence between the suspect's genotype (*sgt*) and that of a crime stain (*tgt*). The primary aim of the proposed models was to assess the strength of the link between the suspect and the crime stain by means of a likelihood ratio. As was seen in Section 7.2, for a fairly general one-stain one-offender case, the likelihood ratio reduces, under certain assumptions, to the inverse of the profile probability. Several variations of this scenario have also been investigated. In Section 7.4, for example, changes in the value of the likelihood ratio have been studied for scenarios in which the alternative proposition was that a sibling of the suspect was the source of the crime stain. In Section 7.6, the population of potential sources of a crime stain was allowed to cover individuals related and unrelated to the suspect.

In this section, scenarios will be addressed in which there is no complete correspondence between the suspect's characteristics and those of a crime stain. In such situations, the suspect is usually considered as excluded as the donor of the crime stain when one assumes that there are no special events that affect profile morphology (such as with traces containing low quantities of DNA). However, as pointed out by Sjerps and Kloosterman (1999), there can be situations in which the two DNA profiles, although different, suggest that the profile a close relative of the suspect might correspond to the crime stain. This may be the case, for example, when the two non-corresponding DNA profiles share several very rare alleles. One could also imagine database searches focusing on individuals that have corresponding alleles in several loci.

Consider a case involving a crime stain with a profile that does not correspond to the profile of the suspect. Suppose further that, as was done in the previous section, the discussion concentrates on a single locus with possible alleles A_1, A_2 and A_x, where the latter is a state that covers all possible outcomes other than A_1 and A_2. Let the genotype of the crime stain (*tgt*) be A_1A_1, for instance. Assume further that the suspect's genotype (*sgt*) is found to be A_1A_2. The pair of propositions under which these observations are evaluated is defined as follows: 'The crime stain comes from the suspect's brother (H_p)' and 'The crime stain comes from an unrelated individual (H_d)'.

The general form of the likelihood ratio for such a setting can be written as follows:

$$V = \frac{Pr(sgt = A_1A_2, tgt = A_1A_1|H_p)}{Pr(sgt = A_1A_2, tgt = A_1A_1|H_d)}.$$

Invoking the third law of probability, this can be re-written in the following form:

$$V = \frac{Pr(tgt = A_1A_1|H_p)}{Pr(tgt = A_1A_1|H_d)} \times \frac{Pr(sgt = A_1A_2|tgt = A_1A_1, H_p)}{Pr(sgt = A_1A_2|tgt = A_1A_1, H_d)}. \tag{7.10}$$

In the absence of knowledge about the genotype of the suspect and a possible brother, the probability of the trace genotype being of type A_1A_1 is the same given H_p and H_d. Thus, the first term on the right-hand side of (7.10) is 1. If the crime stain has in fact been left by the suspect's brother and the trace genotype is A_1A_1, then the genotype of the suspect's brother (bgt) must be of type A_1A_1. Therefore, the numerator of the second term on the right-hand side of (7.10) can be expressed, more shortly, as $Pr(sgt = A_1A_2|bgt = A_1A_1)$. The denominator can be reduced to $Pr(sgt = A_1A_2)$ if an absence of correlation is assumed between the DNA profile of the donor of the crime stain and that of the suspect. The likelihood ratio thus can be approximated by (Sjerps and Kloosterman 1999):

$$V \approx \frac{Pr(sgt = A_1A_2|bgt = A_1A_1)}{Pr(sgt = A_1A_2)}. \tag{7.11}$$

This result can be tracked in several different ways. One possibility would be to re-consider the Bayesian network studied in Section 7.5 (Figure 7.11). In order for this model to be used here, the probability of the state F_p, defined as 'the suspect is the source of the crime stain', is set to zero. The remaining entries of the probability table of the node F, namely F_{d1} ('a random member of the population is the source of the crime stain') and F_{d2} ('a brother of the suspect left the crime stain'), reflect the propositions that are of interest in the current scenario. Allele proportions γ_{A_1} and γ_{A_2} are, once again, set to 0.05.

Assuming equal prior probabilities for F_{d1} and F_{d2}, the following can be considered. On the one hand, if the node tgt is set to A_1A_1 and the node sgt to A_1A_2, then the posterior probabilities for F_{d1} and F_{d2} are 0.16 and 0.84, respectively. This corresponds to a likelihood ratio of 5.25 in support of a brother of the suspect being the donor of the crime stain. On the other hand, the numerator and the denominator of (7.11) can also be evaluated separately. If bgt is instantiated to A_1A_1, then the probability of sgt being A_1A_2 is 0.02625. Initially, the probability of sgt being of type A_1A_2 is given by $2\gamma_{A_1}\gamma_{A_2} = 0.005$. The ratio of these two values is $0.02625/0.005 = 5.25$.

An alternative Bayesian network for evaluating a case involving partially corresponding profiles is shown in Figure 7.13, where consideration is also given to the possibility that a son of the suspect is the offender. There are two principal features. First, a node H is used to evaluate the probability of a brother (H_{p1}), a son (H_{p2}) or an unrelated individual (H_d) being the source of the crime stain. Second, knowledge about the suspect's genotype (sgt) is used to update probabilities of allele and, consequently, conditional genotype probabilities for a suspect's brother (bgt) and son ($son - gt$).

Example 7.6 (*Bayesian network for partially corresponding profiles*) *In order to examine in further detail the model presented in this section (Figure 7.13), consider again a scenario in which only two propositions are retained: either a brother of the suspect or an unrelated individual are the source of the crime stain. The probability of H_{p2} is thus set to zero. Assigning equal prior probabilities to H_{p1} and H_d and using probabilities of $\gamma_{A_1} = \gamma_{A_2} = 0.05$, the Bayesian network yields the same results as those found above, that is if sgt is instantiated to A_1A_2 and tgt to A_1A_1, then the posterior probabilities for H_{p1} and H_d are 0.84 and 0.16, respectively. This corresponds to a likelihood ratio of 5.25. Moreover, evaluating the numerator gives $Pr(sgt = A_1A_2|bgt = A_1A_1)$ as 0.02625, whereas the denominator, $Pr(sgt = A_1A_2)$, is given as 0.005.*

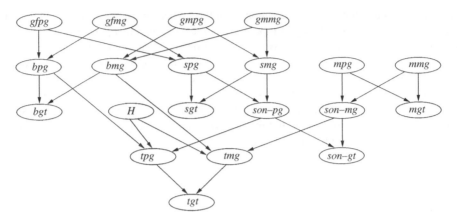

Figure 7.13   Bayesian network for evaluating scenarios in which the suspect's genotype does not correspond to that of a crime stain. Close relatives of a non-corresponding suspect, that is, a brother or a son, are modelled as potential donors of the crime stain. Nodes are *gfpg* and *gfmg*, grandfather's paternal and maternal genes; *gmpg* and *gmmg*, grandmother's paternal and maternal genes; *bpg*, *bmg* and *bgt*, brother's paternal and maternal genes and genotypes; *spg*, *smg* and *sgt*, suspect's paternal and maternal genes and genotype; *mpg*, *mmg* and *mgt*, mother's paternal and maternal genes and genotypes. Nodes *son − pg*, *son − mg* and *son − gt* refer to a son of the suspect for paternal gene, maternal gene and genotype, and nodes *tpg*, *tmg* and *tgt* refer to nodes trace paternal and maternal genes and genotypes. Node *H* has three states, H_{p1}, a brother of the suspect is the source of the crime stain, H_{p2}, a son of the suspect is the source of the crime stain, and H_d, an unrelated individual is the source of the crime stain.

Whenever a non-zero prior probability is used for H_{p2}, a son of the non-corresponding suspect being the donor of the crime stain, the network can be used to discriminate between three potential donors, a situation not considered by (7.11). This is considered in Example 7.7.

Example 7.7 (*Evaluation of partially corresponding profiles and information on additional relatives*) *Another possibility for the use of the proposed network consists in the assessment of the potential of genotypic information on other close relatives to achieve further discrimination between the main propositions. For the purpose of illustration, consider again a scenario where the propositions of interest involve a brother, a son and a person unrelated to the partially corresponding suspect. One of the questions of interest may now be whether or not the analysis of the genotype of the son's mother, for example, could add relevant information, by which is meant that it leads to changes in the posterior probabilities of the three main propositions.*

Assume again a locus described by alleles A_1, A_2 and A_x with $\gamma_1 = \gamma_2 = 0.05$. From Table 7.8, it can be seen that there are situations in which the posterior probabilities of the major propositions can change given information on the son's mother genotype. An extreme example for this is shown in the last column: when the mother's genotype has no allele in common with the genotype of the crime stain, then logically, a son is excluded as a potential donor of the crime stain.

Table 7.8 Posterior probabilities of the major propositions, given genotypic information on the suspect, *sgt*, the crime stain, *tgt*, and the son's mother, *mgt* ('*n.a.*' stands for 'genotypic information not available').

		sgt: A_1A_2	A_1A_2	A_1A_2	A_1A_2
		tgt: A_1A_1	A_1A_1	A_1A_1	A_1A_1
		mgt: n.a.	A_1A_1	A_1A_2	A_2A_2
H :	brother (H_{p1})	0.323	0.025	0.050	0.840
	son (H_{p2})	0.615	0.970	0.941	0.000
	unrelated (H_d)	0.062	0.005	0.009	0.160

Node H has three states, corresponding to the three propositions, a brother, a son or an unrelated individual is the source of the crime stain.

7.8 Mixtures

7.8.1 Considering multiple crime stain contributors

The previous sections have dealt with scenarios involving evidential material assumed to originate from a single individual. Here, the analysis is extended to cases where the crime stain contains material from more than one contributor. The occurrence of mixed stains may be more an issue in some cases than in others, according to particular circumstances. For example, in cases of rape and other physical assaults, traces recovered on a victim may contain material from both the victim and assailant(s).

As noted earlier, for the kind of genetic markers considered here, an individual has at most two alleles. The presence of more than two alleles can thus be taken as an indication that the stain contains biological material from more than one individual. The inferential challenge of evaluating such scenarios consists of regrouping the various competing propositions that may be formulated and the number of combinations of genotypes that enter into consideration.

The approach discussed here assumes, as was done in the previous sections, independence of an individual's alleles both within and across markers (i.e. absence of any sub-population effects). In addition, all contributors to the mixed stain are considered to be unrelated to each other. Therefore, when assessing the denominator of the likelihood ratio, the probability of the findings is not conditioned on the already observed genotypes, such as those of the suspect(s) (Harbison and Buckleton 1998). A further aspect that is also not addressed in this section is that of peak areas and heights as discussed, for example, by Evett et al. (1998a) or Buckleton et al. (2005). A Bayesian network approach to peak area information, restricted to two person mixtures, is presented in Cowell et al. (2007b).

This discussion concentrates on extending the generic modelling approach outlined in the preceding sections to situations in which several individuals may have contributed to a given DNA stain. In order to illustrate such a situation, consider an alleged rape. A vaginal swab is taken and submitted for laboratory analysis. Consider the DNA typing results for a hypothetical marker for which three alleles have been found, denoted A_1, A_2 and A_3. Reference material is also available from the victim and a suspect. The victim was found to be heterozygous A_1A_3 and the suspect is homozygous A_2.

Suppose further that, according to information provided by the victim, the number of contributors can be restricted to two individuals, the victim and the assailant. Assume thus that the propositions put forward by the prosecution and the defence are as follows: 'The crime stain contains DNA from the victim and the suspect (H_p)' and 'The crime stain contains DNA from the victim and an unknown individual (H_d)'.

Denote the genotype of the victim and the suspect by, respectively, vgt and sgt. The profile of the mixed crime stain will be abbreviated csp. The likelihood ratio can then be formulated as follows:

$$V = \frac{Pr(csp, vgt, sgt|H_p, I)}{Pr(csp, vgt, sgt|H_d, I)}$$

$$= \frac{Pr(csp|vgt, sgt, H_p, I)}{Pr(csp|vgt, sgt, H_d, I)} \times \frac{Pr(vgt, sgt|H_p, I)}{Pr(vgt, sgt|H_d, I)}. \qquad (7.12)$$

Considering the victim's and the suspect's genotype to be independent from whether the suspect is or is not a contributor to the mixed stain, (7.12) can be reduced to

$$V = \frac{Pr(csp|vgt, sgt, H_p, I)}{Pr(csp|vgt, sgt, H_d, I)}.$$

The proposition H_d assumes that the victim and a person other than the suspect, that is the true offender (and, hence, the suspect is not the true offender), are contributors to the mixed stain. Assuming independence between the genotypes of the suspect and the true offender, the term sgt can be omitted in the denominator. The likelihood ratio then becomes

$$V = \frac{Pr(csp|vgt, sgt, H_p, I)}{Pr(csp|vgt, H_d, I)}.$$

If H_p is true, $vgt = A_1A_3$ and $sgt = A_2$ then, assuming no intervening special circumstances (such as allelic dropout that may affect traces with low quantities of DNA), then the analysis of mixed stain can be expected to lead to a profile with the three alleles A_1, A_2 and A_3. A probability of 1 can thus be assigned to the numerator.

For the denominator, one needs to consider that the second contributor, an unknown person, may possess one of the following genotypes: A_1A_2, A_2A_2 or A_2A_3, but not A_2A_x for $x \neq 1, 2, 3$ as A_x would then appear in the crime stain profile. Denote these genotypes U_i with

$$U_1 = A_1A_2, U_2 = A_2A_2, U_3 = A_2A_3.$$

Extending the conversation to the possible U_i, and assuming independence between the genotypes of the victim and the suspect, the denominator can be written as follows:

$$Pr(csp|vgt, H_d, I) = \sum_i Pr(csp|vgt, U_i, H_d, I)Pr(U_i|H_d, I).$$

Each of the genotypes U_i, combined with that of the victim, that is A_1A_3, implies an allelic configuration as observed for the crime stain. Thus, again ignoring peak heights, $Pr(csp|vgt, U_i, H_d, I) = 1$ for $i = 1, 2, 3$. Consequently, the denominator reduces to

$$Pr(csp|vgt, H_d, I) = \sum_i Pr(U_i|H_d, I).$$

Assuming that the possible genotypes of the unknown individual do not depend on H_d, the probabilities of the U_i are given by the products of the respective allele probabilities. Let γ_1,

γ_2 and γ_3 denote the probabilities for A_1, A_2 and A_3, respectively. The likelihood ratio can then be written

$$V = \frac{1}{2\gamma_1\gamma_2 + \gamma_2^2 + 2\gamma_2\gamma_3}.$$
(7.13)

7.8.2   Bayesian network for a three-allele mixture scenario

When examining the proposed three-allele mixture scenario in terms of a Bayesian network, the genotypes of the suspect (*sgt*) and the victim (*vgt*) need to be represented in some way. Also, a certain number of intermediate variables are necessary in order to provide a logical connection to a variable representing the observed crime stain profile (*csp*). This can be achieved in much the same way as was done in the previous sections for scenarios involving stains from single donors.

Notice that the probabilistic approach to mixed stains as described in Section 7.8.1 assumes that there to be exactly two individuals contributing to the mixture; that is, on the one hand, the victim and the suspect, and, on the other hand, the victim and an unknown individual. The suspect's profile is known but not considered in the development under the alternative proposition H_d. The mixture can thus be thought of as a combination of two distinct stains, for each of which a given individual is considered as a potential source.

This point of view can be translated into a Baycsian network by choosing two sub-models that are analogous to the graph shown in Figure 7.2 (Section 7.2.2). These sub-models are used to evaluate whether the victim and the suspect are the respective sources of the two underlying stains that make up the DNA mixture. The source nodes are denoted $T1 = v$? and $T2 = s$?. The two stains are unobserved variables and their genotypes arc denoted $T1gt$ and $T2gt$. Further details are given in Figure 7.14.

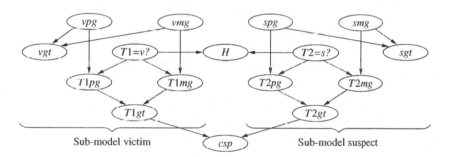

Figure 7.14   Bayesian network for evaluating a scenario where the profile of the crime stain (*csp*) possibly contains more than two alleles. There are four states (source combinations) for the node *H*: *s&v*, suspect and victim; *v&U*, victim and unknown individual; *s&U*, suspect and unknown individual; and *2U*, two unknown individuals. Nodes $T1 = v$? and $T2 = s$? take values *yes* or *no* according to, $T1 = v$?, the victim has contributed to the crime stain, or $T2 = s$?, the suspect has contributed to the crime stain. Nodes *vpg*, *vmg*, *spg* and *smg* denote victim's or suspect's paternal or maternal genes, with *vgt* and *sgt* denoting the victim's or suspect's genotypes. Parameters $T1$ and $T2$ denote the two stains that make up the mixed stain profile *csp*. Nodes $T1pg$, $T2pg$, $T1mg$ and $T2mg$ denote paternal and maternal genes for these stains, with $T1gt$ and $T2gt$ denoting their genotypes. (Reproduced with permission of Academic Press, Ltd.)

Table 7.9 Conditional probabilities assigned to the node H, specifying the contributors to the crime mixed stain.

$T1 = v$?:	Yes		No	
$T2 = s$?:	Yes	No	Yes	No
H : $s \& v$	1	0	0	0
$v \& U$	0	1	0	0
$s \& U$	0	0	1	0
$2U$	0	0	0	1

There are four states (sources) for H, $s\&v$, suspect and victim; $v\&U$, victim and unknown individual; $s\&U$, suspect and unknown Individual; and $2U$, two unknown individuals. Factors $T1 = v$? and $T2 = s$? take values *yes* or *no* according as, $T1 = v$?, the victim has contributed to the crime stain or, $T2 = s$?, the suspect has contributed to the crime stain.

The two sub-models can then be related to each other by adopting two nodes, H and csp, that both are logical combinations of nodes contained in the sub-models. The node H represents a collection of propositions as to how pairs of individuals may contribute to the mixture, that is the suspect and the victim ($s \& v$), the victim and an unknown individual ($v \& U$), the suspect and an unknown individual ($s \& U$) or two unknown individuals ($2U$). The node probability table associated with H may be completed logically with values 0 and 1 (Table 7.9).

The definition of the node csp, denoting the profile of the mixed stain, is crucially dependent on the level of detail of the model's allele and genotype nodes. In order to keep the sizes of the probability tables moderate, and for the ease of argument, consider a hypothetical marker with only three alleles A_1, A_2 and A_3. Let the allele population proportions be denoted γ_1, γ_2 and γ_3. These then are the states of the victim's and the suspect's parental gene nodes (vpg, vmg, spg and smg) as well as the trace parental gene nodes ($T1pg$, $T1mg$, $T2pg$ and $T2mg$). Consequently, there are six states indicated for each of the genotype nodes vgt, sgt, $T1gt$ and $T2gt$: A_1A_1, A_1A_2, A_1A_3, A_2A_2, A_2A_3 and A_3A_3. The probability tables of the genotype nodes are completed analogously to the description provided in Section 7.2 (Table 7.3).

Given the states of the genotype nodes as stated above, there are various possible combinations of alleles that a mixture may contain. For the ease of discussion, restrict the number of states of the crime stain profile node (csp) to $A_1A_2A_3$ and $\overline{A_1A_2A_3}$. The latter state is a global assignment to cover all combinations of alleles other than $A_1A_2A_3$. Despite this reduction, the probability table associated to the node csp is still too large to be given here in full detail. However, one can formulate in a more general way how the table needs to be completed. Notably, a probability of 1 is assigned to $Pr(csp = A_1A_2A_3 | T1gt, T2gt)$ whenever (i) both $T1gt$ and $T2gt$ are heterogeneous and have not more than one allele in common or (ii) one of the stains is heterogeneous and the other is homogeneous and the two genotypes have no allele in common. Otherwise, a zero probability is assigned. Notice that the probability assigned to a state $\overline{A_1A_2A_3}$ of the node csp, given a particular conditioning by $T1gt$ and $T2gt$, is the complement of the probability assigned to state $A_1A_2A_3$.

Example 7.8 *(Bayesian network for mixtures) Consider a numerical example for the Bayesian network shown in Figure 7.14. Imagine a scenario as discussed in Section 7.8.1 involving a victim being heterogeneous A_1A_3, a suspect being homogeneous A_2A_2 and mixed crime stain with alleles $A_1A_2A_3$. The numerical assignments are chosen as follows: $\gamma_1 = 0.2$, $\gamma_2 = 0.3$ and $\gamma_3 = 0.5$. Default probabilities of 0.5 are assigned to the source nodes $T1 = v$? and*

$T2 = s?$. *Notice that the way in which the latter probabilities are chosen is irrelevant for the analysis pursued here. Various instantiations will be made at the node H, and the changes in the observational node csp recorded. The particular choice of the conditional probabilities assigned to the node H implies that, given an instantiation made at the node H, the source nodes $T1 = v?$ and $T2 = s?$ become either true or false. It can then be seen that the assignment of prior probabilities to these variables is merely a technical matter necessary to run the model. Any pair of prior probabilities summing up to 1 may be used.*

An evaluation of the numerator of the likelihood ratio may be obtained through the following instantiations: $vgt = A_1A_3$, $sgt = A_2A_2$ and $H = v$ & s. As may be expected, the effect of these instantiations is that the probability of the crime stain being of type $A_1A_2A_3$ is 1 (node csp). The alternative proposition is that the victim and an unknown individual have contributed to the mixed stain. For evaluating the denominator, the state of the node H must thus be changed to v & U. The change in the conditional H reduces the probability of the crime stain being of type $A_1A_2A_3$ to 0.51. The same results may be obtained using (7.13). The denominator of this formula leads to $2 \times 0.2 \times 0.3 + 0.3^2 + 2 \times 0.3 \times 0.5 = 0.51$. Notice further that, given $H = v$ & U, the actual genotype of the suspect (node sgt) does not affect the probability of the crime stain profile (node csp). This property is entirely in agreement with the stated assumptions.

In this section, a Bayesian network has been set up for the specific scenario involving a mixed stain with three alleles and assuming exactly two contributors. A Bayesian network with the same structure may also serve to evaluate scenarios involving, for example, a mixed stain covering four alleles. This can be achieved by invoking an argument analogous to that used so far, in particular by extending the lists of possible alleles involved in the scenario.

The Bayesian network presented here has initially been described in Mortera (2003) and Mortera et al. (2003). In the latter paper, the interested reader can find further extensions to issues including more than two contributors, missing individuals or silent alleles. Such technicalities have not been developed here in more detail since the major aim was to familiarize the reader with the idea that basic building blocks, or the so-called network fragments, that have separately been set up and validated earlier, can logically be re-used in order to approach further complications that the evaluation of DNA profiling results may entail, such as the occurrence of mixed stains. The paper by Mortera et al. (2003) also discusses alternative network structures in which genotypes are represented indirectly, that is by a collection of Boolean allele nodes, one for each relevant allele. This allows one to avoid large probability tables as a result of an increased number of alleles that may make up a mixed profile.

7.9   Kinship analyses

7.9.1   A disputed paternity

The application of graphical probabilistic models, notably Bayesian networks, to inference problems involving the results of DNA profiling analyses represents a lively area of research. A particular application is kinship analyses as discussed by Dawid et al. (2002). These authors discuss how to derive appropriate graphical structures for Bayesian networks from initial pedigree representations of forensic identification problems. This section focusses on such an example in terms of a classic case of disputed paternity.

A certain male, denoted as the putative father pf, is alleged to be the father of a certain child c. Results are available of measurements taken on genetic markers for the mother m, the putative father pf and the child c. What is thought is the likelihood ratio for the proposition that the putative father *pf* is the true father *tf*, noted *tf* = *pf*?, given knowledge about the child's genotype *cgt* and the mother's genotype *mgt*, when the alternative proposition is that an unknown man, unrelated to the suspect, is the true father.

This scenario can be represented in terms of a basic paternity pedigree as shown in Figure 7.15(i), where squares represent males and circles females. In order to evaluate the scenario in terms of a Bayesian network, Dawid et al. (2002) propose the use of distinct sub-models, as shown in Figure 7.16. The network fragment represented by Figure 7.16(i) describes a child's genotype as a function of a pair of parental genes. These parental gene nodes are unobserved variables as one may not know which of the two alleles of a child's genotype has been transmitted by the mother and which one has been transmitted by the father. The sub-model shown in Figure 7.16(ii) provides a means for evaluating the uncertainty in relation to which of two parental alleles is transmitted to the offspring.

If related to each other in an appropriate way, such network fragments allow available information on the genotypes of the child, the mother and the putative father to be used to draw

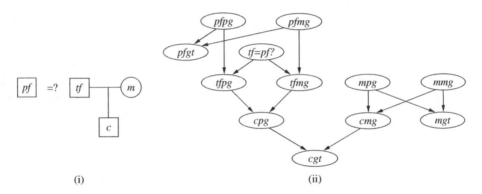

Figure 7.15   Different representations of a case of disputed paternity: (i) paternity pedigree, where squares represent males and circles females; *pf* denotes the putative father, *tf* denotes the true father, *m* denotes the mother and *c* denotes the child and (ii) a Bayesian network. For the Bayesian network, *tfpg*, *tfmg*, *mpg* and *mmg* denote the paternal *p* and maternal *m* (in second place) genes of the true father *tf* and the mother *m* (in first place); *cpg* and *cmg* denote the child's paternal and maternal genes, respectively; *pfpg* and *pfmg* denote the putative father's paternal and maternal genes, respectively; *pfgt*, *mgt* and *cgt* denote the genotypes of the putative father, the mother and the child, respectively, and *tf* = *pf*? takes two values in answer: 'yes' or 'no' as to whether the true father is the putative father.

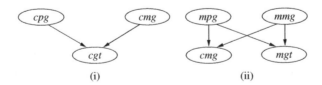

Figure 7.16   Basic sub-models usable for (i) inference about a child's paternal and maternal genes (*cpg* and *cmg*), based on the child's genotype (*cgt*), and (ii) inference about a child's maternal gene (*cmg*), based on the mother's genotype (*mgt*), and the mother's paternal and maternal genes (*mpg* and *mmg*), respectively.

an inference about the proposition according to which the putative father, rather than some other man, is the true father. Dawid et al. (2002) have proposed a structure for a Bayesian network, which is as shown in Figure 7.15(ii). This network provides the same results as that obtained with the classic arithmetic calculus of Essen-Möller under Hardy–Weinberg assumptions of independence.

Example 7.9 *(Bayesian network for kinship analysis) In order to illustrate the Bayesian network for a case of disputed paternity, consider a scenario involving a heterozygous child with genotype A_1A_2. The undisputed mother is homozygous A_1A_1 and the putative father is heterozygous A_1A_2. The numerator of the likelihood ratio is given by the probability of the child's genotype given the genotypes of the mother and the putative father and given the proposition that the putative father is the true father. As the mother will certainly transmit an allele A_1, the event of the child being A_1A_2 depends on whether or not the true father will transmit an allele A_2. The latter event occurs with a probability of 0.5 under the first proposition. Under the proposition of non-paternity, some unrelated individual from the population is the true father. That other individual may be either of genotype A_2A_2 or of A_2A_x (with $x \neq 2$). In the former case, the individual will certainly transmit an allele A_2. This event occurs with probability γ_2^2, that is the probability of that individual being of type A_2A_2. In the event that the individual is A_2A_x, which occurs with probability $2\gamma_2\gamma_x$, an allele A_2 will be transmitted with probability 0.5. The denominator thus is given by $\gamma_2^2 + 0.5 \times 2\gamma_2\gamma_x$. This reduces to γ_2 because $\gamma_x = 1 - \gamma_2$. The probability of the true father, that is an unknown individual different from the putative father, contributing allele A_2 is thus given by the allele proportion in the relevant population, denoted γ_2. The likelihood ratio in favour of paternity thus is given by $1/2\gamma_2$.*

Figure 7.17 shows an illustration of this evaluation assuming the numerical assignments $\gamma_1 = \gamma_2 = 0.05$. As, for purely technical reasons, equal prior probabilities have been assumed for the target node $pf = tf$?, the ratio of the posterior probabilities after instantiating the nodes cgt, mgt and pfgt, equals the likelihood ratio. In the case at hand, both by use of the described Bayesian network and by the formula developed in the previous paragraph, a likelihood ratio of 10 is obtained.

The network shown in Figure 7.17 may also be used for calculating separately the components of the likelihood ratio. For evaluating the numerator, the node $tf = pf$? needs to be set to 'yes', whereas the genotype nodes of the mother and the putative father need to be set to A_1A_1 and A_1A_2, respectively. The probability of the child's genotype being A_1A_2 can then be read from the node cgt, which would display 0.5. In order to obtain an evaluation of the denominator, it is sufficient to change the node $tf = pf$? to 'no'. For the state A_1A_2, the node cgt would then display, as may be expected, a probability of 0.05. This corresponds to the probability of observing an allele A_2. Notice also that for the instantiation of $tf = pf$? to 'no', the actual genotype of the putative father no longer affects the probability distribution for the node cgt.

An extension that may be approached with the network shown in Figure 7.17 is the possibility of mutation. Mutation refers to those events that lead to a change in the sequence of the nucleotides of DNA. In kinship analyses, mutation may lead to situations in which a child is found to carry an allele different from those carried by its parents. Dawid et al. (2002) proposed an approach for evaluating such seeming exclusions in paternity. This approach is based on an additional graphical sub-structure that allows one to make a distinction between an individual's 'actual' parental gene and the parental gene 'originally' carried by the respective

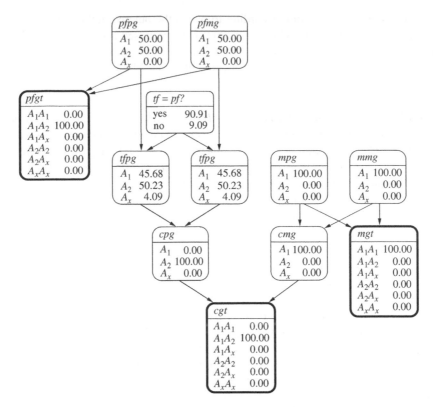

Figure 7.17 Evaluation of a case of disputed paternity. The mother is homozygous $mgt = \{A_1A_1\}$, whereas the putative father ($pfgt$) and the child (cgt) are both heterozygous A_1A_2. The probability of paternity is displayed in the node $tf = pf$?. Nodes $tfpg, tfmg, mpg$ and mmg denote the paternal p and maternal m (in second place) genes of the true father tf and the mother m (in first place); cpg and cmg denote the child's paternal and maternal genes, respectively; $pfpg$ and $pfmg$ denote the putative father's paternal and maternal genes, respectively; $pfgt, mgt$ and cgt denote the genotypes of the putative father, the mother and the child, respectively, and $tf = pf$? takes two values in answer: 'yes' or 'no' as to whether the true father is the putative father. Instantiated nodes are shown with a bold border.

parent. Figure 7.18 summarizes this extension pictorially for an individual's maternally inherited gene. The nodes $mamg$ and $mapg$ denote, respectively, the mother's 'actual' maternal and paternal genes. Notice that the procedure applies analogously to the paternal line. In Figure 7.18, the nodes $comg$ and $camg$ denote, respectively, the child's 'original' and 'actual' maternal genes. Mutation rates are considered when assessing probabilities for original genes mutating into actual genes. Further details are given in Dawid et al. (2002) and Dawid (2003a).

7.9.2 An extended paternity scenario

In cases of disputed paternity, it may regularly happen that scientists need to account for further individuals or of the fact that one or more of the principal individuals, such as the alleged

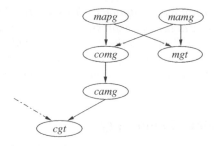

Figure 7.18   Maternal inheritance line in a paternity network with mutation. The nodes *mamg* and *mapg* denote, respectively, the mother's 'actual' maternal and paternal genes. The nodes *comg* and *camg* denote, respectively, the child's 'original' and 'actual' maternal genes.

father for example, are unavailable for DNA typing. It is here that the paternity network discussed in Section 7.9.1 provides a valuable basis, as it may be readily modified and extended as required.

Example 7.10 *(Bayesian network for deficiency kinship analysis) In order to illustrate an extended paternity case, consider the question of whether a certain child is the genetical child of an alleged father. The case is such that the grandmother, the grandfather and the alleged father are deceased. The individuals investigated are a brother and a sister of the alleged father, the child and the (undisputed) mother.*

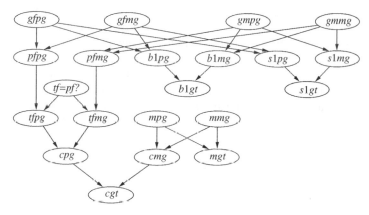

Figure 7.19   Evaluation of a case of disputed paternity where the putative father is not available, but a brother and a sister of the putative father are available. Nodes *gfpg, gfmg, gmpg* and *gmmg* denote the grandfather *gf* and grandmother *gm* paternal *p* and maternal *m* genes. Nodes *pfpg, pfmg, b1pg, b1mg, s1pg* and *s1mg* denote the putative father, *pf*, the brother *b1*, and sister *s1* paternal *p* and maternal *m* genes. Nodes *tfpg, tfmg, mpg, mmg, cpg* and *cmg* denote the true father *tf*, mother *m* (in first place) and child *c* paternal *p* and maternal *m* (in second place) genes. Nodes *b1gt, s1gt, mgt* and *cgt* denote the brother *b1*, sister *s1*, mother *m* and child *c* genotypes. Node *tf = pf*? takes two values: 'yes' if *tf = pf*, the true father is the putative father and 'no' if *tf ≠ pf*, the true father is not the putative father.

Figure 7.19 shows a pictorial representation of this scenario. The extension to siblings of the alleged father is made using arguments expressed in the sub-models shown in Figure 7.16. This extension is also largely analogous to the discussion provided in Section 7.3, where an inference was drawn to the suspect's genotype on the basis of knowledge about a brother's genotype (see also Figure 7.5). The probability tables are completed following the ideas introduced in Section 7.3.

7.9.3   A case of questioned maternity

Consider a case of questioned maternity in which the alleged mother is dead and not available for typing. Typing results are available for the child and for the parents of the alleged mother. The target question is whether the alleged mother is the true mother. The pedigree is shown in Figure 7.20. This case was proposed in the paper challenge 2010 of the English Speaking Working Group of the International Society for Forensic Genetics (ISFG).

Figure 7.21 shows a Bayesian network for approaching this scenario on the level of typing results for a single marker. As a main difference with respect to the paternity network discussed in Section 7.9.2, genotypes are represented here, as outlined earlier in Section 7.2.3 (Figure 7.4), in terms of distinct nodes *gtmin* and *gtmax* that contain the full set of alleles for a given locus. This renders the proposed model more readily applicable in varying casework settings.

Uncertainty about whether or not the alleged mother is the true mother is represented in terms of a sub-structure that is analogous to that used in the paternity network shown in Figure 7.15(i). For the model considered here, the nodes representing the true mother's paternal and maternal genes (*tmpg* and *tmmg*) copy, respectively, the actual state of the alleged mother's paternal and maternal genes (*ampg* and *ammg*) whenever the Boolean[1] node *tm = am*? is in the state 'true' (i.e. the alleged mother is the true mother). Whenever the latter node is the state 'false', then *tmpg* and *tmmg* copy the actual state of, respectively, the paternal and maternal genes of another woman (*ompg* and *ommg*). Although the size of the probability tables of the nodes *tmpg* and *tmmg* may be considerable, in particular when the genetic marker of interest has many alleles, the required assignments of zeros and ones in the node tables can readily be defined through an expression. Considering, for example the node *tmpg*, syntax of the Hugin software allows one to define the probability table in terms of the expression if (C1,C2,C3), where C1 refers to *tm = am*?, C2 to *ampg* and C3

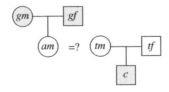

Figure 7.20   Case of questioned maternity proposed in the relationship testing workshop 2010 of the English Speaking Working Group of the International Society for Forensic Genetics (ISFG). Individuals for which typing results are available are shown in grey-shaded nodes. Squares represent males and circles females.

---

[1] This node definition is a slight variation, although conceptually equivalent in output, with respect to the target nodes *tf = pf*? in Sections 7.9.1 and 7.9.2, where a common chance node with states 'yes' and 'no' was defined. The particular choice of a so-called Boolean node is made here in order to facilitate the specification of expressions in Hugin syntax.

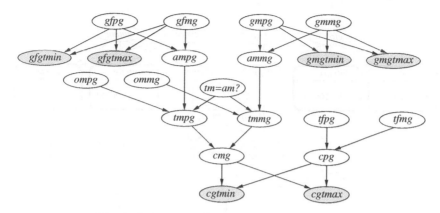

Figure 7.21   Bayesian network for a case of disputed maternity where *gfpg*, *gfmg*, *gmpg*, *gmmg*, *ampg*, *ammg*, *ompg*, *ommg*, *tmpg*, *tmmg*, *tfpg*, *tfmg*, *cpg* and *cmg* denote the paternal and maternal genes of the grandfather *gf*, the grandmother *gm*, the alleged mother *am*, the alternative (i.e. 'other') mother *om*, the true mother *tm*, the true father *tf* and the child *c*. The descriptors *gtmin* and *gtmax* denote, respectively, the smaller and the larger allele that make up an individual's genotype. The node *tm* = *am*? takes two values 'true' and 'false' in answer as to whether the true mother is the alleged mother.

to *ompg*. Notice that without an explicit representation of the paternal and maternal genes of an alternative mother, *ompg* and *ommg*, the probability tables of the nodes *tmpg* and *tmmg* would need to include allele populations proportions, which may be a less efficient way of completing these probability tables.

Observational nodes of the kind *gtmin* and *gtmax* are not included here for the alleged mother and 'other mother (*om*)' because no typing results are available for these individuals. In contrast to this, the genotypes of child and the grandparents are modelled in terms of network fragments that take the structure defined earlier in Section 7.2.3 (Figure 7.4).

Example 7.11 *(Bayesian network for a proficiency exercise in kinship analysis) Consider an application of this Bayesian network for typing results at the locus TH01. In view of U.S. Caucasian population data as published in Butler et al. (2003), the following set of alleles and associated probabilities is considered here:*

$$\{5, 6, 7, 8, 9, 9.3, 10, 11\}, \quad \{0.002, 0.232, 0.190, 0.084, 0.114, 0.368, 0.008, 0.002\}.$$

The child and the grandmother are found to have the genotype $\{7, 9.3\}$, *whereas the grandfather is found to be homozygote* $\{9.3\}$. *Start by considering that, given the genotypes* $\{7, 9.3\}$ *and* $\{9.3\}$ *of the grandparents, the alleged mother could have the genotype either* $\{7, 9.3\}$ *or* $\{9.3\}$, *with probability 0.5 each. If the alleged mother has the genotype* $\{9.3\}$, *then the probability that her child will be of type* $\{7, 9.3\}$ *is* $\gamma_7 = 0.190$. *If, however, the alleged mother has the genotype* $\{7, 9.3\}$, *then her child will have the genotype* $\{7, 9.3\}$ *with probability* $0.5\gamma_{9.3} + 0.5\gamma_7$. *This stems from the fact that the alleged mother may transmit either an allele 7 or an allele 9.3 (each with probability 0.5). The numerator of the likelihood ratio thus is*

$$Pr(c = \{7, 9.3\}|gm = \{7, 9.3\}, gf = \{9.3\}, tm = am) = 0.5\gamma_7 + 0.5(0.5\gamma_{9.3} + 0.5\gamma_7).$$

Using values for γ_7 *and* $\gamma_{9.3}$ *as specified above, the value 0.2345 is obtained.*

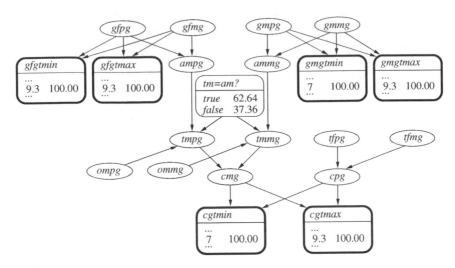

Figure 7.22   Partially expanded representation of the Bayesian network shown in Figure 7.21 for a case of disputed maternity where *gfpg, gfmg, gmpg, gmmg, ampg, ammg, ompg, ommg, tmpg, tmmg, tfpg, tfmg, cpg* and *cmg* denote the paternal and maternal genes of the grandfather *gf*, the grandmother *gm*, the alleged mother *am*, the alternative (i.e. 'other') mother *om*, the true mother *tm*, the true father *tf* and the child *c*. The descriptors *gtmin* and *gtmax* denote, respectively, the smaller and the larger alleles that make up an individual's genotype. The node *tm = am*? takes two values 'true' and 'false' in answer as to whether the true mother is the alleged mother. Instantiated nodes are shown with a bold border. The grandfather's genotype is set to {9.3} and that of the grandmother and the child to {7, 9.3}.

Under the assumption that the alleged mother is not the true mother, the child has two unknown persons as parents. In such a case, the probability of observing the genotype {7, 9.3} can be interpreted as the proportion of individuals that have that genotype. Assuming independence from the genotype of the alleged mother, that probability is given by $2\gamma_7\gamma_{9.3} = 0.13984$. The resulting likelihood ratio thus is $V = 0.2345/0.13984 = 1.68$ (result rounded to two decimals).

This result can also be obtained with the proposed Bayesian network. Figure 7.22 shows such analysis in terms of the posterior probabilities for the node tm = am?, given the observed genotypes {7, 9.3} for the grandmother and the child and {9.3} for the grandfather. Because the initial probabilities of the target node tm = am? were equal, the ratio of the posterior probabilities displayed for this node is equivalent to the likelihood ratio: $0.6264/0.3736 = 1.68$ (result rounded to two decimals).

7.10   Database search

7.10.1   Likelihood ratio after database searching

Throughout this chapter, it has been assumed that a suspect has been found on the basis of information that is completely unrelated to the crime stain. A different situation is one in which the suspect has been selected through a search in a database, and this may lead to a question of the following kind: 'What is the strength of the findings against a suspect who is found as a result of the search of a database?' This practical question, also sometimes referred to as *the database search problem*, has led to considerable discussion within the scientific community, including both forensic scientists and legal practitioners. Different answers, pointing in quite contrary directions, have been offered so far, accompanied with substantial mathematics. This section refers to instances of the most well-supported viewpoint according to which the probative value of a DNA profile that corresponds to a crime stain is greater for the profile which has been found as the result of a database search than for the profile which has not been so found (e.g. Balding 2005; Balding and Donnelly 1995, 1996; Balding et al. 1994; Berry 1994; Buckleton et al. 2005; Dawid 2001; Donnelly and Friedman 1999; Evett et al. 2000a; Evett and Weir 1998; Kaye 2009). For a variation of the classical database search problem that extends to some specified but unobserved relatives of the database members, see Cavallini and Corradi (2006).

Start by considering the basic formulation. In the scenario considered by Balding and Donnelly (1996), a suspect has been found as a result of a search of the DNA profile of a crime stain against a database of N suspects. The suspect's profile was the only profile found to correspond. All the other $(N - 1)$ profiles contained in the database did not correspond. Examining this result suggests there to be two distinct items of information. Let E stand for the correspondence between the profile of the crime stain (E_c) and the profile of the suspect (E_s). Let D denote the proposition that the other $(N - 1)$ profiles in the database do not correspond. Notice that the evaluation of a likelihood ratio incorporating both of these elements of information, E and D, is, in essence, a problem of combining items of information, a topic which will be discussed in more detail in Chapter 8. Part of this discussion is anticipated here by formulating a likelihood ratio for evaluating the combined effect of the two elements E and D. The ratio can be written as follows:

$$V = \frac{Pr(E, D|H_p, I)}{Pr(E, D|H_d, I)}. \tag{7.14}$$

Here, the source level propositions H_p and H_d specify, respectively, the suspect and some unknown person (unrelated to the suspect) as the source of the crime stain. Using the product rule, (7.14) can be re-written as:

$$V = \frac{Pr(E|H_p, D, I)}{Pr(E|H_d, D, I)} \times \underbrace{\frac{Pr(D|H_p, I)}{Pr(D|H_d, I)}}_{V_{db}}. \tag{7.15}$$

Following the analysis of Balding and Donnelly (1996), the first ratio on the right-hand side of (7.15) reduces to $1/\gamma$ approximately, where γ is – very generally speaking – the relevant profile probability. The probability of a correspondence may thus be assumed to be independent of the information that there has been no correspondence amongst the $(N - 1)$ other individuals contained in the database. The variable D in the conditional of the denominator at this point thus is not pursued in further detail.

Further consideration is required for evaluating the second ratio on the right-hand side of (7.15). Evett and Weir (1998) have referred to this fraction as the 'database search likelihood ratio', written V_{db} here. Consider the numerator first. It asks for the probability that none of the $(N - 1)$ suspects correspond, given that the suspect is the source of the crime stain (H_p). Notice that D is not conditioned on any information relating to the observed DNA profile, so the probability of interest is concerned with the potential of the profiling system to differentiate between any two people chosen at random, a property also known as the *discriminating power*. Following a notation used by Evett and Weir (1998), a parameter $\psi_{(N-1)}$ is used to denote the probability that none of the $(N-1)$ innocent individuals has a genotype corresponding to that of any unspecified crime stain.

For evaluating the denominator of V_{db}, one needs to consider that if the suspect is not the source of the crime stain, then necessarily someone else is. Thus, let A denote the proposition that the true source of the crime stain is amongst the $(N-1)$ other members of the database. The denominator can then be extended to

$$Pr(D|H_d) = Pr(D|A, H_d)Pr(A|H_d) + Pr(D|\bar{A}, H_d)Pr(\bar{A}|H_d). \tag{7.16}$$

If the true source of the crime stain is amongst the $(N-1)$ other members of the database, that is A is true, then the probability that there is no correspondence amongst the $(N-1)$ profiles can be considered impossible. Accordingly, the product of the first pair of terms in (7.16) is zero. $Pr(D|\bar{A}, H_d)$ is the probability that the $(N-1)$ profiles do not correspond, given that the true source is not amongst these $(N-1)$ profiles and that the suspect is not the source of the crime stain. Again, this is an assignment that can be made on the basis of $\psi_{(N-1)}$. If ϕ denotes the probability that the true source is amongst the $(N-1)$, then (7.16) reduces to $\psi_{(N-1)}(1-\phi)$ and the database search likelihood ratio V_{db} becomes

$$V_{db} = \frac{Pr(D|H_p)}{Pr(D|H_d)} = \frac{\psi_{(N-1)}}{\psi_{(N-1)}(1-\phi)} = \frac{1}{1-\phi}.$$

It can now be seen that, the greater the size of the database, one should judge it increasingly probable to see that the true source is amongst the $(N-1)$ individuals, thus ϕ is increasing. Consequently, the denominator of V_{db} diminishes and the database search likelihood ratio increases. The primary result of the analysis thus is that, under the stated assumptions, the fact that the suspect has been selected through a search in a database has a tendency to increase the overall likelihood ratio V, defined in (7.15).

In view of capturing this probabilistic approach in terms of a Bayesian network, it is helpful to focus on the three binary variables H, E and D, defined as follows: 'The suspect (H_p) or some other person (H_d) is the source of the crime stain' (variable H), 'The profile of the crime stain corresponds to that of the suspect' (variable E) and 'The other $(N-1)$ profiles of the database do not correspond' (variable D).

Let $\bar{E}$ and $\bar{D}$ denote the negations of the propositions E and D, respectively. In particular, $\bar{D}$ denotes the proposition 'One and only one of the other $(N-1)$ profiles in the database do correspond'. In order to find an appropriate graphical structure that associates these three variables to one another, it is necessary to inquire about the assumed dependencies amongst them. Consider thus (7.14), which states that the variables E and D are both conditioned on the variable H. A graphical representation of these assumptions is that H is chosen as a parental variable for, respectively, E and D so that a diverging connection with $E \leftarrow H \rightarrow D$ is obtained.

Notice that this Bayesian network structure is an explicit representation of the assumption that, knowing H were true, the probability of a correspondence between the suspect's profile and the profile of the crime stain is not influenced by the fact that there has been a search in a database. The conditional probabilities assigned to the variables E and D are presented in Table 7.10.

Table 7.10 Probabilities assigned to the
variables E and D, conditional on the state of H.

H:		H_p	H_d
E:	E	1	γ
	$\bar{E}$	0	$1 - \gamma$
D:	D	$\psi_{(N-1)}$	$(1 - \phi)\psi_{(N-1)}$
	$\bar{D}$	$1 - \psi_{(N-1)}$	$1 - [(1 - \phi)\psi_{(N-1)}]$

Node H has two states: the suspect (H_p) or some other person
(H_d) is the source of the crime stain. Node E has two states:
the profile of the crime stain corresponds or does not
correspond to that of the suspect. Node D has two states: all
other $(N - 1)$ profiles of the database do not correspond or one
and only one of the other $(N - 1)$ profiles of the database
corresponds. The value γ expresses rarity of the corresponding
DNA profile. The probability that the true source of the crime
stain is amongst the other $N - 1$ profiles is denoted ϕ. The
term $\psi_{(N-1)}$ is used to denote the probability that none of the
$(N - 1)$ innocent individuals has a genotype corresponding to
that of any unspecified crime stain.

7.10.2   An analysis focussing on posterior probabilities

An alternative way to approach the 'database search' problem consists in developing an extension to the well-known 'island problem', described along the following lines. A biological
stain is found on a crime scene. It has been typed and found to have genetic profile G_c. It is
assumed here that the method applied for determining the genetic profile of a biological stain
works accurately. The 'island' on which the crime was committed has a population of size N.
Initially, there is no information that directs suspicion to any of the N islanders. Each of them
may be considered to be the source of the crime stain with equal probability. Since the stain is
found to be of type G_c, so must be the person from which the stain comes. A suspect is identified and his blood analysed. For the time being, assume that *no* database search has been
conducted. The genetic profile of the suspect is denoted G_s. It is found to correspond to that
observed for the crime stain, therefore $G_c = G_s$. On the basis of this information, the question
of interest is 'How convinced should one be that the suspect is the source of the crime stain?'

In order to approach this question, information about the occurrence of the corresponding
genetic profile is needed. Suppose that, on the basis of a survey of a comparable population
on another island, the target profile can be taken to occur in about 1% of the population and
that this proportion, written γ for short, can also be considered for the population of the island
on which the crime stain of interest was found (and where the offender is thought to live). It
is further supposed here that knowledge of the suspect's genotype, G_s, does not affect one's
probability that another islander has that profile.

The formal analysis of this inference problem requires some further notation. Within the
population of N individuals, let us index the suspect as person 1 and the remaining individuals
as $2 \ldots N$. Next, let the proposition that a given person i is the source of the crime stain be
denoted H_i. The term H_1 thus stands for the proposition that the suspect is the source of the
crime stain. Analogously, the propositions according to which one of the remaining $N - 1$
people is the source of the crime stain are denoted $H_2, \ldots, H_N$. The initial probability that

a given individual is the source of the crime stain will be written $Pr(H_i) = \pi_i$. Since it is considered, as a starting point, that each of the N persons could be the source with equal probability, one has $\pi_i = 1/N$ and $\sum_{i=1}^{N} \pi_i = 1$. Later on, further notation will introduced in order to allow for the possibility that some of the N individuals are part of a database.

In the setting considered so far, the suspect is the only typed individual amongst the N persons. Write M_1 for the finding that his genotype, G_s, corresponds to that of the crime stain, G_c. The probability that the suspect is the source of the crime stain is then given by Bayes' theorem for discrete observations (data) and multiple discrete propositions [Section 1.1.8, Equation (1.8)]:

$$Pr(H_1|M_1) = \frac{Pr(M_1|H_1)Pr(H_1)}{Pr(M_1|H_1)Pr(H_1) + \sum_{i=2}^{N} Pr(M_1|H_i)Pr(H_i)}. \tag{7.17}$$

Assuming that the suspect will certainly correspond if he is in fact the source of the crime stain, $Pr(M_1|H_1) = 1$, and assuming that he will correspond with probability γ if he is not the source, then, π_1', the posterior probability for H_1, after considering the observation M_1, thus is

$$Pr(H_1|M_1) = \pi_1' = \frac{\pi_1}{\pi_1 + \gamma(1 - \pi_1)}, \tag{7.18}$$

since $Pr(M_1|H_i) = \gamma$ for all i and thus

$$\sum_{i=2}^{N} Pr(M_1|H_i)Pr(H_i) = \gamma \sum_{i=2}^{N} Pr(H_i) = \gamma(1 - \pi).$$

The island problem as described so far can be modified in order to illustrate the effect of a database search. Continue to assume that the variable N represents the size of the total population. However, suppose now that the DNA profiles of the first $1, \ldots, n$ individuals (where the index 1 is that of the suspect) are in a database. The individuals $(n + 1), \ldots, N$ are outside the database. Also, one of the assumptions here is that the profile of the crime stain is compared to all n individuals in the database. This search of the database reveals that only the profile of the suspect corresponds to the profile of the crime stain. This correspondence is denoted, as before, by M_1. Besides, the database search has also revealed that the $2, \ldots, n$ individuals on the database other than the suspect do *not* correspond. The fact that a profile of an individual i (for $i = 2, \ldots, n$) does not correspond to the crime stain is denoted here by X_i. One can thus write $X_2 \& X_3 \& \ldots \& X_n$ for the information that all entries of the database other than that of the suspect do not correspond. The latter two elements need to be jointly evaluated. Following Kaye (2009), write thus the totality of the findings as $E_n = M_1 \& X_2 \& X_3 \& \ldots \& X_n$.

Considering that there are n of the N individuals in a database leads to a minor refinement in the way in which the source level propositions H_i (for $i = 2, \ldots, n$) are formulated. In fact, they can now be framed as 'the individual i in the database is the source of the crime stain'. A more conceptual underpinning of these propositions H_i (for $i = 2, \ldots, n$) is that they refer to individuals who had their DNA profile compared to that of the crime stain. This is a difference with respect to the individuals $(n + 1), \ldots, N$ whose profiles were not compared. On the whole, one can thus think of the population of size N as a splitting into n individuals as database members and $N - n$ that are not. This splitting becomes apparent when rewriting the posterior probability defined earlier in (7.17). Writing this probability for the observations E_n gives

$$Pr(H_1|E_n) = \frac{Pr(E_n|H_1)Pr(H_1)}{\left(\begin{array}{c} Pr(E_n|H_1)Pr(H_1) + \sum_{i=2}^{n} Pr(E_n|H_i)Pr(H_i) \\ + \sum_{i=n+1}^{N} Pr(E_n|H_i)Pr(H_i) \end{array}\right)}.$$

This term can be shown to reduce to (Kaye 2009)

$$Pr(H_1|E_n) = \pi_1' = \frac{Pr(H_1)}{Pr(H_1) + \gamma \sum_{i=n+1}^{N} Pr(H_i)} = \frac{\pi_1}{\pi_1 + \gamma \sum_{i=n+1}^{N} \pi_i}. \tag{7.19}$$

The logic of this result is that the second term in the denominator, $\gamma \sum_{i=n+1}^{N} \pi_i$, is smaller than $\gamma(1 - \pi_1)$ in (7.18). This latter expression involves a sum of prior probabilities for all members of population (with no one except the suspect being in the database) minus the suspect. In (7.19), the sum covers only those members of the population that are not in the database. Stated otherwise, the prior probabilities for the individuals in the database that are found to have profiles different from that of the crime stain are not relevant because of the multiplication with the zero likelihood, that is $Pr(E_n|H_i) = 0$ for $i = 2, \ldots, n$. Because of a smaller denominator, the posterior probability π_1' in (7.19) turns out to be greater than that in (7.18). The selection of a suspect in a database along with an exclusion of other database members by DNA profiling results thus provides more information against the corresponding suspect than if no database search had been conducted.

The result obtained so far can be tracked in a Bayesian network as shown in Figure 7.23(i) (Biedermann et al. 2011a; 2012e). Node N is a numeric node with states 100 and 1000 (other numbers may obviously be chosen) and represents the size of the suspect population, that is the individuals which could have left the crime stain. This node is a parent of node H, which has three states. H_1 represents the proposition according to which the suspect is the source of the crime stain. The proposition according to which one of the individuals $2, \ldots, n$ is the source of the crime stain is represented by the state H_{2_n}. The third state is H_{n+1_N}. It represents the proposition that one of the $N - n$ individuals outside the database is the source of the crime stain. Assuming again prior probabilities of $1/N$ for each of the N individuals, the following node probabilities are specified:

$$Pr(H_1) = 1/N, \quad Pr(H_{2\_N}) = n - 1/N, \quad Pr(H_{n+1\_N}) = (N - n)/N.$$

Node n is a root node and represents the size of the database. Exemplary numerical states could be 2, 10, 100 (other database sizes $n \leq N$ may obviously be chosen). A further root node is γ. It has numeric states that represent the proportion with which the corresponding genetic feature appears in the population. For the purpose of illustration, the values 0.01 and 0.1 are chosen. Notice that this node is not strictly necessary. It would also be possible to specify γ directly in the probability table of the node M_1. A representation of γ in terms of a distinct node is retained here in order to provide a detailed decomposition of the problem at hand.

Part of the findings to evaluate is represented by node M_1, that is the proposition according to which the suspect's (i.e. person 1) profile corresponds to the crime stain profile. The probability table contains the following values:

$$Pr(M_1|H_i, \gamma) = \begin{cases} 1, & i = 1, \\ \gamma, & i \neq 1. \end{cases}$$

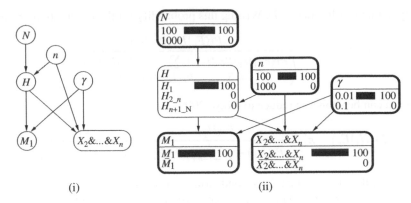

(i)   (ii)

Figure 7.23 (i) Structure of a Bayesian network for evaluating a correspondence (M_1) between the profile of a crime stain and that of a suspect, when the suspect is on a database along with $n - 1$ other individuals. The size of the population of potential offenders is N, where n (with $n \leq N$) of them are on a database. The node H has three states: 'the suspect is the source of the crime stain' (H_1), 'one of the $n - 1$ other individuals in the database is the source of the crime stain' (H_{2_n}) and 'the source of the crime stain is amongst the $N - n$ individuals outside the database' (H_{n+1_N}). The population proportion of corresponding genetic features is γ. The node $X_2\& \ldots \&X_n$ is binary and represents the proposition according to which the profiles of the $n - 1$ individuals in the database, other than the suspect, do *not* correspond to the crime stain. (ii) Expanded representation with observations entered at the nodes $M_1, X_2\& \ldots \&X_n, N, n$ and γ. Instantiated nodes are shown with a bold border.

In turn, node $X_2\& \ldots \&X_n$ represents the proposition according to which the $n - 1$ individuals in the database other than the suspect have profiles that do not correspond to that of the crime stain. The node probability table contains the following assignments:

$$Pr(X_2\& \ldots \&X_n | H_i, n, \gamma) = \begin{cases} (1 - \gamma)^{n-1}, & i = 1, \\ 0, & i = 2, \ldots, n, \\ (1 - \gamma)^{n-1}, & i = n + 1, \ldots, N. \end{cases}$$

Example 7.12 *(Bayesian network for a database search scenario) Figure 7.23(ii) illustrates the logical implications of a probabilistic analysis of a database search scenario. This network captures a case in which a population of $N = 100$ potential sources is assumed. All of these individuals are in a database (i.e. $n = 100$). The findings consist of a correspondence (where the characteristic has population proportion $\gamma = 0.01$) with the suspect (i.e. $M_1 = 1$) and no correspondence with the other $n = 2, \ldots, 100$ individuals (i.e. $X_2\& \ldots \&X_n = 1$). In such a situation, the posterior probability that the corresponding suspect is the source of the crime stain is 1 (proposition H_1). Notice that the value of γ is not relevant in the setting considered here. As may be seen from (7.19), γ would only become relevant if there are individuals in the population not included in the database.*

The model shown in Figure 7.23 can be extended to study the formal requirement of rationality to justify the individualization of person, that is the designation of a person as the source of a crime stain, if that person has been selected through a database search. This aspect is developed in Chapter 11.

7.11   Probabilistic approaches to laboratory error

7.11.1   Implicit approach to typing error

When evaluating DNA profiling results, forensic scientists consider to what degree their findings correspond to their expectations given specified propositions. For example, if a crime stain is found to have the same profile as control material provided by a suspect, then this would be something that one may expect to observe if the suspect were in fact the source of the crime stain. However, a forensic scientist would also be required to state a probability for the event of observing the findings given some alternative proposition were true, for example, that the crime stain comes from an unknown person who is unrelated to the suspect.

If the suspect is assumed not to be the source of the crime stain, then necessarily someone else must be, so one needs to consider the probability that another person would have a corresponding profile. This is an appropriately assigned genotype probability as mentioned earlier. It may be thought of as the probability of a correspondence by coincidence, that is a correspondence between DNA profiles of two different people. This probability has been the main focus of attention of the evaluative procedures discussed so far in this chapter. However, there may be other relevant issues to consider when evaluating DNA profiling results. One of these other issues is the potential for error.

One such error is a false positive, which may be defined as an event that occurs when a laboratory erroneously reports a correspondence between two analysed entities (or materials) that actually have different profiles. A false positive may be due to error in the collection or handling of traces, misinterpretation of test results or incorrect reporting of results (National Research Council 1992; Thompson and Ford 1991; Thompson 1995). The effect that the potential of error through a false positive may have on the value of a reported correspondence may be evaluated in a Bayesian perspective (Thompson et al. 2003). Such an analysis shows that for the evaluation of DNA profiling results, accurate information is necessary not only about the applicable genotype probability but also on the probability of laboratory error.

For the purpose of illustration, consider thus again a single-stain scenario as described in Section 7.1. Let H denote a pair of propositions at the source level ('H_p: the crime stain came from the suspect' and 'H_d: the crime stain did not come from the suspect'). The available information, denoted R, consists of the forensic scientist's report of a correspondence between the DNA profile of the crime stain and that of control material provided by the suspect. According to principles presented so far, a likelihood ratio of the form $V = Pr(R|H_p, I)/Pr(R|H_d, I)$ may be formulated.

In order to account for the potential for error, an intermediate variable M is introduced. The variable M (short for 'match') designates a true correspondence and is differentiated from R, a reported correspondence. As noted by Thompson et al. (2003), this distinction reflects the assumption of the following two possible underlying states of reality: 'The suspect and the crime stain have corresponding DNA profiles (M)' and 'The suspect and the crime stain do not have corresponding DNA profiles ($\bar{M}$)'. It is generally recognized that − for an external observer − it is impossible to know with certainty whether M or $\bar{M}$ is true because the only information available about M and $\bar{M}$ is the laboratory report, which might be mistaken. So both M and H are unobserved states of nature, but beliefs about their truthstate may be revised based on new information, such as R. Thus, when conditioning R on M and H, the likelihood ratio for R can be extended to

$$V = \frac{Pr(R|M, H_p)Pr(M|H_p) + Pr(R|\bar{M}, H_p)Pr(\bar{M}|H_p)}{Pr(R|M, H_d)Pr(M|H_d) + Pr(R|\bar{M}, H_d)Pr(\bar{M}|H_d)}. \tag{7.20}$$

Three assumptions may be made. First, if the analytical characteristics truly correspond (M), then the probability that a correspondence will be reported (R) is not affected by whether the correspondence is coincidental: $Pr(R|M, H_p) = Pr(R|M, H_d) = Pr(R|M)$. Second, if the suspect is the source of the crime stain, then a correspondence is taken to be certain: $Pr(M|H_p) = 1$. Notice that the same assumption has already been made in Section 6.1, where no distinction was made between a true correspondence and a reported correspondence. Third, a situation in which there is no correspondence $(\bar{M})$ can only arise under H_d so that $Pr(R|\bar{M}, H_d)$ can be reduced to $Pr(R|\bar{M})$. Given these considerations, (7.20) can thus be written more succinctly as

$$V = \frac{Pr(R|H_p)}{Pr(R|H_d)} = \frac{Pr(R|M)}{Pr(R|M)Pr(M|H_d) + Pr(R|\bar{M})Pr(\bar{M}|H_d)}. \tag{7.21}$$

The conditional probabilities contained in (7.21) may be assessed as follows. The term $Pr(R|M)$ is the probability that a laboratory will report a correspondence if the suspect and the crime stain have corresponding DNA profiles. Provided that the analysed materials are adequate in quality and quantity, a competent laboratory may be expected to be unlikely to not notice a true correspondence, so that values for $Pr(R|M)$ close to 1 appear appropriate. For present purposes, assume that $Pr(R|M) = 1$. The term $Pr(M|H_d)$ is the probability of a coincidental correspondence. For traces from single sources, $Pr(M|H_d)$ can be interpreted as the assigned genotype probability or the rarity of the corresponding profile in a relevant population, incorporating a sub-population effect. In analogy to the notation used in Section 6.1, this genotype probability is denoted γ. Notice that M and $\bar{M}$ are mutually exclusive and exhaustive events, so that $Pr(\bar{M}|H_d)$ is the complement of γ, that is $1 - \gamma$. The term $Pr(R|\bar{M})$ is the false positive probability, denoted fpp.

Following substitution, the likelihood ratio given by (7.21) takes the following form:

$$V = \frac{1}{\gamma + [fpp \times (1 - \gamma)]}.$$

To account for the probability of error in a Bayesian network, one needs to focus on the number and definition of nodes, as well as their dependencies. From the probabilistic approach described so far, three binary variables may be defined: H, the trace comes from the suspect, M, the suspect and the trace have corresponding DNA profiles, and R, the forensic scientist reports a correspondence between the suspect's profile and the profile of the crime stain.

These three variables need to be combined in a way that appropriately reflects the dependence and independence properties defined by (7.20). In particular, it was assumed that $Pr(R|M, H_p) = Pr(R|M, H_d) = Pr(R|M)$, so that the joint probability of the three variables R, M and H could be written as

$$Pr(R, M, H) = Pr(R|M) \times Pr(M|H) \times Pr(H). \tag{7.22}$$

To represent these dependency assumptions correctly, one needs a serial graph structure: $H \rightarrow M \rightarrow R$. This can be understood by considering the chain rule for Bayesian networks

Figure 7.24 Bayesian network for the problem of laboratory error (implicit inclusion of typing error). The three binary nodes are defined as follows: H, the trace comes or does not come from the suspect, M, the suspect and the crime stain have, or do not have, corresponding DNA profiles, and R, the forensic scientist reports a correspondence or a non-correspondence between the suspect's profile and the profile of the crime stain.

(Section 2.1.8), which yields (7.22) for the three variables R, M and H, and for which the structural relationships are as specified in Figure 7.24. The required conditional probabilities are $Pr(M|H_p) = 1$ and $Pr(M|H_d) = \gamma$ for the node M and $Pr(R|M) = 1$ and $Pr(R|\bar{M}) = fpp$ for the node R.

In the Bayesian networks for the DNA likelihood ratio earlier described in Section 7.2, it has been assumed that the node capturing the observations represents the true state of affairs. For example, by instantiating the node E in the simple network $H \to E$, or, analogously, the nodes sgt and tgt in the network shown in Figure 7.2, the observations are interpreted as an accurate representation of the true state of the respective propositions. In practice, however, care should be taken to examine the conditions under which such an assumption is acceptable. This amounts to making a distinction between a proposition and information that is relevant for that proposition. Figure 7.24 illustrates how the Bayesian networks earlier described in Section 7.2 can be extended to handle this distinction, expressed in terms of the probability of laboratory error.

7.11.2   Explicit approach to typing error

A slightly different approach for combining the probability of laboratory error and the expression of the rarity of the corresponding characteristics has been proposed by Buckleton et al. (2005). Their development involves, on the one hand, E_c, the true type of the profile of the recovered crime stain and, on the other hand, E_s, the profile of the suspect (which is assumed to be determined without error). The target pair of propositions H at the source level is the same as that defined in Section 7.11.1. As a distinctive feature of their approach, Buckleton and Triggs introduce an additional event denoted here Er (the original notation in Buckleton et al. (2005) is $\exists$), defined as the occurrence of a typing error; that is the profile produced in the electropherogram does not correctly represent the true type of the DNA in the crime stain. Accordingly, let $\overline{Er}$ denote the event that the electropherogram correctly reflects the DNA profile of the recovered trace. Incorporating the variable Er in the likelihood ratio introduced earlier in (7.2), and omitting circumstantial information I from the notation, leads to

$$V = \frac{Pr(E_c|E_s, Er, H_p)Pr(Er|H_p) + Pr(E_c|E_s, \overline{Er}, H_p)Pr(\overline{Er}|H_p)}{Pr(E_c|E_s, Er, H_d)Pr(Er|H_d) + Pr(E_c|E_s, \overline{Er}, H_d)Pr(\overline{Er}|H_d)}. \tag{7.23}$$

Several considerations allow one to reduce this formula to simpler terms. If the suspect is truly the source of the crime stain (H_p), and no error has occurred ($\overline{Er}$), then the suspect's genotype certainly corresponds to that of the crime stain. Therefore, $Pr(E_c|E_s, \overline{Er}, H_p) = 1$. The term $Pr(E_c|E_s, Er, H_p)$ is the probability of a false-positive result, given that an error has occurred. Denote this as k. This probability is also assigned to $Pr(E_c|E_s, Er, H_d)$, that is the probability of a correspondence with the suspect when he is not the source of the crime stain, but there is an error. When the suspect is not the source of the crime stain (H_d) and the crime stain profile is determined without error ($\overline{Er}$), then the probability that it will correspond to the profile of the suspect is given by the relevant population proportion γ. Therefore, $Pr(E_c|E_s,\overline{Er},H_d) = \gamma$. Assuming that the occurrence of an error does not depend on whether the suspect or some other person is the source of the crime stain, one can write the probability for Er without the conditioning on the proposition H. More formally, write e for the probability of error so that $Pr(Er) = e$ and $Pr(\overline{Er}) = 1 - e$.

The likelihood ratio in (7.23) may be written as

$$
V = \frac{\overbrace{Pr(E_c|E_s,Er,H_p)}^{k}\ \overbrace{Pr(Er)}^{e} + \overbrace{Pr(E_c|E_s,\overline{Er},H_p)}^{1}\ \overbrace{Pr(\overline{Er})}^{1-e}}{\underbrace{Pr(E_c|E_s,Er,H_d)}_{k}\ \underbrace{Pr(Er)}_{e} + \underbrace{Pr(E_c|E_s,\overline{Er},H_d)}_{\gamma}\ \underbrace{Pr(\overline{Er})}_{1-e}}
$$

$$
= \frac{ke + (1 - e)}{ke + \gamma(1 - e)} = \frac{1 - (1 - k)e}{ke + \gamma(1 - e)}.
$$

These definitional considerations allow one to propose the network structure shown in Figure 7.25. The node Er modelling the event of an error is not conditioned on the main source level proposition H. This is a particular interpretation of the more general formulation in (7.23), where Er is written conditional on H. Accordingly, the node table of Er contains the error probabilities e and $1 - e$ for the states Er and $\overline{Er}$. The node H, too, does not have entering arcs from other nodes. Its node table contains the initial probabilities $Pr(H_p)$ and $Pr(H_d)$. The node E_s, representing the genetic characteristic of the reference material from the suspect, can be restricted to a single state arbitrarily denoted x here. This is sufficient for reproducing the generic development in (7.23), where the focus lies on the characteristic actually observed, with no possibility of an erroneous determination. Table 7.11 summarizes the probability assignments to the states of the variable E_c, the genetic characteristic of the crime stain, given different states of the variables H and Er. In particular, if there is an erroneous determination of the crime stain profile (i.e. Er is true), then E_c will be of type x (i.e. the same as the reference material from the suspect) with probability k. This is the probability

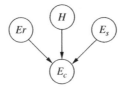

Figure 7.25 Bayesian network for the problem of laboratory error (explicit inclusion of typing error). The nodes H and Er are binary and represent, respectively, the propositions according to which the suspect (some other person) is the source of the crime stain and the event of an error. The node E_c has two states x and $\bar{x}$, which represent the possible analytical characteristics of the crime stain. The node E_s has the single state x, that is the genetic characteristic of the control material pertaining to the suspect.

Table 7.11 Conditional node probability table for the node E_c, the characteristics of the crime stain, which may be of type x or $\bar{x}$.

H:		H_p		H_d	
Er:		Er	$\overline{Er}$	Er	$\overline{Er}$
E_c :	x	k	1	k	γ
	$\bar{x}$	$1-k$	0	$1-k$	$1-\gamma$

The variable H specifies whether the suspect (H_p) or some other person (H_d) is the source of the crime stain. In the event of an error, Er, the variable E_c will take the value x with probability k (i.e. the probability of a false positive). When no error has occurred, that is $\overline{Er}$ holds, E_c will certainly be of type x if the suspect is the source (H_p). Otherwise, that is when H_d is the case (and $\overline{Er}$), variable E_c assumes the state x with probability γ.

Table 7.12 Probabilities for the binary node R, a scientist's report of a correspondence, conditioned on the true occurrence of a correspondence (M), as used in the Bayesian network shown in Figure 7.24.

		Thompson et al.		Buckleton and Triggs	
M:	M	$\bar{M}$		M	$\bar{M}$
R :	R	1	fpp	$1-e+ke$	ke
	$\bar{R}$	0	$1-fpp$	$1-(1-e+ke)$	$1-ke$

The term fpp is the false-positive probability in the notation of Thompson et al. (2003). The terms e and k refer to notation used in Buckleton et al. (2005) to represent, respectively, the probability of an error and the probability of a false-positive correspondence, given that an error has occurred.

of a false-positive correspondence, which applies independently of the truth or otherwise of the proposition H. When no error occurred (i.e. $\overline{Er}$ holds) and the suspect is the source of the crime stain (i.e. H_p holds), then the crime stain will certainly be of the same type, x, as the control from the suspect. When the suspect is not the source of the crime stain, then the crime stain's profile will correspond with probability γ.

It is interesting to note that the approach of Buckleton and Triggs in Buckleton et al. (2005) pursued in this section can be related to that of Thompson et al. (2003) by considering the following. Firstly, one can relax the assumption of no false negatives in Thompson et al. (2003) by changing $Pr(R|M) = 1$ to $1 - e + ke$. This latter expression reflects the view that, when there is in fact a correspondence (M), it will be reported (R) in the event of no error (i.e. $1 - e$) or with probability k in the event of an error. Secondly, one can view the false-positive probability fpp in the development of Thompson et al. (2003) in terms of ke from the development of Buckleton and Triggs in Buckleton et al. (2005). Table 7.12 summarizes these probability assignments. Incorporating these two modified probability assignments in the general Equation (7.20) then leads to the Buckleton and Triggs likelihood ratio (7.23).

7.12   Further reading

7.12.1   A note on object-oriented Bayesian networks

Despite the various compelling capacities of the Bayesian network formalism, the manual model construction may often be a painstaking process, in particular for larger applications. For example, it may be that a model may need to incorporate repetitive sub-models. These may be implemented by 'copy and paste' procedures, but whenever the specification of a sub-model of interest requires changes, all the sub-models of the same type would need to be reviewed as well. This may thus hinder efficient model construction, maintenance and adaptation. As a means of overcoming such difficulties, Bayesian networks have been extended to so-called object-oriented Bayesian networks. The idea behind this approach is to define generic 'classes' of networks, particular nodes of which (so-called instances) can be used, as required, in place of nodes in other networks. The object-oriented Bayesian network language thereby allows one to describe inference problems in terms of inter-related objects. Further discussion of this approach and its application for evaluating results of DNA profiling analyses is presented in Chapter 12.

7.12.2   Additional topics

From a historical point of view, the discussion on the use of graphical probabilistic models for evaluating results of forensic examinations, initiated by Aitken and Gammerman (1989), was continued by Dawid and Evett (1997) who discuss an example which is an advance from that of the single trace. It considers two kinds of traces (i.e. fibres and blood) and the testimonies of two persons. A particular emphasis is made on the various dependence and independence properties implied by a graph structure. It is also emphasized how these relationships may assist the evaluation of different elements within a complex framework of circumstances. Furthermore, it is shown how a graphical structure can help to reduce the number of variables necessary for deriving relevant likelihood ratio formulae. More generally, Dawid and Evett (1997) can be considered the first to focus on Bayesian networks and forensic DNA because it presents an example involving blood staining.

Evett et al. (2002) report an application of Bayesian networks to cases involving small quantities of DNA. Their study addresses recurrent issues related to the evaluation of DNA traces, such as the considerable increase in the sensitivity of DNA analyses that has initiated a tendency of courts to shift from questions of the kind 'whose DNA is this?' to 'how did this DNA reach there?' With regard to such challenges, Evett et al. (2002) show that Bayesian networks are a powerful tool that usefully supplements the evaluation of forensic traces assisted by an already established concept, known as the *hierarchy of propositions*. Evett et al. (2002) discuss two case work examples. The first involves the outcomes of DNA analyses on a cigarette end recovered near the point of entry of a burgled home. The proposed Bayesian network accounts for questions of the following kind: 'Is the suspect the person who smoked the cigarette?', 'Did the person who smoked the cigarette leave sufficient DNA to give a profile?', 'Did DNA from the suspect enter the process by innocent means?', 'Did DNA from some third person enter the process?' These propositions are modelled as independent binary nodes that point towards a common descendant, termed *outcome*, with states 'correspondence', 'mixture/correspondence', 'different' and 'no profile'. In order to study the sensitivity of the outcome to changes in the truthstate of the parental variables (that is the uncertain propositions), beta probability distributions have been used to generate values at random. The authors show that such operations can be used as part of case pre-assessments. A second case example pertains to the outcomes of DNA analyses performed on swabs taken from different areas of a watch (assumed to have been worn by the offender) found on the crime scene. This case is somewhat more complex than the first one, as more intermediate propositions are considered. In addition, the propositions according to which the suspect is or is not the offender have been included. On the whole, the examples illustrate that Bayesian networks allow probabilistic analyses to be made over a large number of variables along with different inter-related issues. This represents valuable support in cases where a full algebraic solution would appear to become increasingly difficult.

Besides the main applications such as inference of source and kinship analyses as discussed throughout the various sections in this chapter, Bayesian networks have also been developed for the study of a variety of further topics that are associated with the evaluation of forensic DNA analyses. One such topic is that of cross-transfer of DNA traces. In fact, existing accounts on 'stain identification' consider only potential transfer of DNA stains in one direction (e.g. a stain from the victim transferred to an offender or a stain transferred by an offender to a crime scene). In practice, however, there may be cases in which a cross-transfer of DNA occurred. For example, DNA traces may be collected on both a victim and a suspect in an assault case. Evaluating DNA profiling results meaningfully in such situations may require a discussion of probabilities for aspects such as transfer, persistence and recovery, innocent acquisition, relevance and innocent presence. Aitken et al. (2003) studied these notions through Bayesian networks.

Further aspects of operational interest are the wrongful designation of a heterozygous genotype as a homozygote, which may lead to a false exclusion (Gill et al. 2009), and the evaluation of likelihood ratios when multiple propositions need to be considered (Buckleton et al. 2006).

A different approach for constructing Bayesian networks for inference from genetic markers has been described by Cowell (2003). This author has developed a software tool, called FINEX (Forensic Identification by Network EXpert systems), where inference problems involving forensic DNA can be expressed through the syntax of a graphical specification language. The program FINEX then uses an algorithm for automatically constructing an appropriate representation in terms of a Bayesian network. Compared to general purpose software, FINEX reduces the time taken in the construction of networks and reduces the potential for error whilst completing large probability tables. In addition, findings for several markers can readily be evaluated in combination.

More generally, Bayesian network models can be used as an integral part of larger IT environments, including, for instance, connections to operational DNA databases. For example, Bruijning-van Dongen et al. (2009) described a system architecture in which Bayesian networks are used for kinship analyses based on DNA profiles. A primary feature of such an implementation consists of its capacity to deal with the analyses of an extensive set of cases. This may be of interest for processes such as victim identification in case of a large disaster.

On the whole, the application of Bayesian networks to inference problems involving the results of DNA profiling analyses is a very lively area of research that expands to several further topics not covered in this chapter. Examples include the estimation of mutation rates (Dawid 2003b; Vicard et al. 2008), the improvement of quantitative mixture interpretation (Cowell et al. 2007a,b, 2008, 2011), the number of contributors to DNA mixtures (Biedermann et al. 2011b, 2012b), uncertainty in allele population proportions (Green and Mortera 2009), evaluation in pedigree structures (Corradi et al. 2003; Corradi and Ricciardi 2013) and X-chromosomal profiling results (Hatsch et al. 2007).

8

Aspects of combining evidence

8.1 Introduction

In the previous chapters, Bayesian networks have been constructed and discussed with the aim of approaching uncertainties that affect inference based on results of various kinds of scientific examinations, conducted on traces, such as fibres and glass, or biological material (DNA). In this chapter, such results are more generally termed *evidence*, or *items of evidence*, because the focus relies on the joint assessment of multiple results pertaining to trace items of, possibly, different natures. This slight shift in wording (from 'findings' to 'evidence') with respect to previous chapters is justified by the fact that the joint assessment of several scientific results is typically operated at an advanced stage in a judicial process (e.g. in Court), where more formal decisions are made as to which scientific results are regarded as 'evidence'.

The networks developed so far in this book provide valuable assistance in addressing some of a wide range of issues that affect a coherent evaluation of probative value. It was emphasized that existing probabilistic solutions proposed in the scientific literature may be used as a guide to elicit appropriate network structures. By providing an explicit representation of the relevant variables together with their assumed dependence and independence properties, Bayesian networks have the potential to clarify the rationale behind a given probabilistic approach, in particular, formulae for likelihood ratios.

As noted by Lindley (2004, p. 24), however, a '(...) problem that arises in courtroom, affecting both lawyers, witnesses and jurors, is that several pieces of evidence have to be put together before a reasoned judgement can be reached (...).' He also adds that '[p]robability is designed to effect such combinations but the accumulation of simple rules can produce complicated procedures. Methods of handling sets of evidence have been developed: for example Bayes nets (...)'. More generally, in the field of fact analysis, probabilistic studies focussed on the various ways in which several items of evidence may interact. This has led to detailed analyses of, for instance, different forms of dissonant and harmonious evidence, as described in Schum (1994).

Bayesian Networks for Probabilistic Inference and Decision Analysis in Forensic Science, Second Edition.
Franco Taroni, Alex Biedermann, Silvia Bozza, Paolo Garbolino and Colin Aitken.
© 2014 John Wiley & Sons, Ltd. Published 2014 by John Wiley & Sons, Ltd.

In forensic contexts, likelihood ratio formulae may attain critical levels of complexity, even for single items of evidence. One often needs to account for particular sources of uncertainty, related to phenomena such as transfer, persistence and background presence. It may thus become increasingly difficult to structure probabilistic analyses properly and discern the relevant variables as well as their relationships. If, in addition, several items of evidence need to be combined, then even further complications may be expected. These, then, are instances where Bayesian networks may assist forensic scientists in constructing coherent and acceptable arguments (Juchli et al. 2012). This chapter aims to illustrate this through a series of examples.

8.2   A difficulty in combining evidence: The 'problem of conjunction'

Unlike the evaluation of single items of scientific evidence, the formal study and analysis of the joint evaluation of several distinct items of forensic evidence has, to date, received some occasional, rather than systematic, attention. Questions about the (i) relationships among a set of (usually unobservable) propositions and a set of (observable) items of scientific evidence, (ii) the joint probative value of a collection of distinct items of evidence and (iii) the contribution of each individual item within a given group of pieces of evidence still represent fundamental areas of research. To some degree, this is remarkable since both forensic science theory and practice require the consideration of multiple items of evidence. A recurrent and particular complication that arises in such settings is that the application of probability theory, that is the reference method for reasoning under uncertainty, becomes increasingly demanding, and even apparently 'simple' questions can lead to fallacious answers. The problem that has become known as the *difficulty of conjunction* provides an illustrative example for this.

The difficulty of conjunction is closely tied with the issue of separating the differences that exist between the probability of the evidence and the probability of an explanatory proposition (Chapter 1). It thus also connects to the problem of posterior probabilities as an expression for the value of the evidence (Taroni and Biedermann 2005). Historically, the difficulty in combining evidence was the subject of a debate between Cohen (1977, 1988) and Dawid (1987). A summary of this debate and the solution proposed by Dawid (1987) can also be found in Aitken and Taroni (2004). In essence, the problem is described as follows: two pieces of evidence, separately supportive of a proposition, when considered in combination seem to produce lower support than when considered separately. As an illustration, let E_1 and E_2 denote two distinct items of evidence. These shall be used to draw an inference concerning some proposition of interest, say H for convenience. H has the two possible outcomes H_p and H_d, denoting the propositions proposed by, respectively, the prosecution and the defence. Imagine further that some evaluator would retain a probability of 0.7 for H_p given the occurrence of either E_1 or E_2, that is $Pr(H_p|E_1) = Pr(H_p|E_2) = 0.7$. The probability of interest is $Pr(H_p|E_1, E_2)$.

If E_1 and E_2 are considered to be independent, given H_p or H_d, their joint probability can be written as the product of the individual probabilities, that is, $Pr(E_1, E_2|H_p) = Pr(E_1|H_p) \times Pr(E_2|H_p)$. It is now tempting to believe that $Pr(H_p|E_1, E_2)$ is obtained analogously, that is by $Pr(H_p|E_1, E_2) = Pr(H_p|E_1) \times Pr(H_p|E_2)$. The apparent contradictory result of this (incorrect) procedure is $0.7 \times 0.7 = 0.49$, which is less than the probability of H_p given either E_1 or E_2.

At this stage, it is useful to consider Bayes' theorem. For two pieces of evidence, E_1 and E_2, and propositions H_p and H_d, the odds form of Bayes' theorem writes

$$\frac{Pr(H_p|E_1, E_2)}{Pr(H_d|E_1, E_2)} = \frac{Pr(E_1, E_2|H_p)}{Pr(E_1, E_2|H_d)} \times \frac{Pr(H_p)}{Pr(H_d)}$$

or

$$\text{posterior odds} = \text{likelihood ratio } (V) \times \text{prior odds}.$$

Assuming equal prior probabilities, $Pr(H_p) = Pr(H_d)$, the target probability $Pr(H_p|E_1, E_2)$ is thus given by $V/(1 + V)$. The likelihood ratio can be obtained as follows:

$$V = \frac{Pr(E_1|H_p)}{Pr(E_1|H_d)} \times \frac{Pr(E_2|H_p)}{Pr(E_2|H_d)}$$

by independence of E_1 and E_2,

$$= \frac{Pr(E_1|H_p)}{Pr(E_1|H_d)} \times \frac{Pr(H_p)}{Pr(H_d)} \times \frac{Pr(E_2|H_p)}{Pr(E_2|H_d)} \times \frac{Pr(H_p)}{Pr(H_d)}$$

since $Pr(H_p) = Pr(H_d)$ so $\dfrac{Pr(H_p)}{Pr(H_d)} = 1$;

$$= \frac{Pr(H_p|E_1)}{Pr(H_d|F_1)} \times \frac{Pr(H_p|E_2)}{Pr(H_d|E_2)}$$

$$= \frac{0.7}{0.3} \times \frac{0.7}{0.3} = \frac{0.49}{0.09}.$$

From this, the probability of interest

$$Pr(H_p|E_1, E_2) = \frac{V}{1 + V} = \frac{0.49/0.09}{1 + 0.49/0.09} = 0.84,$$

which is greater than 0.7. Thus, under the stated assumptions, the combination of the two pieces of evidence yields a higher probability for H_p than when considered separately.

This example illustrates that in cases where two items of evidence are deemed to provide relevant information for the same pair of propositions, the value of the two pieces of evidence in combination cannot readily be determined by the sole use of the posterior probabilities of the respective propositions. This is also one of the reasons why scales of conclusions based on posterior probabilities, as have been proposed for example in the field of shoemark analyses (Katterwe 2003) or handwriting examination (Köller et al. 2004), are inadequate means for the assessment of scientific evidence (Taroni and Biedermann 2005).

Such inferential impasses may be avoided by following established inferential procedures based on the likelihood ratio. By focussing on a likelihood ratio (Section 1.2), one can successively add one piece of evidence at a time and examine the probability of a proposition of interest, H for example, given the available evidence. The posterior odds after considering one item of evidence, E_1 for example, become the new prior odds for the following item of evidence, E_2 say. In a more formal notation, one thus has, for propositions H_p and H_d:

$$\frac{Pr(H_p)}{Pr(H_d)} \times \frac{Pr(E_1|H_p)}{Pr(E_1|H_d)} = \frac{Pr(H_p|E_1)}{Pr(H_d|E_1)} . \tag{8.1}$$

The term on the right-hand side of (8.1) represents the odds in favour of the proposition H_p given E_1. When E_2, a second item of evidence becomes available, then one may proceed as follows:

$$\frac{Pr(H_p|E_1)}{Pr(H_d|E_1)} \times \frac{Pr(E_2|H_p, E_1)}{Pr(E_2|H_d, E_1)} = \frac{Pr(H_p|E_1, E_2)}{Pr(H_d|E_1, E_2)} \; . \tag{8.2}$$

Here, the posterior odds in favour of the proposition H_p incorporate knowledge about both items of evidence, E_1 and E_2. The likelihood ratio for E_2, shown in the centre of (8.2), allows for a possible dependency of E_2 on E_1. More generally, this way of proceeding was concisely summarized by Lindley as 'Today's posterior is tomorrow's prior' (Lindley 2000, p. 301).

8.3   Generic patterns of inference in combining evidence

8.3.1   Preliminaries

When reasoning about a scientific finding or result, one may find that it favours a given proposition rather than others. This can be conceptualized as the ascription of an inferential vector to a given item of scientific information. Such an inferential vector can be characterized by two main aspects, that is an inferential direction and an inferential force. When extending this idea to practical reasoning, however, one comes to realize that one will typically be required to consider several items of evidence and this will generate a whole batch of such vectors.

With respect to inferential directions, two situations can be distinguished. Either the inferential vectors will point towards different propositions or they point towards the same proposition. Following Schum (1994), the first situation involves evidence said to be 'dissonant', whereas the second situation involves evidence said to be 'harmonious'. Using various examples, the sections hereafter lay out the probabilistic underpinnings of these distinctions in some further detail. Figure 8.1 shows the generic Bayesian network structures that underlie these patterns of reasoning.

8.3.2   Dissonant evidence: Contradiction and conflict

8.3.2.1   Contradiction

All dissonant evidence incorporates an inferential divergence, although only some situations of dissonance can properly called *contradictory*. Schum (1994) considers dissonant evidence that is not contradictory as being in 'conflict'. Properly speaking, a contradiction is given only if the occurrence of mutually exclusive events is reported. In order to clarify this, consider a source S_1 stating that the event E occurred. Let this state of S_1 be denoted by E^*. The asterisk in this notation is chosen to refer to a report about the occurrence of event E, which is different from the actual occurrence of E. Next, suppose also a second source S_2 that states E^{c*}, that is 'Event E did not occur'.

Example 8.1 (*Evidential contradiction in questioned document examination*) *In a case involving questioned documents, it may be of interest to learn something about the proposition E, that a given suspect wrote a signature on a handwritten document. Denote by $\bar{E}$, the alternative proposition that the suspect did not write the questioned signature but that someone else did. One cannot directly know whether or not the suspect is the author of the questioned signature. One may therefore rely on an opinion presented by, for example, an eyewitness.*

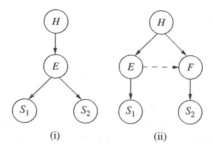

Figure 8.1   Generic Bayesian network structures for dissonant and harmonious evidence. (i) Accounts for contradictory and corroborative settings and (ii) Covers conflicting and converging evidence. The dotted arrow applies whenever one assumes a dependency between the two events E and F conditional upon H.

Let this source of information be denoted by S_1. The report given by this source is written E^*, that is a statement that E occurred. Next, one may also have a further source of information, given by source S_2. This source, too, reports on the proposition E but affirms that its negation, $\bar{E}$, holds. An example for such a second source of information could be another eyewitness or a forensic document examiner.[1]

Given this outcome, a question of interest may be how to draw an inference about a pair of ultimate propositions H_p and H_d, while allowing uncertainty about the true state of the intermediate variable E. For the example introduced above, the variable H could be, for example, the commission of a fraud or any other criminal activity for which the establishment of authorship of the questioned signature at hand is inferentially relevant. For such a situation, the likelihood ratio for the two reports $\{E^*, E^{c*}\}$ takes the following form:

$$V_{E^*, E^{c*}} = \frac{Pr(E^*, E^{c*} | H_p)}{Pr(E^*, E^{c*} | H_d)} \, . \tag{8.3}$$

Assuming a dependency structure as shown in Figure 8.1(i), the likelihood ratio in (8.3) can, as proposed in Schum (1994), be presented in some further detail as follows:

$$V_{E^*, E^{c*}} = \frac{Pr(E^*, E^{c*} | H_p)}{Pr(E^*, E^{c*} | H_d)} = \frac{Pr(E | H_p) + \left[\frac{h_1 m_2}{f_1 c_2} - 1\right]^{-1}}{Pr(E | H_d) + \left[\frac{h_1 m_2}{f_1 c_2} - 1\right]^{-1}} \, , \tag{8.4}$$

where, using to a large extent notation introduced by Schum (1994), h_1 denotes $Pr(E^* | E)$, m_2 denotes $Pr(E^{c*} | E)$, f_1 denotes $Pr(E^* | \bar{E})$ and c_2 denotes $Pr(E^{c*} | \bar{E})$. The extended form of the likelihood ratio shown in (8.4) is reproduced here because it contains the expression $[(h_1 m_2 / f_1 c_2) - 1]^{-1}$. This part of the formula is also referred to as the drag coefficient, as it acts like a drag upon $V_E = Pr(E | H_p) / Pr(E | H_d)$; that is, if the likelihood ratio is thought as a force acting on the probabilities of H_p and H_d to move them from their prior positions to posterior positions, then the size of the force is measured by the value of V_E, and the drag

---

[1] The nature of such expert reports, although entirely realistic, is admitted here only for the purpose of illustration. There is no suggestion here that it is appropriate for forensic document examiners to present categorical statements of the kind 'the suspect is the author of the questioned signature'.

coefficient acts as a drag on this force. As will later be pointed out in Section 8.3.4, the drag coefficient accounts for the credibility of the statements made by the sources of interest. In particular, it will determine the closeness of $V_{E^*,E^{c*}}$ to V_E. Note that the likelihood ratio V_E describes the inference about H on the basis of the intermediate variable E and is given by the fraction of the two likelihoods $Pr(E|H_p)$ and $Pr(E|H_d)$.

The result shown in (8.4) can be further understood by considering local likelihood ratios for drawing an inference about E, on the basis of the distinct items of information $S_1 = E^*$ and $S_2 = E^{c*}$. In other words, one can write a likelihood ratio for item of evidence E^*, written V'_{E^*}, and one for the item of evidence E^{c*}, written $V'_{E^{c*}}$:

$$V'_{E^*} = \frac{Pr(E^*|E)}{Pr(E^*|\bar{E})} = \frac{h_1}{f_1}, \qquad V'_{E^{c*}} = \frac{Pr(E^{c*}|E)}{Pr(E^{c*}|\bar{E})} = \frac{m_2}{c_2}.$$

A prime ($''$) is used here to indicate that the likelihood ratio concentrates on an inference about E only, rather than about the ultimate proposition H.

When taking the inverse of the latter likelihood ratio, then one has an expression of the degree to which E^{c*} favours $\bar{E}$: $V'^{-1}_{E^{c*}} = Pr(E^{c*}|\bar{E})/Pr(E^{c*}|E) = c_2/m_2$. It can now be seen that the overall support of the two statements $S_1 = E^*$ and $S_2 = E^{c*}$ for E depends on the relative magnitude of V'_{E^*} and $V'_{E^{c*}}$. In particular, in all the cases where $(h_1/f_1) > (c_2/m_2)$, the evaluation of the statements will strengthen the proposition E. Conversely, if $(h_1/f_1) < (c_2/m_2)$, then the alternative proposition, $\bar{E}$, will be favoured.

More generally, notice further that the overall inferential force $V_{E^*,E^{c*}}$ is bounded by $V_{\bar{E}}$ and V_E so that $V_{\bar{E}} \leq V_{E^*,E^{c*}} \leq V_E$; that is, V_E represents the capacity of E to discriminate between H_p and H_d, given by $Pr(E|H_p)/Pr(E|H_d)$, whereas $V_{\bar{E}}$ represents the capacity of $\bar{E}$, given by $Pr(\bar{E}|H_p)/Pr(\bar{E}|H_d)$. But usually, one will not have complete knowledge of the occurrence of either E or $\bar{E}$, only evidence in the form of the statements $\{E^*, E^{c*}\}$.

8.3.2.2 Conflict

Situations of evidence in conflict differ from those involving contradicting evidence because 'conflict' relates to events that are not mutually exclusive. This is illustrated in Figure 8.1(ii). For this model, suppose that source S_1 states E^*, that is the occurrence of event E, which is one that favours the proposition H. A second source, S_2, states, F^*, that another proposition F, favouring proposition H_d, occurred. The example given hereafter illustrates this outcome.

Example 8.2 *(Conflicting evidence in questioned document examination) Consider again, as in Example 8.1 given above, a report E^* that event E occurred, that is a given suspect wrote a signature on a questioned document. Imagine further that the questioned document bears ridge skin marks (i.e. fingermarks). Let F denote the proposition according to which the fingermarks come from some person other than the suspect and let F^* denote a scientist's report of such a conclusion.[2] Conversely, let $\bar{F}$ denote the proposition according to which the fingermarks come from the suspect. Assuming that the fingermarks are found in a position on the questioned document where marks from the author of the crime of interest would be expected to be found, the proposition F can be considered relevant in an inference about the*

---

[2] Again, there is no suggestion at this point that this a recommended way of reporting conclusions from fingermark examinations, but it must realistically be conceded that it represents currently the most widespread reporting practice.

proposition H, that is 'the suspect is the author of the fraud'. Clearly, proposition F would favour H_d here because the probability of F can be reasonably be taken to be greater given H_d than given H_p. Stated otherwise, the likelihood ratio for F, written $V_F = Pr(F|H_p)/Pr(F|H_d)$, is smaller than 1. This represents support for H_d. In turn, the proposition E, which relates to the authorship of the questioned signature, provides support for H_p. In fact, following Example 8.1, the likelihood ratio for E is $V_E = Pr(E|H_p)/Pr(E|H_d)$ is greater than 1.

In this example, the evidential values of the reports E^* and F^* by, respectively, source S_1 and source S_2 are given by

$$V_{E^*} = \frac{Pr(E^*|H_p)}{Pr(E^*|H_d)} = \frac{Pr(E^*|E)Pr(E|H_p) + Pr(E^*|\bar{E})Pr(\bar{E}|H_p)}{Pr(E^*|E)Pr(E|H_d) + Pr(E^*|\bar{E})Pr(\bar{E}|H_d)}, \tag{8.5}$$

$$V_{F^*} = \frac{Pr(F^*|H_p)}{Pr(F^*|H_d)} = \frac{Pr(F^*|F)Pr(F|H_p) + Pr(F^*|\bar{F})Pr(\bar{F}|H_p)}{Pr(F^*|F)Pr(F|H_d) + Pr(F^*|\bar{F})Pr(\bar{F}|H_d)}. \tag{8.6}$$

These two likelihood ratios incorporate uncertainty about the actual – but unobserved – state of the events E and F, respectively. This is achieved by writing a given report, for example E^*, conditioned on both E and $\bar{E}$, weighted by the probability of, respectively, E and $\bar{E}$. Note further that the representation of the value of these two pieces of evidence, E^* and F^*, as the product of the separate likelihood ratios requires an assumption of conditional independence given H.

The two likelihood ratios, (8.5) and (8.6), can also be written in a more compact form (Schum 1994):

$$V_{E^*} = \frac{Pr(E|H_p) + \left[\frac{h_1}{f_1} - 1\right]^{-1}}{Pr(E|H_d) + \left[\frac{h_1}{f_1} - 1\right]^{-1}}, \tag{8.7}$$

$$V_{F^*} = \frac{Pr(F|H_p) + \left[\frac{h_2}{f_2} - 1\right]^{-1}}{Pr(F|H_d) + \left[\frac{h_2}{f_2} - 1\right]^{-1}}, \tag{8.8}$$

where $h_1 = Pr(E^*|E), f_1 = Pr(E^*|\bar{E})$, $h_2 = Pr(F^*|F)$ and $f_2 = Pr(F^*|\bar{F})$. As may be seen, the fractions h_1/f_1 and h_2/f_2 represent the evidential values – that is the likelihood ratios – of the reports E^* and F^* for discriminating between the states of the distinct events E and F.

Given the assumption of conditional independence given H, the overall evidential value of the two reports E^* and F^*, that is V_{E^*,F^*}, is given by the product of the individual likelihood ratios: $V_{E^*,F^*} = V_{E^*} \times V_{F^*}$. For Example 8.2, such an assumption seems reasonable. In fact, it appears reasonable to consider ridge skin surface characteristics as independent of handwriting characteristics.

If, however, in a more general case, the events $\{E, \bar{E}\}$ and $\{F, \bar{F}\}$ need to be considered as not necessarily conditionally independent of $\{H_p, H_d\}$, then the overall likelihood ratio will be of the form $V_{E^*} \times V_{F^*|E^*}$. The likelihood ratio for the second report F^* from source S_2 is then conditioned upon knowledge of the first report E^* from source S_1. More formally, this is written as $V_{F^*|E^*}$. While V_{E^*} is as defined above in (8.5), the term $V_{F^*|E^*}$ involves a more extended development that can be shown to reduce to (Schum 1994)

$$V_{F^*|E^*} = \frac{\begin{array}{c} Pr(E|E^*, H_p)[Pr(F|E, H_p) - Pr(F|\bar{E}, H_p)] \\ + Pr(F|\bar{E}, H_p) + \left[\frac{h_2}{f_2} - 1\right]^{-1} \end{array}}{\begin{array}{c} Pr(E|E^*, H_d)[Pr(F|E, H_d) - Pr(F|\bar{E}, H_d)] \\ + Pr(F|\bar{E}, H_d) + \left[\frac{h_2}{f_2} - 1\right]^{-1} \end{array}}. \tag{8.9}$$

Here, $h_2 = Pr(F^*|F)$ and $f_2 = Pr(F^*|\bar{F})$. These two terms represent, respectively, the numerator and the denominator of a local likelihood ratio V'_{F^*} that expresses the degree to which the report F^* discriminates between the intermediate propositions F and $\bar{F}$.

There is a close relationship with respect to (8.8). In fact, when E is irrelevant for the assessment of F conditional on H, then (8.9) reduces to (8.8). When knowledge of E is irrelevant, this relationship is expressed as

$$Pr(F|E, H_p) = Pr(F|\bar{E}, H_p) = Pr(F|H_p) \text{ and}$$

$$Pr(F|E, H_d) = Pr(F|\bar{E}, H_d) = Pr(F|H_d),$$

and this eliminates the terms $Pr(E|E^*, H_p)[Pr(F|E, H_p) - Pr(F|\bar{E}, H_p)]$ and $Pr(E|E^*, H_d)$ $[Pr(F|E, H_d) - Pr(F|\bar{E}, H_d)]$ in, respectively, the numerator and the denominator of the likelihood ratio $V_{F^*|E^*}$.

8.3.3   Harmonious evidence: Corroboration and convergence

8.3.3.1   Corroboration

Schum (1994) distinguishes two main cases of harmonious evidence, notably corroborating evidence and convergent evidence. The former case, corroboration, applies to evidence from sources that comment on the occurrence of the same event. As illustrated in Example 8.3, consider two sources S_1 and S_2. Both state E^*, that event E occurred. Suppose further that $Pr(E|H_p) > Pr(E|H_d)$, that is event E is one that is more probable to occur if the ultimate probandum H_p is true, rather than when the specified alternative, H_d, is true. Using notation introduced so far, this expression of evidential value can also be written as V_E.

Example 8.3 *(Corroborating evidence in handwriting examination) This setting can be considered as a modification of Example 8.1. Consider, for example, a case where two independent handwriting experts each report E^*, that is they provide a report of the event E, defined as 'The suspect is the source of the signature on the questioned document'. These reports represent evidence from two distinct sources. In such a setting, each expert reports the occurrence of the same event E. In turn, E is relevant for an inference about H, the proposition according to which the suspect is the author of a given criminal event of interest. Proposition H may also be called an ultimate probandum as it is a root variable with no entering arcs from other nodes. In a Bayesian network, H is a graphical parent of node E.*

Assuming a dependency structure between the variables as shown in Figure 8.1(i), the likelihood ratio for the reports E_1^* and E_2^* by, respectively, sources S_1 and S_2, follows the general structure defined in (8.3). For the case considered here, the expression can again be developed further and shown to be as follows (Schum 1994):

$$V_{E_1^*, E_2^*} = \frac{Pr(E_1^*, E_2^* | H_p)}{Pr(E_1^*, E_2^* | H_d)} = \frac{Pr(E|H_p) + \left[\frac{h_1 h_2}{f_1 f_2} - 1\right]^{-1}}{Pr(E|H_d) + \left[\frac{h_1 h_2}{f_1 f_2} - 1\right]^{-1}}. \tag{8.10}$$

As may be seen, the overall inferential force of E_1^* and E_2^* depends not only on the value of evidence E for comparison of H_p and H_d, as expressed by the likelihoods $Pr(E|H_p)$ and $Pr(E|H_d)$, but also on the conditional probabilities of the reports given E, that is the local likelihood ratios $V'_{E_1^*} = h_1/f_1$ associated with the first report and $V'_{E_2^*} = h_2/f_2$ associated with the second report.

Notice further that (8.10) can also be extended to multiple, say n, independent sources, assuming conditional independence as before. For such a situation, the likelihood ratio can be shown to take the following form:

$$V_{E_1^*, \dots, E_n^*} = \frac{Pr(E_1^*, \dots, E_n^* | H_p)}{Pr(E_1^*, \dots, E_n^* | H_d)} = \frac{Pr(E|H_p) + \left[\prod_{j=1}^n \frac{h_i}{f_i} - 1\right]^{-1}}{Pr(E|H_d) + \left[\prod_{j=1}^n \frac{h_i}{f_i} - 1\right]^{-1}}. \tag{8.11}$$

Such a setting is typically encountered in so-called testing cases, where n independent examiners work on a well-defined question. As pointed out in Example 8.4, this could be an actual case or an experiment under predefined testing conditions (such as a proficiency test).

Example 8.4 (*Multiple corroborating items of evidence*) *Imagine a situation in which it is of interest to make an inference about a proposition of the kind 'the suspect's photocopier was involved in the production of the questioned document'. Next, consider a group of experts who have all have examined both questioned and known items and provided independent reports on whether or not the suspect's photocopier was involved. For the purpose of illustration, consider publicly available proficiency testing data reported by Collaborative Testing Services Inc. in 2010. As part of their questioned documents test No. 10-521, 149 participants correctly reported a given known source as 'was involved' in the production of a given questioned document. However, there were also 12 participants that incorrectly reported that the known source at hand 'was not involved'.*[3] *If one takes the proposition E, that is 'the suspect's printing device was involved', as evidence (i.e. an intermediate source level proposition) in support of a higher level proposition H, which states the involvement of the suspect in some criminal activity (assuming now the setting of a case, rather than proficiency testing) because if $Pr(E|H_p) > Pr(E|H_d)$, then all statements E^* of the form 'was involved' can be considered as corroborating. They are, however, in contradiction with statements of the kind E^{c*}, that is of the form 'was not involved', following definitions and discussion presented in Section 8.3.2. This proficiency testing scenario is artificial in the sense that the truth of the exercise is known. The general inference structure may, however, also hold in real cases where the same items may be analysed and assessed by different experts independently (i.e. without knowledge each other's conclusions), even though the true state of affairs (i.e. propositions E and H) will typically be unknown.*

[3] Summary Report available at http://www.ctsforensics.com/assets/news/3021_Web.pdf, last accessed May 2014.

In order to obtain overall corroboration with respect to the proposition H_p, it is necessary that $h_i > f_i$ for every source i. This implies that the examination of the credibility of the sources should not be neglected even when confronted with a case of corroboration. Notice further that the likelihood ratio in (8.10) and (8.11) cannot exceed V_E or $V_{\bar{E}}$. The joint value in an inference about H, based on a given number of individual sources that report on E, cannot be higher than that for complete knowledge about E (i.e. a situation in which the actual state of E was known). Alternatively, it can be said that the values of the individual reports for discriminating about H depend on the strength of the individual reports to discriminate between the states of the variable E. For example, if a report E^* is capable of making E certain, then the likelihood ratio for E^*, that is V_{E^*}, would equate to that for E, that is V_E. However, if E^* – which may denote, without loss of generality, a set of independent reports $E_1^*, \ldots, E_n^*$ – leaves some uncertainty about E, then V_{E^*} would be less than V_E.

8.3.3.2   Convergence

A convergence arises when two or more sources state the occurrence of distinct events that support different intermediate propositions that separately support the ultimate proposition. As depicted by Figure 8.1(ii), sources S_1 and S_2 may report the occurrence of the events E and F that are conditionally independent of knowledge of H. This is equivalent to having two independent strains of inference of the kind $E^* \rightarrow E \rightarrow H$, as illustrated in Figure 8.1(i). In such a case, the overall likelihood ratio for the two reports E^* and F^* is given by the product of the likelihood ratios associated with the individual reports. That is, $V_{E^*,F^*} = V_{E^*} \times V_{F^*}$, and (8.5) and (8.6) can again be applied. An illustration for convergence can be obtained by reconsidering Examples 8.1 and 8.2.

Example 8.5 *(Converging evidence, conditionally independent) Suppose a scientist reports E^*, that event E occurred (e.g. a given suspect wrote a signature on a questioned document). In addition, assume further that the questioned document bears ridge skin marks. Let F now denote – distinct from Example 8.2 – the proposition that the fingermarks come from the suspect. Hence, let F^* denote a scientist's report of such a conclusion. Assume again that the fingermarks are found in a position on the document where marks from the offender would be expected to be found. The proposition F can then be considered relevant in an inference about the proposition H, that 'the suspect is the person who committed the fraud'. Consequently, proposition F would now favour H because the probability of F may be taken to be greater given H_p than given H_d; that is, stated otherwise, the likelihood ratio for F, written $V_F = Pr(F|H_p)/Pr(F|H_d)$, is greater than one. Following a likelihood ratio for E, written as $V_E = Pr(E|H_p)/Pr(E|H_d) > 1$, this presents a further element in support of H and thus implies convergence.*

If, however, the events E and F are conditionally dependent on the ultimate proposition H, then (8.5) and (8.9) are relevant. In particular, in the assessment of the probative value of F, it is necessary to account for what has been observed in relation with the first item of evidence. This dependency is expressed by the conditional likelihood ratio $V_{F|E}$. According to the specified probabilistic underpinning, this may lead to the observation that the second observation F has more evidential value when E is already known, compared to a situation in which the outcome of the inspection of the first item of evidence is not known. In such a case, the evidence is called *synergetic*. However, it may also be the case that knowledge about E diminishes the inferential force of F and this would be a situation of also sometimes referred to as *redundancy*. This may go as far as to entail a directional change, rather than only a reduction in the inferential force of F; an event F with $V_F > 1$ that supports H_p, may, with knowledge of E, change to an event $F|E$ that supports H_d with $V_{F|E} < 1$.

Example 8.6 *(Convergent, conditionally dependent, evidence) The assumption of conditional independence in Example 8.5 may not be acceptable in general. The following example pointed out by Lempert (1977, p. 1043) illustrates this point:*

> *(...) the defendant's thumb print was found on the gun the killer used. (...) assume that the factfinder believes that the presence of this evidence is 500 times more likely if the defendant is guilty than if he is not guilty. (...) Now suppose that the prosecution wished to introduce evidence proving that a print matching the defendant's index finger was found on the murder weapon. If this were the only fingerprint evidence in the case, it would lead the factfinder to increase his estimated odds on the defendant's guilt to the same degree that the proof of the thumb print did. Yet, it is intuitively obvious that another five hundredfold increase is not justified when evidence of the thumb print has already been admitted.*

For the construction of a Bayesian network, one can consider three binary variables H, E and F. The variable H stands for the propositions 'The suspect is the killer (H_p)' and 'Some person other than the suspect is the killer (H_d)'. The variables E and F refer to the two finger-marks present on the weapon. For the first mark, consider the event E, defined as 'The friction ridge mark comes from the suspect's thumb (E)' and 'The friction ridge mark came from the thumb of some other person than the suspect ($\bar{E}$)'. For the second mark, let event F denote the possibilities 'The friction ridge mark comes from the suspect's index finger (F)' and 'The friction ridge mark came from the index of some other person than the suspect ($\bar{F}$)'.

Lempert argues that, given marks of the suspect's thumb being found on the murder weapon (event E), the probability of finding marks of his index finger (event F), if he is guilty (event H_p), is not very different from finding the same evidence, if he were not guilty (event H_d). A direct dependency of F on E is thus assumed. Graphically, a structure as shown in Figure 8.1(ii), with an arc pointing from E to F, could be proposed. Note that such a network structure has earlier been discussed in the context of the Baskervilles' case (Figure 2.7, Section 2.1.12).

The evidence E is taken to provide a likelihood ratio of 500 for the proposition of the defendant being guilty, as Lempert (1977, p. 1043) considers that a '(...) mathematically inclined juror might, for example, believe that there is a 0.2 probability that the print would be found if the defendant were guilty (the probability is considerably less than one because guilty people often have taken the trouble to wipe their prints from weapons and, even if they had not, not all prints are identifiable) and a 0.0004 probability that the evidence would be found if the defendant were not guilty.' Consequently, the node probability table for E could contain the probabilities $Pr(E|H_p) = 0.2$ and $Pr(E|H_d) = 0.0004$.

For the probability table associated with the node F, several comments apply. If the first mark did not come from the suspect ($\bar{E}$), then evidence of marks of the suspect's index finger may be given a weight that has the same order of magnitude as that associated with the thumb mark evidence alone. One may thus set $Pr(F|\bar{E}, H_p) = Pr(E|H_p)$ and $Pr(F|\bar{E}, H_d) = Pr(E|H_d)$. In such a case, the values contained in columns 2 and 4 in the body of Table 8.1 may be set equal to the values $Pr(E|H_p) = 0.2$ and $Pr(E|H_d) = 0.0004$ assigned to the node E. In the event of marks of the suspect's thumb being present on the weapon (E), the capacity

Table 8.1 Conditional probabilities assigned to the table of the node F.

H :	H_p		H_d	
E :	E	$\bar{E}$	E	$\bar{E}$
F : F	α	0.2	β	0.0004
$\bar{F}$	$1 - \alpha$	0.8	$1 - \beta$	0.9996

The three conditioning propositions are each binary. Node H has states H_p, the suspect is the killer, and H_d, the suspect is not the killer. Node E, friction ridge marks of the suspect's thumb are, E, or are not, $\bar{E}$, present on the murder weapon. Node F, friction ridge marks of the suspect's index finger are, F, or are not, $\bar{F}$, present on the murder weapon. The probability of finding marks of the suspect's index finger, F, given that he is guilty, H_p, and given that E is also true, the thumb mark evidence is present, is denoted α. The probability of finding marks of the suspect's index finger, F, given that he is innocent, and H_d, given that E is also true, the thumb mark evidence is present, is denoted β. Other probabilities are as explained in the text.

of the second item of evidence to discriminate between H_p and H_d is crucially dependent on the ratio of the probabilities $Pr(F|E, H_p)$ and $Pr(F|E, H_d)$. Denote these two probabilities α and β, respectively. Then, if one believes, as mentioned above, that the probability of finding marks of the suspect's index finger, F, given that he is guilty, H_p (and E is also true, the thumb mark evidence is present), does not greatly differ from the probability of F, given the suspect's innocence, H_d, and E, then $\alpha \approx \beta$ and a likelihood ratio of approximately 1 results for the evidence F given E.

The assessment of the conditional probabilities α and β is a matter of expert judgement and is largely dependent on one's views and beliefs held in a particular case. The following quotation from Lempert (1977, p. 1043) further illustrates this point:

The presence of the second print depends largely on the way the defendant held the gun when he left the thumb print. Unless murderers hold guns differently from non-murderers or are more likely to wipe off some, but not all, of their fingerprints, the finding of the second print is no more consistent with the hypothesis that the defendant is guilty than with its opposite. Indeed, because a murderer is more likely to attempt to wipe off fingerprints from a gun than one with no apprehension of being linked to a murder and since an attempt to wipe off fingerprints might be only partially successful, there is a plausible argument that the presence of the second print should lead jurors to be somewhat less confident that the defendant is the murderer than they would be if only one of the defendant's fingerprints were found.

Note that, strictly speaking, the variables E and F do not denote actual items of evidence. Here, E and F merely represent source level propositions (or events), that is, propositions of the kind 'the friction ridge mark comes from the suspect'. In many contexts, it may be useful to distinguish such events from reports that provide a statement about their occurrence. As suggested by Figure 8.1(ii), these could be represented by, respectively, S_1 and S_2. Propositions S_1 and S_2 could refer, for example, to propositions of the kind 'a certain number of corresponding minutiae are observed in the friction ridge mark found on the weapon and in the prints obtained from the suspect under controlled conditions'.

8.3.4   Drag coefficient

Consider again a situation as in Example 8.1, discussed in Section 8.3.2, where the report E^* of a single expert (source S_1) is used to infer something about the occurrence of an event E. As shown in Figure 8.1 in terms of a path starting at E^*, the event E is, in turn, of interest for an inference about $\{H_p, H_p\}$. The inferential force of the scientist's report is as defined in (8.7):

$$V_{E^*} = \frac{Pr(E|H_p) + \left[\frac{h_1}{f_1} - 1\right]^{-1}}{Pr(E|H_d) + \left[\frac{h_1}{f_1} - 1\right]^{-1}}.$$

Here, the term called *drag coefficient* is given by $[(h_1/f_1) - 1]^{-1}$. It is part of both the numerator and denominator and here written D, for short. As mentioned in Section 8.3.2, D acts like an inferential drag on $Pr(E|H_p)$ and $Pr(E|H_d)$. The drag coefficient is also encountered in other likelihood ratio formulae considered so far in this section, differing only with respect to the probabilities that are incorporated in the expression. The underlying mechanism that generates what may called *inferential drag* is, however, the same.

It is useful to take a closer look at some limiting cases in order to illustrate how D generates inferential drag. Suppose that a given source, for instance S_1, states E^*, but that it has no credibility; stated otherwise, the evidence given by S_1 does not discriminate between E and $\bar{E}$. This is the case whenever the probability of source S_1 providing report E^* given E is the same as given $\bar{E}$ (i.e. the likelihood ratio is 1). Alternatively, one may also say that the so-called hit probability, that is the probability of report E^* when E is in fact true, $Pr(E^*|E)$, equals the 'false positive probability', that is the probability of report E^* when $\bar{E}$ is actually true. In this chapter, these two probabilities have been denoted h_1 and f_1. So, the closer h_1 and f_1 are, the larger the drag coefficient becomes and the more it dominates $Pr(E|H_p)$ and $Pr(E|H_d)$, both of which, being probabilities, are less than 1. Consequently, the likelihood ratio for report E^* becomes closer to 1:

$$V_{E^*} = \lim_{h_1 \to f_1} \frac{Pr(E|H_p) + \left[\frac{h_1}{f_1} - 1\right]^{-1}}{Pr(E|H_d) + \left[\frac{h_1}{f_1} - 1\right]^{-1}} = 1.$$

Hence, the influence of $Pr(E|H_p)$ and $Pr(E|H_d)$ becomes negligible. The drag coefficient dominates the numerator and the denominator so that the likelihood ratio tends towards a value of one. Thus, the failure of E^* to discriminate between E and $\bar{E}$ reduces the ability of E to discriminate between H_p and H_d.

This analysis may be developed further. Consider a situation where S_1 has maximal credibility. That is, its hit probability is one and the probability of a false positive is zero. This is just another way to say that the source S_1 provides perfect evidence for discriminating between E and $\bar{E}$: it always reports E^* when in fact E is true (i.e. $Pr(E^*|E) = h_1 = 1$) and never reports E^* otherwise (i.e. $Pr(E^*|\bar{E}) = f_1 = 0$). It can be seen that in such a case the drag coefficient $\{[(h_1/f_1) - 1]^{-1}\}$ that can be written as $\{f_1/(h_1 - f_1)\}$ tends to zero as f_1 tends to zero, assuming $h_1 \neq 0$. The likelihood ratio then becomes

$$V_{E^*} = \frac{Pr(E|H_p)}{Pr(E|H_d)} = V_E.$$

This result shows that the likelihood ratio for the report E^* of a perfectly credible source equates to the result when the occurrence of E is known for certain.

Finally, imagine yet another situation where S_1 has a hit probability of zero but a false-positive probability of unity. This is a situation in which a given source would systematically report the opposite of what it should state in order to be right. More formally, such a source would report E^* whenever $\bar{E}$ is true (i.e. $Pr(E^*|\bar{E}) = 1$) and report E^{c*} when E is true (i.e. $Pr(E^{c*}|E) = 1$). In such a case, the drag coefficient is $[(0/1) - 1]^{-1} = -1$. Consequently, the likelihood ratio becomes

$$V_{E^*} = \frac{Pr(E|H_p) - 1}{Pr(E|H_d) - 1} = \frac{1 - Pr(E|H_p)}{1 - Pr(E|H_d)} = \frac{Pr(\bar{E}|H_p)}{Pr(\bar{E}|H_d)} = V_{\bar{E}}.$$

This represents the likelihood ratio for knowing the occurrence of $\bar{E}$ for sure.

8.4 Examples of the combination of distinct items of evidence

This section relies on aspects of the joint evaluation of scientific findings as outlined in the previous sections. A sequential procedure is followed in order to illustrate the rationale behind the proposed Bayesian networks. First, network structures are elicited for reasoning separately about each trace item. Then, in a second step, ways are examined in order to combine network fragments logically.

8.4.1 Handwriting and fingermarks

Consider the following scenario, adapted from Stockton and Day (2001, p. 3) with 'credit card' in the original changed to 'chequebook and accompanying bankcard':

> Two people, Mr. Adams and Miss Brown, have been arrested in a shop trying to pass a stolen cheque. Mr. Adams admits to writing the signature on the cheque, but claims he has only just acquired the chequebook and certainly has not written out any more cheques from the book. Miss Brown claims to know nothing of the whole affair. On investigation it is found that sixteen further transactions have taken place using the chequebook and an accompanying bankcard, which was reported stolen a few days ago. The owner, Mr. Constantine, claims the card must have been stolen in the post and his signature does not look anything like the signature on the card. Handwriting samples, including the signatures in the name of Mr. Constantine, are obtained from both Mr. Adams and Miss Brown.

Different analyses have been proposed by Stockton and Day (2001), such as evaluations of single signatures or groups of signatures. Here, an abstraction will be made of the scenario in the sense that consideration will only be given to the evaluation of a single signature present on one of the 16 cheques. This cheque may have been selected because of the particularly high amount of money involved, for example.

Comparative examinations may be performed between the disputed signature and various comparison writings obtained from the suspect, Mr. Adams. The question of authorship may be framed in terms of the propositions 'Mr Adams wrote the signature (S_p)' and 'Mr Adams did not write the signature (S_d)'. As in previously discussed scenarios, the subscripts p and d are used to denote the prosecution's and the defence propositions, respectively.

Different ways may be imagined to describe the findings, according to the desired level of detail. Following Stockton and Day (2001), the findings may be summarized in terms of 'few', 'some' and 'many' similarities. Thus, let E_1 denote the outcomes of the comparative analyses and e_1, e_2 and e_3 stand for, respectively, 'few', 'some' and 'many' similarities. An assessment is then necessary for expressing one's expectation's of observing each of these outcomes, given that the suspect is or is not the author of the disputed signature. These are assessments that are largely dependent on the circumstances of the case at hand. As it may be difficult to find appropriate data from surveys or literature, assessments would consist mostly of an examiner's judgement, taking into account the framework of circumstances of the case at hand. The following provides an idea as to the range of considerations that may be involved when assessing the various outcomes, assuming that the suspect is the author of the questioned signature.

> If Mr. Adams is particularly naive, he might write the samples in exactly the same way as the signature and then we may find many similarities in detail and few differences. However, this is very unlikely. He is more likely to vary his writing in some way (either in his samples or in the questioned signature) so that we may find some similarities but some differences will also be present. Perhaps, most likely is that he will disguise his writing so much that, while we may find a few similarities there will be many differences. (Stockton and Day 2001, p. 4)

If the suspect were not the author of the questioned signature, then one may consider it more probable to observe few similarities than many similarities. Exemplary values for the states of the node E_1 given S_d could thus be $\{0.78, 0.20, 0.02\}$. Given S_p, the following values could be used: $\{0.60, 0.27, 0.13\}$. The values are approximations derived from areas in pie charts proposed by Stockton and Day (2001). Notice that the values retained here mainly serve for the purpose of illustration. Methods exist that focus on the relative order of magnitude of probabilistic assessments rather than particular numerical values. Chapter 13 pursues further discussion of such methods.

So far, the evaluation has been based on a variable E_1, denoting a qualitative expression of the amount of observed similarities, and a variable S, denoting the proposition of authorship of the disputed signature. In terms of a Bayesian network, a basic two-node representation of the kind $S \rightarrow E_1$ could be adopted for this scenario [Figure 8.2(i)]. But often, comparative handwriting analysis may only provide limited information for discriminating between the propositions put forward by, respectively, the prosecution and the defence. This may be in part due to inappropriate comparative material, for example. On other occasions, the disputed handwriting may consist only of a few lines. Difficulties may also arise if the incriminated writing exhibits only a few comparative features, as might be the case with an abbreviated signature. Further elements may thus be needed in order to assess a potential association between the suspect and the incriminating cheque. Fingermarks are one such sort of element. An inferred association between the suspect and the questioned cheque may, in some situations, be considered a relevant item of information for constructing an argument about some higher level probandum (e.g. that the suspect is the author of the signature).

For the scenario considered here, assume, for example, that various kinds of detection techniques have been applied to the questioned document (cheque in this example). During this

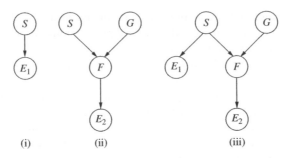

(i)   (ii)   (iii)

Figure 8.2 Bayesian networks for evaluating (i) results of comparative handwriting examinations, (ii) corresponding fingermark features and (iii) both handwriting and fingermarks.
There are two states for node F, 'The suspect is the source of the fingermarks (F)' and 'The
suspect is not the source of the fingermarks ($\bar{F}$)'. Node S has two states, 'The suspect is the
author of the questioned signature (S_p)' and 'The suspect is not the author of the questioned
signature ($\bar{S}_d$)'. Node G has two states, 'The fingermarks have been left by the author of the
questioned signature (G)' and 'The fingermarks have not been left by the author of the questioned signature ($\bar{G}$)'. Node E_1 has three states, the finding of few (e_{11}), some (e_{12}) and many
(e_{13}) similarities given different propositions of authorship for the comparative analysis of
handwriting. Node E_2 has two states, 'There are correspondences between the fingermarks on
the questioned document and the fingerprints of the suspect (e_{21})' and 'There are no correspondences between the fingermarks on the questioned document and the fingerprints of the
suspect (e_{22})'.

process, two fragmentary friction ridge marks became visible. A total of, say, 10 minutiae are
found to correspond to features present in prints obtained under controlled conditions of the
thumb and the index of Mr. Adams' left hand. No unexplainable differences are observed.
Imagine further that the respective position and sequence of the marks present on the document are such that they form what may be called an *anatomical sequence*. Moreover, the marks
are present in a position in which one would naturally expect the author of the questioned
signature to leave such marks. Denote these observations of friction ridge marks by E_2.

Start by considering E_2 separately from the results of handwriting examinations. The current discussion does not aim at addressing aspects of the wide range of issues pertaining to the
process of 'individualizing' particular persons as sources of crime marks. This is treated in
extensive existing literature on the topic (Champod et al. 2004). The discussion here allows for
the general assumption that the probative value of fingermarks can be evaluated and assessed
in probabilistic terms (Champod and Evett 2001).

Observations of friction ridge marks are amenable for inference about the following
pair of propositions, defined here by a variable F: 'The fingermarks come from the suspect,
Mr. Adams (F)' and 'The fingermarks come from an individual other than the suspect ($\bar{F}$)'.
Note that the presentation here distinguishes fingerprints from fingermarks. A fingermark is,
usually, a latent and may be found, for example, at a crime scene or an object thought to be
relevant to the case. A fingerprint is an inked impression of a finger of a suspect, obtained
under controlled conditions.

The scientist may judge the observed correspondences (without any unexplainable differences), written e_{21}, to be more probable, given that they have been left by the suspect

(proposition F), rather than if left by some other individual. Specification of numerical values is avoided here because factors such as the clarity of the marks, the kind of corresponding minutiae, their relative position and so on, are not discussed in further detail here. For the current level of discussion, it suffices to note that one's beliefs may be represented by $Pr(e_{21}|F) \gg Pr(e_{21}|\bar{F})$. This means that the observed correspondences offer clear support for the proposition that the suspect is the source of the friction ridge marks, rather than some other person.

Next, one needs to construct an argument if one seeks to move from the source level proposition F about fingermarks to the propositions S_p and S_d that relate to the authorship of the questioned signature. A Bayesian network described in Section 6.1 (Figure 6.1) may be used for this purpose. Such a network allows for the evaluation of findings in one-stain transfer cases where uncertainties may exist in relation to the relevance of the recovered material. In the context of a scenario involving a disputed signature, the relevance of fingermarks could be taken as an expression of one's belief that the marks had been left by the author of the questioned signature. Accordingly, a binary variable G is defined here as follows: 'The fingermarks come from the author of the questioned signature (G)' and 'The fingermarks do not come from the author of the questioned signature ($\bar{G}$)'. Usually, probability assignments for the variable G would be different from zero and one, although from the position in which the marks are found on the cheque in this case, one could retain a fairly high initial probability for G.

Regarding the conditional probabilities for the variable F, one can proceed in the same way as explained in Section 6.1, using logical values of zero and one as well as an assignment p, that stands for the probability $Pr(F|\bar{G}, S_d)$ that the suspect would have left the friction ridge marks for innocent reasons (e.g. on another occasion).

For a joint evaluation of both, the results of the comparative handwriting examinations and the correspondence in fingermark features, a logical combination needs to be found between the two network fragments shown in Figure 8.2(i) and (ii). Part of a natural way to proceed could consist of recognizing that Figure 8.2(i) and (ii) contain a variable S that has the same definition. The two local networks could thus be combined within a single network by retaining only a single node S, as is shown in Figure 8.2(iii). This form of combination assumes the two observations e_{13} and e_{21} to be independent of knowledge of S. The likelihood ratio for the joint consideration of the two scientific results, V_{12}, may thus be written as the product of the individual likelihood ratios: $V_{12} = V_1 \times V_2$. Here, V_1 denotes the likelihood ratio for the results of the comparative handwriting examinations and V_2 denotes the likelihood ratio for the fingermark comparison. The latter is an abbreviated form of the following:

$$V_2 = \frac{r + \{(1-r)\gamma\}}{(r\gamma) + (1-r)\{p + (1-p)\gamma\}} . \tag{8.12}$$

Note that in deriving (8.12), two assumptions are made. First, $Pr(e_{21}|F) = 1$, an assumption that allows one to simplify the resulting formula. However, strictly speaking, there may be various numbers of corresponding ridge skin features given F, so that the term $Pr(e_{21}|F)$ should actually take values lower than 1 (according to a probability distribution over possible numbers of observed ridge skin features). Second, it is assumed that $Pr(e_{21}|\bar{F}) = \gamma$, where γ is an expression for the rarity of corresponding set of ridge skin features. Note further that the variable r is a shorthand notation for the relevance term $Pr(G)$. Equation (8.12) is a special case of Equation (6.8) with the probability q omitted, which relates to an activity of the suspect at the time of the crime.

8.4.2 Issues in DNA analyses

An important aspect of the procedure described in Section 8.4.1 is that a joint evaluation can be obtained by combining network fragments that, separately, represent existing likelihood ratio formulae. This helps to guide users towards a coherent evaluation that may otherwise become increasingly difficult. This section addresses a further example to illustrate this point.

Imagine a scenario similar to the one described in Section 6.1. A crime has been committed by one offender and a blood stain has been found at the scene. There is a report of a forensic scientist stating that the blood of a suspect shares the same DNA profile as the stain from the scene. Further, suppose that it is also of interest to consider the following aspects: (i) the suspect has been selected through a search in a database containing DNA profiles (i.e. a certain number of individuals with different profiles were observed in the database), (ii) there is a potential of error in the analytical result (iii) there is some doubt as to whether the recovered blood stain truly came from the offender (i.e. the stain being relevant). How is one to draw an inference to the proposition that the suspect is the offender?

On a broad view, this scenario divides into several distinct issues, each of which may be represented by a 'network fragment'. The term *network fragment* is taken here from Neil et al. (2000) who refer to a set of related variables that could be constructed and reasoned about separately from other fragments. In the scenario introduced here, one may distinguish three network fragments. A first network fragment deals with the fact that there is a crime stain with a DNA profile that corresponds to control material provided by the suspect. This is a proposition that depends on whether the suspect is the source of the crime stain. In turn, the latter is a proposition that may be used to infer something about whether the suspect is the criminal, assuming that consideration is given to the relevance of the crime stain. Figure 8.3(i), described in Section 6.1, proposes a representation for this aspect. Second, there is a network fragment for the forensic scientist's report of corresponding DNA profiles. This is a relevant item of information for inferring about the actual occurrence of a correspondence and, beyond this, the proposition according to which the suspect is the source of the crime stain. A network that may be used here is shown in Figure 8.3(ii). It allows one to account for the possibility that the scientist erroneously reports corresponding DNA profiles. Section 7.11 introduced this sub-model in further detail. Third, Figure 8.3(iii) depicts a model for evaluating the results of

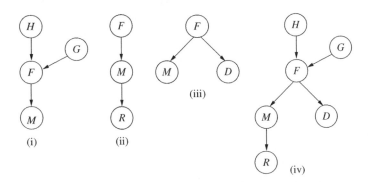

Figure 8.3 Bayesian networks for (i) a one-trace scenario of transfer from the suspect (including the concept of relevance), (ii) the problem of laboratory error, (iii) a database search scenario and (iv) a combination of the aspects represented in (i)–(iii). Table 8.2 summarizes the definitions of the various nodes.

Table 8.2 Definitions of the binary nodes used in Figure 8.3.

Node	Definition
H	The suspect is the offender
F	The crime stain came from the suspect
G	The crime stain came from the offender
M	The suspect's blood and the crime stain have the same DNA profile
R	Forensic scientist's report of corresponding DNA profiles
D	The other $(N - 1)$ profiles of the database do not match

a database search together with the event of a true correspondence between the crime stain's profile and that of control material provided by the suspect. Section 7.10 covers the background of this model.

Table 8.2 summarizes the definition of all propositions that appear in these sub-models. From the number of variables involved and the various possible relevance relationships among them, it can be seen that a formal development of a likelihood ratio formula becomes an increasingly complex task. However, if it can be accepted that the Bayesian network fragments shown in Figure 8.3(i)–(iii) are appropriate representations of the separate issues, then a single Bayesian network for representing all of these issues can be proposed by combining the network fragments such that one obtains a structure as shown in Figure 8.3(iv). This way of combination suggests that separate models can be examined for the presence of nodes with the same definition. A combination between two or more network fragments may then be operated through the node(s) that these fragments have in common. In the resulting network, only one node with the same definition will remain. This stems from the definition of Bayesian networks according to which each variable can only appear once within the same model. Notice however that when combining network fragments, one also needs to examine whether there could be additional links among nodes from the separate networks. In the example considered in this section, this is not the case.

8.4.3   One offender and two corresponding traces

Section 6.3 has already considered scenarios involving more than one stain. In particular, Example 6.5 relates to one offender with two stains found at a scene of a crime. The two traces were of the same kind (e.g. blood) but of different types and, potentially, different degrees of relevance. A question of interest was how to construct an argument for crime level propositions (i.e. that the suspect is the offender). Section 4.3 analysed situations where two traces are found on two distinct crime scenes. The recovered traces were of the same kind and the same type. The aim was to infer something about a potential common source of the two sets of trace material, assuming that no potential source was available for comparative examinations.

The section here considers yet another two-stain scenario. Imagine the following case: during night time, an offender entered a house by forcing a back door. Visibly, several rooms were searched for the presence of objects of value. Drawers were emptied and different items such as clothing and paper were scattered all over the floor. In one of the rooms, a safe had been opened using brute force. All objects of value in the safe had been stolen. Upon examination of the scene, examiners direct their attention to two elements. One is a blood stain found at the point of entry. From the position in which the stain was present, its apparent freshness and abundance, it is believed that it may have been left by the offender while gaining access to the premises. Moreover, the owner asserts that there was no blood there before. A second element of interest is a partial fingermark found on the outer surface of the safe's front door. A few weeks later, the police find a suspect for reasons completely unrelated to these traces. The suspect's blood and the blood trace from the crime scene share the same characteristics. Examiners also observe a correspondence between features in the fingermark lifted on the safe and features in one of the prints obtained from the suspect under controlled conditions. In summary, thus, the scenario here involves two different kinds of traces found on the same crime scene and correspondences observed with a single potential source. This scenario raises the question of how to consider possible dependencies that may exist between arguments affecting the separate evaluation of each trace item. A stepwise construction of a Bayesian network can help to clarify this point.

Start by considering the bloodstain and let E_1 stand for the correspondence between the characteristics of the crime stain and those of control material from the suspect. Section 6.1 pointed out that such a correspondence is a relevant element of information for inferring something about whether the suspect is or is not the source of the crime stain, represented by a proposition F_1. This latter proposition, together with information on the relevance of the crime stain (G_1), allows one to construct a line of reasoning to a crime level proposition H [i.e. 'The suspect is (or is not) the perpetrator of the alleged offence']. This leads to a Bayesian network introduced in Figure 6.1.

In agreement with previous discussion from Section 6.1, conditional probabilities for the node F_1 can be defined as follows. If the suspect is the offender and the crime stain comes from the offender, then necessarily the crime stain comes from the suspect: $Pr(F_1|H_p, G_1) = 1$. If H_p is true and G_1 is false, or vice versa, then the suspect certainly is not the source of the crime stain: $Pr(\bar{F}_1|H_p, \bar{G}_1) = Pr(\bar{F}_1|H_d, G_1) = 1$. Probabilities different from 0 and 1 may be applicable for the state F_1 if it is thought that there could be a possibility for the suspect leaving a stain even though he is innocent of the crime. Denote such an assessment p_1. Consequently, the state $\bar{F}_1$ will assume a probability equal to $1 - p_1$.

Next, consider the second trace, E_2, the correspondence between features of the friction ridge mark found on the safe and features in one of the prints obtained from the suspect under controlled conditions. One can extend the existing Bayesian network by duplicating the structure developed so far. Besides E_2, this creates two new nodes. One is G_2 with states 'The fingermark has been left by the offender (G_2)' and 'The fingermark has not been left by the offender ($\bar{G}_2$)'. The other is F_2 with states 'The suspect is the source of the fingermark (F_2)' and 'The suspect is not the source of the fingermark, some other person is ($\bar{F}_2$)'.

The Bayesian network constructed so far is that shown in Figure 8.4(i). Next, one of the questions that the scientist should consider is whether this is a satisfactory structure or whether there are any properties for which this network does not yet account. Notice that the structure shown in Figure 8.4(i) assumes that the two traces are independent given knowledge of H. A more close examination of the probability table for node F_2 will show whether this represents an appropriate assumption. Start by considering the following three assignments. First, if the suspect is the offender and the fingermark has been left by the offender, then certainly the fingermark comes from the suspect: $Pr(F_2|H_p, G_2) = 1$. Second, if the suspect is the offender but the fingermark has not been left by the offender, then certainly the suspect is not the source of the fingermark: $Pr(F_2|H_p, \bar{G}_2) = 0$. Third, if the suspect is not the offender and

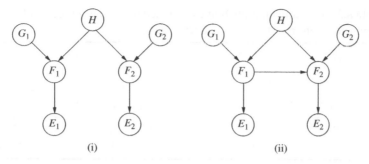

(i)   (ii)

Figure 8.4   (i) and (ii) Bayesian networks for the joint evaluation of two trace items found on the same crime scene. Node H has two states, 'The suspect is the offender (H_p)' and 'The suspect is not the offender (H_d)'. Node F_1 has two states, 'The suspect is the source of the blood stain (F_1)' and 'The suspect is not the source of the blood stain $(\bar{F}_1)$'. Node F_2 has two states, 'The fingermark comes from the suspect (F_2)' and 'The fingermark does not come from the suspect $(\bar{F}_2)$'. Node G_1, relevance of the blood stain, has two states, 'The stain came from the offender (G_1)' and 'The stain did not come from the offender $(\bar{G}_1)$'. Node G_2, relevance of the fingermark, has two states, 'The mark came from the offender (G_2)' and 'The mark did not come from the offender $(\bar{G}_2)$'. Node E_1 has two states, 'The blood stain and the control material provided by the suspect have corresponding analytical features (E_1)' and 'There is no such correspondence $(\bar{E}_1)$'. Node E_2 has two states, 'There is a correspondence between the friction ridge mark at the crime scene and a print provided by the suspect (E_2)' and 'There is no such correspondence $(\bar{E}_2)$'.

the fingermark comes from the offender, then certainly the suspect is not the source of the fingermark: $Pr(F_2|H_d, G_2) = 0$.

These appear reasonable assumptions and knowledge about any of the nodes used for modelling inference about the blood stain (i.e. nodes G_1, F_1 and E_1) should not affect these assessments. There is, however, one situation where a need may be felt for a dependency: when the suspect is not the offender and the second trace (fingermark) did not come from the offender. In such a situation, the suspect could have left the stain for innocent reasons. Here, knowledge about whether the suspect has left the first trace appears to be potentially informative. For example, one might be inclined to be more confident in the suspect being the source of the second trace for innocent reasons if he were also known to be the source of the first trace. Notice that it is still assumed here that the suspect is not the offender, so were he the source of the first trace, then he must be so for innocent reasons. This stems from the probability 0 assigned to $Pr(F_1|H_d, G_1)$.

So, if it is believed that for some assessments of F_2 knowledge about the truthstate of F_1 is relevant, then one should draw a directed edge between these two nodes. This situation is shown in Figure 8.4(ii). Table 8.3 shows the node probability table associated with node F_2. According to this table, knowledge about F_1 affects one's assessment of the probability that the suspect is the source of the second trace for innocent reasons, that is when both H_p and G_2 are false (i.e. a case of asymmetric independence, Section 2.1.12). In particular, one can account for a probability denoted p_2, used to describe the expectation that the suspect is the source of the second trace (for innocent reasons), given that he is the source of the first trace: $Pr(F_2|H_d, \bar{G}_2, F_1) = p_2$. Moreover, a probability denoted p'_2 can be defined to describe the

Table 8.3 Conditional probabilities assigned to F_2, the source level propositions referring to the second trace item.

H :	H_p				H_d			
G_2 :	G_2		$\bar{G}_2$		G_2		$\bar{G}_2$	
F_1 :	F_1	$\bar{F}_1$	F_1	$\bar{F}_1$	F_1	$\bar{F}_1$	F_1	$\bar{F}_1$
F_2 : $\quad F_2$	1	1	0	0	0	0	p_2	p'_2
$\bar{F}_2$	0	0	1	1	1	1	$1 - p_2$	$1 - p'_2$

Factors F_i (for $i = \{1, 2\}$) have two states, 'The suspect is the source of trace i (F_i)' and 'The suspect is not the source of trace i ($\bar{F}_i$)'. Factor H has two states, 'The suspect is the offender (H_p)' and 'The suspect is not the offender (H_d)'. The variable G_2, relevance of the second trace, has two states, 'The second trace came from the offender (G_2)' and 'The second item did not come from the offender ($\bar{G}_2$)'. The probability that the suspect is the source of the second trace, given that he is the source of the first trace, he did not commit the crime and the second trace did not come from the offender is p_2. The probability that the suspect is the source of the second trace, given that he is not the source of the first trace, he did not commit the crime and the second trace did not come from the offender is p'_2.

expectation that the suspect is the source of the second trace (for innocent reasons) given that he is not the source of the first trace: $Pr(F_2|H_d, \bar{G}_2, \bar{F}_1) = p'_2$. The former probability may be thought of as a conditional probability of innocent acquisition. This means that the object on the crime scene has acquired a stain from the suspect in a situation unrelated to the crime under investigation. The particular nature of this assessment is its conditioning on knowledge about the presence of another stain left by the suspect for innocent reasons. Notice however that for the link $F_1 \rightarrow F_2$ to have some meaning, the probability of innocent acquisition of the first item of evidence, denoted p_1, must be different from 0. In addition, the probabilities of innocent acquisition of the second item of evidence, p_2 and p'_2 (Table 8.3), must have different values. Whenever one of the latter two conditions is not satisfied, then the link is superfluous in the sense that the network would yield the same result as the one shown in Figure 8.4(i).

An interesting aspect of the Bayesian network for combining conditionally independent trace items [Figure 8.4(i)] is its logical relationship with a basic diverging connection of the kind $E_1 \leftarrow H \rightarrow E_2$, for which it provides a natural extension. Compare, for example, the properties of these two networks for particular specifications. That is, for the network shown in Figure 8.4(i), suppose $Pr(E_1|F_1) = 1, Pr(E_1|\bar{F}_1) = 0.01, Pr(F_1|H_p, G_1) = 1, Pr(F_1|H_p, \bar{G}_1) = Pr(F_1|H_d, G_1) = Pr(F_1|H_d, \bar{G}_1) = 0$ (analogously for E_2 and F_2). For the network $E_1 \leftarrow H \rightarrow E_2$, suppose $Pr(E_1|H_p) = 1, Pr(E_1|H_d) = 0.01$ (analogously for E_2). If, in addition, the nodes G_1 and G_2 in Figure 8.4(i) are instantiated to G_1 and G_2, respectively, then the joint occurrence of E_1 and E_2 supports the proposition H_p by a likelihood ratio of 10^4. This is the product of the likelihood ratios provided by the individual trace items. Verbally, it may be stated that, for example, the occurrence of E_1 (i.e. correspondence between the characteristics of the crime stain and the control material provided by a suspect) supports F_1 (the suspect being the source of the crime stain) versus the alternative $\bar{F}_1$ by a factor of $Pr(E_1|F_1)/Pr(E_1|\bar{F}_1) = 1/0.01 = 100$. As it is assumed that the crime stain came from the offender ($Pr(G_1) = 1$) and the suspect could not have left the stain for innocent reasons ($Pr(F_1|H_d, \bar{G}_1) = 0$), a trace item which is such as to increase the odds in favour of the suspect being the source of the crime stain (i.e. proposition F_1) by a factor of 100 exerts the same increase in the odds in favour of the suspect being the offender. The same holds for E_2. The network $E_1 \leftarrow H \rightarrow E_2$ also provides a result of this magnitude. Here, the likelihood ratio for the joint occurrence of the two trace items is just given by $[Pr(E_1|H_p)/Pr(E_1|H_d)] \times [Pr(E_2|H_p)/Pr(E_2|H_d)] = 100 \times 100 = 10^4$.

What then is the aim of using the network shown in Figure 8.4(i)? One of the features that this structure offers over the diverging connection $E_1 \leftarrow H \rightarrow E_2$ is that the relevance of each stain can be specified in a range of values between 0 and 1 (i.e. by leaving the two nodes G_1 and G_2 uninstantiated). As mentioned above, in the special case where $Pr(G_1) = Pr(G_2) = 1$, the network in Figure 8.4(i) provides the same result as the network $E_1 \leftarrow H \rightarrow E_2$. But whenever the relevance of a trace item i (for $i = \{1, 2\}$) is smaller than 1, then the degree to which the finding of that trace will affect (or induce) changes in the probability of H through the associated source level proposition F_i is reduced. In the special case where a relevance node G_i assumes the value 0 (i.e. the stain of interest did not come from the offender), the likelihood ratio reduces to 1. Notice, however, that the degree of support that a trace item i provides for its associated source node, F_i, would remain unaffected by whatever value is assumed by the relevance node. Imagine, for example, a situation where knowledge is available on the occurrence of E_1. This supports the proposition F_1, the suspect being the source of the crime stain, compared to the alternative $\bar{F}_1$, by a factor of 100 (i.e. through a likelihood ratio of $Pr(E_1|F_1)/Pr(E_1|\bar{F}_1) = 1/0.01 = 100$). However, when assuming that the crime stain did not come from the offender, then this would not affect the probability of the proposition H_p, the suspect being the offender. All of these are reasonable results.

8.4.4 Firearms and gunshot residues

8.4.4.1 Marks present on fired bullets

Section 6.2.5 presented a graphical model for evaluating results of comparative mark examinations that may be encountered, for example, with fired bullets. The basic idea follows an approach initially described by Evett et al. (1998b) in the context of footwear marks. Notice solely that the model described in Section 6.2.5 (Figure 6.7) does not go into the details of explicit incorporation of particular criteria retained in comparative examinations. On the contrary, the model is intentionally kept on a general level. This should allow for an evaluation of subjective degrees of beliefs, leaving room for scientists to retain those criteria that they consider most relevant to a case at hand. An example of a characteristic that can help to inform a scientist's opinion is the count of consecutive matching striations (CMS) between marks on two bullets, a statistic suggested by Biasotti (1959) and Biasotti and Murdock (1997). Bunch (2000) reviews this method, whereas Nichols (1997, 2003) gives general reviews of criteria for evaluating results of comparative examinations in the area of firearms and toolmarks.

Many discussions on firearms and toolmark examination focus rather directly, and often rather exclusively, on the question of evaluating corresponding striations. Agreement between features of manufacture is regarded only as a preliminary requirement for the extension of comparative analyses to acquired features. However, features of manufacture bear their own probative value as they allow for a reduction in the size of the population of firearms, which show features that are compatible with features observed on a fired bullet that is related to a crime. With the proposed Bayesian network, this variation in treatment can be overcome because this model allows for a logical combination of information pertaining to acquired characteristics and information pertaining to features of manufacture.

The combination of these pieces of information may become important, for example, when only poor information is available on acquired characteristics. There may be cases, for instance, where bullets are severely damaged so that only limited information is available on the number and quality of corresponding striations. This then is an instance where information on features of manufacture, such as calibre, the numbers of lands and grooves and their twist may be useful to enable some reduction in the size of the population of firearms that could have fired the questioned bullet. Ideally, scientists can inform their beliefs on aspects based on appropriate statistics.

8.4.4.2 Gunshot residues

The discussion in the previous paragraph focussed on different levels of detail encountered within the same trace category, that is marks present on fired bullets. Their observation during comparative analyses can lead to findings that form the basis for inference about whether or not a suspect's gun has been used to fire a particular questioned bullet.

Besides characterizing the relationship that may exist between a suspect's gun and a bullet of interest, firearms examiners may also be called on to conduct range-of-fire evaluations. For this purpose, firearms specialists may examine the distribution pattern of particle matter, more generally also termed *gunshot residues* (*GSR*), in the area of the entrance hole. It is thought that, for a given weapon and ammunition, there is a dependence between the distribution of such particles and the distance from which the firearm has been discharged (Lichtenberg 1990). In the discussion, the general term GSR is used but it is acknowledged that there is discussion within the field of what ought to be regarded as GSR (see also discussion presented in Section 10.6). When residue patterns are poorly visible, for example when present on a dark target surface, various techniques may be applied for visualizing residue patterns. In addition, examiners would also test fire the suspect's weapon towards appropriate targets arranged at different distances. Next, they would visually compare aspects such as size and density of the patterns thus obtained against the pattern in question (for which the distance of firing is unknown). Such experiments are thought to provide assistance in evaluating the distance from which the questioned pattern was shot.

There is some common terminology that firearms examiners use for describing various ranges of firing (Rowe 2000a). A distant shot, for example, refers to a distance that is such that no residue would reach the target. Close-range shots refer to distances close enough for residue particles to reach the target. When firing a shot at a range of 1 inch (2.5 cm) or less, it may be referred to as a *near-contact shot* and when the shot is fired with the muzzle in contact with the target surface, it may be referred to as a *contact shot* (Rowe 2000a). Certain kinds of observations are typically encountered with near-contact and contact shots, such as the singeing of hairs or textile fibres. The examination of gunshot wounds during autopsy may also reveal distinctive defects, such as the splitting of tissue.

As an example, consider the Modified Griess Test (MGT), a technique used for visualizing nitrite compounds formed when smokeless powder burns or partially burns (Dillon 1990). Using a suspect's gun and ammunition in combination, examiners can reproduce residue patterns at known distances. When 'approximating' a questioned pattern by patterns obtained from firings at known distances, some degree of subjectivity is involved. Generally, factors such as size and density may serve well for summarizing a given pattern. For the purpose of the current discussion, patterns will be described by a more simple and general measure, such as the number of visualized nitrite residues. These will appear in an orange colour subsequent to the application of the MGT.

Table 8.4   Mean and standard deviation of the number of visualized
nitrite residues for firings at known distances (hypothetical data).

Distance (cm):		30	50	70	90
Number of residues:	mean	500	350	120	55
	standard deviation	80	60	35	20

The number of firings at each distance is 10.

Imagine a scientist fires a suspect's weapon at distances of, say, 30, 50, 70 and 90 cm.
Usually, several firings would be made at each distance so as better to assess a pattern's variability. Conditions for firing would be chosen as closely as possible to the circumstances
assumed to have existed at the moment the questioned pattern was shot. A hypothetical set of
data are shown in Table 8.4. From personal experience, it is reasonable to assume a Normal
probability density function with parameters μ and σ^2 for the number of particles, denoted
$N(\mu, \sigma^2)$. The parameters are estimated by the sample mean and variance. For example, let
D_{30} denote the number of GSR particles found for firings made at a distance of 30 cm. The distribution may be written $(D_{30}|500, 80^2) \sim N(500, 80^2)$, or $D_{30} \sim N(500, 80^2)$ for short, with
similar notation for the other distances. Y refers to a questioned pattern, with y denoting a
particular outcome, that is a certain number of visualized particles. For the purpose of the
current example, let y be 400. It is accepted here that the approximation of the distribution of
a discrete random variable (that is, the number of particles) by a continuous random variable
is reasonable because of the large number of particles involved.

Next, imagine that a scientist seeks to evaluate the questioned pattern given particular
distributional assumptions. For example, the scientist may wish to assess the finding of 400
particles in a questioned pattern with respect to an assumed distance of firing of 30 cm. But,
as usual, it is necessary to consider the observation of interest given at least one alternative
proposition. Suppose, for instance, 50 cm is proposed as an alternative distance of firing. More
formally, denote the competing propositions as follows: 'The questioned pattern was shot from
a distance of 30 cm (H_p)' and 'The questioned pattern was shot from a distance of 50 cm (H_d)'.

Notice that the approach described here does not lead scientists to propose any specific
value for the distance from which the questioned pattern was shot. Rather, observations serve
the purpose of discriminating between a chosen pair of discrete propositions. The choice of
these propositions may follow from the positions put forward by, respectively, the prosecution
and the defence. An advantage of this approach is that the scientist may conduct carefully
designed experiments, that is firings under controlled conditions (i.e. at particular distances),
so as to obtain suitable data for subsequent evaluation of observations.

A likelihood ratio V can be formulated in order to express the degree to which observations
favour one proposition over another. For the scenario considered here, V can take the form of
a ratio of two Normal probability densities, denoted $f_i(y|\mu, \sigma^2)$, where μ and σ^2 are replaced
by estimates appropriate to the propositions H_i, where $i = \{p, d\}$:

$$V = \frac{f_p\left(y|\bar{x}_{D_{30}}, \sigma^2_{D_{30}}\right)}{f_d\left(y|\bar{x}_{D_{50}}, \sigma^2_{D_{50}}\right)} = \frac{f_p(400|500, 80^2)}{f_d(400|350, 60^2)} = \frac{0.0023}{0.0047} \approx 0.49. \tag{8.13}$$

The numerator represents the probability density at the point $y = 400$ when the distribution
is $D_{30} \sim N(500, 80^2)$. Analogously, the denominator is given by the probability density at the

point $y = 400$ when the distribution is $D_{50} \sim N(350, 60^2)$. In the current example, the particle count supports, weakly, H_d, the proposition according to which the questioned pattern was shot from a distance of 50 cm. The likelihood ratio supporting H_d is V^{-1}, or, approximately 2.

More generally, the likelihood ratio for a given observation $Y = y$ (assuming Y is normally distributed) and propositions $H_1 : N(\mu_1, \sigma_1^2)$ and $H_2 : N(\mu_2, \sigma_2^2)$ may be written:

$$\frac{\sigma_2}{\sigma_1} \exp \frac{1}{2} \left[\left(\frac{y - \mu_2}{\sigma_2} \right)^2 - \left(\frac{y - \mu_1}{\sigma_1} \right)^2 \right].$$

An evaluation of alternative scenarios shows that the value of V depends crucially on the specified propositions, the parameters of the associated probability density functions and on the observations (i.e. the number of visualized GSR particles). It is, for example, little surprising that the likelihood ratio given by (8.13) is small. The distributions of D_{30} and D_{50} partially overlap, and the number of GSR particles of the questioned pattern lies in the interval of overlap. For illustration, contrast this finding with the value of the likelihood ratio when the number of particles found in the questioned pattern is 250. The two propositions compared are as before. The likelihood ratio becomes

$$V = \frac{f_p\left(250|\bar{x}_{D_{30}}, \sigma_{D_{30}}^2\right)}{f_d\left(250|\bar{x}_{D_{50}}, \sigma_{D_{50}}^2\right)} = \frac{f_p(250|500, 80^2)}{f_d(250|350, 60^2)} = \frac{3.78 \times 10^{-5}}{1.66 \times 10^{-3}} = 0.0228.$$

Here, the findings still support H_d, but the value has increased to $1/0.0228 \approx 44$. When, for example, the propositions H_p and H_d considered are D_{50} and D_{70}, respectively, V becomes

$$V = \frac{f_p\left(250|\bar{x}_{D_{50}}, \sigma_{D_{50}}\right)}{f_d\left(250|\bar{x}_{D_{70}}, \sigma_{D_{70}}\right)} = \frac{f_p(250|350, 60^2)}{f_d(250|120, 35^2)} = \frac{1.66 \times 10^{-3}}{1.15 \times 10^{-5}} = 144.$$

Note that it is not always necessary to have two discrete propositions for comparison.

8.4.4.3   Bayesian network for evaluating residue particles

The procedure described in Section 8.4.2 for evaluating residue particles may be implemented in a Bayesian network (Biedermann and Taroni 2006). Consider this in terms of two variables D and Y that provide the nodes for such a network. The node D is discrete and represents the various propositions relating to the distance from which a questioned pattern may have been shot. In this example, there are four states: D_{30}, D_{50}, D_{70} and D_{90}, denoting distances of firing of, respectively, 30, 50, 70 and 90 cm. The node Y is continuous and represents the quantity of particles observed in a questioned pattern.

The node table of D contains probabilities that are not conditioned on any other variable (except on the general framework of circumstances). They represent prior beliefs held about the distance from which the questioned pattern was shot. Assume that, a priori, there is no preference for any of the four distances of interest so that $Pr(D_{30}) = Pr(D_{50}) = Pr(D_{70}) = Pr(D_{90}) = 0.25$. Note that this is a discrete probability distribution.

For the time being, the continuous variable Y is introduced here without providing any further details. Chapter 9 covers further details on the technicalities associated with continuous variables. The table of the node Y requires means and variances given each of the possible states of the parent node D. Table 8.5 proposes exemplary values for illustration, in agreement

Table 8.5   Means, μ_i, and variances, σ_i^2, of the probability density functions assigned to the variable Y, the number of particles of gunshot residue for various values of D, represented by D_i, where i is the distance in centimetres from which the gun was fired.

i		30	50	70	90
D:		D_{30}	D_{50}	D_{70}	D_{90}
Y:	mean (μ_i)	500	350	120	55
	variance (σ_i^2)	6400	3600	1225	400

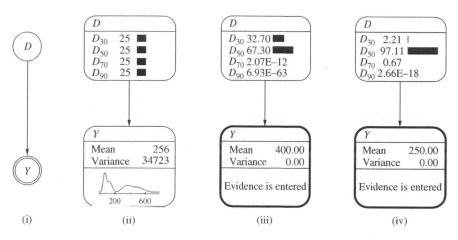

(i)   (ii)   (iii)   (iv)

Figure 8.5   Bayesian network for inference about the distance of firing: (i) abstract representation of the graphical structure (double borders are used to represent a continuous variable and a single circle is used to represent a discrete variable); (ii) initial state of the numerically specified network; note that the marginal distribution of Y is multimodal; and (iii) and (iv) states of the network after specifying observations at the node Y (shown with a bold border), that is, observation of 400 and 250 GSR, respectively.

with the values from Table 8.4. As the quantity of visualized GSR particles is thought to depend on the distance of firing, one can choose node D as a parent node for the node Y. Figure 8.5(i) illustrates this graphically.

Figure 8.5(ii) displays the Bayesian network in its initial state, that is when no evidence is entered. All four possible states of the variable D show the same probability. The variable Y indicates the mean number of GSR particles that may be expected to be found, given the specified prior beliefs. This value is given by

$$E(Y) = \mu_{30}Pr(D_{30}) + \mu_{50}Pr(D_{50}) + \mu_{70}Pr(D_{70}) + \mu_{90}Pr(D_{90}) \,. \tag{8.14}$$

This is a particular example of a general result that given two dependent random variables Y and X, then $E(Y) = E(E(Y|X))$. Further use of this result is made in Chapter 9 (Equation (9.8)). Note also that whilst the distribution of Y, conditional on a particular value of D, is Normal, the marginal distribution of Y is multimodal.

Consider some of the properties of the proposed Bayesian network when specifying particular observations. When instantiating the variable D to any of its four possible states, then the node Y displays the parameters of the respective Normal distribution specified in Table 8.5. For example, if $Pr(D = D_{90}) = 1$, then the node Y indicates a Normal distribution with a mean of 55 and a variance of 400. The value 55 is also obtained by (8.14), which reduces to $E(Y) = \mu_{90} = 55$ [due to $Pr(D = D_{90}) = 1$].

Next, consider an inference in the opposite direction. Figure 8.5(iii) shows a situation with 400 observed particles. This finding is entered at the node Y. The node D displays the posterior probability of each hypothesis given the observation. For example, the probability associated with the state D_{30} is the posterior probability of the distance of firing being 30 cm, given that 400 particles have been observed, formally written $Pr(D = D_{30}|y = 400)$. The latter probability can be obtained as follows:

$$Pr(D = D_{30}|y = 400) = \frac{f(y|D_{30})Pr(D_{30})}{\sum_i f(y|D_i)Pr(D_i)}, \quad \text{with } i = (30, 50, 70, 90)$$

and $f(y|D_i)$ being the probability density function for y at distance D_i. The value for $Pr(D = D_{30}|y = 400)$ thus obtained is 0.3270. Analogously, one can obtain the posterior probabilities for D_{50}, D_{70} and D_{90}.

Consider, for example, the ratio of the posterior probabilities of, respectively, D_{30} and D_{50} given y, which yields $0.327/0.673 = 0.49$. This value corresponds to the likelihood ratio obtained in (8.13). Notice however that this agreement requires equal prior probabilities for D_{30} and D_{50}. Another likelihood ratio found above focussed on 250 visualized particles. Such a situation is shown in Figure 8.5(iv).

8.4.4.4   Combining results of comparative examinations of marks and visualized gunshot residues

The procedure described in Section 8.4.3 focusses on discriminating between a particular set of discrete hypotheses. Roughly speaking, the procedure combines observed data, that is the quantity of visualized GSR particles, with previous knowledge about various possible distances of firing. It thus helps rank hypotheses according to their credibility.

Consider now a scenario where a forensic firearms specialist would be called on to conduct examinations on submitted items. Imagine that one of the available elements is a bullet extracted from a dead body. Assume further that only one bullet has been fired during the incident and that the particles around the entrance hole on the victim's clothing originate from that firing (i.e. they are GSR). Besides, a gun is available from a suspect. The scientist uses this weapon to fire bullets under controlled conditions and then compares these bullets with the bullet in question (found in the victim's body). Suppose also that the scientist judges the quality of the observations to be sufficient for reaching the (subjective) opinion that the incriminated bullet has been fired from the suspect's gun. The discussion introduces such a categoric conclusion at this point only for illustrating further argument in the following, as well as possible limitations.

In a subsequent part of the examinations, the scientist may proceed with experiments designed to inform about the range of firing. Given available ammunition thought to come from the same lot as that used to fire the incriminated pattern, the scientist can use the suspect's weapon because it is believed to have fired the questioned pattern (as noted in the previous paragraph), for firings at known distances and under controlled conditions. After that, the scientist would compare the GSR patterns thus obtained with the pattern surrounding the entrance hole on the victim's clothing. Throughout such an examination, a procedure described in terms of the model in Figure 8.5 can assist the scientist with the inferential task associated with these experiments.

So far, the evaluation of the results of the mark comparison was considered quite independently from the evaluation of the observed quantity of particles, and vice versa. Usually, scientists would not proceed with an assessment of a range of firing until they have reached a given subjective degree of belief that some specified weapon is actually the one used to fire an incriminated pattern. However, this opinion is essentially probabilistic in nature, in the sense that there remains, usually, some uncertainty about whether the suspect's weapon is in fact the one used during the shooting incident. Arguably, it may be felt that this uncertainty should be accounted for in a range-of-fire evaluation. This represents an interesting issue in the joint evaluation of findings concerning marks and residue particles.

Consider the Bayesian network shown in Figure 8.5(i). For each of the potential distances of firing (node D), the node Y models the number of visualized particles on the target by a probability distribution. An implicit assumption of this approach is that the particle count is assessed only under selected propositions for distances of firing. For example, a GSR count y is assessed under the proposition that the distance of firing was 30 cm ($D = D_{30}$) and under the proposition that the distance of firing was 50 cm ($D = D_{50}$). However, both settings suppose that the suspect's weapon is in fact the one that fired the questioned pattern. This is an assumption that may not necessarily reflect the defence position. More generally, the proposition that the suspect's weapon has fired the incriminated pattern may be doubted to varying degrees, in particular when the results for the comparative examinations of marks provide only weak support.

This inconvenience may be overcome with the addition of a further node, F say. F is a binary node that takes two values F_s and $\bar{F}_s$. These represent, respectively, the propositions 'The bullet was fired by the suspect's weapon' and 'The bullet was fired by another weapon'. The node F acts as a parent for the node Y so that the Bayesian network shown in Figure 8.5(i) becomes $D \rightarrow Y \leftarrow F$ as shown in Figure 8.6(i). Consequently, the probability table of the node Y extends as given in Table 8.6. When the first four columns of this table correspond to

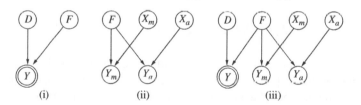

(i)   (ii)   (iii)

Figure 8.6 Bayesian networks for (i) evaluating distances of firing based on observed particles, (ii) evaluating marks given source level propositions and (iii) the joint evaluation of GSR and marks. Node Y is a continuous node representing the number of residue particles found from a firing. Node D is a discrete node for the firing distance, with four states, representing distances 30, 50, 70 and 90 cm. Node F has two states, F_s and $\bar{F}_s$, the bullet was or was not fired from the suspect's weapon, respectively. Factors Y_m, Y_a, X_m and X_a denote the observations made on, respectively, the incriminating bullet (Y_m, Y_a) and the bullets (X_m, X_a) from a known source. The subscripts m and a indicate that the marks originate from the firearm's features of manufacture and acquired characteristics, respectively.

Table 8.6   Parameters of the Normal distributions associated with the variable Y.

F:	F_s				$\bar{F}_s$			
i	30	50	70	90	30	50	70	90
D:	D_{30}	D_{50}	D_{70}	D_{90}	D_{30}	D_{50}	D_{70}	D_{90}
Y: mean (μ_i)	μ_{30}	μ_{50}	μ_{70}	μ_{90}	μ'_{30}	μ'_{50}	μ'_{70}	μ'_{90}
variance (σ_i)	σ^2_{30}	σ^2_{50}	σ^2_{70}	σ^2_{90}	σ'^2_{30}	σ'^2_{50}	σ'^2_{70}	σ'^2_{90}

Means, μ_i, μ'_i, and variances, σ^2_i, σ'^2_i, of the probability density functions assigned to the variable Y, the number of particles of gunshot residue for various values of D, represented by D_i, where i is the distance in centimetres from which the gun was fired. Proposition F has two states, F_s and $\bar{F}_s$, the bullet was or was not fired from the suspect's weapon.

Table 8.5, then, with F instantiated to F_s, the Bayesian network $D \rightarrow Y \leftarrow F$ yields strictly the same results as the network $D \rightarrow Y$ [Figure 8.5(i)].

The parameters specified for Y given $\bar{F}_s$ characterize the Normal distributions for Y when a firearm other than that associated with the suspect was used to fire the incriminating bullet. Note that these parameters are labelled with a ' ' ' as they may be different from the parameters assumed for Y given F_s. The parameters for Y given $\bar{F}_s$ tend to affect the evaluation of some observation y if, for a given distance D_i of firing, the parameters of the Normal distributions specified for Y given F_s and $\bar{F}_s$ differ, and the probability of node F being in state F_s is different from one. If F_s is assumed known to be true, then the evaluation is not affected by whatever value is specified for Y given $\bar{F}_s$. This is how uncertainty about the variable F enters the process of evaluation.

To illustrate this further, write f_0 and f_1 for the prior probabilities for, respectively, F_s and $\bar{F}_s$. Then, the mean number of GSR particles expected to be found, given prior beliefs specified for the variables D and Y, is given by

$$E(Y) = E(E(Y|D))f_0 + E(E(Y|D))f_1$$
$$= \sum_i \mu_i Pr(D_i)f_0 + \sum_i \mu'_i Pr(D_i)f_1$$
$$= \sum_i Pr(D_i)(\mu_i f_0 + \mu'_i f_1), \quad \text{for } i = (30, 50, 70, 90) \ .$$

If one can accept that uncertainty may exist with respect to the variable F, the proposition according to which the suspect's weapon has fired the incriminated bullet, then it may be desirable to inform one's belief about the truth or otherwise of that variable, based on other forensic findings. The results of the comparative mark examinations represent a suitable element for this task. Imagine thus that the scientist performs comparative examinations between the questioned bullet and bullets fired (under controlled conditions) through the barrel of the suspect's weapon. Such an examination process leads to observed similarities and differences. Following the previous discussion, one can assess such observations given source level propositions using the Bayesian network constructed and discussed in Section 6.2.5 [partially reproduced in Figure 8.6(ii)]. The nodes Y_m, Y_a, X_m and X_a in this model refer to the observations made on, respectively, the incriminating bullet (Y_m, Y_a) and the bullets (X_m, X_a) from a known source, such as a suspect's weapon. The subscripts m and a indicate that the marks originate from, respectively, the firearm's features of manufacture and acquired characteristics.

Figure 8.6(iii) depicts a combination of the networks for evaluating GSR and marks evidence. The two network fragments are combined by the binary node F, defined as 'The questioned bullet has been fired by the suspect's weapon'. As mentioned earlier, the importance of incorporating F when evaluating observed residue particles is that one may account for the uncertainty about whether the weapon used for the firings at known distances, that is the suspect's weapon, is in fact the one used to fire the incriminated pattern. A particular feature of this network thus is that one's beliefs about the truthstate of F does not necessarily need to be a categoric assumption. In fact, the model allows one to rely coherently on knowledge available from comparative examinations of marks.

The Bayesian network shown in Figure 8.6(iii) can cope with the analysis of quite different positions that may be held by the prosecution and the defence. For example, the prosecution's case may be that the suspect's weapon has fired the incriminated bullet (F_s) and that the distance of firing was about 30 cm. The defence may argue that a weapon other than that of the suspect was used ($\bar{F}_s$) and that the distance of firing was about 70 cm. In such a setting, the likelihood ratio in this model for some observation $Y = y$ is given by

$$
\frac{\sigma'_{70}}{\sigma_{30}} \exp \frac{1}{2} \left[\left(\frac{y - \mu'_{70}}{\sigma'_{70}} \right)^2 - \left(\frac{y - \mu_{30}}{\sigma_{30}} \right)^2 \right].
$$

The proposed Bayesian network clarifies the potential scenarios as to how the questioned pattern may have been produced, and the assumptions made by the expert during the evaluative process. A difficulty however is that experts may only have limited knowledge about the possible outcomes in the various scenarios. In particular, it may be difficult for experts to specify values for the parameters assigned to Y given $\bar{F}_s$ in Table 8.5, which relates to the kind of particle patterns produced by firearms other than that of the suspect. In some situations, scientists may however be able to make reasonable assumptions about, for example, values for μ' and σ' in order to describe the particle deposition of an alternative weapon. This is typically the case when such a weapon is available for firings under controlled conditions. Otherwise, they may make the explicit assumption that it was the suspect's weapon that has been used to fire the incriminated pattern.

8.4.5 Comments

In the examples discussed through Sections 8.4.1–8.4.3, two trace items have in each example been used to infer something about an ultimate probandum. This represents a main difference with respect to the scenario involving GSR and marks treated in Section 8.4.4, where analyses focussed on inferences about different propositions; that is, results of mark examinations are used to infer something about whether the suspect's weapon has been used to fire an incriminating bullet. In turn, information on the GSR pattern is used to infer something about the distance of firing. The evaluation is interrelated because of propositions that are relevant for both trace types.

It may be felt that the counting of single GSR particles is too laborious a task to be used in practice. Also, scientists may consider that the phenomenon of particle transfer on a target is not directly amenable to a particular numerical description. As far as chromophoric tests are concerned, examiners may prefer a visual inspection based on characteristics such as the spread and density of an incriminated pattern. However, the general questions that need to be addressed to evaluate such findings follow the same logic. Scientists may need to assess whether their observations are more 'compatible' with one distance of firing rather than with another. Here, experts would rely on an essentially qualitative expression of their findings. Bayesian networks can also address inferences on such a level of detail. Further details on this topic are given in Section 13.1.1.

The approach to distance evaluation discussed in Section 8.4.4, based on a Bayesian network, avoids categorical statements about particular distances of firing. In the proposed model (Figure 8.6), the probative value of observations (i.e. a GSR count) depends on specified distributional assumptions. The output of this procedure consists of an indication as to which propositions are favoured by the observations, which is a common way to look at evidence in Bayesian data analysis (D'Agostini 2003).

9

Networks for continuous models

So far events and counts and the probability of their occurrence have been discussed. These ideas may be extended to consider measurements about which there may be some uncertainty or randomness. In certain fairly general circumstances, the way in which probability is distributed over the possible values for the measurements can be represented, mathematically, by functions known as *probability distributions*. The most well-known distribution for measurements is that of the Normal distribution. This will be described in Section 9.1.1. Before Normal probability distributions can be discussed here, however, certain other concepts have to be introduced.

9.1 Random variables and distribution functions

The concept of a *random variable* [or *random quantity* or *uncertain quantity*, Lindley (1991)] needs some explanation. There has been some discussion on this earlier in Sections 1.1.9 and 2.1.2. A statistician draws conclusions about a population. This may be done by conducting an experiment or studying a population. The possible outcomes of the experiment or the study are known as the sample space Ω. The outcomes are unknown in advance of the experiment or the study and, hence, are said to be variable. It is possible to model the uncertainty of these outcomes because of the randomness associated with them. Thus, the outcome is considered as a random variable. Each random variable has a set of possible states which might be categorical (the random variable has category labels as possible values) or numerical (the random variable has numbers as possible values). A random variable, in a rather circular definition, is something which varies at random. For example, there is variation in the elemental concentrations between glasses, or in the alcohol concentration in blood.

When considering data in the form of counts, the variation in the possible outcomes is represented by a function known as a *probability mass function*, often denoted $Pr(x)$, which shows how the probabilities associated with the random variable are spread over the sample space. The variation in measurements x, which are continuous, may also be represented

---

Bayesian Networks for Probabilistic Inference and Decision Analysis in Forensic Science, Second Edition.
Franco Taroni, Alex Biedermann, Silvia Bozza, Paolo Garbolino and Colin Aitken.
© 2014 John Wiley & Sons, Ltd. Published 2014 by John Wiley & Sons, Ltd.

mathematically by a function, known as a *probability density function*, often denoted $f(x)$. Both probability mass functions and probability density functions are examples of probability models. Note that a value $f(x)$ of a probability density function f is not a probability. In order to consider a probability for a measurement x (with associated random variable X), another function of x is needed. This is the probability distribution function, conventionally denoted $F_X(x)$, and defined by

$$F_X(x) = Pr(X \leq x) = \int_{-\infty}^{x} f(x)dx.$$

Thus, the probability that a random variable X takes values in any interval (a, b) can be computed as

$$Pr(a \leq X \leq b) = Pr(X \leq b) - Pr(X \leq a) = F_X(b) - F_X(a) = \int_{a}^{b} f(x)dx.$$

This can be seen as the area underlying the density function $f(x)$ corresponding to the interval (a, b), as it is shown in Figure 9.1.

Note that from this result $Pr(X = a) = Pr(a \leq X \leq a) = 0$, the probability that a continuous random variable X is equal to a given value x is null. In fact, the probability of a given point x can be seen as the area underlying the density function corresponding to an interval of zero length, which is zero. Often, the function representing the variation of a random variable is referred to as its probability distribution (with the thought of the probability being *distributed* over its sample space).

Any function of x can be a probability density function, provided it satisfies two conditions: first, that it is nonnegative, $f(x) \geq 0$ for all x, and second, that it integrates to 1 over the sample space of x, so that $\int_{-\infty}^{\infty} f(x)dx = 1$. Note that these conditions allow for a probability density function to take values greater than 1. Fortunately, the variation in measurements, or simple transformations of them, of many natural phenomena can be represented by one of a few well-studied functions, such as the Normal probability function (Section 9.1.1).

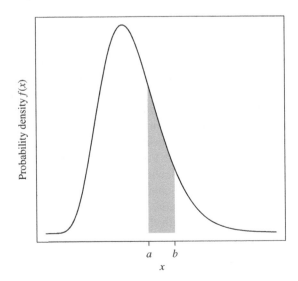

Figure 9.1 Probability density function of the random variable X. The grey shaded area represents the probability of the interval $[a, b]$.

Notation is useful in the discussion of random variables. Rather than writing out in long-hand phrases such as 'the concentration of silicon in a fragment of glass' the phrases may be abbreviated to a single uppercase Roman letter. For example, let X be short for 'the concentration of silicon in a fragment of glass' and the phrase 'the probability the concentration of silicon in a fragment of glass is less than 30' may be written as $Pr(X < 30)$, or more generally as $Pr(X < x)$ for 'the probability the concentration of silicon in a fragment of glass is less than x' for a general value x of the concentration of silicon.[1]

The mean of a random variable is the corresponding population mean. In the examples above, this would be the mean refractive index of the population of all fragments of glass or the mean concentration of silicon in glass (both conceptual populations). The mean of a random variable is given a special name, the *expectation*, and for a random variable, X say, it is denoted $E(X)$. Similarly, the variance of a random variable is the corresponding population variance. For a random variable X, it is denoted $Var(X)$. The standard deviation of a random variable is the positive square root of the variance: being expressed in the same measurement unit of the random variable, it is easier to interpret. The mean may be thought of as a measure of location to indicate the size of the measurements. The standard deviation may be thought of as a measure of dispersion to indicate the variability in the measurements.

9.1.1   Normal distribution

The most well-known probability function is the *Normal* or *Gaussian* distribution, named after the German mathematician Carl Friedrich Gauss (1777–1855). The variation of a random variable with a Normal distribution is such that it may be represented by a probability density function which is unimodal, symmetric and bell shaped. Two and only two characteristics of the measurement are required to define the Normal probability density function. These are the mean, or expectation, θ, and the variance, σ^2. Given these parameters, the Normal probability density function for x, $f(x|\theta, \sigma^2)$, is given by

$$f(x|\theta, \sigma^2) = \frac{1}{\sqrt{2\pi\sigma^2}} \exp\left\{ -\frac{(x-\theta)^2}{2\sigma^2} \right\}. \tag{9.1}$$

The function $f(x|\theta, \sigma^2)$ is a function which is symmetric about θ. It takes its maximum value when $x = \theta$, it is defined on the whole real line for $-\infty < x < \infty$ and is always positive. The area underneath the function is 1 from the law of total probability, since x has to lie between $-\infty$ and ∞. In general, a Normally distributed measurement, X say, with mean θ and variance σ^2, is denoted

$$(X|\theta, \upsilon^2) \sim N(\theta, \sigma^2),$$

or with the abbreviated notation where the dependence of X on θ and σ^2 is not made explicit

$$X \sim N(\theta, \sigma^2). \tag{9.2}$$

The first symbol within parentheses on the right-hand side of the expression conventionally denotes the mean, the second conventionally denotes the variance. Explicit mention of the dependence of X on θ and σ^2 is not always necessary, and such abbreviated notation will be used often. As an example of a probability model for a continuous measurement,

---

[1] Upper case letters denote a random variable and lower case letters denote what is known as a *realization* of the random variable, which is a particular value of the random variable.

consider the estimation of the quantity of alcohol in blood. From experimental results, it has been determined that there is variation in the measurements, x (in g kg^{-1}), provided by a certain procedure.

Example 9.1 *(Alcohol concentration in blood) In some countries, a person is considered to be under the influence of alcohol if the alcohol level in blood is greater than 0.8 g kg^{-1}. The variability inherent in a measurement, X, with a particular type of instrument, of the quantity of alcohol is known from previous experiments to be such that it is Normally distributed with variance σ^2 of 0.005, about the true value θ of the blood alcohol level of a person. The distribution of X is, denoted X $\sim$ N(θ, 0.005). The true value θ is taken to be the expectation of X. The variation σ^2 in the measurements is taken to be constant over all instances of use of the measuring instrument; it is independent of θ and over time. Also, the instrument is assumed to be unbiased since the probability model assumes E(X) = θ. Note that a random variable whose expectation is not equal to the parameter for which it is an estimate is said to be biased. If the measurements from the instrument had a mean value greater or less than the true value the instrument would be said to be biased. Next, consider a person whose actual quantity, θ, of alcohol in the blood is 0.7 g kg^{-1}. The probability density function $f(x|\theta, \sigma^2)$ for the measurement of the quantity of alcohol in the blood is then obtained from (9.1) with the substitution of 0.7 for θ and 0.005 for σ^2. The function is illustrated in Figure 9.2. Note the labelling of the ordinate as 'Probability density $f(x)$'. This figure provides an illustration of a probability density function which takes values greater than 1.*

There is a special case of the Normal distribution in which the mean is zero ($\theta = 0$) and the variance is one ($\sigma^2 = 1$). The Normal probability density function is then

$$f(z|0, 1) = \frac{1}{\sqrt{2\pi}} \exp\left(-\frac{z^2}{2}\right),$$

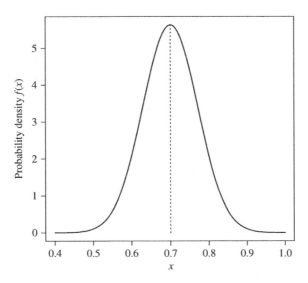

Figure 9.2 Probability density function $f(x)$ for a Normal random variable X (with realization x), with mean 0.7 and variance 0.005.

where z is used instead of x to denote the special nature of a Normally distributed random variable with zero mean and unit variance. A random variable with this distribution is said to have a standard Normal distribution. If a random variable Z has a standard Normal distribution, it is denoted as

$$Z \sim N(0, 1),$$

a special case of (9.2) where the conditioning on $\theta = 0$ and $\sigma^2 = 1$ on the left-hand side has been omitted for clarity.

The determination of probabilities associated with general Normally distributed random variables is made possible by a process known as *standardization*, whereby a general Normally distributed random variable $X \sim N(\theta, \sigma^2)$ is transformed into one which has a standard Normal distribution. Let

$$Z = (X - \theta)/\sigma.$$

Then $E(Z) = 0$ and $Var(Z) = 1$ and the random variable Z has a standard Normal distribution.[2] Note that standardization requires variability, as represented by σ, to be taken into account. For example, the division by σ ensures that the resulting statistic is dimensionless. Consider the measurements on blood alcohol concentration in units of gram per kilogram. Both the numerator and the denominator of $(X - \theta)/\sigma$ have these units, so the resultant statistic Z is dimensionless.

The standard Normal probability density function is so common that it has special notation. Let $Z \sim N(0, 1)$. Then the probability distribution function $F_Z(z) = Pr(Z < z)$ is denoted $\Phi(z)$. However, the probability that a Normally distributed random variable lies in a certain interval cannot be determined analytically. Reference has to be made to statistical packages such as R (R Core Team 2013), from which probabilities for the general Normal distribution may be determined directly, or to tables of probabilities of the standard Normal distribution, from which probabilities for the general Normal distribution can be determined. Certain values of z are used commonly in the discussion of significance probabilities, particularly those values for which $1 - \Phi(z) = Pr(Z > z)$ is small, and some of these are tabulated in Table 9.1.

Corresponding probabilities for negative or absolute values of Z may be deduced from the tables by use of the symmetry of the Normal distribution. By symmetry,

$$\Phi(-z) = Pr(Z < -z) = 1 - Pr(Z < z) = 1 - \Phi(z).$$

Thus,

$$
\begin{aligned}
Pr(|Z| < z) &= Pr(-z < Z < z) \\
&= Pr(Z < z) - Pr(Z < -z) \\
&= \Phi(z) - \Phi(-z) \\
&= 2\Phi(z) - 1.
\end{aligned}
$$

---

[2] Note that this result requires the intermediate result that a linear function of a Normally distributed random variable is also Normally distributed (i.e. the family of Normal probability distributions is said to be closed under linear transformations). A linear function $W = aX + b$ of a Normal random variable X, with $X \sim N(\mu, \sigma^2)$, is still Normally distributed with mean equal to $a\mu + b$ and variance equal to $b^2\sigma^2$, that is, $W \sim N(a\mu + b, a^2\sigma^2)$. As an example to illustrate that such a result is not always the case, consider a binomial random variable Y which takes integer values on the set $\{0, 1, 2, \ldots, n\}$. A linear function $aY + b$ need not necessarily be binomially distributed. For example, it may be neither integer nor nonnegative.

Table 9.1   Values of probability distribution function $\Phi(z)$ and its complement $1 - \Phi(z)$ and absolute values for the standard Normal distribution for given values of z.

z	$\Phi(z)$	$1 - \Phi(z)$	$Pr(\|Z\| < z)$ $= 2\Phi(z) - 1$	$Pr(\|Z\| > z)$
1.6449	0.950	0.050	0.90	0.10
1.9600	0.975	0.025	0.95	0.05
2.3263	0.990	0.010	0.98	0.02
2.5758	0.995	0.005	0.99	0.01

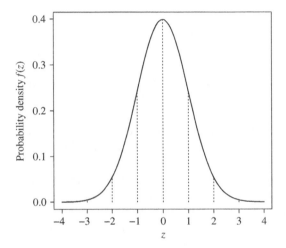

Figure 9.3   Standard Normal probability density. The dotted vertical lines indicated the positions of the mean $z = 0$ and the symmetric probability intervals of approximate width $0.68(-1, 1)$, $0.95(-2, 2)$ and $0.997(-3, 3)$.

Furthermore, the following standard results are often implemented in practice. Given a Normal random variable $X \sim N(\theta, \sigma^2)$, then

$$Pr(\mu - \sigma < X < \mu + \sigma) = Pr(-1 < Z < 1) \simeq 0.68$$

$$Pr(\mu - 2\sigma < X < \mu + 2\sigma) = Pr(-2 < Z < 2) \simeq 0.95$$

$$Pr(\mu - 3\sigma < X < \mu + 3\sigma) = Pr(-3 < Z < 3) \simeq 0.997.$$

Figure 9.3 shows the standard Normal probability density along with these key features.

Example 9.2 *(Alcohol concentration in blood, continued) Consider again Example 9.1 where the alcohol concentration X in blood is assumed Normally distributed such that $X \sim N(\theta, \sigma^2 = 0.005)$, where θ is the true unknown blood alcohol level of the person. A person is stopped because of suspicion of driving under the influence of alcohol. Assume that the true level of alcohol in this person's blood is $1g\ kg^{-1}$. What is the probability that a measurement obtained* *from some of the blood of this person, submitted to a forensic laboratory, is greater than 0.8 g kg⁻¹? This can be computed by the process of standardization, that is,*

$$Pr(X > 0.8) = Pr\left(\frac{X - \theta}{\sigma} > \frac{0.8 - 1}{\sqrt{0.005}}\right)$$
$$= Pr(Z > -2.828) \simeq 0.998.$$

This probability value can be easily obtained using the software R, *by typing the command*

$$\text{pnorm}(x,m,s,lower.tail=FALSE),$$

where x *is equal to 0.8,* m *is equal to 1, and* s *is equal to* $\sqrt{0.005}$ *to give,* pnorm(0.8, 1, sqrt(0.005),lower.tail=FALSE). *Note that* sqrt(0.005) *is used directly. An alternative approach by which* $\sqrt{0.005}$ *is calculated separately and a rounded result entered explicitly into the* pnorm *command can lead to a rounding error in the final result.*

It should also be noted that the result for Pr(X > 0.8) is not what is sought in practice. The value for Pr(X > 0.8) is the probability of a person whose true blood alcohol level θ is 1 g kg⁻¹ giving a reading X greater than 0.8. In practice, the court is interested in the probability that a person whose reading X is greater than 0.8 has a true blood alcohol level θ greater than 0.8. This requires a prior probability model for θ with considerable subjective input. Use of Bayes theorem with the prior probability distribution for θ and the distribution of the measurement X, given θ, gives a posterior probability distribution of θ, given X, from which a probability that θ > 0.8 may be obtained.

As well as representing the variation of the measurements (or transformation of the measurements, e.g. with a logarithmic transformation) of many natural phenomena, the Normal distribution is important because of its role in a theorem known as the *Central Limit Theorem* (CLT). The CLT states that for a random variable X with a mean θ and a variance σ^2, the distribution of a random variable X, corresponding to the mean of a sample of size n from the original distribution such that $\bar{X} = \sum_{i=1}^{n} X_i/n$, tends to a Normal distribution with mean θ and variance σ^2/n, regardless of the original distribution of X.

9.1.2 Bivariate Normal distribution

In many experimental situations, the value of more than one random variable may be observed. Multiple observations could arise because several characteristics are measured on each person or item of interest. For example, in blood taken from a person suspected of driving under the influence of alcohol (with associated measurement of blood alcohol concentration X), other, possibly illegal, substances may also be measured [e.g. the quantity of tetrahydrocannabinol (THC)]. Thus, a new variable, Y, the quantity of THC may be defined. A two-dimensional random vector (X, Y) is then associated with each examined (blood) item. The most commonly used distribution for multivariate data is the multivariate Normal (MVN) distribution. The central role played by the MVN distribution can be explained because of its tractability and its role in the CLT . The CLT shows that the random variable corresponding to the mean of a sample of size n from an arbitrary multivariate distribution, for which a mean and a variance exist, has a multivariate normal distribution (Press 1982).

Some general results are introduced here for use later. Let (X, Y) be a bivariate random vector with a Normal probability distribution, mean vector $\theta = (\theta_x, \theta_y)$ and covariance matrix

$$\Sigma = \begin{pmatrix} \sigma_x^2 & \rho\sigma_x\sigma_y \\ \rho\sigma_x\sigma_y & \sigma_y^2 \end{pmatrix},$$

where σ_x and σ_y are the standard deviations of X and Y, respectively, and ρ is the correlation coefficient between X and Y. The associated probability density function $f(x, y)$ is a bivariate function and is given by

$$f(x, y) = \frac{1}{2\pi\sigma_x\sigma_y\sqrt{1 - \rho^2}} \exp\left\{ -\frac{1}{2(1 - \rho^2)} \left[\left(\frac{x - \theta_x}{\sigma_x}\right)^2 \right.\right.$$
$$\left.\left. -2\rho\left(\frac{x - \theta_x}{\sigma_x}\right)\left(\frac{y - \theta_y}{\sigma_y}\right) + \left(\frac{y - \theta_y}{\sigma_y}\right)^2 \right] \right\}. \quad (9.3)$$

The marginal distributions are easily derived with use of the result that the marginal distribution of any subvector of a Normally distributed random vector, is also Normally distributed. In particular, starting from the probability density function (Equation 9.3), X and Y have Normal distributions given by

$$X \sim N(\theta_x, \sigma_x^2), \quad (9.4)$$
$$Y \sim N(\theta_y, \sigma_y^2). \quad (9.5)$$

The conditional distribution of X, given $Y = y$, is also a Normal distribution with mean $\theta_{x|y}$, a linear function of y, such that

$$\theta_{x|y} = \theta_x + \rho\frac{\sigma_x}{\sigma_y}(y - \theta_y), \quad (9.6)$$

and variance $\sigma_{x|y}^2$ is given by

$$\sigma_{x|y}^2 = \sigma_x^2(1 - \rho^2). \quad (9.7)$$

9.1.3   Conditional expectation and variance

As explained previously, it is possible to have a situation in which one random variable, Y, for example, is dependent on another random variable, X, for example. This situation is different from the more familiar linear regression model in which there is one random variable which is dependent on a fixed variable whose values are known exactly and may be fixed by an experimenter.

Consider Y is dependent on X. It may be of interest to know the unconditional expectation and variance of Y, which is unconditional on the value of the random variable X. The unconditional expectation and variance are given by the following formulae:

$$E(Y) = E_X(E(Y|X)) \quad (9.8)$$

and

$$Var(Y) = E_X(Var(Y|X)) + Var_X(E(Y|X)). \quad (9.9)$$

Note that the subscript X in $E_X(E(Y|X))$ and $E_X(Var(Y|X))$ is introduced to underline that the first expectation $E_X(\cdot)$ is computed with respect to the variable X. The same argument is valid for the subscript X in $Var_X(\cdot)$.

Examples will be given to illustrate these results. Proofs are not given here but may be found in books on probability theory (e.g. O'Hagan 1988). Note that (9.8) has earlier been used in Section 8.4.4.

9.2   Samples and estimates

For any particular type of forensic trace material, the distribution of the characteristic of interest is important. This is so because scientists often need to assess the rarity or otherwise of any particular observation. For the refractive index and elemental composition of glass, for example, the distributions of the refractive index and elemental composition measurements are important. These distributions relate to variability within and between pieces of glass from the same or different populations, such as window glass and car head lamp glass. Sometimes, the population may be a conceptual population only (e.g. the population of all possible fragments of window glass). In practice, these distributions are not known exactly. They are estimated from empirical data, a sample from some background relevant population.

A characteristic of interest from the population is known as a *parameter* (e.g. a population mean). The corresponding characteristic from the corresponding sample, that is, items or members drawn from the population of interest, is known as an *estimate*. It is hoped that an estimate will be a good estimate in some sense. Different samples from the same population will almost certainly produce different estimates. Different results from different samples do not mean that some are wrong and others are right. They merely indicate the natural variability in the distribution of the characteristics amongst different members of the population. Variation is accounted for by including a measure of the variation, known as the *standard deviation*, in the process. The square of the standard deviation is the *variance*. A sample variance may be denoted s^2, and the corresponding population variance σ^2. These terms have been used earlier in Section 8.4.4.

A notational convention uses Roman letters for functions evaluated from measurements from samples and Greek letters for the corresponding parameters from populations. Thus, a sample mean may be denoted $\bar{x}$ and the corresponding population mean θ. A sample standard deviation may be denoted s and the corresponding population standard deviation σ. Another more general notation is the use of the symbol to indicate an estimate of a parameter. Thus an estimate of a population mean θ may be indicated as $\hat{\theta}$, read as 'theta-hat'.

A *statistic* is a function of the data. Thus, the sample mean and the sample variance are statistics. A particular value of a statistic which is determined to estimate the value of a parameter is known as an *estimate*. The corresponding random variable is known as an *estimator*. An estimator, X say, of a parameter, θ say, which is such that $E(X) = \theta$ is said to be unbiased. If $E(X) \neq \theta$, the estimator is said to be biased. The applications of these concepts are now briefly discussed in the context of probability distributions for measurements.

9.2.1   Summary statistics

Consider a population of continuous measurements with mean μ and standard deviation σ. Given a collection of observations $(x_1, x_2, \ldots, x_n)$ of measurements from this population (called here sample data), μ and σ may be estimated from the sample data as follows. The sample mean, denoted $\bar{x}$, is defined by

$$\bar{x} = \sum_{i=1}^{n} x_i / n,$$

(9.10)

where $\sum$ denotes summation such that $\sum_{i=1}^{n} x_i = x_1 + \dots + x_n$. The sample mean $\bar{x}$ can be taken as an estimate of the population mean θ, that is, $\hat{\theta} = \bar{x}$. The sample standard deviation, denoted s, is defined as the square root of the sample variance, s^2, which is itself defined by

$$s^2 = \sum_{i=1}^{n} (x_i - \bar{x})^2 / (n - 1). \tag{9.11}$$

Expression (9.11) can also be calculated as

$$s^2 = \left\{ \sum_{i=1}^{n} x_i^2 - \left(\sum_{i=1}^{n} x_i \right)^2 / n \right\} / (n - 1). \tag{9.12}$$

The sample variance s^2 can be taken as an estimate of the population variance σ^2, that is, $\hat{\sigma}^2 = s^2$. As an example of the calculations, consider the following three measurements of the concentrations of silicon in a fragment of glass ($n = 3$),

$$
\begin{array}{ccc}
x_1 & x_2 & x_3 \\
27.85 & 27.83 & 28.27
\end{array}
$$

Then, from (9.10) with $n = 3$

$$\bar{x} = \sum_{i=1}^{n} x_i / n = (27.85 + 27.83 + 28.27)/3 = 83.95/3 = 27.983.$$

From (9.12)

$$s^2 = \left\{ \sum_{i=1}^{n} x_i^2 - \left(\sum_{i=1}^{n} x_i \right)^2 / n \right\} / (n - 1)$$

$$= \{(27.85^2 + 27.83^2 + 28.27^2) - 83.95^2/3\}/2$$

$$= (2349.32 - 83.95^2/3)/2 = 0.062$$

and the sample standard deviation is

$$s = \sqrt{0.062} = 0.248.$$

Note that the sample mean and standard deviation are quoted to one more decimal place than the original measurements. It is left as an exercise to check that

$$s^2 = \sum_{i=1}^{n} (x_i - \bar{x})^2 / (n - 1)$$

$$= \{(27.85 - 27.983)^2 + (27.83 - 27.983)^2 + (28.27 - 27.983)^2\}/2$$

$$= 0.062.$$

9.2.2   The Bayesian paradigm

Assuming the distribution underlying the generation of observations is known, one of the main purposes of statistical methods is to make an inference about a parameter of interest θ, starting from the collected observations x, whilst probabilistic modelling provides the probability distribution of hypothetical data set, before any observation is taken, conditional on θ. A general description of the inversion of probabilities is given by Bayes' theorem, presented earlier in Section 1.1.6 for the discrete case. A continuous version of Bayes' theorem states that given two random variables X and Y, with conditional distribution $f(x|y)$ and marginal distribution $g(y)$ of Y, the conditional distribution of Y, given X, is

$$g(y|x) = \frac{f(x|y)g(y)}{\int f(x|y)g(y)dy},$$

where the marginal distribution $f(x)$ of X is written as $\int f(x|y)g(y)dy$.

The fundamental element of the Bayesian paradigm states that all uncertainties characterizing a problem must be described by probability distributions (Bernardo and Smith 2000). Statistical inference about a quantity of interest is described as the modification of one's uncertainty about the true value of the quantity given data, and Bayes' theorem specifies how this should be done. Hence, under the Bayesian paradigm, the uncertainty about a parameter θ is modelled through a probability distribution π for θ, called a *prior distribution*, that summarizes the knowledge that is available on the values of θ before the data are obtained. Bayes' theorem allows the prior information on the parameter θ to be updated by incorporating the information contained in the observations (x). Inference is then based on the posterior distribution, denoted $\pi(\theta|x)$, the distribution of θ conditional on x:

$$\pi(\theta|x) = \frac{f(x|\theta)\pi(\theta)}{\int_{\Theta} f(x|\theta)\pi(\theta)d\theta} = \frac{f(x|\theta)\pi(\theta)}{f(x)}, \tag{9.13}$$

where $f(x)$ is the marginal distribution of x. Statistical inference about the parameter θ is based on the modification of the uncertainty about its value in the light of evidence represented by data (measurements) x. It can be observed that if $f(x|\theta)$ is multiplied by a term independent of θ, which may be a function of x, then that term will cancel in the calculation in the right-hand side of (9.13), and the same posterior density for θ will be obtained. Thus, the denominator of (9.13) is often omitted and the following expression is used:

$$\pi(\theta|x) \propto f(x|\theta)\pi(\theta).$$

Consider data in the form of n independent observations, $x_1, \ldots, x_n$ from a Normal distribution, $X \sim N(\theta, \sigma^2)$. The sample distribution can be written as the product of n Normal probability densities, that is,

$$f(x_1, \ldots, x_n|\theta) = \prod_{i=1}^{n} \frac{1}{\sqrt{2\pi}\sigma} \exp\left[-\frac{1}{2\sigma^2}(x_i - \theta)^2\right]. \tag{9.14}$$

Suppose that some background information is available such that a Normal prior distribution may be used for the parameter θ, say $\theta \sim N(\mu, \tau^2)$, with both μ and τ known. Its density is thus given by

$$\pi(\theta) \propto \exp\left[-\frac{1}{2\tau^2}(\theta - \mu)^2\right]$$

where the multiplier $1/\sqrt{2\tau^2}$ has been omitted since it does not include θ. The posterior distribution $\pi(\theta|x_1, \ldots, x_n)$ can be obtained by substituting in (9.13), the Normal likelihood (9.14) and the prior density $\pi(\theta)$, so that

$$\pi(\theta|x_1, \ldots, x_n) \propto \exp\left[-\frac{1}{2\sigma^2}\sum_{i=1}^{n}(x_i - \theta)^2\right]\exp\left[-\frac{1}{2\tau^2}(\theta - \mu)^2\right],$$

where the multipliers $1/\sqrt{2\sigma^2}$ and $1/\sqrt{2\sigma^2}$ have been omitted as they do not include θ. Completing the squares, it follows with a small effort (e.g. Berger 1985; Press 2003) that

$$\pi(\theta|x_1, \ldots, x_n) \propto \exp\left[-\frac{1}{2}\frac{\frac{\sigma^2}{n} + \tau^2}{\frac{\sigma^2}{n}\tau^2}\left(\theta - \left(\frac{\frac{\sigma^2}{n}}{\frac{\sigma^2}{n} + \tau^2}\mu + \frac{\tau^2}{\frac{\sigma^2}{n} + \tau^2}\bar{x}\right)\right)^2\right].$$

The posterior distribution is still Normal $N(\mu_x, \tau x^2)$ (i.e. the distributions are conjugate) with mean

$$\mu_x = \frac{\frac{\sigma^2}{n}}{\frac{\sigma^2}{n} + \tau^2}\mu + \frac{\tau^2}{\frac{\sigma^2}{n} + \tau^2}\bar{x}, \tag{9.15}$$

and variance

$$\tau_x^2 = \frac{\frac{\sigma^2}{n}\tau^2}{\frac{\sigma^2}{n} + \tau^2}, \text{ or precision} \tag{9.16}$$

$$\frac{1}{\tau_x^2} = \frac{1}{\tau^2} + \frac{1}{\sigma^2/n}.$$

The posterior mean, μ_x, is a weighted average of the prior mean μ and the sample mean $\bar{x}$, where the weights are the variance σ^2/n of the sample mean $\bar{x}$ and the variance τ^2 of the prior distribution, respectively, such that the component (sample mean or prior mean) which has the smaller variance has the greater contribution to the posterior mean. The mean of the distribution with lower variance (higher precision) receives greater weight. The posterior precision $1/\tau_x^2$ is the sum of the precisions of the prior and the likelihood. Note that this result also holds (with $n = 1$) if only one measurement x is taken.

9.3   Continuous Bayesian networks

Propagation in discrete networks is exact, as particular variables within the network become instantiated, and probabilities can be updated accordingly. Discrete distributions typically supported by Bayesian network software, such as Hugin, include the binomial, geometric, negative binomial and Poisson distributions. In practical applications, however, one may wish to be able to use models in which some or all of the variables in the system of interest are measurements and can take values within a continuous range. First, a decision has to be made on how to represent a continuous entity in a network. Several options are possible for Normal and non-Normal random variables.

The theory underlying Bayesian networks is able to provide exact results for Normal random variables, and the methods used are described below. For now, the case in which all nodes within the network are continuous is considered. An extension of the univariate Normal distribution to MVN distribution lends itself extremely well to the field of continuous networks, since exact updates can be made in the light of data as a direct result of the properties of the MVN distribution (see Section 9.1.2). The MVN distribution thus has advantages for modelling nodes in a continuous network in that the results are exact. If the Normality assumption appears inappropriate, the modeller may consider transforming the data. For example, it is often the case that logarithms of the measurements have a Normal distribution in which case the original measurements are said to have a lognormal distribution (Aitken 2005). It is also possible to present a continuous entity as a variable with states representing intervals for the continuous entity (Section 9.3.3). Moreover, some continuous distributions may be approximated using a mixture of Normal distributions (Section 9.4.1).

In order to illustrate the propagation within a network containing continuous nodes, consider the case illustrated in Example 9.2 concerning a drunk-driving case. The network is illustrated in Figure 9.4. Nodes for discrete variables are denoted with single circles, nodes for continuous variables are indicated with double borders. The left-hand node θ represents the true level of alcohol in the blood of someone suspected of driving whilst under the influence of alcohol. In the Bayesian paradigm, a prior distribution for θ is required. In this example, it is taken to be a Normal distribution. Denote the mean of this distribution by μ and the variance by τ^2. The right-hand node X in the network represents the measured level of alcohol in the person suspected of driving whilst under the influence of alcohol. The arrow pointing from θ to X indicates that the measured level of alcohol is dependent on the true level of alcohol. It is assumed that the measured level of alcohol has a Normal distribution, $X \sim N(\theta, \sigma^2)$, and that the measuring device is unbiased in that the expected value of X is θ, the true level of alcohol in the blood. The arrow in Figure 9.4 illustrates this dependency of X on θ. There is uncertainty in the measurement, X, and this is represented by the variance of X, denoted σ^2.

Note that it is common practice in many forensic laboratories to analyze the concentration of alcohol in blood by performing several independent analytic procedures. For a given procedure, one obtains several measurements, say $x = (x_1, \ldots, x_n)$, with a mean $\bar{x} = \sum_{i=1}^{n} x_i$. The distribution of the corresponding random variable for the sample mean $\bar{X} = \sum_{i=1}^{n} X_i/n$, conditional on θ, denoted $(\bar{X}|\theta)$ is $N(\theta, \sigma^2/n)$. Assume that $\theta \sim N(\mu, \tau^2)$, that is, θ has itself a distribution. The unconditional (marginal) distribution of X can then be shown to be $N(\mu, \sigma^2/n + \tau^2)$. It can further be shown that the posterior distribution of θ, given a vector of measurements $x_1, \ldots, x_n$ for X such that $x = \frac{1}{n} \sum_{i=1}^{n} x_i$, and given σ^2, μ and τ^2 is still Normal

$$(\theta|\bar{x}, \sigma^2, \mu, \tau^2) \sim N(\mu_x, \tau_x^2),$$

with posterior mean μ_x and posterior variance τ_x^2 obtained as in (9.15) and (9.16), respectively.

Figure 9.4   A Bayesian network with two continuous nodes, θ and X where θ is the parent of X. Both nodes have a double border to indicate that they are continuous nodes.

Example 9.3 *(Alcohol concentration in blood, continued) Consider a case in which a person is stopped because of suspicion of driving under the influence of alcohol. Blood is taken from that individual and submitted to a forensic laboratory, where a measurement x is obtained. Consider the values 0.82 for x, 0.005 for σ^2, 0.7 for μ and 0.01 for τ^2. The prior beliefs about the true value are thus given by a 95% interval over (0.5, 0.9), that is, $0.7 \pm 2 \times 0.1$, where 0.1 is the standard deviation (square root) of the variance 0.01. The variance σ^2 is the sensitivity of the instrumentation as assessed on the basis of analyses performed under controlled conditions. From (9.15) and (9.16) it can be verified that the posterior mean μ_x for the distribution of θ is $0.78 g\ kg^{-1}$, the posterior variance τ_x^2 is 0.0033 and the posterior standard deviation τ_x is 0.0574.*

Figure 9.5(i) provides a schematic illustration of the marginal distributions for the true and measured blood alcohol levels, and Figure 9.5(ii) illustrates the posterior distribution for the true blood alcohol level, given a measured level of $0.82\ g\ kg^{-1}$. Thus, given a reading of $0.82\ g\ kg^{-1}$, the probability the true blood alcohol level is greater than $0.80\ g\ kg^{-1}$ is $1 - \Phi((0.80 - 0.78)/\sqrt{(0.0033)}) \simeq 1 - \Phi(0.34) = 1 - 0.63 = 0.37$.

This probability (0.37) for the posterior cumulative probability that the actual level of alcohol in the blood is larger than $0.8\ g\ kg^{-1}$ may be thought rather low, given that the observed value is $0.82\ g\ kg^{-1}$. This is explained by the choice of a Normal prior distribution centered on 0.7 to represent the personal prior beliefs about the unknown quantity θ. A sensitivity analysis may be performed to quantify the robustness of the decisions taken on the basis of different specifications of the prior distribution. If conclusions are sensitive to the prior specification, then it would be advisable to check the elicitation process, or to increase – if possible – the number of observations, as it will be discussed in Section 13.2.4.

This example provides a relatively simple illustration of propagation with Bayes' theorem. Both nodes represent Normally distributed random variables, the laboratory measurements (X) determined to be Normally distributed from results from previous experiments, and the true but unknown concentration of alcohol in blood (θ) where the (subjective) uncertainty is assumed to be represented by a Normal distribution.

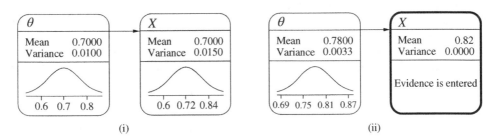

Figure 9.5   (i) Marginal distributions of θ (which is the prior distribution) and the measured value X of blood alcohol in gram per kilogram where the marginal mean of X is $\mu = 0.70$ and the marginal variance of X is $\tau^2 + \sigma^2/n$ (with $n = 1$) $= 0.015$. (ii) Posterior distribution of the true value θ of blood alcohol concentration, given a measured value X of blood alcohol concentration of 0.82 g kg^{-1}. This instantiation of node X is shown with a bold border. Note that the variance of X is now recorded as zero as X is assumed to be known and thus has no uncertainty or variation associated with it.

9.3.1 Propagation in a continuous Bayesian network

Computation and propagation become more involved as the number of nodes increase as in the Bayesian network in Figure 9.6(i) (Cowell et al. 1999). The three variables X, Y and Z are all random variables. Their probability density functions are taken to be Normally distributed. Thus, X represents a parent node and has a marginal distribution, taken here for illustrative purposes to be $N(0, 1)$. Random variable Y has a distribution dependent on X. As a child of X its distribution is represented as $(Y|X = x)$ and taken to be $(Y|X = x) \sim N(x, 1)$. Random variable Z has a distribution dependent on Y. Because Z is a child of Y, the distribution of Z is taken to be $(Z|Y = y) \sim N(y, 1)$. The variances are all taken to be 1 for ease of explanation of the underlying principles. The mean of the distribution of Y, conditional on X, is given by a value x of X, and the mean of the distribution of Z, conditional on Y, is given by a value y of Y. Thus,

$$f(x) = \frac{1}{\sqrt{2\pi}} \exp\left\{-\frac{1}{2}x^2\right\},$$

$$f(y|X = x) = f(y|x) = \frac{1}{\sqrt{2\pi}} \exp\left\{-\frac{1}{2}(y - x)^2\right\},$$

and

$$f(z|Y = y) = f(z|y) = \frac{1}{\sqrt{2\pi}} \exp\left\{-\frac{1}{2}(z - y)^2\right\}.$$

At this stage, it is convenient to introduce the notion of a probability *potential* that is often encountered and used when working with probabilistic networks since it simplifies computations. A probability potential, often denoted by the Greek letter ϕ, is a nonnegative function defined on the product space over the domain of a set of variables (e.g. $\phi(x)$ denotes a potential defined in the domain of the random variable X, dom(X)) (Kjærulff and Madsen 2008). A probability potential is turned into a probability distribution $f(x)$ through a Normalization, that is,

$$\frac{\phi(x)}{\sum_{\text{dom}(X)} \phi(x)}.$$

A potential can be specified for each clique $\{X, Y\}$ in the junction tree that is characterized by a conditional distribution $f(x \mid y)$.

In the network in Figure 9.6(i), two cliques are identified, namely, $\{X, Y\}$ and $\{Y, Z\}$. It is important to know the so-called clique potentials $\phi(x, y)$ and $\phi(y, z)$. The clique potentials are derived as follows, where $f(x)$ denotes the probability density function for variable X

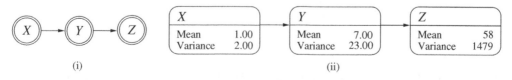

X		Y		Z	
Mean	1.00	Mean	7.00	Mean	58
Variance	2.00	Variance	23.00	Variance	1479

(i)   (ii)

Figure 9.6 A Bayesian network with three continuous nodes, X, Y and Z, such that X is a parent node for Y, Y is a child of X and a parent of Z and Z is a child of Y. (i) shows the network in collapsed form, and (ii) shows partially expanded nodes with the marginal distributions for continuous variables Y and Z as discussed in Example 9.4.

and $f(x, y)$ denotes the bivariate Normal probability density function for variables X, Y. For propagation, it is sufficient to know that $f(x, y)$ may be written as $f(y|x)f(x)$. Then,

$$\phi(x, y) \propto f(x, y)$$

$$= f(y|x)f(x)$$

$$= \frac{1}{\sqrt{2\pi}} \exp\left\{-\frac{1}{2}(y - x)^2\right\} \frac{1}{\sqrt{2\pi}} \exp\left(-\frac{1}{2}x^2\right)$$

$$= \frac{1}{2\pi} \exp\left\{-\frac{1}{2}(y^2 + 2x^2 - 2xy)\right\}. \tag{9.17}$$

It can be checked that this is a bivariate Normal probability density with zero mean $\theta_x = \theta_y = 0$, $\sigma_x^2 = 1$, $\sigma_y^2 = 2$ and correlation coefficient $\rho = 1/\sqrt{2}$. For example, this can be done with the substitution of these values into (9.3) to obtain the probability density (9.17). The clique potential $\phi(y, z)$ may be derived similarly. By definition, $\phi(y, z) \propto f(y, z) = f(z|y)f(y)$. Hence, in order to derive $f(y, z)$, it is first necessary to determine the marginal density function $f(y)$. The marginal distribution of any subvector of a Normally distributed random vector is Normal. The marginal probability density $f(y) = N(0, \sigma_y^2 = 2)$ may be derived from the joint distribution $f(x, y)$ in (9.17), as illustrated in (9.3) and (9.5).[3]

Therefore, the potential $\phi(y) \propto \exp(-y^2/4)$. Now that the separator potential (i.e. the potential for y that separates the two cliques) $\phi(y)$ has been derived, the potential $\phi(y, z)$ may be derived.

$$\phi(y, z) \propto f(y, z)$$

$$= f(z|y)f(y)$$

$$= \frac{1}{\sqrt{2\pi}} \exp\left\{-\frac{1}{2}(z - y)^2\right\} \frac{1}{\sqrt{2\pi}} \exp\left\{-\frac{1}{4}y^2\right\}$$

$$= \frac{1}{2\pi} \exp\left\{-\frac{1}{2}(\frac{3}{2}y^2 + z^2 - 2yz)\right\}. \tag{9.18}$$

---

[3] In general, given a continuous random variable (X, Y), the marginal distribution of variable Y can be obtained by taking the integral of the joint probability function $f(x, y)$ with respect to the possible values of X. For the case at hand, it can be checked that

$$f(y) = \int f(x, y)dx$$

$$= \frac{1}{2\pi} \int \exp\left\{-\frac{1}{2}(y^2 + 2x^2 - 2xy)dx\right.$$

$$= \frac{1}{2\pi} \exp\left(-\frac{1}{2}y^2\right) \int \exp\{-(x^2 - xy)\}dx$$

$$= \frac{1}{\sqrt{2}\sqrt{2\pi}} \exp\left\{-\frac{1}{4}y^2\right\}.$$

This is the probability density function of a Normally distributed random variable with expectation 0 and variance 2. These values of 0 and 2 may also be obtained from (9.8) and (9.9). The distribution of $Y|X$ is $N(x, 1)$ and the distribution of X is $N(0, 1)$. Thus, $E(Y|X = x) = x$, and the quoted result that $E(Y) = E_X(E(Y|X))$ gives $E(Y) = E(X) = 0$. Also $Var(Y|X) = 1$ so $E_X(Var(Y|X)) = 1$, and $E(Y|X = x) = x$ so $Var(E(Y|X)) = Var(X) = 1$. Thus, $Var(Y) = E_X(Var(Y|X)) + Var_X(E(Y|X)) = 2$. The distributional result is obtained from the result that if $(Y|X)$ and X are Normally distributed then Y is Normally distributed.

This is a bivariate Normal probability density with zero mean $\theta_y = \theta_z = 0$, $\sigma_y^2 = 2$, $\sigma_z^2 = 3$, $\rho = \sqrt{2/3}$, that is,

$$f(y, z) \propto \exp\left\{ -\frac{1}{2\frac{1}{3}}\left(\frac{y^2}{2} - 2\frac{\sqrt{2}}{\sqrt{3}}\frac{y}{\sqrt{2}}\frac{z}{\sqrt{3}} + \frac{z^2}{3} \right) \right\}.$$

Now, consider the scenario in which the information that variable $Y = 1.5$ is obtained. This information may be propagated through the network in order to obtain posterior potentials reflecting the new beliefs for both X and Z. Note that all updated posterior potentials are marked by the superscript * (e.g. ϕ^*). Firstly, $\phi(x, y)$ becomes $\phi(x, 1.5)$, which is then the posterior potential function on X, $\phi^*(x)$. Then, $\phi^*(x) = \phi(x, 1.5) \propto \exp(1.5x - x^2)$. Thus, the potential function is proportional to the corresponding posterior marginal probability density for X, $f^*(x)$, which is still Normally distributed. In fact, recalling notions of conditional distributions from Section 9.1.2, it can easily be observed that the conditional distribution for X, given $Y = 1.5$, is Normal with mean $\theta_{x|y} = \frac{1}{\sqrt{2}\sqrt{2}}1.5 = 0.75$ (Equation (9.6)) and variance $\sigma_{x|y}^2 = 1 - 1/2 = 1/2$ (9.7), that is, $(X|Y = 1.5) \sim N(0.75, 0.5)$. It is interesting to note that the prior belief on X was $X \sim N(0, 1)$. Learning that Y has taken the value 1.5 shifts the expectation on the value of X to 0.75 and decreases the variability from 1 to 0.5, as one would expect from the dependence structure described in Figure 9.6(i).

Belief in the variable Z may be updated by the calculation of the posterior potential $\phi^*(z)$ in an analogous way such that $\phi^*(z) = \phi(1.5, z) \propto \exp(1.5z - 0.5z^2)$. Recalling that $(Z|Y = y) \sim N(y, 1)$, then $(Z|Y = 1.5) \sim N(1.5, 1)$, without decreasing the variability. Therefore, the distributions of X and Z have been updated after obtaining information on the value of variable Y.

Alternatively, instead of instantiating the value of the clique separator Y, now consider the situation in which the value of variable Z has been learnt. Assume that the data received is $Z = 1.5$. This information has to be passed back through the network, exploiting the dependence structure to update beliefs on the variables X and Y accordingly. The marginal potential for Y may be updated using (9.18): $\phi^*(y) = \phi(y, 1.5) \propto \exp(1.5y - 0.75y^2)$. This potential function is proportional to the corresponding posterior marginal probability density for Y. As above, the conditional distribution of Y, given $Z = 1.5$, is Normal with mean $\theta_{y|z} = \sqrt{\frac{2}{3}}\sqrt{\frac{2}{3}}(1.5) = 1$ and variance $\sigma_{y|z}^2 = 2(1 - \frac{2}{3}) = \frac{2}{3}$. Thus, $(Y|Z = 1.5) \sim N(1, \frac{2}{3})$.

The update of beliefs on X is more complex because there is no direct causal link between X and Z. The information about Z is passed to X through Y. Variables X and Z are conditionally independent, given a value for Y. It is of interest to determine the potential function that is proportional to the conditional probability density function $f(x|z)$. Firstly, derive the updated clique potential $\phi^*(x, y)$ which is proportional to $f(x, y|z)$. The following calculations are, therefore, required (the notations $f(y|z)$ and $\phi^*(y)$ are essentially analogous).

$$\phi^*(x, y) \propto f(x, y|z)$$
$$= f(x|y, z)f(y|z)$$
$$= f(x|y)f(y|z) \text{ because of conditional independence}$$
$$= \frac{f(x, y)}{f(y)}f(y|z).$$

Thus,

$$\phi^*(x, y) \propto \phi(x, y) \frac{\phi^*(y)}{\phi(y)}.$$

Using this relation, it is possible to calculate the updated clique potential on $\{X, Y\}$, $\phi^*(x, y)$:

$$\phi^*(x, y) \propto \exp\left\{-\frac{1}{2}(y^2 + 2x^2 - 2xy)\right\} \frac{\exp(1.5y - 0.75y^2)}{\exp(-0.25y^2)}$$

$$= \exp(-x^2 + xy - y^2 + 1.5y).$$

Then, integration over y provides $\phi^*(x)$:

$$\phi^*(x) \propto \int \exp(-x^2 + xy - y^2 + 1.5y) dy$$

$$= \exp(-x^2) \int \exp\{-y^2 + (x + 1.5)y\} dy$$

$$= \exp(-x^2) \exp\left\{\left(\frac{x + 1.5}{2}\right)^2\right\} \int \exp\left\{-\left(y - \frac{x + 1.5}{2}\right)^2\right\} dy$$

$$\propto \exp(-x^2) \exp(0.25x^2 + 0.5625 + 0.75x)$$

$$\propto \exp(-0.75x^2 + 0.75x)$$

$$\propto \exp\{-0.75(x - 0.5)^2\}.$$

Note that $\int \exp\{-(y - [(x + 1.5)/2])^2\} dy = 1$.

Therefore, the posterior density function for X is of the form

$$f^*(x) = \frac{1}{\sigma\sqrt{2\pi}} \exp\left\{-\frac{1}{2\sigma^2}\left(x - \frac{1}{2}\right)^2\right\},$$

where σ^2 is taken to be equal to $\frac{2}{3}$ to ensure that $f^*(x)$ is indeed a density (proportional to $\phi^*(x)$). Thus, $(X|Z = 1.5) \sim N(\frac{1}{2}, \frac{2}{3})$.

The choice of a standard Normal distribution for the parent node X as well as the variance equal to 1 for all-continuous nodes makes computations quite feasible. However, whenever any Normal distribution is taken for the parent node, and the variance is allowed to take any value greater than zero, the computation and propagation become more cumbersome. In general, if $X \sim N(\theta, \sigma^2)$ and $(Y|X = x) \sim N(x, \tau^2)$, then $Y \sim N(\theta, \sigma^2 + \tau^2)$. A further generalization is that if $X \sim N(\theta, \sigma^2)$ and $(Y|X = x) \sim N(\alpha x + \beta, \tau^2)$ then $Y \sim N(\alpha\theta + \beta, \alpha^2\sigma^2 + \tau^2)$.

Consider again the Bayesian network in Figure 9.6(i). Set $X \sim N(\theta, \sigma^2)$, $(Y|X = x) \sim N(\alpha x + \beta, \tau^2)$ and $(Z|Y = y) \sim N(\gamma y + \delta, \eta^2)$. Then, repeated use of the results above gives the marginal distributions of Y and Z as

$$Y \sim N(\alpha\theta + \beta, \alpha^2\sigma^2 + \tau^2), \tag{9.19}$$

$$Z \sim N(\gamma(\alpha\theta + \beta) + \delta, \gamma^2(\alpha^2\sigma^2 + \tau^2) + \eta^2). \tag{9.20}$$

Example 9.4 *(Propagation in a continuous Bayesian network) Consider the following hypothetical example were $X \sim N(1, 2)$, $(Y|X = x) \sim N(3x + 4, 5)$ and $(Z|Y = y) \sim N(8y + 2, 7)$. Substitution of these values for $\alpha, \beta, \tau, \theta, \sigma, \gamma, \delta$ and η in (9.19) and (9.20) gives the following results.*

$$Y \sim N(7, 23),$$

$$Z \sim N(58, 1479).$$

These results may be checked and compared with those provided by an appropriate software tool. Using the program Hugin (Vers. 8.0), for example, label the nodes X, Y and Z as in Figure 9.6(i). Then, double click on the node labelled X, and set the mean θ equal to 1 and the variance σ^2 equal to 2. Double click on the node labelled Y. Three cells appear for which entries are required, labelled Intercept, X and Variance, as the structure of the network is that Y is dependent on X. In the cell labelled 'Intercept', enter the value for β, which is 4 in this example. In the cell labelled 'Variance', enter the value for τ^2, which is 5 in this example. In the cell labelled 'X', enter the value for α, which is 3 in this example. Double click on the node labelled Z. Three cells appear for which entries are required, labelled 'Intercept', 'Y' and 'Variance', as the structure of the network is that Z is dependent on Y. In the cell labelled 'Intercept', enter the value for δ, which is 2 in this example. In the cell labelled 'Variance', enter the value for η^2, which is 7 in this example. In the cell labelled 'Y', enter the value for γ, which is 8 in this example. Figure 9.6(ii) provides a schematic illustration of the marginal distributions for Y and Z computed with Hugin.

Example 9.5 *(Blood alcohol analysis, continued) Consider again Example 9.1 in which the alcohol concentration measured in the blood of a person suspected of driving under the influence of alcohol was assumed to be Normally distributed. A Bayesian network with two continuous nodes was proposed in Section 9.3, where the node X represented the measured level of alcohol in blood and the node θ the true but unknown level of alcohol in blood. It might be observed that, usually, measurements are taken on blood that is preserved in a vial and, in fact, the quantity of alcohol that is contained in the blood from a vial may not necessarily be equivalent to (although being dependent on) the quantity of alcohol present in the body of a person at the time when the blood was taken. Thus, one may propose a three-node continuous Bayesian network as shown in Figure 9.7(i), in which the node X represents the measured level of alcohol in the vial, the node θ_v represents the true level of alcohol in the vial and the node θ represents the true unknown level of blood alcohol. Thus, θ is represented by a parent node and has a marginal distribution that is taken to be $N(\mu = 0.7, \tau^2 = 0.01)$, as in Example 9.3. The random variable θ_v representing the true level of alcohol in the vial has a distribution dependent on θ that is taken to be $(\theta_v|\theta) \sim N(\theta, \eta^2 = 0.01)$. The random variable X representing the measured level of alcohol in the vial has a distribution dependent on θ_v that is taken to be $(X|\theta_v) \sim N(\theta_v, \sigma^2 = 0.005)$. Figure 9.7(ii) provides a schematic illustration of the marginal distributions for the true level of alcohol in blood θ, the true level of alcohol in the vial θ_v and the measured blood alcohol levels X. Figure 9.8 illustrates the posterior distribution for the true blood alcohol level, given a measured level of 0.82 g kg^{-1}. Thus, given a reading of 0.82 g kg^{-1}, the probability the true blood is greater than 0.80 g kg^{-1} is $1 - \Phi((0.8 - 0.748)/\sqrt{\{0.006\}}) \simeq 0.25$. The additional level of uncertainty produces a decrement of this probability from 0.37 (as in Example 9.3) to 0.25.*

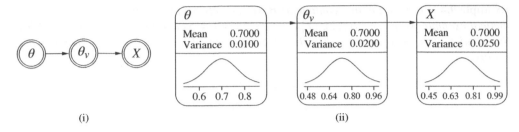

(i)   (ii)

Figure 9.7   (i) A Bayesian network with three continuous nodes, θ, θ_v and X where θ represents the true level of alcohol in blood, θ_v the true level of alcohol in the vial and X the measured alcohol level. (ii) Marginal distributions of the true level of alcohol θ in blood (the prior distribution), the true level of alcohol in the vial θ_v and the measured value X of blood alcohol in gram per kilogram where the marginal mean is equal to 0.7 and the marginal variance is equal to 0.025.

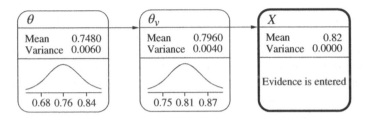

Figure 9.8   Posterior distribution of the true value θ of blood alcohol concentration, given a measured value X of blood alcohol concentration of 0.82 g kg^{-1}. The instantiated node X is shown with a bold border. Note that the variance of X is now recorded as zero as X is assumed known and thus has no uncertainty or variation associated with it.

9.3.2   Background data

Examinations in real-world circumstances are often affected by inefficiencies. Imagine a scenario where n distinct images captured from the recordings from a surveillance camera are available of a male individual and where it is of interest to estimate the height of that individual. A measured height of the individual obtained by the available camera is denoted Y and is assumed to have a Normal distribution. These measured heights are biased because of errors introduced by the context in which the images were taken, such as the posture of the individual, the presence or absence of headgear and the angles of the camera relative to the individual. Denote the bias by ξ. There are two sources of variance and these are assumed independent. The first is the precision of the measurement devices (assumed constant across devices), denote their variance by σ^2. The second is the variation associated with the context and is denoted δ^2. Hence, $Y \sim N(\theta + \xi, \sigma^2 + \delta^2)$.

The uncertainty about θ is modelled through a Normal distribution with mean μ and variance τ^2, $\theta \sim N(\mu, \tau^2)$, whilst hyperparameters ξ and δ^2 are assumed to be known from the scenario of interest. This appears to be admissible because the assumption is made that the case at hand is basic in the sense that several images of an individual are available from video recording of a given surveillance camera. That is to say, parameters ξ and δ^2 could be obtained through an ad hoc reconstruction. Whenever recordings would be available from different surveillance cameras on different locations, then the scenario is one in which it would be necessary to model the prior mean and variance.

It can be shown that, given a set of measurements $y = (y_1, \dots, y_n)$, the posterior distribution is still a Normal density with mean

$$\mu_y = \frac{\tau^2(\bar{y} - \xi) + \mu\left(\frac{\sigma^2 + \delta^2}{n}\right)}{\tau^2 + \frac{\sigma^2 + \delta^2}{n}}$$

and variance

$$\tau_y^2 = \frac{\tau^2\left(\frac{\sigma^2 + \delta^2}{n}\right)}{\tau^2 + \frac{\sigma^2 + \delta^2}{n}}.$$

The variance is higher with respect to an error-free setting, whilst more weight is given to the prior mean μ, according to the amplitude of the error component. This is illustrated in Example 9.6. This can be implemented using a Bayesian network, as in shown Figure 9.9, where the continuous node Y represents the sample mean of the observations, the continuous node θ represents the true height of the individual and, finally, the discrete node n represents the number of available camera recordings.

Example 9.6 (*Evaluating height using surveillance camera recordings*) *Imagine a video recording is made by a surveillance camera during a bank robbery. The recordings depict an individual appearing in $n = 10$ images. Measurements on the available recordings $(y_1, \dots, y_{10})$ yield $\bar{y} = 178$ cm. It is of interest to infer the height of that individual. The precision of the measurement procedure (independent of the complexity of the scenario) is known, and it is set to $\sigma^2 = 0.1$. Parameters μ and τ^2 are chosen on the basis of the following arguments. There is an eyewitness, suggesting the mean μ is equal to 175 cm. Values less than 170 cm and greater than 180 cm are considered extremely unlikely; therefore, the variance τ^2 is fixed and is equal to 2.67. Finally, repeated measurements are obtained in experiments under controlled conditions (i.e. a reconstruction), which enables values to be chosen for the hyperparameters of the Normal distribution of the error. These values are taken to be $\xi = 1$ and $\delta^2 = 1$, respectively. Then,*

$$\mu_x = \frac{2.67(178 - 1) + 175\left(\frac{0.1 + 1}{10}\right)}{2.67 + \frac{0.1 + 1}{10}} = 176.92\ cm.$$

Figure 9.10(i) provides a schematic illustration of the prior distribution for the true and the measured height, and Figure 9.10(ii) illustrates the posterior distribution for the true height, given a measured mean of 178 cm.

Figure 9.9   A Bayesian network with two continuous nodes, Y and θ, and one discrete node n. The continuous node Y represents the sample mean of the observations, the continuous node θ represents the true height of the individual and, finally, the discrete node n represents the number of available camera recordings.

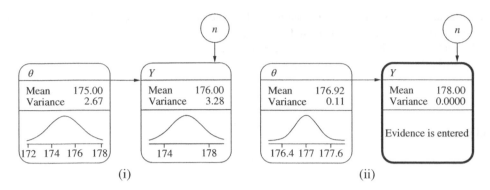

Figure 9.10 (i) Prior distributions of the true value θ and the measured height Y in centimetres, where the unconditional mean of Y is $\mu + \xi$ and the unconditional variance of Y is $(\sigma^2 + \delta^2)/n$. (ii) Posterior distribution of the true height θ, given $n = 10$ observations and a sample mean of 178 cm.

9.3.3   Intervals for a continuous entity

It is possible to represent a continuous entity as a discrete variable with states representing intervals for the continuous entity. An example is given here of the use of a beta distribution in conjunction with a binomial distribution. The derived model is used in the context of consignment inspection, as described in Aitken (1999). The problem is to determine the number of units from a consignment to inspect in order to make an inference about the proportion of units in the consignment have a particular characteristic. The application is to a limited case in which the units in the consignment are homogeneous in all respects except one. The characteristic in which the units are not homogeneous is a binary characteristic, that is, one with only two responses, say positive and negative. It is of interest to know the proportion of units in the consignment which are positive. A certain number of units (i.e. a sample) are taken from the consignment, and their examination reveals the numbers of positives and negatives. These data are then to be used to make an inference about the consignment as a whole (i.e. the proportion of units in the consignment that are of a certain kind). Three examples will help illustrate the problem. First, consider a consignment of white tablets, homogeneous with respect to colour, texture, shape and size. The requirement is an inference about the proportion of tablets which is illicit. Second, consider a consignment of compact discs (CDs). Here the requirement is an inference about the proportion which is pirated. Third, consider a file of computer images. Here the requirement is an inference about the proportion of images are pornographic.

A statistical model may be proposed for these three examples. The structure is the same in all the examples as long as reasonable assumptions are made. First, each unit in the consignment is assumed homogeneous in all respects except for the characteristic of interest. The characteristic of interest can take one and only one of two possible values (e.g. illicit or not, pirated or not, pornographic or not). The probabilities that a unit, when inspected, may have each of the two characteristics are assumed constant over all units in the consignment. Denote these as θ and $1 - \theta$. These probabilities are independent from the inspection of one unit to another.

Let X be the number of positive units selected in a sample of size n from a consignment sufficiently large that inspection may be considered to be with replacement. The model for this situation is a binomial model. The distribution of X is the binomial distribution with associated probability function

$$Pr(X = x|n, \theta) = \binom{n}{x} \theta^x (1 - \theta)^{n-x}, \quad 0 < \theta < 1; \quad x = 0, \ldots, n.$$

The conjugate prior distribution for θ is a beta distribution which, with parameters α and β, has a probability density function

$$f(\theta|\alpha, \beta) = \frac{\Gamma(\alpha + \beta)}{\Gamma(\alpha)\Gamma(\beta)} \theta^{\alpha-1} (1 - \theta)^{\beta-1}, \quad 0 < \theta < 1; \quad \alpha, \beta > 0,$$

where Γ is the gamma function, $\Gamma(z) = \int_0^\infty t^{z-1} e^{-t} dt$, with $\Gamma(z + 1) = z\Gamma(z)$ for integer $z > 0$, $\Gamma(1) = 1$, $\Gamma(1/2) = \sqrt{\pi}$.

The posterior density function for θ can be obtained by Bayes' theorem and is

$$f(\theta|x, \alpha, \beta) = \frac{\Gamma(\alpha + \beta + n)}{\Gamma(\alpha + x)\Gamma(\beta + n - x)} \theta^{x+\alpha-1} (1 - \theta)^{\beta+n-x-1},$$

where $0 < \theta < 1$, the parameters $\alpha, \beta > 0$ and $x = 0, \ldots, n$. It can be observed that this is also a beta distribution with updated parameters $\alpha^* = \alpha + x$ and $\beta^* = \beta + n - x$, that is, $(\theta|x) \sim Be(\alpha^*, \beta^*)$.

From this result, it is possible to determine the probability that θ is greater than a certain proportion, θ_0 say, from the equation

$$Pr(\theta > \theta_0|n, x, \alpha, \beta) = \int_{\theta_0}^1 f(\theta|\alpha + x, \beta + n - x)d\theta.$$

The application of this result for selecting the number units to be inspected is described in detail in Aitken (1999) with a summary in Aitken and Taroni (2004).

Example 9.7 (*Inspection of units from a consignment of pills*) *The use of the method is illustrated here for the example of inspecting units from a consignment of white tablets. A criterion for the number of units to be inspected is specified such that if all inspected tablets are illicit (positive), then one can be 95% certain that at least 50% of the consignment is illicit (Aitken 1999). Thus, $x = n$ and $n - x = 0$. It is also specified that the prior distribution is uniform, that is, $\alpha = \beta = 1$ and $f(\theta|1, 1) = 1$. Then the number n of units to be inspected is the smallest integer value of n such that $Pr(\theta > 0.5|n + 1, 1) = 0.95$, that is,*

$$\int_{0.5}^1 f(\theta|n + 1, 1)d\theta = 0.95.$$

Use of a suitable statistical package, such as R (R Core Team 2013), shows that the solution of this equation is $n = 4$. Results for different levels of certainty associated with different proportions, different levels of the numbers of units that are negative and different values of α and β are available. See, for example, Aitken (1999) and Aitken and Taroni (2004). Guidelines have also been issued by the ENFSI Drugs Working Group (2004) and Unites Nations Office on Drugs and Crime (2009).

The question of how many units to inspect can also be approached using Bayesian networks. Figure 9.11 shows a Bayesian network useable for inference about a proportion

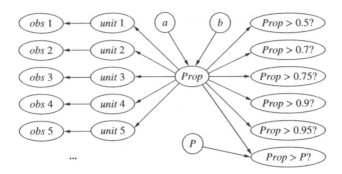

Figure 9.11 A Bayesian network for inference about a proportion of a large consignment. The definitions of the nodes are as given in Table 9.2.

of a large consignment (Biedermann et al. 2007; Taroni et al. 2010). The proportion θ of positives in a consignment is modelled here with a discrete chance node *Prop* with intervals $0 - 0.05, 0.05 - 0.1, \dots, 0.95 - 1$. The number of discrete intervals as well as their size may be varied, according to the analyst's needs. The probabilities assigned to the intervals of *Prop* are determined by a beta distribution whose parameters (α and β) are provided by the nodes *a* and *b*, which are parents of *Prop*. The states of nodes *a* and *b* consist of numbers $0.5, 1, 2, \dots, 10$. Again, the choice of these numbered states is a subjective one and depends on the analyst's needs for the case at hand. Thus, other values (> 0) might be defined as required. The value of 0.5 is included here because the nodes *a* and *b* instantiated to this value allow the node *Prop* to have a $Be(0.5, 0.5)$ distribution, which is sometimes used to represent prior beliefs according to which either no (or nearly none of the) units or all (or nearly all) units are positive.

A particular aspect of the Bayesian network shown in Figure 9.11 consists in the way in which the inspection procedure is modelled. Instead of a variable x ($x \le n$) for the overall number of positive units amongst the n inspected units (as done in Section 10.5.2), separate nodes are used to represent the target characteristic ('positive' or 'negative') of each inspected unit. In addition, a distinction is made between the true, but unknown, condition of an unit (i.e. containing or not containing an illegal substance) and what is observed in the course of an experiment designed to 'detect' the presence or absence of that target characteristic. Note that in Section 10.5.2, the observation of a 'positive' (measurement) result is equated with the examined unit being truly positive, which is a simplification of the underlying states of reality. Such distinctions are advocated, for example, in the context of DNA profiling analyses (Thompson et al. 2003), and are also challenged in other forensic disciplines (Saks and Koehler 2005).

The structure of the model shown in Figure 9.11 is described as follows. Binary nodes labelled *unit n* with states 'positive' and 'negative' represent propositions according to which the nth unit may or may not contain illegal substance. The result of an examination or analysis of the nth unit is modelled by a binary node *obs n*, again with states 'positive' and 'negative'. The outcome of a given analysis depends directly on the presence or absence of the respective characteristic. Directed edges thus connect the nodes *obs n* and *unit n*. Five pairs of nodes *unit n* and *obs n* are incorporated in the current model. For routine use, more trials may be needed and additional nodes can be added analogously.

There may be circumstances shedding doubt on the result of an analysis. The condition of a unit chosen for inspection or erroneous experimental settings are possible reasons for this.

Table 9.2   Definitions of nodes used in the Bayesian network shown in Figure 9.11.

Node	Definition	States
a, b	Parameters α and β of the beta distribution defined for the node *Prop*	$0.5, 1, 2, \ldots, 10$
Prop	Proportion of positives in the consignment (θ)	$0 - 0.05, \ldots, 0.95 - 1$
P	Lower limit for evaluating cumulative probabilities of the proportion of positives in the consignment	$0, 0.05, \ldots, 0.95, 1$
Prop > P?	Is the proportion of the positives in the consignment greater than P?	*yes, no*
Prop > 0.5 (0.7,...)	Is the proportion greater than 0.5 (0.7, ...)?	*yes, no*
obs 1 (2,...)	Outcome of test conducted in order to determine the characteristic of unit 1 (2, ...)	*positive, negative*
unit 1 (2,...)	True (but unknown) characteristic of unit 1 (2, ...)	*positive, negative*

Thus, an analysis may not always turn out to be 'positive' when the inspected unit is truly positive or may turn out to be 'positive' when the inspected unit is truly negative. Generally, two probabilities can be used to describe the accuracy: the probability of a positive result when the unit is truly positive, $Pr(obs\ n = positive | unit\ n = positive)$, and the probability of a negative result when the unit is actually negative, $Pr(obs\ n = negative | unit\ n = negative)$. Sometimes, these two values are referred to as *sensitivity* and *specificity* (e.g. Balding 2005; Kaye 1987; Lindley 2006; Robertson and Vignaux 1995). They can be used to complete the probability table of the nodes *obs n*. Note that a result is taken to indicate the presence or absence of a unit's target characteristic with certainty if one assumes that $Pr(obs\ n = positive | unit\ n = positive) = Pr(obs\ n = negative | unit\ n = negative) = 1$. For the purpose of the current discussion, a hypothetical value of 0.99 is chosen for both of these probabilities.

Before inspection of a unit, the probability that it is found to contain or not to contain an illegal substance depends directly on the proportion of units in the consignment that contain illegal substances. Directed edges are thus drawn from the node *Prop* to the nodes *unit n*. The probability tables of the nodes *unit n* can be completed, for example, through the expression Distribution(Prop,1-Prop) (syntax used in Hugin).

The model also contains auxiliary nodes, namely, *P*, *Prop > P?*, *Prop > 0.5?*, *Prop > 0.7?*, *Prop > 0.75?*, *Prop > 0.9?* and *Prop > 0.95?*, which define a substructure from which cumulative probabilities may be evaluated. The definition of these variables is given in Table 9.2.

Example 9.8 *(Inspection of pills suspected to contain drugs, continued) Consider again, as in Example 9.7, a scenario in which $n = 4$ units are inspected and found to be positive. In such a situation, following discussion presented in Aitken (1999), there is – assuming a uniform prior probability distribution for θ – a probability greater than 0.95 so that the proportion of positive units θ is greater than 0.5.*

Figure 9.12 (i) depicts the Bayesian network described above with nodes expanded and instantiations made at the relevant nodes. The parameters for the beta distribution of the *node Prop are set by instantiating both nodes a and b to 1. The nodes unit 1 to unit 4 are set to 'positive'. This represents the observation of the four units found to be positive. Instantiation of nodes 'unit n' rather than 'obs n' follows from the assumption that the determination of the characteristics of an inspected unit is made without error. The node Prop > 0.5? displays the target probability $Pr(\theta > 0.5|n = y = 4, \alpha = 1, \beta = 1)$. The value 0.97 agrees with results reported in Aitken (1999). Notice also that the node Prop shows the updated (posterior) probability distribution for the proportion of positives in the consignment. As may be read from the graph, the result is that higher proportions now are more probable than before the inspection, based on the assumption of the uniform prior. Cumulative probabilities for various intervals of proportions (other than > 0.5) are displayed in the nodes at the far right-hand side.*

The proposed Bayesian network readily allows an examination of a setting in which the determination of the analytical characteristics cannot be assumed to be error free. This is shown in Figure 9.12 (ii). Here, the nodes '*obs n*' are instantiated instead of the nodes '*unit n*'. By defining a value of 0.99 for sensitivity and specificity of the analytic procedure, each observation of a positive result provides a likelihood ratio of 99 for the proposition according to which the unit of interest is in fact positive. The effect of this uncertainty on the value of the target cumulative probability is weak as the probability changes by less than 1%.

In a more general case, one could enter one observation at the time and subsequently observe the changes in the probability distributions for the nodes of interest. Notice further that one is not limited to consider the proposed Bayesian network only for evaluating findings that have actually been obtained. One may also evaluate the probability with which future trials, given previous observations, can be expected to yield positive and negative results, respectively. This is illustrated in Figure 9.12 where a probability of approximately 0.83 is indicated for the fifth unit being positive.

Besides considering the probability of future trials resulting in positive or negative findings, one may also evaluate the information that future findings can be expected to provide. More can be learned about such a question by instantiating, for example, the node *obs* 5 (not shown in Figure 9.12). If a positive result is obtained, one can find that the probability of the proportion being greater than 0.5 would increase a further 2% to approximately 0.98. In case of a negative finding, this probability would decrease by approximately 6%.

9.4 Mixed networks

When attempting to construct graphs to model a real-life system, it is often the case that both discrete and continuous variables will be involved within the system. Hence, it is important to consider the case of *mixed graphs*, in which both discrete and continuous nodes are of interest. Propagation through a mixed network is very similar in form to the propagation examples already covered. The junction tree is again formed and subsequent efficient propagation can be made through this junction tree. However, there is one very important restriction in the modelling conditions under which a mixed network can be used to propagate information. Exact mixed node propagation can only be implemented through the junction tree if the cliques of the directed acyclic graph (DAG) follow the strong running intersection property

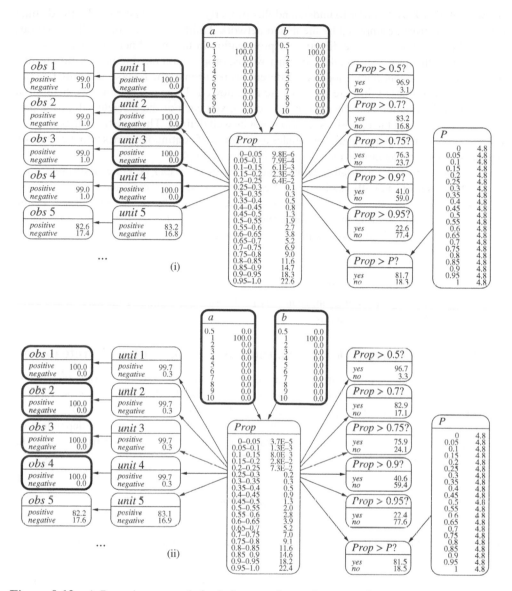

Figure 9.12   A Bayesian network for inference about a large consignment of discrete units (Example 9.8). The definitions of the nodes are in Table 9.2. (i) displays the analysis of a setting assuming a uniform prior distribution for the parameter θ (node *Prop*) and four inspected units, all of which are found to be positive (assuming error-free analyses). (ii) shows an evaluation of the scenario allowing for error in the analysis of the inspected units. Instantiated nodes are shown with a bold borderline.

(Cowell et al. 1999). In order to understand this property, one must first consider the distinct difference in structure imposed by the mixed network. Unlike the cases of an all-continuous or an all-discrete network, the direction of propagation in the mixed network will influence the form that this propagation takes. In other words, propagation passing information from a discrete node to a continuous node is a different proposition to the reverse scenario. Continuous parent variables of discrete variables are not allowed. There is a natural specification of conditional distributions for the continuous node, given each of the discrete values the parent(s) could take, but the opposite scenario is not as well defined. The logistic distribution is a possible way of incorporating discrete nodes as children of continuous nodes into the network but subsequent propagation is then considerably more complex.

Consider a mixed Bayesian network where each node v can be either a discrete node with a finite set of mutually exclusive and exhaustive states or a continuous node. Denote with $\mathcal{Z} = \{\mathcal{X}_\Delta, \mathcal{Y}_\Gamma\}$ the set of random variables represented by the set of nodes V of the DAG, with $\mathcal{X}_\Delta$ representing the set of discrete nodes V_Δ and with $\mathcal{Y}_\Gamma$ representing the set of continuous nodes V_Γ. The joint distribution over all the variables has the form

$$\prod_{v \in V_\Delta} f(X_v | X_{PA(v)}) \prod_{v \in V_\Gamma} f(Y_v | Z_{PA(v)}),$$

where $X_{PA(v)}$ are the parents of X_v and can only be discrete and $Z_{PA(v)}$ are the parents of Y_v and can be either discrete or continuous.

9.4.1   Bayesian network for a continuous variable with a discrete parent

In order to illustrate the propagation within a Bayesian network containing both discrete and continuous nodes, consider the network shown in Figure 9.13, where X is a discrete random variable which takes the value x_i with probability p_i, for $i = 1, \ldots, k$ and such that $\sum_{i=1}^{k} p_i = 1$, and Y is a continuous random variable with a Normal probability distribution dependent on X, $(Y|X = x_i) \sim N(\mu_i, \sigma_i^2)$. For each configuration x_i of node X, the joint distribution has the form $p_i N(\mu_i, \sigma_i^2)$.

The marginal distribution of Y can be computed as

$$f(y) = \sum_{i: p_i > 0} p_i \times N(\mu_i, \sigma_i^2). \tag{9.21}$$

The unconditional expectation (mean) of Y, $E(Y)$, can be computed according to the formula given in (9.8), that is,

$$E(Y) = E_X(E(Y|X)) = \sum_{i: p_i > 0} p_i \mu_i. \tag{9.22}$$

The computation of the unconditional variance of Y, $Var(Y)$, requires slightly more effort. In fact, one needs to compute

$$Var(Y) = E_X(Var(Y|X)) + Var_X(E(Y|X)).$$

Figure 9.13   A Bayesian network with one discrete (X) and one continuous (Y) node.

Now, since $Var(Y|X = x_i) = \sigma_i^2$,

$$E_X(Var(Y|X)) = \sum_{i:p_i>0} p_i\sigma_i^2.$$

The variable $E(Y|X)$ may be considered as a discrete variable, T, say which takes the value μ_i with probability p_i. Then $E(T) = \sum_{i:p_i>0} p_i\mu_i$ and

$$Var(T) = E(T^2) - \{E(T)\}^2$$

$$= \sum_{i:p_i>0} p_i\mu_i^2 - \left(\sum_{i:p_i>0} p_i\mu_i\right)^2,$$

and

$$Var(Y) = E_X(Var(Y|X)) + Var_X(E(Y|X))$$

$$= \sum_{i:p_i>0} p_i\sigma_i^2 + \sum_{i:p_i>0} p_i\mu_i^2 - \left(\sum_{i:p_i>0} p_i\mu_i\right)^2. \tag{9.23}$$

It can be observed that the marginal distribution of Y in (9.21) is a mixture of Normal distributions (in other words, a linear combination of k Normal probability distributions where all coefficients are nonnegative and sum up to one) that can be used to approximate a continuous distribution. This option is interesting as in this way it is possible to approximate any probability distribution. A Gamma distribution, for example, can be approximated using a two-component mixture of Normal distributions, as it is illustrated in Example 9.9.

Example 9.9 *(Gamma approximation) Imagine a case where a network contains a node with a Gamma distributed variable, say $Y \sim Ga(3, 2)$. This distribution is illustrated in Figure 9.14 (continuous line). This distribution can be approximated using a two-component mixture of Normal distributions*

$$f(x) = 0.5N(0.95, 0.55^2) + 0.5N(1.9, 0.95^2).$$

Figure 9.14 shows the result for approximating the $Ga(3, 2)$ distribution with the above two-component mixture of Normal distribution (dashed line). It can be verified using relations in (9.22) and (9.23) that

$$E(X) = 0.5 \times 0.95 + 0.5 \times 1.9 = 1.425,$$

$$Var(X) = (0.5 \times 0.55^2 + 0.5 \times 0.95^2) + (0.5 \times 0.95^2 + 0.5 \times 1.9^2) - 1.425^2 = 0.8281,$$

whilst the expected mean $E(Y)$ for a $Ga(\alpha = 3, \beta = 2)$ distribution is given by $\alpha/\beta = 3/2 = 1.5$ and the variance $Var(Y)$ is given by $\alpha/\beta^2 = 3/4 = 0.75$.

Let us see how this can be implemented through a Bayesian network with a structure as given earlier in Figure 9.13. The node Y represents the variable whose distribution is not Normal and will be approximated as in Equation (9.21). The discrete selector variable X with m states is introduced as a parent of Y. Each component $p_iN(\alpha_i, \gamma_i)$ in the mixture is represented by a state $x_i \in X$. The prior distribution on node X is $Pr(X = x_i) = p_i$, whilst variable Y has a conditional Normal distribution, $(Y|X = x_i) \sim N(\alpha_i, \gamma_i)$. Figure 9.15 provides a schematic illustration of the marginal distribution of the approximated random variable Y.

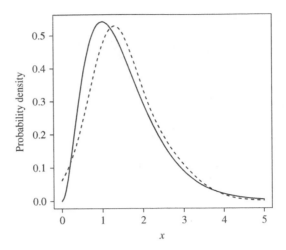

Figure 9.14 A two-component approximation based on the two Normal distributions (dashed line) of the Gamma(3,2) distribution (continuous line).

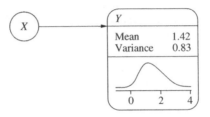

Figure 9.15 A Bayesian network with one discrete (X) node and one continuous (Y) nodes. Node Y is shown in expanded form with the marginal distribution of random variable Y.

9.4.2 Bayesian network for a continuous variable with a continuous parent and a binary parent, unmarried

Consider the Bayesian network in Figure 9.16 such that Y is a child of both X and W, where W and Y are Normally distributed random variables and X is a discrete variable with a set of k mutually exclusive and exhaustive states.

The probabilistic structure is

$$W \sim N(\mu, \sigma^2),$$

$$Pr(X = x_i) = p_i, \qquad i = 1, \ldots, k$$

$$(Y|X = x_i, W) \sim N(\alpha_i w + \beta_i, \tau_i^2),$$

where $\sum_{i=1}^{k} p_i = 1$. From these distributional forms, the mean and the expectation of Y may be derived. First, consider the expectation. When $X = x_i$, $E_W(E(Y|W, X)) = \alpha_i \mu + \beta_i$, thus

$$E(Y) = E_X(E_W(E(Y|W, X))) = \sum_{i=1}^{k} p_i(\alpha_i \mu + \beta_i). \qquad (9.24)$$

The results of Section 9.4.1 may be used, with μ_i replaced by $\alpha_i w + \beta_i$ to derive the expression for the variance, although the computations are more complex here than in Section 9.4.1. It can be verified that if X is a binary discrete node, then

$$Var(Y)$$
$$= Var(E(Y|X)) + E(Var(Y|X))$$
$$= p_1 p_2 [(\alpha_1 - \alpha_2)^2(\sigma^2 + \mu^2) + (\beta_1 - \beta_2)^2 + 2(\alpha_1 - \alpha_2)(\beta_1 - \beta_2)\mu]$$
$$+ (\alpha_1 p_1 + \alpha_2 p_2)^2 \sigma^2 + \tau_1^2 p_1 + \tau_2^2 p_2. \tag{9.25}$$

The following numerical example may be used to check the results given by the above formulae with those given by the program Hugin. Set $X = x_1$ and $X = x_2$ for X, $\beta_1 = 2$ and $\beta_2 = 5$ for 'Intercept', $\tau_1^2 = 3$ and $\tau_2^2 = 6$ for 'Variance' and $\alpha_1 = 1$ and $\alpha_2 = 4$ for W. The other values used are $\mu = 8, \sigma^2 = 10, p_0 = 0.2, p_1 = 0.8$. Then

$$E(Y) = (8 + 2)0.2 + (32 + 5)0.8 = 31.6$$

and

$$Var(Y) = 0.16[3^2(10 + 64) + 3^2 + 2(-3)(-3)8]$$
$$+(0.2 + 3.2)^2 10 + (3 \times 0.2) + (6 \times 0.8)$$
$$= 252.04.$$

Figure 9.16(ii) provides a schematic illustration of the marginal distribution of the continuous random variable Y.

Example 9.10 *(Blood alcohol analysis, continued) Consider again Example 9.1 where the alcohol concentration measured in the blood of a person suspected of driving under the influence of alcohol was assumed Normally distributed. A Bayesian network with two continuous nodes was proposed in Section 9.3 (Figure 9.4), where the node X represented the measured level of alcohol in blood and the node θ the true but unknown level of alcohol in blood. Suppose that the laboratory can perform two independent analytical procedures, denoted here as HS and ID, and that procedure HS has greater precision than the alternative procedure ID but is more expensive. Assuming measurements are Normally distributed, then the conditional distribution of measurements X, given the unknown level of alcohol concentration θ and the implemented procedure P, is $(X|\theta, P) \sim N(\theta, \sigma_p)$, where the standard deviation σ_p has different values for $p = \{HS, ID\}$. The standard deviations σ_p are assumed to be known as they have*

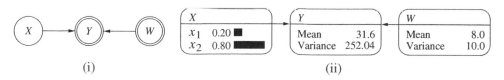

(i)   (ii)

Figure 9.16 Collapsed (i) and expanded (ii) representation of a Bayesian network for a continuous variable, Y with a continuous parent, W, and a binary parent, X, unmarried. The numerical specification and the form of the marginal distribution of the continuous random variable Y is as defined in Section 9.4.2.

been evaluated on the basis of previous experiments conducted under controlled conditions by both procedures and are equal to 0.0229 g kg^{-1} for procedure HS and equal to 0.0463 g kg^{-1} for procedure IS. Thus, one may propose a Bayesian network with three nodes as shown in Figure 9.17(i), where the node P is a discrete node with states HS and ID, representing the implemented procedure, and is a parent node of X. The marginal distribution of measurement X is represented in Figure 9.17(ii), where the marginal mean is equal to 0.7 and is obtained as in (9.24)

$$E(X) = E_P[E_\theta(E(X|\theta, P))] = E_P[\mu] = \mu = 0.7,$$

and the marginal variance is equal to 0.0113 and is obtained as in (9.25) with $\alpha_1 = \alpha_2 = 1$, $\beta_1 = \beta_2 = 0$, $\sigma_1^2 = \sigma_{HS}^2 = 0.0229^2$, $\sigma_2^2 = \sigma_{ID}^2 = 0.0463^2$ and $\tau^2 = 0.01$, $p_1 = p_2 = 0.5$ and is

$$Var(Y) = 0.01 + 0.0229^2 \times 0.5 + 0.0463^2 \times 0.5 = 0.013.$$

Note that it is a common practice to perform multiple measurements with a given implemented procedure and to report the mean of the observed measurements. The Bayesian network proposed in Figure 9.17 can be easily extended to allow for multiple measurements, as shown in Figure 9.18. Note solely that node X has been replaced by node $\bar{X}$, representing the sample mean of the measurements, whilst a discrete node n has been included to take into account the number of observations. The conditional distribution of the mean of the measurements is $(\bar{X}|\theta, n, P) \sim N(\mu, \sigma_p^2/n)$. As an example, suppose two measurements are obtained with the procedure HS, with the result $\bar{x} = (x_1 + x_2)/2 = (0.6066 + 0.5778)/2 = 0.5922$. The posterior distribution of θ is still Normal with posterior mean μ_x equal to 0.5950 and posterior variance τ_x^2 is equal to 0.0003, as can be seen in Figure 9.18(i). Recalling (9.15) and (9.16), it can easily be verified that

$$\mu_x = \frac{(0.0299^2/2)0.7 + 0.01(0.5922)}{0.0229^2/2 + 0.01} = 0.5950$$

$$\tau_x^2 = \frac{(0.0229^2/2)0.01}{0.0229^2/2 + 0.01} = 0.0003.$$

Note that these results can be also obtained in a sequential Bayesian updating procedure for θ, based on individual measurement results X. This is illustrated in Figure 9.18(ii) where nodes

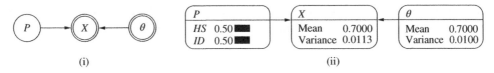

	P		X		θ
HS	0.50	Mean	0.7000	Mean	0.7000
ID	0.50	Variance	0.0113	Variance	0.0100

(i)   (ii)

Figure 9.17 Collapsed (i) and expanded representation (ii) of a Bayesian network with two continuous nodes θ, representing the true level of a concentration of interest, and X, representing the measured concentration, and one discrete node P with two states HS and ID, representing two distinct analytical procedures. Nodes θ and P are parent nodes for X. (ii) shows the initial state of the network (i.e. without any instantiations) with values for the marginal distributions as discussed in Example 9.10.

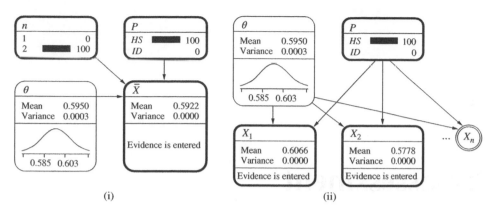

(i)   (ii)

Figure 9.18   (i) Bayesian network with two continuous nodes θ, representing the true level of alcohol concentration in blood, and $\bar{X}$, representing the mean of measured concentrations, and two discrete nodes, P, with two states HS and ID (representing two distinct analytical procedures), and n (representing the number of measurements). Node θ shows the posterior distribution of the true level of blood alcohol concentration, given a mean $\bar{X}$ ($= 0.5922$) of two values of blood alcohol concentration measurements obtained with the procedure IS. The instantiated nodes X and n are shown with a bold border. (ii) Bayesian network with nodes θ and P that have the same definition as the network (i) whilst nodes n and $\bar{X}$ are replaced by continuous nodes $X_1, X_2, \ldots, X_n$, representing results of $1, 2, \ldots, n$ distinct concentration measurements performed with a given method (as defined by node P). The network shows a situation in which results from two measurements are available, that is, $X_1 = 0.6066$ and $X_2 = 0.5778$ g kg^{-1}.

n and $\bar{X}$ from Figure 9.18(i) are replaced by continuous nodes $X_1, X_2, \ldots, X_n$, representing results of $1, 2, \ldots, n$ distinct concentration measurements performed with a given method. This is equivalent to duplicating node X n times in Figure 9.17(i). Figure 9.18(ii) reflects the situation introduced above, where two separate results from measurements are available, notably 0.6066 and 0.5778 g kg^{-1}. Entering these findings in nodes X_1 and X_2 leads to the same posterior distribution for node 0 as in Figure 9.18(i).

10

Pre-assessment

10.1 Introduction

An evaluation process starts when the scientist first meets the case. It is at this stage that the
scientist thinks about the questions that are to be addressed and the outcomes that may be
expected. Together with the submitting agency or party, the scientist should attempt to frame
propositions of interest that can help to think about the value of expected findings (Evett et al.,
2000b). There is a wide tendency, however, to consider evaluation only as a final step in casework examination, notably at the time of preparation of the formal report. This is so even if
an earlier interest in the process would enable the scientist to make better choices about the
allocation of resources. One of the first approaches to strategic management in an operational
forensic science unit was proposed by Cook et al. (1998b). It is based on a model embody-
ing the likelihood ratio as a measure of probative value. The proposed model is intended to
enhance the cost-effectiveness of casework activity from initial contact with the person or
party requesting particular forensic services. The aim is to offer support in making better
informed choices about which analyses to undertake.

In routine work, forensic scientists thus often require answers to questions such as 'What
are the magnitudes of the likelihood ratios that may be obtained?', or 'How probable is it that
a likelihood ratio of a given magnitude will be obtained?' A submitting party may wish to
know about such information prior to deciding about how to proceed in a given case.

For the purpose of illustration, imagine a case of disputed paternity. The alleged father is
unavailable, but a cousin of the alleged father could potentially be considered and analysed. In
such a setting, the two propositions of interest may be of the usual form 'The analysed person
is a cousin of the true father', versus 'The analysed person is unrelated to the child'. In such a
scenario, two issues may be raised. One relates to whether one can obtain an evidential value
such that the first hypothesis is favoured over the other. The other pertains to the question of
how a laboratory or a customer can take a rational decision on the necessity of performing
DNA analyses, after an assessment of possible values for the likelihood ratio. The former of
these two questions refers to a process known as *case pre-assessment* developed mainly in

Bayesian Networks for Probabilistic Inference and Decision Analysis in Forensic Science, Second Edition.
Franco Taroni, Alex Biedermann, Silvia Bozza, Paolo Garbolino and Colin Aitken.
© 2014 John Wiley & Sons, Ltd. Published 2014 by John Wiley & Sons, Ltd.

the late 1990s in the UK's Forensic Science Service (Cook et al. 1998b) and is treated here throughout this chapter. The latter question is one that refers to decision analysis as discussed, for example, in Taroni et al. (2005, 2007). Decision theory is more formally introduced in this book in Chapters 1 and 11.

10.2   General elements of pre-assessment

In order to follow a logical procedure, scientists need to take into account the circumstances of the case. These form an integral part of the framework in which examinations will be carried out. They also help to provide a description of the expected outcomes (Cook et al. 1998b), but they have to be known before any examination of the submitted items is made so as to propose relevant propositions.

The choice of a propositional level according to the scheme described by Cook et al. (1998a), described earlier in Chapter 3, is important and provides a basis for consistency for scientists. It encourages them to consider carefully factors such as circumstantial information and data that are to be used for the evaluation.

Example 10.1 *(A generic glass case) The scientist should proceed by considering an assessment of the probability of whatever results will be obtained, given each proposition of interest. Consider, for example, a case potentially involving glass fragments due to a smashed window. Assume that the prosecution and defence propose activity-level propositions such as 'the suspect is the man who smashed the window' (H_p) and 'the suspect has nothing to do with the incident' (H_d). The examination of the suspect's pullover leads to the recovery of a quantity Q of glass fragments. For the purpose of illustration, let the possible states for Q cover the descriptions 'none', 'few' and 'many'.*

Then, for assessment of the numerator of the likelihood ratio, the question to be answered is 'What is the probability of finding a quantity Q of "matching" glass fragments if the suspect is the man who smashed the window?' Similarly, for the denominator of the likelihood ratio, one needs to address a question of the following kind: 'What is the probability of finding a quantity Q of "matching" glass fragments if the suspect is not the man who smashed the window and has nothing to do with the case?'

Initially, the scientist is asked to assess a total of six distinct probabilities. These relate to the three states of the variable Q, given, respectively, H_p and H_d. The various assessments may be informed by different sources of information, such as surveys or relevant publications on the subject matter, paired with personal judgment by the scientist (Taroni et al. 2001). These probabilities may not be easy to assign, however, because of possibly incomplete information. It may be difficult, for example, to assess transfer probabilities when the scientist has no or only vague indications about parameters pertaining to the modus operation. In particular, it may not be known how the window was smashed and how close the perpetrator was with respect to that window at the moment it was smashed.

Consider this in terms of a scenario described by Cook et al. (1998b). These authors propose exemplary probability distributions, summarized here in Table 10.1, for finding quantities Q of glass fragments, with corresponding analytical features, under the two competing propositions H_p and H_d. These values allow one to find likelihood ratios as a measure of the value of finding a quantity Q of glass fragments. In view of these assignments, a pre-assessment could take the following wording:

Table 10.1 Probabilities of finding quantities Q of glass with corresponding analytical characteristics.

Q	$Pr(Q|H_p)$	$Pr(Q|H_d)$	V
None	0.05	0.95	1/19
Few	0.30	0.04	7.5
Many	0.65	0.01	65

The propositions are H_p and H_d, the suspect is the man who smashed the window, and, respectively, he had nothing to do with the breaking of the window. The values of the likelihood ratio V for the various findings are indicated in the right-hand column. *Source*: Adapted from Cook et al. (1998b)

[...] on the basis of this assessment, if the suspect is indeed the person who committed the burglary, there is a 65% chance that the result of the examination will provide moderate support for that proposition; and a 30% chance that it will provide weak support. If, on the other hand, the suspect is truly not *the offender then there is 95% chance of moderate evidence to support his innocence although there is a 5% chance of evidence which will tend falsely to incriminate him. (Cook et al. 1998b, p. 155)*

On the basis of such an assessment, it may be thought that the scientist is in a position to help the customer take an informed decision. Whilst this stage is a fundamental one in a perspective that seeks to inform about expected findings, it does not offer clear criteria of a genuine decision-making process. For example, the assessment does not offer an explicit framework to address the question of performing or not a given examination (e.g. an analysis in the laboratory). Section 11.1 offers further discussion of this critical point which is essential because scientists routinely take decisions, either individually or on behalf of requesting parties, and they do so generally under uncertainty. A more extended logical framework, based on decision theory, should be employed for this task.

10.3   Pre-assessment in a fibre case: A worked through example

10.3.1   Preliminaries

Scenarios proposed earlier in Chapter 5 focused on evaluation in situations where a group of fibres has been found during the investigation of an offence. Retaining offence-level propositions, the discussion distinguished between two main kinds of situations. One of them considers recovered fibres as potentially coming from the offender under both propositions of interest. The other treats recovered fibres as being present by chance alone in the denominator. A formal development shows that amongst the factors of interest are transfer, persistence and recovery, as well as the presence by chance (probability of background). The rarity of the fibre's characteristics in a relevant population, denoted γ, represents a further relevant factor.

The following sections describe a more detailed study of a scenario involving textile fibres, initially described by Champod and Jackson (2000). The proposed hypothetical case allows one to illustrate the interplay between the tasks of defining the information scientists may need together with relevant propositions used to assess findings, the actual conduct of a case pre-assessment, the selection of an examination strategy, the assessment of a likelihood ratio and the evaluation of the effect when changing propositions.

The scenario is as follows. Two armed and masked men burst into a post office and threaten the staff. The takings for the day were handed over and the men left. Witnesses report that one of the men was wearing a dark green balaclava mask and the other man was wearing a knotted stocking mask. Witnesses further declare that the two men left the scene in a car driven by a third man. Some way along the presumed getaway route, a dark green balaclava was found. The following day Mr U is arrested. He denies all knowledge of the incident. The scientists seize a few hairs from the suspect's head, take a swab for DNA and take combings from his head hair. Mr U has not yet been charged with the robbery because there is very little evidence against him.

10.3.2   Propositions and relevant events

At some stage during the investigative proceedings, it may be of interest to know whether Mr U has worn the incriminated mask. As a starting point, the scientist may formulate propositions at the source level, such as 'Hairs in the mask came from Mr U' versus 'Hairs in the mask came from someone else'. Other pairs of propositions could be 'Saliva in the mask came from Mr U' versus 'Saliva in the mask came from someone else' or 'Fibres in U's hair combings came from the mask' versus 'Fibres in U's hair combings came from some other garment or fabric'.

Another level of propositions, often more useful for investigative authorities, focuses on an activity. A typical example for such a hierarchical level is a pair of propositions, such as 'Mr U wore the mask at the time of the robbery' (H_p) versus 'Mr U has never worn this mask' (H_d). Such propositions can be more relevant for the court because they offer a closer link with the alleged offence. However, scientists will need more detailed information on the circumstances pertaining to the incident, such as the time the alleged offence took place, conditions under which the suspect was arrested, the time when traces were searched, and so on. Such data enable the scientist better to assess the factors that relate to transfer and persistence of any recovered trace material. Besides the circumstances, scientists should also focus on transfer, persistence and recovery of hairs, fibres and saliva of individuals wearing masks and survey data on masks.

Published data on hairs and saliva are very limited. Data on transfer of fibres to head hair, including persistence, are more readily available. So if the above requirements regarding available information are met, the scientist may consider fibres first. If, however, the background information is not sufficient for considering trace material (fibres, saliva, hairs) under activity-level propositions, then scientists may be obliged to remain with source-level propositions. In the scenario described above, a feasible strategy would be to offer an assessment, given activity-level propositions for fibres, but to evaluate the other trace material, given source-level propositions.

A second step in a case pre-assessment consists of thinking about the possible findings. Considering fibres on hair combings, scientists may encounter results that can be described in terms of the following categories: no fibres are recovered, a small number of fibres are recovered (i.e. 1–3) or a large number of fibres are recovered (i.e. more than 3). The definition of such categories is flexible and may depend on the available data. It is also possible that more than one group of fibres could eventually be found.

An assessment for activity-level propositions requires the definition of further variables. In order to help scientists to determine relevant variables, for a case at hand, it may be useful to consider a question of the kind: 'What could happen if Mr U wore the mask at the time of the robbery?' Such a question leads to three main events: T_0, no fibres have been transferred (arguably, there is no persistence and no recovery of fibres that are crime related), T_s, a small number of fibres has been transferred (i.e. the fibres have persisted and have successfully been recovered) and T_l, a large number of fibres has been transferred (i.e. the fibres have persisted and have successfully been recovered). These events are suitable for definitions of the states of a variable in a Bayesian network. Later in this chapter, such a variable will be denoted 'transfer', or T for short.

Two main propositions can account for the presence of fibres. One refers to a crime-related transfer, as noted in the previous paragraph. Another is the presence by chance prior to discovery. Such a possible background presence can be defined in terms of the two events P_0, no group of fibres is present by chance, and P_1, one group of fibres is present by chance. A group of fibres present by chance may be either small or large. Let these events be denoted as follows: S_s, the group of fibres present by chance is small, and S_l, the group of fibres present by chance is large. Note that the discussion here refers to head hair, which is a particular target area for the search of fibres. It is reasonable in this case to expect either no or at most one group of fibres. This assumption may not be appropriate for other target surfaces where textile fibres are much more prevalent (e.g. the surface of a garment).

Note that in Chapter 5, outcomes for P_i and $S_{i,j}$ have been aggregated within a single variable $b_{g,m}$. Probabilities of background $b_{g,m}$ account for the chance of occurrence of g group(s) of m foreign fibres on the receptor. This probability can be considered as a combination of probabilities p_i and $s_{i,j}$ as is currently practised in the evaluation of glass fragments [e.g. Curran et al. (2000)]. Values for p_i denote the probabilities by which i (≥ 0) groups of fibres are present. The term $s_{i,j}$ denotes the probability that group i of recovered material is of size j, where j may take a positive integer value. Notice that j may also be replaced by l or s in order to refer to a large or small group, respectively. Literature on the evaluation of glass fragments considers that (i) there is no association between the number of groups found on surfaces of interest and the sizes of those groups and (ii) there is no association between the occurrence of a given type of glass fragments with either the number of groups or the size of the group (Curran et al. 2000). In the context of fibres, such assumptions may be more difficult to maintain. Properties such as transfer and persistence may be affected by physical characteristics of the type of fibres involved. To keep the development at a tractable level, the present discussion will, however, consider similar assumptions as acceptable.

When comparing a recovered group of fibres of unknown origin with fibres from a known source, that is, a control, two outcomes are possible. Either the recovered fibres are found to correspond to the control with respect to some features of interest, an event denoted M, or the recovered fibres are found to have 'different' characteristics, denoted $\bar{M}$. Note that the discussion here will make the assumption that for the case of discrete variables, the probability of a correspondence in analytical characteristics is 1 when a given item (or trace), known to originate from a particular source, is compared with a reference item from that source.

10.3.3   Expected likelihood ratios

When searching hair combings for the presence of fibres, and when analysing fibres thus recovered, scientists could reasonably meet any of the four situations summarized in Table 10.2. Note that this listing of possible outcomes does not take into account other scenarios, such as one group of fibres being present due to a transfer and a second group of fibres being present as background. The current discussion thus considers the possible presence of at most one group of fibres.

Table 10.3 provides a summary of the various events, given, respectively, H_p and H_d, and the associated probabilities. Note that $Pr(E|H_p)$ and $Pr(E|H_d)$ associated with outcomes O_3 and O_4 correspond to (5.6), Section 5.1.6.

Table 10.2   Major possible outcomes of a procedure for searching and analysing fibres present in hair combings.

Outcome	Number of groups	Number of non-corresponding groups	Number of corresponding groups	Size of corresponding groups
O_1	0	0	0	–
O_2	1	1	0	–
O_3	1	0	1	Small
O_4	1	0	1	Large

Table 10.3   Events and probabilities relating to findings under H_p and H_d, propositions referring to the mask's wear by the suspect.

Outcome from Table 10.2	Events to occur if H_p is true	$Pr(E|H_p)$
O_1	T_0, P_0	$t_0 p_0$
O_2	$T_0, P_1, \bar{M}$	$t_0 p_1 (1 - m)$
O_3	T_0, P_1, S_s, M or T_s, P_0	$t_0 p_1 s_s m + t_s p_0$
O_4	T_0, P_1, S_l, M or T_l, P_0	$t_0 p_1 s_l m + t_l p_0$

Outcome from Table 10.2	Events to occur if H_d is true	$Pr(E|H_d)$
O_1	P_0	p_0
O_2	$P_1, \bar{M}$	$p_1 (1 - m)$
O_3	P_1, S_s, M	$p_1 s_s m$
O_4	P_1, S_l, M	$p_1 s_l m$

Events $T_{\{0,s,l\}}$ relate to transfer, persistence and recovery, whereas $P_{\{0,1\}}$ and $S_{\{s,l\}}$ relate, respectively, to the presence by chance and the size of a group of fibres (with subscripts s and l referring to 'small' and 'large'). Corresponding probabilities are denoted with lower case letters $t_{\{0,s,l\}}$, $p_{\{0,1\}}$ and $s_{\{s,l\}}$. Variable M has two states, the recovered fibres correspond, M, or do not correspond, $\bar{M}$, to the control fibres with respect to some features of interest, with associated probabilities, m and $1 - m$.

The next step in the process of pre-assessment focuses on the assignment of values for these probabilities using data from published literature, case-specific experiments or personalized assessments based on the scientist's experience. Start by considering probabilities for the event of transfer. In order to find appropriate values for t_0, t_s and t_l, it may be useful to answer questions of the kind 'If the suspect wore a mask, what is the probability of, respectively, no fibres, a small or a large number of fibres being transferred, having persisted and successfully being recovered?' An appropriate assessment should also consider information pertaining to the suspect (e.g. type and length of his hair), the material involved (e.g. the sheddability of the mask), the methods used to search and collect fibres as well as the circumstances of the case (e.g. alleged activities, time delays).

The probabilities of background are expressed by p_0, p_1, s_s and s_l. These component values serve the purpose of assessing situations involving the presence, by chance, of no fibres or one group of fibres (which could be small or large, as previously defined), assuming that no fibres have been transferred or, alternatively, that the suspect never wore the mask. Notice that if the alternative proposition changes, for example, to one that the suspect wore a similar mask 2 days before the alleged facts, then probabilities for background change and new assessments may be required.

The probabilities abbreviated by m represent, in some way, an assessment of the rarity of extraneous fibres found in the head hair of an innocent person, falsely accused of wearing a mask. These are fibres that correspond by chance to the control fibres originating from the mask. This rarity may be assessed in various ways. For example, the scientist may refer to studies on textile fibres recovered from head hair and consider the relative proportions of fibres presenting the features of interest. Eventually, they may also consult so-called target fibre studies. Note, however, that such studies lead to other kinds of probabilities, such as probabilities of observing by chance one target group of fibres that correspond to the control: $Pr(P_1, S_s, M|H_d)$ and $Pr(P_1, S_l, M|H_d)$, previously referred to as γ_b (Section 5.1.2). To some extent, databases may also be used, assuming that the potential sources of fibres are hats, neckwear, bedding or jumpers, so that the scientist is able to assess occurrence of the corresponding fibres in these populations.

Example 10.2 *(Case involving fibres and a mask) When using the probabilities* $\{t_0 = 0.01, t_s = 0.04, t_l = 0.95\}$ *for events of transfer,* $\{p_0 = 0.78, p_1 = 0.22\}$ *for the number of groups,* $\{s_s = 0.92, s_l = 0.08\}$ *for group size and m = 0.05 for the observation of corresponding features, as proposed in Champod and Jackson (2000), the likelihood ratios for the outcomes defined in Table 10.2 are 0.01 for each of the two outcomes O_1 and O_2, 3.0 for outcome O_3 and 842 for outcome O_4.*

'Is it useful to conduct analyses of textile fibres?' Likelihood ratios obtained as a result of the pre-assessment of textile fibres can assist in the provision of an answer to such a question. The numerical example shows that all considered outcomes offer likelihood ratios that allow one to discriminate between the competing propositions of interest (i.e. likelihood ratios are different from 1). If no fibres at all are recovered or if a group of fibres is recovered and this group can be distinguished from the control object, then the likelihood ratios support the proposition of the defence. On the other hand, if one group (small or large) of fibres is recovered and this group corresponds to the control group (i.e. shares the same analytical features), then a likelihood ratio greater than 1 is obtained, and the prosecution's case is supported.

Table 10.4 Conditional probabilities assigned to the outcome node O of Figure 10.1.

T:	T_0				T_s				T_l			
P:	P_0		P_1		P_0		P_1		P_0		P_1	
S:	S_s	S_l	S_s	S_l	S_s	S_l	S_s	S_l	S_s	S_l	S_s	S_l
$O:$ O_1	1	1	0	0	0	0	0	0	0	0	0	0
O_2	0	0	0.95	0.95	0	0	0	0	0	0	0	0
O_3	0	0	0.05	0	1	1	0	0	0	0	0	0
O_4	0	0	0	0.05	0	0	0	0	1	1	0	0
O_5	0	0	0	0	0	0	1	1	0	0	1	1

Node O has five states, the discovery of no group of fibres, O_1; a group of non-corresponding fibres, O_2; a small group of corresponding fibres, O_3; a large group of corresponding fibres, O_4 and a generic state O_5 for all possible situations other than O_i (for $i = 1, \ldots, 4$). Variable T has three states, no transfer, T_0; transfer of a small amount of fibres, T_s, and transfer of a large amount of fibres, T_l. Variable P has two states, P_0 and P_1 corresponding to no (0) or one (1) group of fibres being present by chance. Variable S has two states, S_s and S_l, corresponding to the group of fibres being present by chance being small (s) or large (l), respectively. The value 0.05 is the probability m assigned for the event of observing corresponding analytical features on either a small or a large group of fibres. The value $1 - m = 0.95$, in turn, corresponds to the probability of observing non-corresponding features on a group of fibres of any size.

10.3.4   Construction of a Bayesian network

Following discussion presented in Section 10.3.2, four variables appear relevant: H (with states H_p and H_d), T (with states T_0, T_s and T_l), P (with states P_0 and P_1) and S (with states S_s and S_l). These propositions define nodes for a Bayesian network along with a fifth node O, for 'outcomes' as defined in Section 10.3.3. The node O has the states O_i (for $i = 1, \ldots, 4$) as defined in Table 10.2, but this list is not exhaustive. For example, theoretically, there may have been transfer and a presence of background. For this reason, let there be an additional state O_5 for all outcomes other than those defined in Table 10.2. This will assure that the node probabilities for O sum up to one.

On a structural account, one can consider that the phenomenon of transfer depends on whether or not the proposition H is true. In turn, the variable T acts as a conditioning for the outcome O, along within P and S. In combination, this leads to a network structure as shown in Figure 10.1.

With this Bayesian network, one can obtain the same likelihood ratios as in Example 10.2 (Section 10.3.3). It is solely necessary to specify the same numerical assignments for phenomena of transfer and the number and the size of fibre groups. In addition, conditional probabilities for the node O as presented in Table 10.4 need to be adopted.

10.4   Pre-assessment in a cross-transfer scenario

10.4.1   Bidirectional transfer

As a basic approach to inform a decision-making process, case pre-assessment can also be applied in more challenging cases. In order to illustrate this, imagine a case involving a cross-transfer, also known as *two-way transfer* (Aitken and Taroni, 2004; Champod and Taroni, 1999). The scenario of interest is that of an assault involving one victim and one

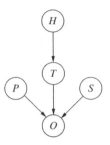

Figure 10.1 Bayesian network useable for pre-assessment in cases involving transfer material. Node O, defined as 'outcome', has five states, the discovery of no group of fibres, O_1; a group of non-corresponding fibres, O_2; a small group of corresponding fibres, O_3; a large group of corresponding fibres, O_4, and a generic state O_5 for all possible situations other than O_i (for $i = 1, \ldots 4$). Node T has three states, no transfer, T_0; transfer of a small amount of fibres, T_s, and transfer of a large amount of fibres, T_l. Node P has two states, P_0 and P_1 corresponding to no (0) or one (1) group of fibres being present by chance, with associated probabilities p_0 and p_1, respectively. Node S has two states, S_s and S_l, denoting the events that to the group of fibres being present by chance is small (s) or large (l), with associated probabilities s_s and s_l, respectively. Node H has two states, H_p, Mr U wore the mask at the time of the robbery, and H_d, Mr U has never worn the mask.

criminal. The criminal and the victim have, in some way yet to be defined in further detail (see Example 10.3), been in contact with each other. The scientific data under consideration are DNA profiles corresponding to, respectively, the victim and the criminal.

In a rape case, the biological material could consist of semen and vaginal fluids, for example. The material could also be blood in an assault case. It needs to be remembered, however, that transfer can occur also in one direction only. Suppose, for example, that a victim has been killed with a knife and there are no signs of a fight, that is, the offender has not been injured. In such a setting, one may consider that the probability of a transfer of blood from the criminal to the victim is low. In turn, the probability of a transfer of blood from the victim to the criminal may be high. More generally, the two sets of recovered traces have to be considered as dependent, at least under the prosecution's proposition because it assumes that the alleged criminal activity was committed by the suspect. In fact, if a transfer has occurred in one direction and the expert has recovered traces characterizing this transfer, then the expert would generally expect – still under the assumption of the prosecution's case – to find trace material characterizing transfer in the other direction. The presence of one set of transfer material can provide information about the presence of the other set of transfer material. Obviously, the absence of the other set of transfer material could by itself be significant as discussed in Sections 5.2 and 10.4.

The dependence between the two sets of materials plays an important role in the assessment of a case and allows scientists to think in a hierarchical manner. In particular, scientists can update an assessment by following a staged approach. In a scenario involving fibres, for example, they may examine the victim's pullover first and then, in a second step, proceed with an examination of the suspect's garment (Cook et al. 1999). The results of the examination of one of the garments can offer relevant information for a scientist's client who may need to decide whether or not the second garment should be examined.

Analogous reasoning could also be applied in other kinds of cases. For example, if the crime involves the smashing of a sheet of glass and if clothing is submitted from a suspect, then the phased approach can help select an order in which the garments are examined. If, for instance, the examination of the jacket reveals no glass, then this information may be useful for assessing the option of examining the trousers or shoes. Section 10.4.3 will approach such a scenario.

Example 10.3 *(Cross-transfer and textile fibres) Consider a case described by Cook et al. (1999). Mr A disturbed a man who had broken into his home. He attempted to restrain the intruder and struggled with him without being able to prevent him from making his escape. Sometime later, a man, Mr B, is arrested.*

Given the extent of available case-related information, police investigators are interested in the following two activity-level propositions: 'Mr B is the man who struggled with Mr A (H_p)', and 'There has been no physical contact between Mr B and Mr A (H_d)'. For clarity of discussion, it may be useful to add that in the latter case, it is supposed that the suspect has nothing to do with the case at hand (e.g. because the two men do not know each other and have never met in the past). This allows one to avoid considerations of potential fibre presence due to a transfer not related in any way to the alleged incident.

The possible findings consist of a quantity, say Q, of transferred fibres. The quantity Q concerns, independently, both Mr A and Mr B. For the sake of clarity, let Q_A and Q_B denote the quantities found on Mr A and Mr B, respectively. Then, when scientists find a quantity Q_B of fibres on the surface of the clothing of Mr B, indistinguishable from fibres in the fabric of Mr A's garment, then the scientist needs answers to questions such as 'What is the probability of finding a quantity Q_B of fibres on the clothing of Mr B if he is the man who struggled with Mr A?' and 'What is the probability of finding a quantity Q_B of fibres on the clothing of Mr B if there has been no physical contact between him and Mr A?' In a general approach to this case, scientist may define the variable Q in rather broad qualitative terms such as 'none', 'few' and 'many' fibres.

With respect to Mr A, they may choose to describe potential findings as a quantity Q_A of fibres that has been transferred, has persisted and was successfully recovered from the surface of Mr A's garment. Also part of such a definition is that the scientist finds the fibres to be of the same type as control material taken from, for example, the jumper of the suspect, Mr B. Regarding the examination of the suspect's clothing, scientists may define the potential outcome as a quantity Q_B of fibres that has been transferred, has persisted and was successfully recovered from the surface of Mr B's garment. Likewise, they may add that observation indicates that the fibres are of the same type as control material taken from clothing of the victim, Mr A.

It is important to keep in mind that, as mentioned by Cook et al. (1999), one may allow for different intensities of contact – described here as a 'struggle' – between the victim and the aggressor. In fact, the extent of contact is a major factor affecting fibre transfer. It may be incorporated in the analysis as a distinct source of uncertainty. A variable 'intensity of contact' can account for this need by assuming states referring to, for example, 'light', 'medium' and 'heavy' contact.

10.4.2 A Bayesian network for a pre-assessment of a cross-transfer scenario

When building a Bayesian network for pre-assessment in a case where a cross-transfer is a potential issue, aspects such as the number of variables, their states and the relationships amongst them depend closely on the level of detail and the scope of considerations that are required in the work of the scientists. As a guiding example, the discussion in the following will refer to Example 10.3 introduced in the previous section.

Example 10.4 (*Cross-transfer and textile fibres – continued, Bayesian network construction*) *The main propositions of a cross-transfer scenario as described in Example 10.3 can serve the purpose of the definition of nodes for a Bayesian network with a structure as shown in Figure 10.2.*

The node H accounts for the two main propositions, namely, 'Mr B is the man who struggled with Mr A' (H_p) and 'there has been no physical contact between Mr B and Mr A' (H_d). Related to this is the node C that represents the event 'contact'. It depends directly on the main propositions. Conditional probabilities thus need to be assigned for each state of C, given, respectively, H_p and H_d. Cook et al. (1999) suggested probabilities of 0.1, 0.3 and 0.6 for light, medium and heavy contact, respectively. Notice that if H_d is true, that is, the proposition according to which there has been no physical contact between Mr B and Mr A, then the variable C cannot be light, medium or heavy but should assume a different state, denoted 'no contact' here. The variable C is thus described by a total of four states. The latter state, 'no contact', takes probability 1 if H_d is true, and 0 otherwise.

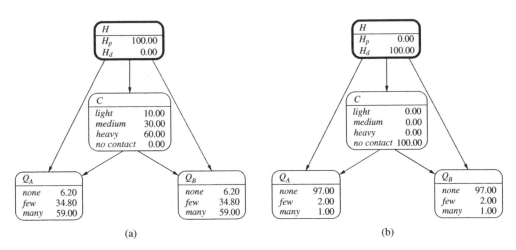

Figure 10.2 Assessment of the expected quantities Q of recovered fibres, given (a) H_p is true and (b) H_d is true. Node C represents the level of contact. Node H represents the prosecution proposition, H_p, that Mr B is the man who struggled with Mr A and the defence proposition, H_d, that there has been no physical contact between Mr B and Mr A. Nodes Q_A and Q_B represent the quantity of fibres recovered on, respectively, Mr A's and Mr B's garments. The instantiation of node H is shown as a bold border.

Table 10.5   Probability distribution for the quantity Q of fibres recovered, given contact, C, and the prosecution proposition, H_p, that Mr B is the man who struggled with Mr A.

	C:	Light	Medium	Heavy	No contact
Q:	None	0.2	0.1	0.02	1
	Few	0.6	0.4	0.28	0
	Many	0.2	0.5	0.70	0

The node Q_A represents the quantity of fibres recovered on Mr A's garment. These fibres correspond, in terms of the compared analytical features, to those composing Mr B's garment. For the sake of simplicity, the node Q_A (and Q_B here below) is allowed to have a broad definition, including phenomena of transfer, persistence and recovery. The direct dependency of node Q_A on C, the intensity of contact, is expressed by the arc $C \rightarrow Q_A$. Node H also exerts influence on Q_A, but in two distinct ways. One influence is direct through $H \rightarrow Q_A$, whilst the other is indirect, in terms of the path $H \rightarrow C \rightarrow Q_A$. The probability table for the node Q_A thus is not dictated by C alone. Cook et al. (1999) made the particular assumption that both garments have comparable shedding and retentive properties. Therefore, given that for both garments, the extent of contact was the same and the lapse of time between the alleged incident and fibre recovery is approximately similar, it may be acceptable to assign the same probabilities to both nodes Q_A and Q_B (see Table 10.5). This assumption applies when considering H_p as a conditioning state. However, when H_d is true, then the variable 'intensity of contact' is not relevant and the scientist is concerned only with the presence by chance of a quantity Q of corresponding fibres. For the purpose of illustration, consider conditional probabilities of 0.97, 0.02 and 0.01 for none, few and many fibres, respectively. Notice that these values represent an assessment of the rarity of the extraneous fibres found on the receptor; they are fibres that correspond by chance to the control fibres. This rarity may be assessed in various ways. The scientist may, for example, consider target fibre studies. However, as noted earlier in this chapter, it should be remembered that such studies lead to different probability assignments, notably probabilities of observing by chance no or one (small or large) target group of fibres that correspond to the control.

Besides Q_A, there is a further observational node Q_B. This variable has the same states as Q_A, with the sole difference that the fibres with corresponding features are found on Mr B's garment.

10.4.3   The value of the findings

One way to conduct a pre-assessment consists of considering the option of a phased approach. That is, the case is reassessed once a first item has been examined. In the case introduced in Example 10.3, the scientist could, for instance, start by considering the black sweater of the victim Mr A, leaving the garment from the suspect aside for the time being. There could be red fibres on Mr A's garment, which might correspond to the red fibres of the sweater worn by the suspect, Mr B. Notice that the scenario is simplified here in the sense that it is assumed that at most one group of foreign fibres may be recovered. The scientist thus focuses on red fibres without interest in any other possible background fibres.

For whatever quantity Q_A of fibres that may be found, the scientist can evaluate a likelihood ratio, that is,

$$\frac{Pr(Q_A|H_p)}{Pr(Q_A|H_d)},$$

where the competing propositions H_p and H_d refer to a physical aggression as defined earlier in Example 10.3. Example 10.5 explores this through a numerically specified Bayesian network. Subsequently, Example 10.6 continues the pre-assessment for Mr B's garment, conditional on outcomes from the examination of the victim's clothing.

Example 10.5 (*Pre-assessment for fibres in a case of possible cross-transfer – continued*) *What is the probative value of a quantity Q_A of fibres found on a victim – here, Mr A. –of a physical assault? The Bayesian network constructed earlier in Section 10.4.2 can help to address this issue. Figure 10.2 shows an expanded view of this network with a display of the outputs of the two competing scenarios. The model shown in Figure 10.2(i) indicates the conditional probabilities of finding quantities Q_A of fibres, given H_p is true (i.e. that Mr B struggled with Mr A): $Pr(Q_A = none|H_p) = 0.062$, $Pr(Q_A = few|H_p) = 0.348$ and $Pr(Q_A = many|H_p) = 0.590$. In turn, Figure 10.2(ii) displays the expected quantities Q_A of fibres when H_d is true (i.e. Mr B has nothing to do with the case): $Pr(Q_A = none|H_d) = 0.970$, $Pr(Q_A = few|H_d) = 0.020$ and $Pr(Q_A = many|H_d) = 0.010$.*

Using these values, likelihood ratios V for finding quantities Q_A (i.e. none, few and many) fibres on Mr A's garment can be obtained as follows:

- $V_{Q_A=none} = \frac{Pr(Q_A=none|H_p)}{Pr(Q_A=none|H_d)} = \frac{0.062}{0.970} = 0.064$. *This scenario provides moderate support for the defence's case, that is, $1/0.064 \approx 16$.*

- $V_{Q_A=few} = \frac{0.348}{0.020} = 17.4$. *In this setting, there is moderate support for the prosecution's case.*

- $V_{Q_A=many} = \frac{0.590}{0.010} = 59.0$. *This result also provides moderate support for the prosecution's case.*

This analysis enables the scientist, even before an examination of the first garment (i.e. that of Mr A), to inform the customer about expected outcomes and their probative value. The scientist could say that, for example, if the suspect, Mr B, is the man who struggled with Mr A (H_p), then there is a probability of less than 0.1 (0.062) that a result will be obtained that will provide moderate support for the innocence of Mr. B. There is also a probability of about 0.35 (0.348) that a result will be obtained that will provide moderate support for the prosecution's case, that is, through a likelihood ratio of approximately 17. Moreover, it is more probable than not (probability 0.59) that a result will be obtained that will provide moderate support for the prosecution's case through a likelihood ratio of 59.

If, on the other hand, Mr B is truly innocent (H_d), then there is a 0.97 probability that the examination will result in a finding (i.e. a quantity $Q_A = none$) that offers moderate support for the defence's case. Besides, there is also a probability of 0.03 to obtain a result that will, falsely, incriminate Mr B. Notice that the value 0.03 is the cumulative probability of the conditional probabilities $Pr(Q_{A=few}|H_d)$ and $Pr(Q_{A=many}|H_d)$. In addition, recall that both outcomes 'few' and 'many' fibres are results with associated likelihood ratio in support of the prosecution's case.

Example 10.5 is to be noted because it focuses on pre-assessment when no other scientific results are yet available. This can be contrasted with a pre-assessment for findings on the suspect's clothing, assuming knowledge of results of a search conducted on the victim's clothing. Such a pre-assessment, based on some initial results on other examined items, is also known in the context as *staged (or, phased) pre-assessment*. Example 10.6 provides a numerical example.

Example 10.6 (*Staged pre-assessment for potential cross-transfer and textile fibres – continued*) *Imagine that the scientist started by examining Mr A's garment but did not find any fibres with corresponding analytical features to the clothing of Mr B. At this stage, the scientist may contact the customer in order to discuss whether or not to proceed with the examination of Mr B's red sweater. In such a situation, evaluation of $Pr(Q_B|Q_A = none, H_p)$ and of $Pr(Q_B|Q_A = none, H_d)$ becomes relevant as it allows the scientist to assess the expected probative value for different outcomes on Mr B's sweater. The Bayesian network defined in Example 10.4 can be used to find these target conditional probabilities.*

By instantiating node H to H_p and node Q_A to none, the probabilities of the states of the variable Q_B change to 0.117 (previously 0.062) for none, 0.441 (previously 0.348) for few and 0.442 (previously 0.59) for many corresponding fibres recovered on Mr B's garment. Figure 10.3(i) graphically illustrates this evaluation, and Figure 10.3(ii) analogously for a setting in which H_d is assumed to be true (along with $Q_A = none$). The likelihood ratios for finding quantities Q_B of fibres on Mr B's garment, given that no fibres have been found on Mr A's garment ($Q_A = none$), can then be found as follows:

- $V_{Q_B=none|Q_A=none} - \dfrac{Pr(Q_{B=none}|Q_{A=none},H_p)}{Pr(Q_{B=none}|Q_{A=none},H_d)} = \dfrac{0.117}{0.970} = 0.121.$

- $V_{Q_B-few|Q_A=none} = \dfrac{0.441}{0.020} = 22.05.$

- $V_{Q_B=many|Q_A=none} = \dfrac{0.441}{0.010} = 44.1.$

On the basis of this assessment, the scientist may contact the mandating party for a follow-up of the pre-assessment. If Mr B is the man who struggled with Mr A (H_p), then there is an approximate probability of 0.11 that the findings on Mr B's garment will tend to support his innocence. Since the likelihood ratio is smaller than 10, this support is considered as weak ($V^{-1} = 1/0.121 \approx 8$). Thus, knowledge of $Q_A = none$ has approximately doubled the probability (from 0.062 to 0.117) of obtaining results supporting Mr B's innocence. However, the degree of support decreases from a likelihood ratio of approximately 16 to one of about 8.

Further, likelihood ratios of approximately 22 and 44 can be obtained, each with a probability of about 0.44. These likelihood ratios provide moderate support for the prosecution's case. Thus, the effect of knowing $Q_A = none$ is to reduce the overall probability of obtaining evidence incriminating Mr B.

The scientist could also advise the client that if the suspect is truly innocent (H_d), then there is a 0.97 probability of obtaining results that provide moderate support for the defence proposition and that there is a 0.03 probability of obtaining results that will falsely incriminate the suspect.

It is also possible to allow for uncertainty about the truthstate of H (i.e. leave H uninstantiated) and assess the probabilities of the various outcomes in node Q_B, given knowledge about an observed quantity Q_A. For example, when assuming $Pr(Q_A = none) = 1$, then the probabilities of finding quantities Q_B of fibres are, respectively, 0.919, 0.045 and 0.036. This would imply, for instance, that the probability of finding no matching fibres on Mr B's garment, given that no matching fibres have been found on Mr A's garment is high (0.919). Such an assessment can provide the customer with further information to support a decision.

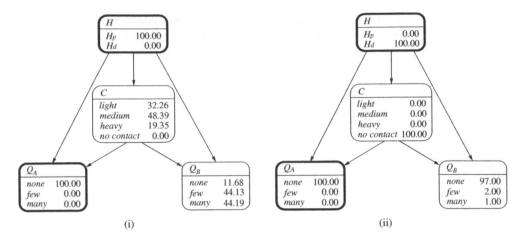

Figure 10.3 Assessment of expected quantities Q_B of fibres on Mr B's garment, given (i) H_p is true and (ii) H_d is true. It is further assumed that no fibres have been found on Mr A's garment. Node C represents the level of contact. Node H represents the propositions, the prosecution proposition, H_p, that Mr B is the man who struggled with Mr A and the defence proposition, H_d, that there has been no physical contact between Mr B and Mr A. Nodes Q_A and Q_B represent the quantity of fibres recovered on, respectively, Mr A's and Mr B's garments. The instantiation of the nodes H and Q_A is shown with a bold border.

10.5   Pre-assessment for consignment inspection

10.5.1   Inspecting small consignments

Operational forensic science laboratories commonly encounter consignments or collections of discrete units whose attributes may be relevant within the context of a criminal investigation. The kind of consignments considered here involves individual units that are all of the same kind. Examples include tablets or CDs within a population of, respectively, tablets or CDs. The individual units may or may not have a given target characteristic, such as illegal content. A small consignment, as considered here, is one that typically contains less than 50 units.

The analyses pursued here below follow the Bayesian approach outlined in Aitken (1999). It involves a number M of examined units, Z of which are found to be positive, and allows one to (i) re-evaluate one's uncertainty about θ (the proportion of the total number of units in the consignment that are positive), represented in terms of a beta probability distribution, and (ii) derive a probability distribution for Y, the number of positive units amongst those not analysed, which is binomial. The latter becomes a beta-binomial distribution when combined with the posterior beta distribution for θ. Section 10.5.2 describes a Bayesian network, shown in Figure 10.4, that can approach inferences about both θ and Y (Biedermann et al. 2008b). Table 10.6 summarizes the node definitions.

In this context, the notion of 'sample' is typically used to refer to a selection of units from a given population of units (or more generally speaking, the extraction of part of a whole). It is useful to emphasize that this term has a meaning that differs from that of the notion of 'specimen'. The latter notion also refers to part of a whole, yet has a fundamentally different meaning. In a great majority of forensic applications, a specimen represents a single (possibly

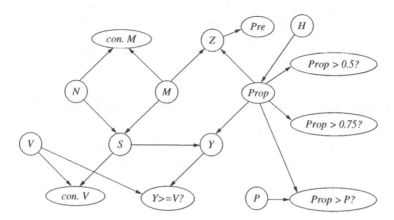

Figure 10.4 Bayesian network for inference about small consignments. Node definitions are given in Table 10.6.

Table 10.6 Definition of nodes used in the Bayesian network shown in Figure 10.4.

Node	Definition	States
N	Consignment size	$n_i(i = 0, 1, \dots, 20)$
M	Number of analysed units	$m_j(j = 0, 1, \dots, 20)$
S	Number of units not inspected	$s_k(k = 0, 1, \dots, 20)$
Z	Number of positives amongst the inspected units	$z_l(l = 0, 1, \dots, 20)$
Y	Number of positives amongst the uninspected units	$y_f(f = 0, 1, \dots, 20)$
H	Prosecution's (H_p) and defence's (H_d) case	H_p, H_d
$Prop$	Proportion of positives in the consignment	$0 - 0.05, \dots, 0.95 - 1$
V	Lower limit for evaluating cumulative probabilities of the number of positives amongst the uninspected units	$v_w (w = 0, 1, \dots, 20)$
P	Lower limit for evaluating cumulative probabilities of the proportion of positives in the consignment	$0, 0.05, \dots, 0.95, 1$
$Y \geq V?$	Are there at least V positives amongst the uninspected units?	yes, no
$Prop > P?$	Is the proportion of the positives in the consignment at least P?	yes, no
$con.M,$ $con.V$	Constraints on, respectively, M and V	yes, no
Pre	Outcomes (number of positives amongst the inspected units)	none (0) few (1–5) some (6–10) many (11–20)
$Prop > 0.5$ $(0.75)?$	Is the proportion greater than 0.5 (0.75)?	yes, no

degraded or even contaminated) item, such as a stain found on a crime scene. A characteristic feature of an item (or trace) of this kind is that it may not offer the same qualities as a sample. For example, it may not be representative and/or replicable. The use of the term 'sample' for such units thus is not appropriate, but in more loose and common language, the distinction is often not properly made.

10.5.2   Bayesian network for inference about small consignments

A scenario in which it is of interest to enquire about the number of units to be analysed usually starts with the scientist considering some sort of consignment. Let the individual units of the consignment consist of, for example, bags containing powder of unknown composition. It may be that the powder in a certain proportion of these units (i.e. the bags) is illegal drugs. More is to be learned about that proportion by inspecting the composition of the content of some of the units.

The first variable considered here is N, the consignment size. The Bayesian network shown in Figure 10.4 incorporates this variable as a discrete chance node. N is modelled as a root node because it is not thought to depend on any other variable. For implementation of the network in the Hugin environment, a node subtype termed 'numbered' should be chosen for this variable. Its states have numeric values that are amenable to, for instance, expression building and the specification of standard discrete distribution functions. Here, N is allowed to cover a total of 21 states, that is, $0, 1, \ldots, 20$. A lower case letter n_i with $i = 0, 1, \ldots, 20$ is used to denote a particular state of N. The Bayesian network will thus enable its user to cope with scenarios involving up to 20 units.

The number of inspected units, M, is defined in the same way as N, that is, in terms of a numbered discrete chance node with states $0, 1, \ldots, 20$, abbreviated as m_j ($j = 0, 1, \ldots, 20$). The probabilities to be specified for the tables of the nodes N and M do not require particular discussion because they are merely a technical matter necessary to run the model. When using the Bayesian network, the nodes N and M will be instantiated so that the probabilities initially assigned in the tables of these nodes become irrelevant. The definition of Bayesian networks solely requires that $\sum_{i=0}^{20} Pr(N = n_i) = \sum_{j=0}^{20} Pr(M = m_j) = 1$.

A method is required, however, to assure that only pairs of values $\{n_i, \quad m_j\}$ as instantiations of the nodes N and M will be allowed such that $n_i \geq m_j$ ($i, j = 0, 1, \ldots, 20$). The Bayesian network should allow for assumptions of states of nature that are impossible; one cannot inspect more units (M) than there are units in the consignment (N). Such a constraint is incorporated in the Bayesian network through definition of a discrete chance node $con.M$ which is a direct descendant of both N and M. The node $con.M$ is binary with possible states labelled, for instance, 'yes' and 'no'. The probability table is completed as follows:

$$Pr(con.M = yes | N = n_i, M = m_j) = \begin{cases} 1, & i \geq j, \\ 0, & i < j, \end{cases} \quad (i, j = 0, 1, \ldots, 20).$$

The node $con.M$ is not strictly necessary for the Bayesian network shown in Figure 10.4. One could use the model correctly if one is sure to instantiate nodes N and M only in ways such as to agree with the constraint mentioned above. The reason of adopting a node C_M thus is one of convenience. It can help the user to avoid fixing values for n_i and m_j that are logically incompatible.

Knowledge about the consignment size N together with information about the number of inspected units M allows one to determine the state of a variable S, the number of units in the consignment that are not inspected. In analogy to N and M, S is defined as a discrete chance node of type 'numbered' with states s_k ($k = 0, 1, \ldots, 20$). The node table of S contains the logical values 1 and 0, that is, 1 for the states $s_{k=i-j}$ of S for all parental configurations n_i, m_j satisfying the constraint defined by $con.M$. For all parental configurations that do not conform to $con.M$, the value 1 is assigned to the state s_0 by default.

A node of principal interest in the network shown in Figure 10.4 is *Prop*, a placeholder for θ, the proportion of units in the consignment that are positive. Strictly speaking, θ is a continuous beta-distributed variable that can assume values between 0 and 1. In the network here, it is represented by a discrete chance node whose states represent disjoint intervals between 0 and 1. Some Bayesian network programs, such as Hugin, offer a subtype of discrete chance nodes named 'interval' that supports such a definition. In the model here, a node *Prop* with a total of 20 intervals is defined, that is, 0–0.05, 0.05–0.1, ..., 0.95–1. Notice that other intervals of different size could obviously be used as well.

For the time being, ignore node H in the Bayesian network displayed in Figure 10.4 and take *Prop* as a root node. In order for the node *Prop* to represent the beta-distributed variable θ, the probabilities assigned to the various interval states of *Prop* need to be chosen appropriately. In Hugin language, this can be obtained by defining the expression Beta(alpha,beta), where alpha (> 0) and beta (> 0) are the shape parameters of the beta distribution. These may be varied to obtain a distribution that provides a good representation of the initial beliefs held by the evaluator. Note solely that alpha=beta=1 provides a uniform distribution over the values from 0 to 1. An alternative, more explicit way of modifying the shape of the beta distribution associated with *Prop* consists of modelling the parameters alpha and beta as distinct discrete chance nodes named a and b, with each having numbered states in the range of, for example, $1, 2, \ldots, 10$. The nodes a and b are then adopted as parental nodes for *Prop* so as to obtain the local structure $a \rightarrow Prop \leftarrow b$. In addition, the expression Beta(a,b) would be specified for *Prop*. The user can then make instantiations at the nodes a and b, and these will provide the parameters of the beta distribution according to which the Bayesian network will distribute the probabilities over the states of the node *Prop*. Such an example is discussed in Section 10.6.2.

Generally, the definition of the properties of the prior probability distribution for *Prop* is outside the province of the forensic scientist. The topic of prior probabilities is, however, interesting in itself as their formulation provides a regular source of debate. Within the environment of a Bayesian network, diverging prior opinions can be accommodated in an elegant way. This is illustrated in Figure 10.4 in terms of a node H, introduced as a parent node for *Prop*. There could be as many states for H as there are diverging opinions. For the purpose of the present discussion, let there be two, that is, H_p and H_d where the subscripts p and d could reflect the standpoints of the prosecution and defence. Within the local network structure $H \rightarrow Prop$, one can then define the probability distributions $Pr(Prop|H_p)$ and $Pr(Prop|H_d)$. For illustration, Beta(1,1) is specified for the former and Beta(2,5) for the latter. Note that the expression Beta(2,5) assigns higher probabilities to proportions below 0.5, whereas Beta(1,1) provides, as mentioned earlier in this section, a 'flat' distribution. An interesting property of this local network structure is that it allows one to assign probabilities to H_p and H_d, so as to regulate the extent to which the opinions subsumed under H_p and H_d contribute to form the shape of the beta distribution for the node *Prop*. When choosing $Pr(H_p) = Pr(H_d) = 0.5$, for example, one conveys the idea that each of the diverging opinions is given the same weight.

Associated to *Prop* are three summary nodes, *Prop* > 0.5?, *Prop* > 0.75? and *Prop* > *P*?, where 0.5, 0.75 and *P* (the latter being user-defined) are lower limits for the evaluation of cumulative probabilities for the proportion of positives in the consignment. For example, the discrete node *Prop* > 0.75? with states labelled 'yes' and 'no' provides the probability that θ, represented by the node *Prop*, is greater than 0.75, given particular instantiations made at other nodes in the Bayesian network. The Hugin expression `if(Prop>0.75,Distribution(1,0), Distribution(0,1))` provides the probabilities for the node table of *Prop* > 0.75? and similarly for the nodes *Prop* > 0.5? and *Prop* > *P*? by replacing 0.75 in this expression by, respectively, `0.5` and `P`.

Besides *Prop*, a further variable of interest in discussions about consignment inspection is *Y*, the number of positives amongst *S*, the units in the consignment that are not inspected. The distribution of *Y*, given θ and *S*, that is, $(Y|\theta, S)$, is binomial (Aitken 1999). In a Bayesian network, the relationship between *Y*, θ and *S* can be expressed by adopting a numbered discrete chance node *Y* with entering arcs from the nodes *S* and *Prop* (see also Figure 10.4). Node *Y* has states numbered 0, 1, ... , 20 and the node table is completed according to the expression `Binomial(S,Prop)`. In analogy to the previously defined variables, lower case letters y_f with subscript $f = 0, 1, ... , 20$ are used to denote particular outcomes of *Y*.

It may be that an analyst seeks more general statements about *Y*, the number of positive units amongst those not inspected. For this purpose, a summary node, denoted by *Y* >= *V*? here, could be adopted. It is binary with states labelled 'yes' and 'no'. The role of this node is to provide a probability for the event that the number of positives amongst the units not inspected (node *Y*) is at least, or greater than, *V*, a user-specified integer number. Node *V* is a discrete chance node with v_w ($w = 0, 1, ... , 20$) denoting numbered states in the range 0, 1, ... , 20. Both *V* and *Y* are parental variables of the node *Y* >= *V*?. The probability table of the node *Y* >= *V*? contains logical values 0 and 1 and completes as follows:

$$Pr(Y >= V? = yes | V = v_w, Y = y_f) = \begin{cases} 1, & f \geq w, \\ 0, & f < w, \end{cases} \quad (f, \ w = 0, 1, ... , 20).$$

The node *V*, in turn, would be instantiated whilst using the Bayesian network so that no particular requirement for the respective unconditional probabilities is required (except that the values sum up to 1). In order to ensure that illogical queries cannot be addressed through the node *V* – for example, inquiring about whether the number of positives *Y* amongst the uninspected units *S* is greater than *S* – the highest numbered state of *V* that one should be allowed to select should depend on the actual state of *S*. This constraint on *V* can implemented through the use of a node *con.V*, in much the same way as was done previously in this section for the node *con.M*. Here, *con.V* is a discrete chance node with two states labelled 'yes' and 'no' and a probability table containing the logical values 0 and 1 (as defined by the constraint).

A node comparable to *Y* is *Z*, a node representing the number of positives amongst *M*, the number of inspected items. The node *Z* is a variable with a binomial distribution with arguments provided by *M* and *Prop*. This allows one to define, in Hugin language, the expression `Binomial(M,Prop)` to complete the node probability table. The numbered states of *Z* are abbreviated as z_l ($l = 0, 1, ... , 20$).

For pre-assessment, it might not be of primary interest, however, to know – prior to the analysis – the probability of finding exactly one, two, four or any other given number ($\leq M$) of positives. Rather, one might be interested to reason about an inference problem in broader terms. For example, imagine that one wishes to inquire about the probability of finding 'some' or 'many' positives amongst the analysed units. Such general propositions can help the scientist to evaluate expected findings prior to the examination stage. A node that can help approach such questions is *Pre*, a discrete chance node with states labelled 'none (0)', 'few (1–5)', 'some (6–10)' and 'many (11–20)'. *Pre* is a summary node because its states accumulate the probabilities associated with specified groups of states of the node *Z*. For example, the probability of the state 'few (1–5)' of the node *Pre* corresponds to $\sum_{l=1}^{5} Pr(Z = z_l)$. Note that the intervals defined in *Pre* are chosen arbitrarily and may be modified as required. The category 'many', for example, may have different meanings in cases with different numbers of analysed units. The probability table of *Pre* is assigned the value 1 for the state 'none (0)' if *Z* is in the state 0 (otherwise a probability of 0 is assigned). Analogous assignments apply to the state 'few (1–5)' when *Z* takes one of the states $1, 2, \ldots$ or 5, to the state 'some (6–10)' if *Z* takes one of the states $6, 7, \ldots$ or 10 and to the state 'many (11–20)' if *Z* takes one of the states $11, 12, \ldots$ or 20. The probabilities that can be obtained for the various states of *Pre* can inform a process of pre-evaluation when inspecting small consignments, as discussed in the next section.

10.5.3   Pre-assessment for inspection of small consignments

Consider a case that involves a small consignment of size 15 and suppose that it is of interest to gather further information for inference about the proportion of positives in the consignment. A given customer, such as an investigating magistrate, may consider the option of paying for the analysis of 5 units but may remain undecided about whether or not to proceed. In particular, the official may be unclear about the impact that various findings may have on the case. The forensic scientist may thus be requested to evaluate – prior to the examination – the probabilities for various outcomes and their associated degree of support in favour of the customer's proposition, as opposed to a given alternative. Such information can help officials in their decision to allocate monetary resources.

Example 10.7 *(Bayesian network for pre-assessment when inspecting a small consignment) The Bayesian network described throughout Section 10.5.2 assists with the calculation of probabilities for observations of $Z = 0, 1, \ldots, 5$ positives amongst 5 inspected units, taken from a consignment of 15 units, given particular assumptions about the unknown proportion θ. For the purpose of illustration, suppose that the investigating magistrate's proposition of interest is that the proportion of positive units in the consignment is greater than 0.75. In turn, assume that the alternative proposition is that the proportion of positive units is lower than 0.5. It may be that less than 0.5 is not worth bothering about, greater than 0.75 opens up the possibility of a custodial sentence and in between is the possibility of a non-custodial sentence.*

Table 10.7 summarizes the probabilities for observing $Z = 0, 1, \ldots, 5$ positives amongst 5 analysed units (from a consignment covering 15 units) when the prosecution's case is that the proportion of positives in the consignment is greater than 0.75 (column three) and when the defence's case is that the proportion of positives in the consignment is lower than 0.5 (column four). The values can be found with the Bayesian network described in the previous section with instantiations made at the appropriate nodes. In particular, the prosecution's case is evaluated by fixing node N to n_{15}, node M to m_5 and Prop > 0.75? to 'yes'. The defence's case is evaluated by fixing N to n_{15}, M to m_5 and Prop > 0.5? to 'no'. Figure 10.5 illustrates this latter situation graphically. Notice that the values compiled in Table 10.7 are based on a uniform prior distribution for the unknown parameter θ (node Prop).

Table 10.7 Values for the likelihood ratio V (V^{-1}) in support of the prosecution's (defence's) case when the findings (i.e. number of positives amongst $M = 5$ analysed units from a consignment of $N = 15$) are $Z = \{0, 1, \ldots, 5\}$ or, more generally, $Pre = \{none\ (0), few\ (1 - 5)\}$.

Node	State	$Prop > 0.75$	$Prop < 0.5$	V	V^{-1}
Z:	0	0.0002	0.3281	0.0006	1640.5
	1	0.0031	0.2969	0.0104	95.774
	2	0.0251	0.2188	0.1147	8.7171
	3	0.1130	0.1146	0.9860	1.0175
	4	0.3107	0.0365	8.5123	0.1175
	5	0.5480	0.0052	105.38	0.0095
Pre:	*none* (0)	0.0002	0.3281	0.0006	1640.5
	few (1 − 5)	0.9998	0.6719	1.4880	0.6720

It is assumed that the propositions retained by the prosecution and the defence are that the proportion of positives in the consignment is greater than 0.75 and lower than 0.5, respectively.

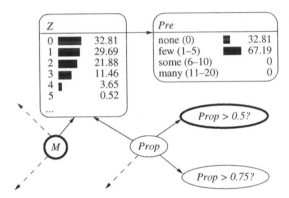

Figure 10.5 Pre-assessment for the defence's case as described in Example 10.7 on the basis of a partial representation of the Bayesian network shown in Figure 10.4, with the nodes Z (the number of positives units amongst the analysed items) and *Pre* (expected outcomes in terms of grouped numbers of positives) displayed in expanded form. The node M denotes the number of analysed items (instantiated here to the state $m = 5$, not shown) and *Prop* the proportion of positives in the consignment. The nodes *Prop* > 0.5   (0.75) are binary and reply to the question whether the proportion is greater than 0.5   (0.75). The node *Prop* > 0.5 is instantiated to 'no' to represent the defence's proposition according to which the proportion is inferior to 0.5. The borders of instantiated nodes are shown with a bold line.

Columns five and six in Table 10.7 list the likelihood ratio V (and V^{-1}) that each possible outcome provides for the propositions put forward by, respectively, the prosecution and defence. For example, there is a probability of 0.3107 of observing four positives amongst the 5 analysed units if H_p is true, and a probability of 0.0365 for the same event if H_d is true. This finding would thus provide a likelihood ratio V of approximately 8.5 (8.5123 to four decimal places) in favour of the prosecution's case.

Further, the results presented in Table 10.7 show that the defence can expect to obtain higher likelihood ratios than the prosecution. The highest possible value of the likelihood ratio in support of the defence case is about 1640, whilst the prosecution can expect a maximum likelihood ratio of approximately 105. However, there is a difference in the probabilities with which these likelihood ratios may be obtained. The prosecution can expect to obtain the highest possible likelihood ratio (about 105) for their case – given their proposition were true (i.e. Prop > 0.75) –with a probability of 0.55, whereas the defence can expect the highest possible likelihood ratio, given their proposition being true (i.e. Prop < 0.5) with a probability of 0.33.

The two lines at the bottom of Table 10.7 list the probabilities of the states none (0) and few (1 − 5) of the node Pre conditional on, respectively, Prop > 0.75 and Prop < 0.5. Such cumulative probabilities can be of interest in a perspective of pre-assessment when the focus is not that of individual outcomes of the kind Z = {0, 1, 2, … , M} (where M is the number of analysed units, here 5) but that of a more coarse categorization. In the case here, an evaluation at the node Pre shows that the prosecution has – if their proposition is true – an overwhelming probability of obtaining evidence in the form of few (1–5) 'positives' that will support their case. However, the associated likelihood ratio is low.

The cumulative classes (for groups of the number of positives) may be more informative when the consignment is larger and more units are analysed, so that further categorizations can be adopted as states of the node Pre. This helps to illustrate the general idea that propositions in a case pre-assessment can be expressed at various levels of detail.

10.6   Pre-assessment for gunshot residue particles

10.6.1   Formation and deposition of gunshot residue particles

The discharge of a firearm – that is, the firing of a cartridge in a firearm – is accompanied by the generation of discharge residues that consist of a variety of materials. These may stem, for instance, from the primer, from the propellant, from the bullet (i.e. its core or its jacket) and possibly from the cartridge case (e.g. Rowe 2000b). Discharge residues are complex heterogeneous mixtures that are known according to the context as firearm discharge residues (FDR), cartridge discharge residues (CDR) or gunshot residues (GSR). The third of these terms will be used throughout this book, but it should be recognized that there is discussion about what actually ought to be considered a GSR and on what basis (Romolo and Margot 2001; Romolo et al. 2013). In the discussion here, it will be assumed that the scientist can provide sound observational support for what are referred to as *distinct particles*.

The formation of GSR is a complex process that involves hot gases and high pressure that are produced during the discharge of a cartridge. The residues escape from the firearm essentially through the barrel although other openings may also be involved, such as ejection ports of self-loading or automatic firearms or the space between the barrel and the chamber in revolvers. This means that GSR particles may deposit on surfaces in the vicinity of the discharging firearm – typically, a shooter's or bystander's hands, clothing, face and head hair (e.g. Zeichner and Levin 1995). Searches of interesting particles are, therefore, regularly conducted on individuals (or items belonging to individuals) when it is suspected that they have been exposed to the discharge of a firearm. There is a second major forensic application where the analysis of GSR plays an important role. That application focuses on GSR particles present on a target and seeks to assess the range from which a shot was fired (e.g. Haag 2006). For discussion of this usage of GSR, see Section 8.4.4.

When a certain quantity (which may be nil) of particles is recovered from the backs of the hands of a suspect, a question that may be of interest to a forensic scientist's client is the degree to which such a finding allows one to discriminate between competing propositions of the kind 'the suspect has shot a firearm' versus 'the suspect has not shot a firearm'. The latter kind of proposition is typically used – as in many other contexts focusing of potentially transferred particle materials – as a way to separate the suspect from the criminal event, if the circumstances allow one to do so (e.g. the proposition is in agreement with what the suspect says). The scientist can then reason about the quantities of particles that may be found on individuals with no recent exposure to the discharge of a firearm. Other alternative propositions may readily complicate the evaluation because of scarce empirical data. For example, a more tentative pair of propositions that may be used is 'the suspect is associated (is not associated) with primary exposure to GSR' (primary transfer). This formulation is sometimes used with the aim of introducing a distinction with respect to particles present due to secondary transfer. Secondary transfer may occur, for example, when suspects are seated in contaminated vehicles or detained in offices or interview rooms with contaminated equipment, such as tables or chairs (Berk et al., 2007). In order to approach such questions of evidential assessment, forensic scientists need an understanding of the occurrence of GSR particles in various case circumstances [e.g. Jalanti et al. (1999)]. For example, the amount of GSR particles that may be found depends crucially on the nature and intensity of the activities of a person (exposed to a GSR deposit) as well as the time elapsed between the alleged firearm discharge and the particle search. These considerations arise from a number of studies (e.g. Andrasko and Maehly 1977, Jalanti et al. 1999, Kilty 1975, Knechtle and Gallusser 1996, Nesbitt et al. 1976, Tillman, 1987). Additional circumstantial information, such as the place where the shooting took place (indoors or outdoors) and meteorological conditions should also be considered (Zeichner and Levin, 1995). Due to such factors and circumstances, the evaluation of target particles in firearm discharge cases represents a particularly challenging topic and it is not rare that after incorporating case-related information, the probative value of particular findings may be considerably reduced. This, however, should not be considered as a drawback of the Bayesian approach. On the contrary, it is through a Bayesian approach that such conclusions can be arrived at and explained transparently.

10.6.2   Bayesian network for grouped expected findings (GSR counts)

In a case pre-assessment, scientists may prefer a general classification scheme for expected findings – in the context here, a particle count – rather than a highly detailed description as given by, for example, an individual count variable with states $\{0, 1, 2, 3, \ldots\}$. As in Section 10.5 on consignment inspection, it may be of interest to scientists to describe potential findings (i.e. a GSR count) in summary statements such as 'none', 'few', 'some' and 'many'. The Bayesian network shown in Figure 10.6 exemplifies this in terms of a node *Pre* with states 'none (0)', 'few $(1-4)$', 'some $(5-8)$' and 'many (> 8)' (see also Table 10.9). Notice that the definition here for the qualitative categorizations is confined to rather small categories (i.e. in terms of numbers of particles) essentially because practical experience often shows that few numbers of particles are detected when searching the surface of hands.

The node *Pre* is a sink node that represents a starting point for modelling a cascaded backward-reasoning process (i.e. bottom-up reasoning against the direction of the network's arcs) that aims to discriminate between particular competing propositions. This model, shown

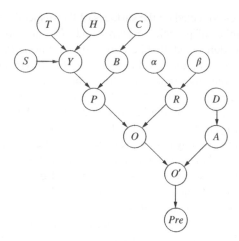

Figure 10.6 Bayesian network for evaluating detected GSR particles. Node definitions are given in Table 10.9.

in Figure 10.6, consists of several generic inferential levels. These cover several components, described below for a setting in which particles are searched for on the surface of a suspect's hands.

The parental node of *Pre* is O', representing the overall particle count on a lifting device after application to the surface of the suspect's hands. This is a node with numbered states. The range of possible counts o'_l is defined here for $l = 0, 1, 2, \ldots, 70$, but the reader may choose other ranges as required. Node O' is defined as the logical combination of the two nodes O and A, that is, respectively, the number of GSR particles collected on the surface of the suspect's hands and the number of GSR particles present on the lifting device prior to the particle collection. The two nodes O and A cover, respectively, the range of numbered states o_k (for $k = 0, 1, 2, \ldots, 60$) and a_i (for $i = 0, 1, 2, \ldots, 10$). The probability table of O' is defined as follows:

$$Pr(O' = o'|A = a, O = o) = \begin{cases} 1, & o' = a + o, \\ 0, & o' \neq a + o. \end{cases}$$

The presence of particles on the particle lifting device prior to its application is an aspect modelled in terms of the two nodes A and D, with numbered states a_i (for $i = 0, 1, 2, \ldots, 10$) for A that represent the number of particles present as contamination on the lifting device prior to its use and with states 'clean' and 'not clean', denoting the condition of the device for D, a binary node. The unconditional probabilities assigned to the states of the node D allow one to express personal beliefs about the condition of a particle collection device, according to the circumstances of a case at hand. In turn, the probability table of the node A contains the logical value 1 for $Pr(A = 0|D = clean)$. For the conditioning $D = not\,clean$, the user specifies a discrete probability distribution over the various states that the node A can take. This distribution expresses the user's personal probability for there being $0, 1, 2, \ldots$ particles on a contaminated device. It could take, for example, the form of a Poisson distribution (see also Chapter 4, Example 4.1) if the scientist can demonstrate that the choice of a particular value for the Poisson parameter is appropriate for the case at hand.

The number of particles collected on the surfaces of the suspect's hands (node O) depends not only on the total number of particles present (node P) but also on the performance of the particle recovery procedure, modelled here in terms of a node R. This node represents a proportion that covers a range of interval values between 0 and 1 (including these end points). Conceptually, this is a continuous variable, but for technical reasons, it is implemented here in terms of an interval node that covers a total of 20 discrete states – in the same way as the node *Prop* defined earlier in Section 10.5.2. Uncertainty about the value of the proportion R is modelled by a probability distribution from the beta family. The parameters α and β for this beta distribution are provided by two discrete nodes labelled α and β with numeric states $\{0.1, 0.5, 1, 2, \ldots, 10\}$. The choice of α and β as parental nodes for R allows one to specify the node table of R through the expression `Beta(alpha,beta)` (in Hugin syntax).

In combination, the nodes R (proportion of lifted particles) and P (the true but unknown total number of particles present on the suspect's hands) act as parental variables for the node representing the number of actually collected particles (node O). Node O has the same number of states as node P, that is, $\{0, 1, \ldots, 60\}$. The node probability table contains binomial probabilities that can be defined by the expression `Binomial(P,R)` (in Hugin syntax).

The total number of particles present on the surfaces of the suspect's hands (P) is a quantity that is a logical combination of particles present as background (B) and particles present as the result of the discharge of a firearm (Y). The nodes B and Y are both numerical and define the node table for P in the following way:

$$Pr(P = p \mid Y = y, B = b) = \begin{cases} 1, & p = y + b, \\ 0, & p \neq y + b. \end{cases}$$

The quantity of GSR particles that do not originate from the firearm discharge of interest is a variable that may be considered as a representation of 'background' (B). Assuming that the occurrence of background GSR particles is a rather rare event, it is feasible to retain only a limited number of numeric states for this node (e.g. $\{0, 1, 2, \ldots, 10\}$). The number of background GSR particles is modelled here as a Poisson-distributed variable with parameter λ_b. A categorization of the quantity of the background GSR particles is introduced here in terms of an additional variable C with states 'none', 'low', 'medium' and 'high' (other states may be chosen as required). Several distinct values for the parameter λ_b can then be considered. For the purpose of the current discussion, suppose the following choice: $Pr(B = b \mid C = \{low, medium, high\}) \sim Pn(\lambda_b)$ where λ_b can take values 0.25, 2 and 5 depending on the state 'low', 'medium' and 'high', respectively. For $C = none$, the probability table of node B contains the following values: $Pr(B = 0 \mid C = none) = 1$ and $Pr(B \neq 0 \mid C = none) = 0$. In turn, the probability table of the node C contains unconditional probabilities that reflect the scientist's expectations about various degrees of background presence of particles. In a less elaborate version of the network, node C could be left out and personal probabilities assigned directly to the states of node B.

A key node in the network is Y, representing the particles on the suspect's hands due to the firearm discharge in the event of H_p and conditioned on (i) particular settings S (e.g. type of weapon, number of shots) and (ii) the time elapsed between the incident and the examination of the suspect. The assignment of conditional probabilities to the table of node Y is amongst the most challenging aspects of the network specification, essentially because the number of possible situations is potentially large and published data are very scarce. One example is given

Table 10.8 Mean number of particles retrieved from hands of shooters after various intervals of time in two different experimental settings (e.g. in terms of the weapons used), as reported by Cardinetti et al. (2006).

Experimental setting	Mean GSR particle count after						
	2 h	3 h	4 h	5 h	6 h	8 h	10 h
1	16.17	9.8	8.60	6.50	5.25	3.67	2.50
2	23.80	5.75	5.20	5.20	4.00	4.00	3.50

by Cardinetti et al. (2006). To illustrate the possible use of such data, the probability assignments for the node table of Y are chosen here in terms of Poisson distributions with parameters equating the reported mean number of particles (see the study summarized in Table 10.8). For example, the conditional probabilities for the node Y, given H_p, the experimental setting 1 and a time interval of 5 h would be defined as $Pr(Y = y|H_p, S = 1, T = 5) \sim Pn(6.5)$. In turn, under H_d, the node table of Y contains the values $Pr(y = 0|H_d) = 1$ and $Pr(y \neq 0|H_d) = 0$. This does not mean that there could be no particles (e.g. as background), given H_d, but this is accounted for in other parts of the network (e.g. node B and P).

Note that the Bayesian network described in this section has a generic structure that allows one to explain the wide range of considerations that may have a bearing on the evaluation of detected GSR particles. The numerical assignments for node probability tables as chosen here have only an illustrative character. For the analysis of a given case, assessments should be made such as to suit the framework of circumstances. In the particular case of the node Y, probabilities have been assigned here on the basis of data from the scientific literature in order to illustrate the feasibility of informing a network with such data. Scientists may wish to conduct their own experiments in case the experimental settings in published reports or studies do not adequately fit the needs for a case at hand. For example, it may be that particles have not (only) been searched on the surfaces of a suspect's hand but (also) on his head (e.g. face, hair) or clothing, and this may affect the definition of the node Y. In that respect, the Bayesian network is valuable as a help in the specification of the kind of (empirical) data that is necessary for a probabilistic evaluation and, thus, in the consideration of the design of experimental studies to collect the appropriate data.

10.6.3 Examples for GSR count pre-assessment using a Bayesian network

The Bayesian network described in the previous section can support scientists in several distinct operations in which pre-assessment is involved. One particular example is the assignment of probabilities for the various potential findings (or possible outcomes), given each of the competing propositions of interest. In Figure 10.6, the possible outcomes are defined in terms of the qualitative apportionments 'none (0)', 'few (1–4)', 'some (5–8)' and 'many (> 8)' that make up the states of the node *Pre* (short for pre-assessment). The examples here below discuss

Table 10.9   Definition of the nodes of the Bayesian network shown in Figure 10.6.

Node	Definition	States
A	Number of GSR particles present on lifting device prior to its use	$a_i (i = 0, 1, 2, \dots , 10)$
B	Background content of GSR particles on the suspect's hands	$b_j (j = 0, 1, 2, \dots , 10)$
C	Degree of background content of GSR particles on the suspect's hands	none, low, medium, high
D	Condition of stub	clean, not clean
H	Competing propositions of interest	H_p, H_d
O	Number of GSR particles collected on the suspect's hands	$o_k (k = 0, 1, 2, \dots , 60)$
O'	Overall GSR count on lifting device (after application to the surface of the suspect's hands)	$o'_l (l = 0, 1, 2, \dots , 70)$
Pre	Outcomes (quantity of detected particles)	none (0), Few (1–4), Some (5–8), many (>8)
P	Total number of GSR particles on the suspect's hands (due to the shooting of a firearm and background content of GSR particles)	$p_m \quad (m = 0, 1, 2, \dots , 60)$
R	GSR particle recovery rate (lifting efficiency), modelled as a beta random variable	$r_n = 0 - 0.05, \dots , 0.95 - 1$
S	Situational (or experimental) setting	$s = 1, 2$
T	Time interval (hours) between firearm discharge and collection of GSR particles	$t = 2, 3, 4, 5, 6, 8, 10$
Y	Number of GSR particles on the suspect's hands due to the shooting of a firearm	$y = 0, 1, 2, \dots , 50$
α, β	Parameters of the beta-distributed variable R	$0.1, 0.5, 1, 2, 3, \dots , 10$

the evaluation of conditional probabilities for these states for hypothetical cases with varying circumstantial settings.

Example 10.8 *(Pre-assessment for a GSR count) Consider a scenario in which the scientist swabbed the surfaces of the hands of a suspect 5 h after a shooting incident. The scientists wish to conduct a pre-assessment with respect to discriminating between propositions of the following kind: 'The suspect is the shooter (H_p)', 'An unknown person is the shooter and the suspect has nothing to do with the case at hand (H_d)'.*

It is assumed that there is circumstantial information that suggests that background contents of GSR particles are present to an extent that can be quantified as 'medium'. Further, the scientist considers that the distributional assumptions about the number of GSR particles on the suspect's hands due to the discharge of a firearm (as assumed under H_p) approximate the experimental setting 1 described earlier in Table 10.8. The proportion of retrieved GSR particles is described by a Beta(10,1) distribution. The particle collection device is assumed to be clean prior to its use. This outset leads to the following node instantiations in the Bayesian

Table 10.10 Probabilities for the states of the node *Pre* given, respectively, H_p (column three) and H_d (column four) and further settings as described in Examples 10.8 and 10.9. The column on the far right-hand side provides the likelihood ratio (V) values in favour of H_p.

	Pre	$Pr(Pre\|H_p)$	$Pr(Pre\|H_d)$	V
Example 10.8:	*none* (0):	0.000653	0.165478	0.004
	few (1 − 4):	0.125563	0.795902	0.158
	some (5 − 8):	0.503890	0.038485	13.1
	many (> 8):	0.369895	0.000134	2760
Example 10.9:	*none* (0):	0.086156	0.196828	0.4
	few (1 − 4):	0.346800	0.627355	0.6
	some (5 − 8):	0.314141	0.166266	1.9
	many (> 8):	0.252903	0.009552	26.5

network shown in Figure 10.6: $t = 5$, $C = medium$, $S = 1$, $\alpha = 10$, $\beta = 1$, $D = clean$. Given these settings and each of the two competing propositions (i.e. H_p and H_d), Table 10.10 summarizes the resulting conditional probabilities for the states 'none (0)', 'few (1–4)', 'some (5–8)' and 'many (> 8)' of the node Pre (representing qualitative categorizations of numbers of GSR particles), as well as their combination that leads to a likelihood ratio (last column in Table 10.10).

According to this analysis, the scientist would expect – if H_p is true – to obtain a finding that will moderately support H_p (i.e. by a likelihood ratio of about 13) with a probability of about 0.5 (see results for 'some' in Table 10.10). He would also expect to obtain findings that will strongly support the same proposition (i.e. by a likelihood ratio of approximately 2700) but a somewhat lower probability (i.e. 0.37, actually). On the other hand, if H_d holds, then the scientist assesses, for example, a probability of almost 0.8 for a result that will weakly support H_d (i.e. by a likelihood ratio of $0.158^{-1} = 6.3$). A finding that will moderately support H_d (i.e. by a likelihood ratio in the order of $0.004^{-1} = 250$) is assigned a probability of 0.17. In addition, the scientist would also reserve some probability but only to a very limited extent (i.e. about 0.03), for a finding that would incorrectly support H_p.

Example 10.9 *(Pre-assessment for a GSR count – continued) As a modification of Example 10.8, consider a situation in which the background contents are assumed to be more substantial (i.e. $C = high$). Moreover, circumstances are such that the scientists need to assume that the particle collection device may be affected by contamination (i.e. $D = not\ clean$). In addition, the scientist prefers to remain vague with respect to the efficiency of the applied method for particle recovery and express this in terms of the parameters $\alpha = 1$, $\beta = 1$ for the beta-distributed variable R. In such a setting, still assuming a time interval of 5 h (i.e. $t = 5$) and experimental conditions $S = 1$ (as in Example 10.8), the resulting likelihood ratio values concentrate around more moderate values, in either direction (see also Table 10.10). This tendency appears intuitively reasonable because the settings for the nodes D and C (i.e. 'not clean' and 'high', respectively) reflect that the scientist gives increased weight to alternative sources of residue particles. In a cascaded inference as defined by the Bayesian network shown in Figure 10.6, these intervening variables tend to weaken the discriminative* *capacity that observations at the node Pre have with respect to the propositions summarized by the node H. This modified example also illustrates that the various intervening factors can have a substantial bearing on the overall conclusion. It is thus important that scientists investigate the sensitivity of the overall inference due to distributional assumptions made at the various hierarchical levels and support the quantitative model specification by sound empirical data.*

11

Bayesian decision networks

One of the main threads of argument pursued so far in this book considers Bayesian networks as a widely and flexibly applicable formalism that supports sound reasoning when the available information is incomplete. On the basis of observations, of the kind typically encountered in forensic science, one can construct probabilistic arguments to propositions of interest. As such, Bayesian networks provide a coherent environment wherein beliefs about target propositions can be revised upon receipt of newly acquired information. This represents a fundamental preliminary requirement for an additional step, that is, the coherent use of beliefs in action. This is also more commonly known as *decision making under uncertainty* (Lindley 2006). Chapter 2 showed that Bayesian networks can be extended to incorporate the basic ingredients necessary to perform so-called Bayesian decision analyses. In essence, this extension consists of decision and utility nodes (Section 2.1.9). They represent, broadly speaking, actions available to the scientist – or, in a wider sense, to the decision maker – and values for possible consequences of these actions, respectively. This chapter emphasizes the relevance of these extensions for forensic science by focusing on selected examples.

11.1 Decision making in forensic science

There are many situations in forensic science in which decisions need to be made under circumstances of uncertainty. It is routinely asked, for example, which if any forensic examination or analysis ought to be conducted. Recall also that one of the more important questions discussed in Chapter 10, on pre-assessment, focused on the extent to which a scientist expects to obtain a likelihood ratio that will support either H_p or H_d in a given scenario. It was acknowledged that such inquiry should be extended to questions of how a laboratory or customer can take a rational decision on the necessity to perform examinations, after an initial assessment of possible likelihood ratio values. As previously noted, the former category of question refers to the process of case pre-assessment, whereas the latter category of question refers to decision making, the main topic pursued in this chapter.

Bayesian Networks for Probabilistic Inference and Decision Analysis in Forensic Science, Second Edition.
Franco Taroni, Alex Biedermann, Silvia Bozza, Paolo Garbolino and Colin Aitken.
© 2014 John Wiley & Sons, Ltd. Published 2014 by John Wiley & Sons, Ltd.

In the early 1950s, a debate was initiated as to how people should make decisions involving money that were in some sense rational and as to how people in fact made monetary decisions, and whether these could be regarded as rational (Edwards 1954; Lindley 1985; Luce and Raiffa 1958; Raiffa 1968; Savage 1951; Savage 1972; Smith 1988). This debate – through the use of the notion of a so-called 'utility function' – was extended to the context of financial decision making. Subsequently, it has also been proposed to develop these ideas in legal and forensic contexts, involving scientific findings (Kaye 1988). With regard to forensic science, decision theory can be used to develop general approaches for determining optimal choices given a certain set of items of information and values. The perspective is typically that of an individual decision maker, who is either the customer or acts on behalf of the customer (such as the forensic scientist) and who is interested in defining an optimal course of action, using formal modelling.

Forensic scientists may need to decide about whether or not to use a given analytical instrument, apply a particular chemical reagent or search for a particular category of trace material. Examples of questions requiring a decision could be, for instance, whether to process a counterfeit banknote for fingermarks or DNA, what kind of analyses to perform (e.g. nuclear DNA or mitochondrial DNA in a case of contested kinship) or how many of such analyses (e.g. the number of loci in a DNA trace to be analysed or the number of units in a consignment). The component of Bayesian inference in such questions thus presents an instance for extending considerations to Bayesian decision theory, a shift of the focus from inference to particular practical actions.

11.2   Examples of forensic decision analyses

11.2.1   Deciding about whether or not to perform a DNA analysis

Figure 11.1 reproduces part of a larger influence diagram introduced earlier in Figure 2.6, and provides a short reminder of the main ingredients of an influence diagram. Figure 2.6 can readily be used to illustrate the decision problem of the identification or exclusion of a potential source (such as a person or an object) as the source of a particular trace (Biedermann et al. 2008a). Losses, or utilities, in node L (or U) are defined with respect to whether or not the suspected source is truly the origin of the recovered trace material, that is, propositions θ_1 and θ_2 in node Θ, and what conclusion the decision maker declares in terms of identification (d_1) or exclusion (d_2). These conclusions are described by node D.

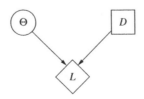

Figure 11.1   Generic Bayesian decision network structure over three nodes. Node Θ is a discrete node with states θ_1 and θ_2 representing the states of nature. Node D is a decision node. Its states d_i, for $i = 1, 2, \ldots, n$, correspond to the n elements of the decision space. The diamond-shaped node L defines the loss function, that is, a loss value for each state combination of the parental nodes H and D.

The influence diagram shown in Figure 11.1 depicts a very general framework for decisions that leaves room for a broad interpretation, in particular for so-called one-stage decision problems. To take this idea further, consider the question of whether or not to perform DNA analyses in a case of questioned kinship (Taroni et al. 2005). Imagine two individuals, say A and B, are uncertain about whether they are full sibs (proposition H_1) or unrelated (proposition H_2). The two questions of interest before performing DNA profiling analyses in this scenario are, first, 'Can a value of the likelihood ratio be obtained that will be such as to support H_1 or H_2?' and, second, 'How can the laboratory take a rational decision on the need to perform DNA profiling analyses?' Case pre-assessment is an effective tool to help scientists approach the former question (see Chapter 10). Decision analysis deals with the latter question.

11.2.1.1   One-stage decision approach

Fundamental to pre-assessment is the specification of appropriate distributions for findings, given each of the two competing propositions. For this purpose, simulation techniques can be employed. For example, given an appropriate population genetic model and relevant allele population proportions (at different loci) from relevant databases, two artificial databases can be generated by simulation. One database can be generated to consist of the simulated DNA profiles of what would be a large number of pairs of siblings and one can be generated to consist of the simulated DNA profiles of a large number of pairs of unrelated individuals. Then, for a given pair of siblings, A and B, in the first database, a likelihood ratio, V, of the following general form is assigned:

$$V = \frac{Pr(Agt, Bgt|H_1)}{Pr(Agt, Bgt|H_2)},\qquad(11.1)$$

where Agt and Bgt represent the genotypes of individuals A and B, respectively. The same procedure is performed for pairs of individuals coming from the database of unrelated individuals. By calculating likelihood ratios for many pairs in this way, one can obtain two distributions of likelihood ratio values. The first assesses the support for full sibship when the individuals are related (e.g. brothers, H_1), the second assesses the support for full sibship when the individuals are unrelated (H_2). Thus, they provide information for the question 'Could we obtain a value supporting the hypothesis H_1 or H_2 in this scenario?' As shown by Figure 11.2, such simulations can lead to informative results.

In the context of paternity and kinship analyses, it may be acceptable to make a digression from consideration solely of the likelihood ratio V and to consider the probability that individuals A and B are full sibs, that is, the probability that H_1 is true. The reason for this is that in some judicial systems, this is commonly requested by judicial rules and jurisprudence. This probability is known as the *probability of sibship*. Hummel (1971, 1983) provides a verbal scale, given here on Table 11.1, columns 1 and 2. Hummel's scale can be used to characterize states of nature (θ_j) in the decision-making approach, because legal decision in kinship cases is closely related to this scale by jurisprudence.

Note that states of nature are specified from a particular point of view. That may be, for example, the perspective of full siblings, where an individual is interested in proving the sibling relationship. On the contrary, if an individual is interested in proving unrelatedness, then the order of column 3 in Table 11.1 should be inverted. That is, θ_1 represents the state 'not useful', and so on. The one-stage decisions of interest are 'Perform DNA profiling analyses' and 'Do not perform DNA profiling analyses', denoted here d_1 and d_2, respectively.

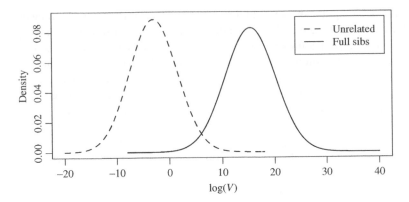

Figure 11.2   Distribution of the logarithm of the likelihood ratio (11.1), log(V), for simulated pairs of full siblings (continuous curve) and unrelated individuals (dashed curve).

Table 11.1   Hummel's scale (columns 1 and 2) and states of nature (column 3) defined for the purpose of the current example.

Probability of sibship	Likelihood of sibship	States of nature
Greater than 0.9979	Practically proved	θ_1
0.9910–0.9979	Extremely likely	θ_2
0.9500–0.9909	Very likely	θ_3
0.9000–0.9499	Likely	θ_4
0.8000–0.8999	Undecided	θ_5
Less than 0.8000	Not useful	θ_6

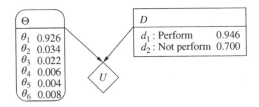

Figure 11.3   Kinship problem represented in terms of a decision network. The decision node D has two states, perform or do not perform DNA analyses. The chance node Θ has six states, ranging from 'practically proved' (θ_1) to 'not useful' (θ_6). The utility node, U, has 12 states corresponding to the 12 combinations of the decision and chance node states.

Figure 11.3 illustrates that the decision node D and the chance node Θ condition the utility node U, as is expressed by U's entering arcs. The utility node describes the value of the consequence C_{ij}, for decisions d_i, $i = (1, 2)$, and states of nature θ_j, $j = (1, \dots, 6)$. The number associated with it is denoted $U(d_i, \theta_j)$. Utilities equal to 1 (for $U(d_1, \theta_1)$) and 0 (for $U(d_1, \theta_6)$) are assigned to states 'practically proven' and 'not useful', respectively, considered to be the best and worst consequences in case decision d_1, 'Perform DNA analyses'. Utilities u assigned to intermediate consequences C_{1j} (for $j = 2, \dots, 5$) are determined from answers to the following question: Does the decision maker prefer the intermediate consequence or does he/she prefer the best consequence ('Practically proven') with probability equal to u? For the example here, the following utility assignment is proposed for consequences of decision d_1: $U(C_{1j}) = \{1, 0.5, 0.1, 0.1, 0.1, 0\}$, for $j = 1, \ldots , 6$. Consider, for example, $C_{ij} = C_{13} = 0.1$, the consequence of deciding d_1 (performing analyses) in connection with the state of nature θ_3 (conclusion 'very likely'). Here, the decision maker specifies that he/she is indifferent between, on the one hand, performing the DNA analyses and learning that the kinship is very likely (plausibility between 0.9500 and 0.9909) and, on the other hand, accepting that if he/she performs the analyses the result that kinship is practically proved (plausibility greater than 0.9979) will be obtained with probability 0.1 (the value of u assigned to θ_3 for d_1).

The utility of the consequence under d_2 is constant, for the sake of illustration, equal to 0.7. That is, $U(C_{2j}) = 0.7$, $\forall j$, written more shortly as C_2. This expresses the assumption that the decision maker is indifferent between, on the one hand, not performing analyses and, on the other hand, performing analyses and obtaining the best consequence ('practically proven') with probability 0.7. The utility of not performing the test is a direct consequence of the overall cost of performing it. This cost may be purely monetary. More generally, it may combine monetary costs with other non-financial burdens, such as inconvenience, intrusiveness and so on. Different decision makers may state very different values for $U(C_{1j})$ (for $j = 1, \ldots , 6$) and $U(C_2)$, depending on the kind of interest there is in determining the level of parentage and on their adversity to risk.

The probability table associated with the chance node (Fig. 11.3, node θ) presents probabilities for $\theta_1, \ldots , \theta_6$ obtained through simulation techniques as previously presented for two individuals under the proposition H_1 of full sibship for a fixed prior probability of sibship equal to 0.1 (Taroni et al. 2005). Note that there is no parent node for the decision node (i.e. no are pointing to this node). In fact, the example here is such that no state of nature is known at the time the decision is made, so that no link is required. Note further that the network is 'static' in the sense that no instantiation is made. It merely serves as a representation of the expected utilities at a given instance of time, and given a particular view (here, results supposing full sibship). The network as shown in Figure 11.3 allows the decision maker to check expected utilities for the two decisions to be able to choose the more suitable decision. In the case at hand, the network shows that if a pair of individuals are full siblings, DNA analyses will generally confirm it and d_1 should always be used ($EU(d_1) = 0.946$ and $EU(d_2) = 0.7$). The situation becomes more complicated depending on the chosen prior probabilities and if DNA analyses are requested to investigate a different level of parentage, such as half-sibship versus unrelatedness or full sibship versus half-sibship (Taroni et al. 2005). Example 11.1 illustrates possible extensions to Figure 11.3 in order to deal with alternative assumptions.

Example 11.1 *(Extended influence diagram for a case of questioned kinship) The structure of the influence diagram shown in Figure 11.3 was kept limited for the sake of clarifying the general characteristics of the inference problem at hand. In particular, the probability distribution for the node Θ was specified for a prior probability of full sibship of 0.1, and analytical results under the assumption of full sibship. A second 'static' assumption was adopted for the utility $U(C_2)$, which was assigned the value 0.7. In forensic pre-assessment, it may be important to remain flexible and investigate the effect of changes in assumptions. One way to do this for the current scenario is to extend the influence diagram as shown in Figure 11.4. This extension features an additional node H with states H_1 and H_2, allowing the user to specify whether the distribution to be considered in node Θ reflects the assumption of full sibship (H_1) or of being unrelated (H_2). A further node is H', with states $\{0.1, 0.2, \ldots , 0.9\}$ (other values*

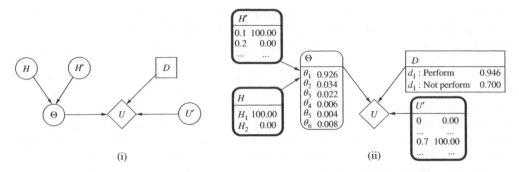

Figure 11.4 Collapsed (i) and partially expanded (ii) form of an extended influence diagram for the kinship problem. The decision node D has two states, perform (d_1) or do not perform DNA analyses (d_2). The chance node Θ has six states, ranging from 'practically proved' (θ_1) to 'not useful' (θ_6). The node U specifies utility values for all state combinations of the parental nodes Θ, D and U'. The latter node, U', specifies values for $U(C_{2j})$, one of which is to be selected by the analyst. Situational instantiations are also required for nodes H and H'. The former specifies possible views of the analyst (i.e. H_1, full sibship, or H_2 being unrelated). The node H' specifies the prior probability (to be selected amongst the numerical states $\{0.1, 0.2, \ldots, 0.9\}$) for the perspective specified in node H.

in the range between 0 and 1 may be chosen as required), representing the prior probabilities for the perspective specified in node H. Accordingly, the node Θ will contain probabilities for states θ_j ($j = \{1, \ldots, 6\}$), conditional on the decision maker's perspective (node H) and the prior probability for this perspective (node H'). Such values can be found through simulation as illustrated, for example, in Tables 5 and 6 in Taroni et al. (2005, p. 900).

As a third extension, Figure 11.4 contains a node U' with states $\{0, 0.1, \ldots, 1\}$, representing utility values for $U(C_2.)$, the utility of consequences in case decision d_2 (not performing the DNA analysis) is adopted. Different ranges of values may be defined, depending on the needs of the case at hand. It is supposed that, for a given analysis, the decision analyst will instantiate the node to one particular value (out of the defined range). This represents a more flexible approach than the one implemented in Figure 11.3 where the table of node U contained a fixed value of 0.7 for consequences of decision d_2. Note that in the model here, node U will copy the state instantiated in node U' for all utility assignments $U(C_{2j})$, $j = \{1, \ldots, 6\}$.

It is clearly seen that the influence diagram in Figure 11.4 represents a coherent extension of the previous diagram (Figure 11.3) when instantiations are made in the former (as shown in Figure 11.4(ii)) that reflect the situation captured in the latter. That is, instantiations refer to the view H_1 and the prior probability of 0.1 for H_1. In addition, $U(C_{2j})$ is specified by instantiating U' to 0.7. The agreement in results with respect to Figure 11.3 is established for utilities for actions in node D.

11.2.1.2   Two-stage decision approach

A particular feature of the approach pursued above is that it focuses only on a single decision (i.e. whether or not to perform DNA analyses) and degrees of support (i.e. likelihood ratios or their translation into levels of a posterior conclusion scale), interpreted as states of nature, given particular views held by the decision analyst. The expected utility of a decision to perform DNA analyses is evaluated with respect to possible posterior states of belief about sibship. This is a particular way of specifying the decision problem, influenced largely by the framework of thinking outlined in Chapter 10 on pre-assessment. This one-stage approach is pragmatic in the sense that it exploits simulation results for specifying target probabilities, thus avoiding an exhaustive consideration of all genotypic configurations that may be observed for two individuals.

The question of whether or not it is in the interest of a decision maker to conduct DNA analyses can also be investigated in terms of the classical two-stage decision model, outlined earlier in Section 2.1.9. This perspective is different in the sense that it distinguishes two decisions. One decision is that of deciding whether or not two persons are brothers. This can be called the *ultimate* or *terminal decision*. Another decision relates to whether or not to perform DNA analyses, that is, a decision to generate results that may provide data that allows the decision maker to revise his/her beliefs in the propositions according to which the two persons of interests are related (e.g. brothers) or unrelated, prior to choosing an ultimate decision. According to such a perspective, the expected value of performing DNA analyses is less immediate, as it requires a comparison with the expected utility of deciding about brotherhood in a situation in which no analyses are performed.

Figure 11.5 sketches part of an influence diagram for this purpose. The proposed structure covers a node H for the target propositions, that is, particular states of kinship, a node D_2 for the ultimate decision about kinship, a node D_1 for the preliminary decision about performing DNA analyses and a utility node U. Node U specifies utilities for various consequences of deciding about kinship, given the actual state of kinship. The structural relationship of these nodes follows the decision structure previously described in Figure 2.6. The influence diagram shown in Figure 2.6 is simple in the sense that the possible findings upon deciding to conduct analyses are given by a single-chance node conditioned only on the node for the preliminary decision and the node for the target state H. In the scenario considered here, the findings consist of observed genotypes for two individuals A and B. Because the list of possible pairs of genotypes, even for a single locus, is potentially long, the genotypes of persons A and B are modelled here in terms of nodes for single alleles, following methodology outlined earlier in Section 7.2.3, using nodes *Agtmin* (*Bgtmin*) and *Agtmax* (*Bgtmax*) for, respectively, the minimum and the maximum of the associated two parental gene nodes *Apg* (*Bpg*) and *Amg* (*Bmg*). For a situation in which person B is a brother of person A, the paternal and maternal gene of person B is inherited from the parents of A, represented by the nodes *fpg* (father paternal gene), *fmg* (father maternal gene), *mpg* (mother paternal gene) and *mmg* (mother maternal gene). In case person B is a half-brother (*hb*), a cousin (*c*) or unrelated (*u*) to person A, the nodes *Bpg* and *Bmg* copy the states of, respectively, nodes *hbpg* and *hbmg*, *cpg* and *cmg* and *upg* and *umg* (where *pg* and *mg* denote paternal and maternal gene, respectively). Due to limitations of space, the inheritance structure for a cousin c of person A, based on the grandparents of A, is not shown in full detail (abbreviated in Figure 11.5 by '[...]'). The structure is given by standard modelling methodology outlined in Section 7.9 and can be shown to provide results that agree with the computation procedure for conditional genotypes discussed in Section 7.4.2. The nodes *AgtminObs* (*BgtminObs*) and *AgtmaxObs* (*BgtmaxObs*) copy the current state of the corresponding parent nodes *Agtmin* (*Bgtmin*) and *Agtmax* (*Bgtmax*) given decision $d_{1,1}$ ('Perform DNA analysis') of node D_1. For state $d_{1,2}$ ('Do not perform DNA analyses') of node D_1, the nodes with extension *Obs* take the dummy state '99' (short for 'no result'). This modelling feature has been exemplified earlier in Table 2.10. Further, observe that nodes with extension *Obs* have information links to the decision node D_2.

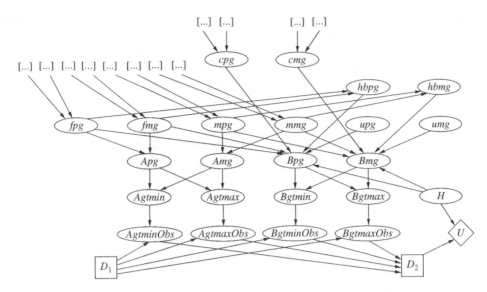

Figure 11.5 Influence diagram for a two-stage decision approach to deciding about whether or not to perform a DNA analysis that informs about the genotypes of two persons A and B. The genotypes of A and B are represented by nodes *Agtmin*, *Agtmax*, *Bgtmin* and *Bgtmax*, defined as the minimum (*gtmin*) and maximum (*gtmax*) of the two parental gene nodes *pg* and *mg*, inherited via the paternal and maternal line, respectively. Nodes with the same names but with extension *Obs* refer to the observed genotypes in situations in which DNA analyses are performed. Nodes *cpg*, *cmg*, *fpg*, *fmg*, *hbpg*, *hbmg*, *mpg*, *mmg*, *upg*, *umg*, *Apg*, *Amg*, *Bpg* and *Bmg* are gene nodes, where *c* denotes cousin, *f* father, *hb* half-brother, *p* paternal, *m* mother (if in first place) or maternal (if in second place) and *u* unknown person. Node H represents the scope of possible relationships with states 'Brothers (H_1)', 'Half-brothers (H_2)', 'Cousins (H_3)' and 'Unrelated (H_4)'. Node D_2 represents the ultimate decision with two states $d_{2,1}$ ('A and B are brothers') and $d_{2,2}$ ('A and B are not brothers'). Node D_1 represents the decisions to perform ($d_{1,1}$) and not to perform ($d_{1,2}$) DNA analyses. Node U contains the utility function for outcomes $(d_{2,i}, H_j)$, $i = 1, 2, j = 1, \ldots, 4$.

A key aspect of the model shown in Figure 11.5 obviously is the definition of the nodes H, U and D_2. For the purpose of the current scenario, it appears sensible to define node H in terms of the following reasonable coverage of the possible states of nature: H_1 ('Persons A and B are brothers' or, more generally and without the loss of generality, siblings), H_2 ('Half-brothers'), H_3 ('Cousins') and H_4 ('Unrelated'). Note that to obtain an exhaustive set of propositions, H_4 covers all relationships that are more distant than cousins. They are subsumed here under the generic state named 'unrelated'. Assuming that the only decisions of interest are 'A and B are brothers' and 'A and B are not brothers', regardless of whatever state of relatedness may hold given this latter state, one can define node D_2 in terms of these two states, written $d_{2,1}$ and $d_{2,2}$, respectively. If the main focus is on correct conclusions of brotherhood, one can specify the following utility function for node U:

$$U(d_{2,i}, H_j) = \begin{cases} 1, & i, j = 1, \\ 1, & i = 2, \quad j \neq 1, \\ 0, & \text{elsewhere.} \end{cases}$$

This function assigns utility 1 to all correct conclusions of 'brotherhood' and 'not brotherhood', and a utility 0 in all cases in which 'brotherhood' is concluded when in fact A and B are not brothers (i.e. states of nature H_i for $i = 2, 3, 4$).

Without information about the genotypes of persons A and B, the calculus of the expected utility for decisions $d_{2,i}$ ($i = 1, 2$) is based on the prior probabilities for propositions H_j ($j = 1, 2, 3, 4$). Taking these probabilities as equal, for the sole purpose of illustration, one can find that the expected utility of decision $d_{2,1}$ ('A and B are brothers') reduces to

$$EU(d_{2,1}) = U(d_{2,1}, H_1)Pr(H_1) = 0.25$$

because all utilities $U(d_{2,1}, H_j)$ for ($j \neq 1$) are zero. In turn, the expected utility of decision $d_{2,2}$ is given by

$$EU(d_{2,2}) = U(d_{2,2}, H_2)Pr(H_2) + U(d_{2,2}, H_3)Pr(H_3) + U(d_{2,2}, H_4)Pr(H_4) = 0.75.$$

The term $U(d_{2,2}, H_1)$ does not appear in the formula because it takes the value 0. Thus, for the prior probabilities as assumed in this hypothetical setting, the decision analyst should decide $d_{2,2}$, that is, 'Persons A and B are not brothers'. Note that, more generally, this result is also given by $\max_i \sum_{j=1}^{4} U(d_{2,i}, H_j)Pr(H_j)$.

To decide whether or not to perform a DNA analysis on persons A and B, one needs to know how much should be paid for such an 'experiment'. A quantity of interest in such a context is the notion of expected value of perfect information (EVPI, Section 1.4.6). It expresses how much it would be worth to know which proposition is true. It requires the calculation of the expected utility with perfect information (EUPI), obtained by multiplying the maximum utility for each hypothesis by the probability of that hypothesis, and summing all these products for the four propositions, that is, in the case here:

$$\text{EUPI} = \sum_{j=1}^{4} \max_i U(d_{2,i}, H_j)Pr(H_j), \quad \text{for} \quad i = 1, 2, \quad j = 1, 2, 3, 4.$$

In the case here, it is readily seen that the EUPI is 1 because it is optimal to decide $d_{2,1}$ if H_1, a consequence having utility of 1, and decide $d_{2,2}$ in all other cases, that is, consequences also having a utility of 1. The difference between this result and the expected utility of the *a priori* optimal decision defines the EVPI

$$\text{EVPI} = \sum_{j=1}^{4} \max_i U(d_{2,i}, H_j)Pr(H_j) - \max_i \sum_{j=1}^{4} U(d_{2,i}, H_j)Pr(H_j)$$

and leads to $1 - 0.75 = 0.25$ in the scenario here. It represents the maximum price that one should be willing to pay for this perfect information, assuming that the analyst can map the quantification of the consequences of the decisions on the same scale as the cost of experimentation.

In a subsequent step, one can consider the expected utility of deciding about brotherhood once experimental results E are available. This leads to posterior probabilities for the propositions H_j, for $j = 1, 2, 3, 4$, and allows one to find the so-called a posteriori optimal decision, defined as $\max_i \sum_{j=1}^{4} U(d_{2,i}, H_j)Pr(H_j|E)$, also called the *Bayes action*. This expression differs from the expression of the *a priori* optimal action by conditioning the probabilities of the propositions H_j by the particular result E.

Usually, however, there may be various possible outcomes for E, and the decision analyst might be interested to inquire about the expected utility before performing the analysis producing E, so as to compare this gain with respect to the cost of obtaining the results. Following theory presented earlier in Chapter 1 (Section 1.4.6), the interest thus lies on maximizing the expected loss for any possible result and is expressed by the expected utility with partial information (EUpI) . It is obtained by multiplying the maximum expected utility for each possible result of the DNA analysis by the probability of that result and then summing over all these products. Write the individual results as E_k, for $k = 1, \ldots, m$, where m is the number of possible genotypes that may be observed for two individuals A and B. Then one can write the EUpI as follows:

$$\text{EUpI} = \sum_{k=1}^{m} \max_i \sum_{j=1}^{4} U(d_{2,i}, H_j) Pr(E_k|H_j) Pr(H_j).$$

In order to obtain the EVpI, one needs to find the difference between the EUpI and the expected utility of the *a priori* optimal action:

$$\text{EVpI} = \sum_{k=1}^{m} \max_i \sum_{j=1}^{4} U(d_{2,i}, H_j) Pr(E_k|H_j) Pr(H_j) - \max_i \sum_{j=1}^{4} U(d_{2,i}, H_j) Pr(H_j).$$

The various expressions above are given here in a partially generic form because they depend on the number of alleles that are considered at the genetic marker of interest and that define the states for the nodes representing individual alleles in Figure 11.5. For a given person and a locus with a alleles, there will be $a(a + 1)/2$ different genotypes. Hence, for two persons A and B, there will be $(a(a + 1)/2)^2$ different genotypic configurations that can be observed.

The EVpI allows the decision maker to address the question of whether or not to conduct DNA analyses because it provides an indication of the maximum price that one should be willing to pay for obtaining the partial information (i.e. the DNA profiling results). This consideration supposes, however, that the decision analyst can interpret the value given by the EVpI with respect to the financial cost of the analysis.

11.2.2   Probability assignment as a question of decision making

11.2.2.1   Scoring probability statements: Proper scoring rules

Generally, proper scoring rules enable a clear definition of probability in the context of an assignation of a probability to an event E. To apply such a proper scoring rule, a person – in the context here an expert or a scientist – is asked what probability he/she assigns to a given event E. This question is accompanied with the information that the probability statement thus elicited will be scored, with respect to the actual truth or falsity of E, denoted by, respectively, 1 and 0. This constraint is added as an incentive for persons to report their true beliefs, rather than a deliberately chosen value (i.e. one that is different from the one they have in mind).

This topic is of interest in forensic contexts as it often happens that scientists report categorical conclusions rather than genuine evaluations of probability. As an example, consider the notion of relevance that expresses the relationship between a given stain or trace and the offender. According to Stoney, the probability of relevance '(...) may range from very likely to practically nil (...)' (Stoney 1994, p. 18). Such a probability assignment relates to the proposition of the kind 'the stain or mark comes from (or was left by) the offender' and appears as a factor in various likelihood ratio developments, such as for the evaluation of footwear marks (Evett et al. 1998b). Typically, one should maintain a belief p different from 0 and 1 about the origin of the stain because one is unsure about whether or not a given stain comes from the offender. But, for the ease of argument, it is not uncommon for evaluators to suppress uncertainty by declaring a probability of relevance $p' = 1$. A probability smaller than 1 is rounded up to 1. Scoring rules are a device to illustrate that this rounding up is not advisable.

In the context here, the term score refers to the square of the difference between the probability for the event E, as declared by the scientist, and the actual truth value of E. Hence, the name 'quadratic scoring rule'. To illustrate this rule, suppose that a scientist states the value 0.8 as his/her probability for the event E. Then, one can distinguish two cases. One is the case in which E is true and thus assumes the truth value 1. This situation leads to a score of $(1 - 0.8)^2 = 0.2^2$. The other case arises when the event E is not true and thus has the truth value 0. This leads to the score $(0 - 0.8)^2 = (-0.8)^2$. But these score calculations, leading to the expressions of actual penalty, are only hypothetical. Given the fact that the scientist is uncertain about whether or not E is true or not, the scientist cannot know the actual score. He/she can only consider a so-called prevision of the scoring and seek a way of proceeding that minimizes the expected penalty. Here, the term *prevision of the scoring*, also called *expected penalty* or *expected loss*, is taken to mean combining the possible scores according to the probability with which they may be produced. This general concept was previously introduced in Section 1.4.3.

A particular aspect of the quadratic scoring rule is that it implies a minimization of the expected penalty when the scientist states his/her actual belief. That is, if he/she would have stated, for example, 0.6 or 0.9, whilst actually thinking 0.8, the expected score would have been greater. Thus, under such a rule, scientists are directed to state their true beliefs as probabilities, provided that they accept that it is in their interest to minimize the expected score. Historically, the quadratic scoring rule appears in many writings of the subjective probabilist de Finetti [e.g. de Finetti (1962), de Finetti (1982)] although, after a paper by Brier (1950) on an application in meteorology, it is also known as the *Brier's rule*. The attribute 'proper' refers to the rule's property of ensuring that the probability assessor's best policy is that of stating his/her actual belief. Conversely, for rules that do not have this property, one may have a greater interest in stating something else than one's true belief. The quadratic scoring rule is often retained because it is one of the simplest proper scoring rules. Other proper scoring rules exist. Good (1952), for example, proposed one based on the logarithm. Scoring rules have also been studied for evaluating and comparing probability assignments from different experts who were asked to present their subjective probabilities for various events (DeGroot and Fienberg 1982). This leads to additional notions for assessing the quality of the models for the assessment of uncertainty, such as calibration, which are not covered here.

Note also that the quadratic scoring rule can also be used for the evaluation of conditional probabilities, that is, a person's belief about the occurrence of an event E, given that a particular event H has occurred. This is achieved by supposing that the penalties only apply if the conditioning event H holds [e.g. de Finetti (1972)].

11.2.2.2 Illustrating probability scoring using an influence diagram

The general idea of quadratic scoring for evaluating probability assignments can be represented in an influence diagram as shown in Figure 11.6 (Biedermann et al. 2013). This model implements a representation of the target event of interest in terms of the Boolean node E,

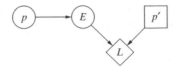

Figure 11.6 Influence diagram for scoring reported probabilities. Nodes p and p' represent, respectively, an individual's true beliefs and reported probability. They have the numerical states $\{0, 0.1, 0.2, \ldots, 0.9, 1\}$. Node E is Boolean and denotes the occurrence of an event of interest. Each reported probability p', along with a particular state of node E (i.e. its truth or falsity), defines a so-called 'consequence'. Each consequence is assigned a value (i.e. a quadratic score) in node L.

with states 'true' and 'false'. One option would be to have this node as a root node and assign one's belief directly for the two states without a conditioning on other nodes. Another option, shown in Figure 11.6, amounts to specifying an entering arc from a node p. Such an arc is not thought to represent a 'causal' relationship because an individual's belief cannot influence the occurrence of event E. The relationship is of a purely technical nature and serves the purpose of copying an individual's actual belief, represented in terms of one of the numerical states of node p, to the state 'true' of node E. Conversely, the state 'false' of node E takes the value 1 minus the value of the current state of node p. Thus, the node p only serves as a device to ease the investigation of alternative scenarios by making instantiations rather than editing the node table of E manually. The scope of numerical states of p is restricted here to $\{0, 0.1, 0.2, \ldots, 0.9, 1\}$. The reader may choose other increasing sequences of values from the interval $[0, 1]$. These states represent an individual's possible beliefs about the occurrence of event E. The choice of such a discrete collection of values is sufficient for the purpose of illustration pursued here. Indeed, in many applications (also beyond forensic science), such as forecasting, it might be standard practice that experts will state values from a predefined collection of values (DeGroot and Fienberg 1982), and there is no need to strive for larger sets of values with an increasing precision that scientists may feel difficult to support. For technical reasons, the model requires the specification of probabilities for each of these numbered states. However, when using the model for the purpose described here, one of the states will be instantiated, so that the initially specified probabilities will no longer be of importance. For this reason, no particular distribution is discussed here. The only requirement is that the specified probabilities sum to one. Conceptually, these probabilities have no actual meaning because there is no intention here to specify a probability for a probability, a standpoint also supported in literature [e.g. Lindley (2006)].

Node p' is a decision node and represents the various numerical values amongst which the scientist may choose one as his/her reported belief. For technical convenience, the action space is defined in terms of the same collection of numerical values used for representing the reasoner's actual beliefs p, that is, $\{0, 0.1, 0.2, \ldots, 0.9, 1\}$. It is now straightforward to see that the comparison between p and p' allows one to define loss values in node L, which is defined as a descendant of nodes E and p'. The possible combinations of states for E and p' represent the so-called consequences, and losses for each consequence are defined by the quadratic scoring rule. When E is true and the individual chooses to state a value $p' = i$ (for $i = \{0, 0.1, 0.2, \ldots, 0.9, 1\}$), then node L takes the penalty $(1 - i)^2$. In turn, when E is false, then this node takes the penalty i^2.

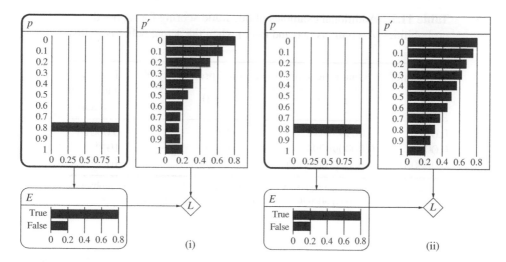

Figure 11.7 Expanded representation of the influence diagram shown in Figure 11.6. In the case shown here, the probability assessor's true belief is $p = 0.8$. In (i), node L defines a quadratic score. In (ii), node L uses the simple difference between the true state of E and the reported probability p' as the score. In both diagrams, node p' displays the expected penalty for each of the possible probability reports $p' = \{0, 0.1, 0.2, \dots, 0.9, 1\}$, given a supposed actual belief $p = 0.8$ (this is shown with a bold node border).

Example 11.2 *(Probability assignment using quadratic scoring) Imagine a case in which an expert's actual belief p about the occurrence of an event E is 0.8. A moderately sized value of this kind could be of interest in situations where scientists are uncertain about the relevance of a given crime stain. The value of 0.8 is proposed here in order to keep the presentation visual, because examples with very low probabilities may involve calculations with many zeros. Some complementary examples of the latter kind are tabulated later on (Table 11.2). As shown in Figure 11.7(i), the expert's true belief 0.8 is communicated to the model by instantiating the state 0.8 of the node p. The effect of this is that node E will display the value 0.8 as the probability with which the expert believes that the event E is true. The question of interest then is 'What is the optimal value that the scientist should report, given that he/she acts under a quadratic scoring rule?' This question is answered by the decision node p'. For each of the possible actions $p'_i = \{0, 0.1, 0.2, \dots, 0.9, 1\}$, the node p' displays the expected penalties given by*

$$EL(p'_i) = L(p'_i, E = \text{true})Pr(E = \text{true}) + L(p'_i, E = \text{false})Pr(E = \text{false}),$$

or $(1 - p')^2 p + p'^2(1 - p)$ for short. One can readily see that the expected penalty is minimal for $p' = p = 0.8$, where it is 0.16 (Figure 11.7, node p', state 0.8):

$$(1 - p')^2 p + p'^2(1 - p) = (1 - 0.8)^2 \times 0.8 + 0.8^2 \times (1 - 0.8)$$

$$= 0.16.$$

It is thus in the scientist's interest to report his/her actual belief. The reader can readily verify that for any reported value p' different from p = 0.8, the expected penalty would be larger,

Table 11.2   Additional examples of quadratic scoring for reported probabilities p' (columns two and three) when the actual belief is p (first column).

True belief	Quadratic score for reported probability p'		Expected additional penalty
p	$p' = p$	$p' = 0$	
0.01	0.0099	0.01	0.0001
0.0001	0.00009999	0.0001	10^{-8}
0.000001	0.000000999999	0.000001	10^{-12}

The values in the last column represent the expected additional penalty when a probability $p' = 0$ (column three) is reported instead of an honest report of probability $p' = p$ (column two).

as is illustrated visually by the expanded node p'. This points out the nature of the quadratic scoring rule, which implies that whatever one's true belief p, one should sincerely choose that value as one's reported probability, because this is the choice with the minimal expected penalty. This feature makes the quadratic scoring rule a proper scoring rule. Other scoring rules may not necessarily exhibit this property. Example 11.3 points this out for the simple difference between the true state of E and the reported probability p'.

Table 11.2 summarizes some further examples to illustrate how the quadratic scoring rule works, in particular for probabilities of interest p that tend towards 0. Columns two and three contain the expected penalties for, respectively, honest probability reports (i.e. when $p' = p$) and fixed reports of a zero probability (i.e. $p' = 0$). The difference between these two scores leads to the expected additional penalty indicated in column four. More generally, it can be shown that a decision strategy that rounds off a non-zero belief to a zero-reported probability has an expected penalty equal to p.

Although the expected additional penalties for the situations summarized in Table 11.2 become increasingly smaller, this should not be taken to imply that the practice of rounding a small probability to 0 could be regarded as an acceptable alternative to honest probability reporting. As noted by Lindley (1985, p. 25), the reason for not reporting extreme values is that they will result in extreme penalties if the ground truth does not lie in the direction one has thought. Honest probability reporting proves to be the consistently better strategy and thus reflects the idea that it pays to be cautious.

Example 11.3 *(Probability assignment, continued) It may be asked why not use a measure other than the squared difference. One is obviously free to do so, but the point is that under an alternative measure, it may no longer be of interest for the scientist to state his/her true beliefs. To illustrate this point, consider a score given by the simple difference, rather than the squared difference. To investigate this choice, one can use the influence diagram introduced above but change the definition of the loss function in node L. For situations in which E is true, the score for reporting p' is only $1 - p'$, rather than $(1 - p')^2$ as used for the quadratic scoring rule. In turn, the score is p' for settings in which E is falsified, rather than the quadratic score p'^2. An influence diagram with such a definition for the node L is shown in Figure 11.7(ii). In this figure, it is again supposed that the expert's true belief is $p = 0.8$. The modified scoring* *rule now leads to the following expected penalty for reporting $p' = 0.8$:*

$$(1 - p')p + p'(1 - p) = (1 - 0.8) \times 0.8 + 0.8 \times (1 - 0.8) = 0.32.$$

It may now be asked if honest probability reporting may be recommended under this scoring rule. As shown visually in the expanded node p', reporting $p = p' = 0.8$ is no longer the optimal decision for the scientist. In fact, the expected penalty is lower for reports of values $p' > p$. Actually, the expected penalties for reporting 0.9 and 1 are, respectively, 0.26 and 0.2. Therefore, under a simple scoring rule, it may not be the best strategy for the scientist to report sincerely his/her actual belief p.

11.2.3   Decision analysis for consignment inspection

As noted previously in Section 10.5, the question of how many units to inspect when confronted with a consignment of units represents a topic of ongoing interest to the forensic science community essentially because of its crucial role in laboratory planning and working protocols. Forensic literature covers thorough (Bayesian) probabilistic approaches for this and they are now widely implemented in practice. They allow one, for instance, to obtain probability statements that parameters of interest (e.g. the proportion of a seizure of units that present particular features, such as illegal content) satisfy particular criteria (e.g. a threshold or an otherwise limiting value). Besides probability statements, it may also be asked how a forensic decision maker ought actually to *decide* about a proportion or the number of units to inspect. Often, such questions are not formally explored in current forensic applications. Methodology based on decision theory may help to cope usefully with such issues in consignment inspection. They help scientists to address a variety of concepts such as the value of sample information, the (expected) value of sample information or the (expected) decision loss (Biedermann et al. 2012a). These are all aspects that directly relate to questions encountered in casework.

The decision problem considered in this section focuses on a unknown proportion, denoted θ here. It could refer, for example, to the proportion of a consignment of individual units that are of a certain kind (e.g. contain an illegal substance). The parameter θ may take any value in the range $[0, 1]$, including the end points. For the ease of discussion, the presentation here uses the term 'positive' to denote an individual unit that has a characteristic of interest. Conversely, a 'negative' (result) refers to a unit that does not have the target trait of interest. The actual decision problem the decision maker faces could cover a variety of options, but the discussion here will concentrate on two decisions, denoted d_0 and d_1. The former, d_0, amounts to accepting the view that the proportion θ of positive units in the consignment is not greater than some specified value θ^*, for example, $\theta^* = 0.9$. The latter, d_1, is the view according to which the proportion θ of positive units in the consignment is greater than the specified value θ^*. Notice that the two actions d_0 and d_1 can be conceptualized as decisions to accept one of the two composite hypotheses $H_0 : \theta \leq \theta^*$ and $H_1 : \theta > \theta^*$.

A decision d is correct if the true value of the unknown proportion (θ) lies in the range of values defined by such a decision. Otherwise, it is an incorrect decision. The discussion here will consider that correct decisions do not have an associated positive 'loss' (or, more generally, a penalty). A positive loss intervenes, however, if a conclusion is erroneous. This is the case whenever the true proportion θ does not lie in the range specified by a particular decision d. More generally, let $L(d_i, \Theta_j)$ denote the loss associated with the decision d_i, for $i = 0, 1$, whilst Θ_j, for $j = 0, 1$, is the true state of affairs, written in more detail as $\Theta_0 = [0, \theta^*]$ and

$\Theta_1 = (\theta^*, 1]$. Thus, the consideration that correct decisions do not have an associated positive loss is represented by $L(d_i, \Theta_j) = 0$, for $i, j \in \{0, 1\}$ and $i = j$. Further, write l_i for the loss $L(d_i, \Theta_j)$ associated with an erroneous conclusion when action d_i is taken, $i, j \in \{0, 1\}$ and $i \neq j$. There are different ways to conceptualize a loss l_i, notably in terms of a variable that takes a numerical value incorporating both monetary and non-monetary components. For the purpose of the current discussion, the losses l_i will be interpreted as monetary values. Consequences of incorrect decisions will be assigned constant and equal values; in such a way a constant and symmetric loss function is specified. It is not always the case that loss functions are of the type $0 - l_i$ and symmetric. Other forms of loss functions are possible, such as linear or quadratic [e.g. Press (2003)]. In the analysis pursued here, imagine that the person in charge of making a decision is an examining magistrate who typically faces the practical problem of high workload as well as multiple cases. One aspect of coping with this is to concentrate efforts (e.g. in terms of time and money) on cases where the proportion of seized units is above a certain threshold and to prioritize these cases. Alternatively, one could also imagine that a given specialized investigative unit has, as an organizational requirement, the mission of focusing only on the so-called 'high-profile' or 'priority' cases. In such settings, it may be of interest to focus on a limit above which the actual proportion of illegal units of a given consignment should be. For such an outset, l_0 represents the loss that incurs when the decision maker falsely regards a case as one in which $\theta \leq \theta^*$. Only financial losses are being considered. Here, the loss could consist of the funds or monetary value of property that could have been confiscated by the government as a penalty and given to the public treasury or to the investigative unit. In turn, l_1 represents the loss from falsely considering the proportion θ of a consignment to be greater than θ^*. In such a situation, the magistrate or investigative unit would in fact pursue a case that actually does not meet the requirements specified by the investigative unit's mission statement. As a consequence of this, an investigative unit may deploy efforts and generate expenses (e.g. due to expensive investigative techniques) which, when compared to the reduced funds that may be seized in a case with $\theta \leq \theta^*$, could represent a net loss. The loss could also represent the amount of compensation to be allocated to an individual found erroneously guilty and subsequently exonerated. The choice of particular values for l_i depends on the characteristics of the case at hand, but caseworkers may be reluctant to articulate their choices. It is recognized here that this is a challenging task. For the remainder of this section, an equal and constant value $l_0 = l_1 = 100000$ (e.g. €) is adopted as a broad choice and conceivable, for example, in the area of investigations of drug crimes. There is no suggestion that the chosen value is to be considered as generally representative of losses in criminal investigations. The purpose at this juncture is to agree on some exemplary value in order to illustrate and provide insight to the forthcoming analyses. This may encourage readers to think about and provide their own numerical assignments. Examples 11.4–11.8 illustrate the potential of influence diagrams to help with various decisional issues arising in the context of attribute inspection.

Example 11.4 *(Deciding about a proportion without inspecting any units) Consider the decision problem outlined so far and imagine that a decision d needs to be taken with regard to the proportion θ of illegal units. Should one decide to act as if the proportion of illegal units in a consignment at hand is not greater than θ^* or should one decide to act as if the proportion is greater than θ^*? These two possible decisions were defined above as, respectively, d_0 and d_1. For the purpose of the example here, suppose that the question of how to choose between d_0 and d_1 is considered under the assumption that no units are inspected. This point is also sometimes termed* prior analysis *and contrasts with 'posterior analysis' that refers to* *a setting in which data are available from an analysis by a laboratory. See Example 11.7 for an illustration of this.*

Let θ^* assume the value 0.9, as noted previously, and consider, initially, that one considers all possible proportions θ equally plausible. This is expressed in terms of the so-called uniform prior probability distribution. Then, the probability of the composite hypothesis $H_0 : \theta \leq 0.9$ is given by the integral of this uniform prior density function (in this case, a beta distribution with parameters $\alpha = \beta = 1$), with end points 0 and 0.9:

$$Pr(\theta \leq 0.9) = \int_0^{0.9} f(\theta | \alpha = 1, \beta = 1) d\theta = 0.9. \tag{11.2}$$

The beta distribution is a widely used distribution for representing uncertainty about a parameter such as a proportion θ (Bernardo and Smith 1994). It also has many applications in forensic statistics (e.g. Aitken 1999; Aitken and Taroni 2004; Biedermann et al. 2009, Taroni et al. 2010). A uniform distribution (beta distribution with $\alpha = \beta = 1$) is taken to translate the initial belief according to which no proportion θ is thought to be more probable than any other. The aim here is to express a generic point of view, and readers may obviously choose other combinations of values to express their own view. Note that the above implies that $Pr(\theta > 0.9) = 0.1$, the complement of $Pr(\theta \leq 0.9)$. In the example here, Bayesian networks will be used to obtain such cumulative probabilities, which will help readers to obtain results for their personal choices of values for θ^* and parameters for the beta distribution.

On the basis of (11.2), and its complementary result, one can calculate the prior expected loss of a decision d_i, for $i = 1, 2$, written, for short, $EL^0(d_i)$, where the superscript 0 refers to the 'initial' point of time. For instance, if one decides $d_0 : \theta \leq 0.9$, then there is a 0.9 probability of a zero loss and a 0.1 probability of a loss l_0:

$$EL^0(d_0) = Pr(\theta \leq 0.9) \times L(d_0, \Theta_0) + Pr(\theta > 0.9) \times L(d_0, \Theta_1)$$
$$= 0.9 \times 0 + 0.1 \times l_0 = 0.1 \times l_0. \tag{11.3}$$

In the same way, the prior expected loss of decision $d_1 : \theta > 0.9$ is obtained

$$EL^0(d_1) = Pr(\theta \leq 0.9) \times L(d_1, \Theta_0) + Pr(\theta > 0.9) \times L(d_1, \Theta_1)$$
$$= 0.9 \times l_1 + 0.1 \times 0 = 0.9 \times l_1. \tag{11.4}$$

The optimal decision d_{opt}^0 will be the one for which $EL^0(d_i)$ is minimal. Following the losses $l_0 = l_1 = 100000$ defined above, one thus obtains

$$EL^0(d_0) - 10000, \quad EL^0(d_1) = 90000, \quad \Rightarrow d_{opt}^0 = d_0$$

The conclusion thus is that in a setting in which the decision maker does not wish to inspect any units (i.e. the decision is to be based solely on the prior beliefs about θ), it is advisable to conclude d_0, that is, the proportion is smaller than 0.9, because this action has the smaller expected loss.

Since this recommendation for the decision d_0 is obtained as a function of the decision maker's prior beliefs about the proportion θ and the choice of the loss function, it is worth mentioning that the recommendation may change as these input values change. For the purpose of illustration, imagine that the prior beliefs of the decision maker are represented by a beta$(3, 0.5)$ distribution. This distribution reflects increased beliefs for high values of θ. In *particular, it implies that more than 50% of prior belief weight is given to values of θ that are greater than 0.9:*

$$Pr(\theta > 0.9) = \int_{0.9}^{1} f(\theta | 3, 0.5) d\theta = 0.55.$$

Accordingly, $Pr(\theta \le 0.9) = 0.45$ and the prior expected losses of decisions d_i are as follows:

$$EL^0(d_0) = 0.55 \times 100000 = 55000,$$

$$EL^0(d_1) = 0.45 \times 100000 = 45000, \quad \Rightarrow d_{opt}^0 = d_1.$$

Because a decision maker with a beta(3, 0.5) prior distribution considers that it is more probable that the true value of the parameter θ is above 0.9, rather than below 0.9, it is advisable for this person to select d_1 (i.e. the decision according to which θ > 0.9). The expected loss associated with this decision is lower than that for d_0.

It may appear from the preceding that only the probability for the states of nature is important since in both examples, the decisions only depend on the probabilities of the states on nature. Notice, however, that this is only so at this point because the loss function is taken to be symmetric. More generally, one should take action d_0 when the associated expected loss is minimal, that is, when

$$Pr(\theta \in \Theta_1) \times l_0 < Pr(\theta \in \Theta_0) \times l_1.$$

Rearranging terms, it is advisable to decide d_0 any time $Pr(\theta \in \Theta_1) < l_0/(l_0 + l_1)$. It thus becomes clear that whenever the losses associated with erroneous conclusions are equal, the preferred action is the one in which the associated parameter values have the higher probability.

Example 11.5 *(Influence diagram for deciding about a proportion) The prior analysis in Example 11.4 can be translated into an influence diagram as shown in Figure 11.8(i). The definition of the nodes is summarized in Table 11.3. The part of the network covering the discrete chance nodes $\alpha, \beta, \theta, \theta^*$ and $\theta > \theta^*$? follows the definitions given in Biedermann et al. (2008b). It represents uncertainty about θ by means of a beta distribution with parameters α and β, as described earlier in Section 9.3.3. The node $\theta > \theta^*$ represents probabilities for values of the proportion θ that are greater than the threshold value defined by the node θ^*. This model fragment is extended here with a decision node D (covering the two available actions d_0 and d_1) and two discrete chance nodes l_0 and l_1. The chance nodes cover a single numeric state, which provides the losses of interest l_i, $i = \{0, 1\}$. Additional states may be chosen as required in order to allow the model to cope with scenarios involving other losses. In the same way, the user may also define other ranges of numerical values for the nodes α, β and θ^*. The loss node L has its state defined according to the loss function defined in Example 11.4, that is, a zero loss for correct decisions and losses $l_0 = l_1 = 100000$ for erroneous decisions (i.e. a constant symmetric loss function). Technically speaking, the node L takes, respectively, the value 0, or the numerical values defined for l_0 and for l_1, depending on the actual configurations of the parental nodes of L, that is, actions d_i, the true states of nature Θ_j and the specified limiting value θ^*.*

Figure 11.8(ii) shows a partially expanded representation of the influence diagram with θ^* instantiated to 0.9 and α and β instantiated to 1. The settings for α and β (the parameters of the

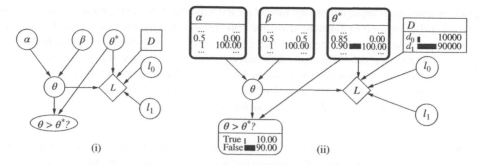

Figure 11.8 Collapsed (i) and partially expanded (ii) representation of an influence diagram for deciding about a proportion when no units are inspected. The node θ represents the uncertain proportion, the diamond-shaped node L the loss and the squared node D the available actions. The remaining nodes are defined according to Table 11.3. Instantiated nodes are shown with a bold border.

Table 11.3   Definitions of the nodes used in the influence diagram shown in Figure 11.8.

Node	Description	State(s)
α, β	Parameters of the beta-distributed variable θ	$0.1, 0.5, 1, 2, \ldots, 10$ (e.g.)
θ	Proportion θ of 'positive' units in the population (i.e. seizure or consignment)	$0{-}0.05, \ldots, 0.95{-}1$
θ^*	Target value for the unknown proportion	$0.85, 0.9, 0.95$ (e.g.)
$\theta > \theta^*$?	Is the true proportion θ greater than the specified target value θ^*?	$true, false$
D	Decision about the proportion (available actions)	d_0, d_1
l_0	Loss for erroneously deciding d_0 : $\theta \le \theta^*$ when Θ_1 : $\theta > \theta^*$ is the true state of affairs	100000 (e.g.)
l_1	Loss for erroneously deciding d_1 : $\theta > \theta^*$ when Θ_0 : $\theta \le \theta^*$ is the true state of affairs	100000 (e.g.)
L	Decision loss	$L(d_i, \Theta_j), \quad i, j = 0, 1$

The states of the nodes $\alpha, \beta, \theta^*, l_0$ and l_1 are chosen according to the requirements of the case under investigation. Other ranges of values may be chosen as required.

beta distribution) reflect the initial starting point defined earlier in Example 11.4. The value displayed for the state 'true' of the node $\theta > \theta^*$? corresponds to the cumulative probability that θ is greater than θ^*, for $\theta^* = 0.9$, found in (11.2). The node D displays the expected prior losses of the two available actions d_0 and d_1. In agreement with the results from Example 11.4, these are, respectively, 10000 and 90000 with d_0 being the a priori *optimal decision.*

Example 11.6 (*Expected value of perfect information in a decision problem about a proportion) The expected loss of the* a priori *optimal action, as calculated in Examples 11.4 and 11.5, is also sometimes called the* expected value of perfect information *(EVPI) about the true state of θ. Although, in practice, it may not be feasible for a decision maker to obtain perfect information about the true state of nature, the EVPI is a measure of appreciable conceptual and practical relevance. As noted earlier in Section 1.4.6, the EVPI allows one to indicate the* *maximum amount of money that one should be willing to pay for (expert) information such that it would allow one to know or learn the true state of nature with certainty. In the current scenario, where the prior beliefs are given by a beta$(1, 1)$ distribution and the losses are $l_0 = l_1 = 100000$, one should accept additional information about the true proportion only if the cost for that information does not exceed $EL^0(d_{opt}) = EVPI = 10000$. If the loss of the a priori optimal action (EVPI) is smaller than the cost of additional information, then it is more reasonable for the decision maker not to consider that information.*

Example 11.7 *(Influence diagram for posterior analysis) As an extension of the previous examples in this section, focus now on finding the optimal action for a situation in which information is available on some of the units from the consignment. For illustrating the procedure, consider a setting in which 21 units are chosen randomly from a large consignment and analysed. The discussion here assumes that a consignment is sufficiently large that the sampling process may be treated as sampling with replacement. Suppose further that the analyses of all selected units yielded positive results. This is a special situation because, theoretically, with 21 inspected units, there can be 22 different results, that is, 0, 1, ... , 21 positive units. This particular setting is chosen here because according to current forensic consignment inspection criteria (Aitken 1999), a set of 21 units all found to be positive is required to be 90% certain that the proportion is greater than the specified value $\theta^* = 0.9$ (under the assumption of a beta$(1, 1)$ prior distribution). More formally, the number of positive units required to be $100p\%$ certain that θ is greater than θ^* is given by the smallest integer greater than $[\log (1 - p)/ \log (\theta^*)] - 1$.*

Assuming again the loss function defined earlier in Example 11.4, the question of interest thus is 'Given the information on the analysed units, how is one to decide about the proportion in an optimal way?' As for the prior analysis, the aim is that one is going to choose that of the two actions d_0 and d_1 which presents the minimum expected loss. However, one will focus now on expected losses that are calculated by weighting the possible losses of the actions by posterior probabilities $Pr(\theta \in \Theta_j | \alpha, \beta, s, n)$, rather than the prior probabilities $Pr(\theta \in \Theta_j | \alpha, \beta)$ as used in the prior analysis. Note that the terms s and n denote, respectively, the number of positives and the number of inspected units (i.e. 21 in the case here).

If the prior distribution for θ is a beta(α, β) distribution with $\alpha = \beta = 1$ as defined earlier, then as a standard result in Bayesian statistics, the corresponding posterior probability of H_0 (i.e. the composite hypothesis $H_0 : \theta \leq \theta^*$) given n inspected units, with s units found to be positive (here $n = s = 21$), is (e.g. Lee 2004; Taroni et al. 2010)

$$Pr(\theta \leq 0.9 | \alpha, \beta, s, n) = \int_0^{0.9} f(\theta | \alpha + s, \beta + n - s) d\theta$$

$$= \int_0^{0.9} f(\theta | 22, 1) d\theta = 0.10.$$

The Bayesian updating rule here consists of adding the number of positives (s) to the first parameter and the number of negatives $(n - s)$ to the second parameter of the beta prior distribution. Accordingly, the cumulative probability for $\theta > 0.9$, that is, $Pr(\theta > 0.9 | \alpha, \beta, s, n)$, is given by $1 - Pr(\theta \leq 0.9 | \alpha, \beta, s, n)$, giving 0.90 in the example here.

With the posterior probabilities for the Θ_j now being available, the expected loss for each decision can be found by weighting the possible loss of a selected decision by the probability *of the occurrence of that loss. Unlike (11.3) and (11.4), the expressions of the expected losses here are written only for the losses different from zero, that is,*

$$EL'(d_0) = Pr(\theta > 0.9|\alpha, s, \beta, n)L(d_0, \theta > 0.9) \tag{11.5}$$

$$= 0.90 \times 100000 = 90000,$$

$$EL'(d_1) = Pr(\theta \leq 0.9|\alpha, s, \beta, n)L(d_1, \theta \leq 0.9)$$

$$= 0.10 \times 100000 = 10000. \tag{11.6}$$

Because the expected posterior loss of the decision d_1 is smaller than that of d_0, one ought to decide d_1 and adopt the viewpoint according to which $\theta > 0.9$. It may be convenient to rewrite this as $d'_{opt} = d_1$. This result thus states that in view of the predominant part of the (posterior) probability being assigned to values of the proportion greater than $\theta^* = 0.9$, it is advisable to conclude that the true proportion θ lies in that area. In fact, if this should indeed be the case, then the loss associated with this consequence (i.e. a correct decision) would be zero. There is only a probability of approximately 0.1 that the true proportion is lower than 0.9, and this leads to an expected loss that is lower than that for the decision d_0 : $\theta \leq \theta^*$. Actually, deciding d_0 : $\theta \leq \theta^*$ is associated with a probability of approximately 0.9 for the loss l_0 (and, thus, a greater expected loss). Again, loss functions of type $0 - l_i$ are not necessarily symmetric, and the decision rule derived at the end of Example 11.4 applies here. Action d_0 is optimal whenever $Pr(\theta \in \Theta_1|\alpha, \beta, s, n) < l_0/(l_0 + l_1)$.

To calculate the posterior probabilities for the unknown parameter θ, and the associated expected losses for the available decisions d_i, one can reuse the influence diagram previously described in Figure 11.8. It is only necessary to add a minor extension to account for the information about the inspected units. This can be done by adding two additional nodes n and s as shown in Figure 11.9 (Biedermann et al. 2008b). These two nodes allow one to specify the number of inspected units n amongst which s units are found to be positive (in the context also known as a binomial sample). In the network shown in Figure 11.9, the nodes n and s are both instantiated to the value 21, to reflect the available new information. The node θ will account for the resulting posterior distribution, based on a beta(1, 1) prior distribution (nodes α and β instantiated to 1), whilst the (cumulative) posterior probability that the proportion θ is greater than θ^* (here θ^* is set equal to 0.9) is provided by the node $\theta > \theta^*$. The node D shows the posterior expected losses of the two decisions d_i. These values agree with the results obtained above in (11.5) and (11.6). The proposed network structure is generic in the sense that results for other numbers n of inspected units and numbers of units found to be positive s, can readily be specified and the effect on the expected losses in decision node D can be evaluated.

There is an additional element, other than the consequences of decisions, that must be taken into account. That element appears in the form of experimental cost, that is, the cost of taking observations, the size of which depends mainly on the total number of observations, n. An *overall loss function* that takes into account both aspects can be defined by considering the sum of the decision loss and the cost of the observations and it is denoted $L_c(d, \Theta, n)$ (Berger, 1988). The subscript c is used to indicate that the loss function includes the cost of the experimentation. Therefore,

$$L_c(d, \Theta, n) = L(d, \Theta) + \sum_{i=1}^{n} k_i,$$

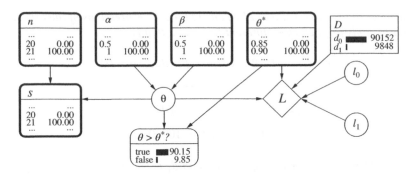

Figure 11.9 Partially expanded representation of the Bayesian decision network introduced earlier in Figure 11.8, extended here by two nodes n (for the number of inspected units) and s (for $s \leq n$, representing the number of positive units amongst the inspected units) in order to support the calculation of the posterior distribution for the parameter θ (the uncertain proportion). The diamond-shaped node L represents the loss function and the squared node D the available actions. The remaining nodes are defined according to Table 11.3.

with k_i denoting the ith observational cost. Note that the aim is to find a coherent balance between the decision loss and the cost of experimentation, since to lower the decision loss, it will generally be necessary to run a larger experiment, which is associated with higher experimental costs. The expected loss $EL'_c(d_i)$ of a given decision d_i, $i = 0, 1$, can, therefore, be computed as

$$EL'_c(d_i) = \sum_{j=0,1} Pr(\Theta_j | \alpha, \beta, s, n) L_c(d_i, \Theta_j, n)$$

$$= \sum_{j=0,1} Pr(\Theta_j | \alpha, \beta, s, n)(L(d_i, \Theta_j) + kn)$$

$$= \sum_{j=0,1} Pr(\Theta_j | \alpha, \beta, s, n)L(d_i, \Theta_j) + \sum_{j=0,1} Pr(\Theta_j | \alpha, \beta, s, n) \times kn$$

$$= \sum_{j=0,1} Pr(\Theta_j | \alpha, \beta, s, n)L(d_i, \Theta_j) + kn$$

$$= Pr(\Theta_j | \alpha, \beta, s, n)L(d_i, \Theta_j) + kn \qquad \text{for } i \neq j,$$

since $L(d_i, \Theta_j) = 0$ for $i = j$.

Example 11.8 (*Influence diagram for posterior analysis including cost of analyses*) *A minor modification of the Bayesian decision network shown in Figure 11.9 allows one to approach situations in which the cost of analyses needs to be taken into account. One aspect of the required modification is an additional node k whose numeric states represent the cost of inspecting individual units. The node k is a parent node of the loss node L. The user may define several distinct values for k, depending on the case at hand. Here, 150 and 0 are chosen in order to perform – given appropriate instantiations made at this node – calculations of posterior losses with or without consideration of the cost of analysis, assuming equal costs for each experimental unit. Another part of the extension is a directed edge that connects the nodes n and L. Nodes k and n are needed to define the component of cost in the loss*

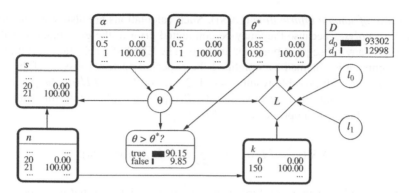

Figure 11.10 Partially expanded representation of the Bayesian decision network shown in Figure 11.9, extended here by a node k, representing the cost for inspecting individual units (instantiated here to 150, assuming equal costs per experimental unit). The node θ represents the uncertain proportion, the diamond-shaped node L the decision loss and the squared node D the available actions. The node n defines the number of analysed units and s the number of positives found after analysing all n units. The remaining nodes are defined according to Table 11.3. Nodes are instantiated, shown with a bold border, according to the scenario discussed in Example 11.8. The node D displays the expected posterior losses.

function. That is, for correct decisions, there is no longer necessarily a zero loss, because in cases where n units were inspected, there will be a loss given by $n \times k$. Similarly, for incorrect decisions, the loss will not necessarily be $l_0 = l_1$, but $l_{i=0,1} + n \times k$. Figure 11.10 shows the Bayesian network that incorporates these extensions. The partially expanded form illustrates that information on the inspected units is specified in the nodes n and s, along with the costs of analysis fixed at 150 (per analysed unit) in node k. The node D displays the expected posterior losses, which correspond to (with rounding)

$$EL'_c(d_0) = Pr(\theta > 0.9 | \alpha, \beta, s, n)L(d_0, \theta > 0.9) + k \times n$$

$$= (0.90 \times 100\ 000) + (150 \times 21) = 93150,$$

$$EL'_c(d_1) = Pr(\theta \leq 0.9 | \alpha, \beta, s, n)L(d_1, \theta \leq 0.9) + k \times n$$

$$= (0.10 \times 100\ 000) + (150 \times 21) = 13150.$$

As may be seen, the expected loss of the a posteriori optimal action, $EL'_c(d_{opt})$, has increased by the cost of analysing n units at a cost of $k = 150$ per unit, that is, by a total of $k \times n = 3150$. This leads to the expected loss of now 13150. Compare this result to the value of 10000 found in Example 11.7, shown also in Figure 11.9 as 9850 where the difference is due to rounding errors in the calculations shown in the text.

The influence diagrams presented in the above examples support a wide range of computations relating to notions useful in the context of consignment inspection. For example, before starting laboratory examinations on n units selected within a consignment, one does not know how many, s, positive units ($s = 0, 1, 2, \ldots, n$) will be observed. The decision maker is in a position in which he/she assigns probabilities to the number $s = 0, 1, 2, \ldots, n$ of positive units which are expected to occur. These probabilities enable calculation of the so-called *expected value of sample information* (EVSI). Such calculations are also sometimes termed *pre-posterior analyses* because they may be conducted prior to the examination of the individual units and any subsequent posterior analysis. The EVSI of a selection of n units drawn from a consignment is obtained by weighting the expected posterior losses (of the posterior optimal action) for each possible result $s = 0, 1, 2, \ldots, n$, that is, $EL'(d_{opt}|\alpha, \beta, s, n)$, by the probabilities of obtaining the respective results, $Pr(s|n)$, and subtracting this sum from the expected loss of the *a priori* optimal action, $EL^0(d_{opt})$:

$$\text{EVSI}(n) = EL^0(d_{opt}) - \sum_{s=0}^{n} EL'(d_{opt}|s, n)Pr(s|n).$$

This definition can be taken a step further in order to find the *expected net value of sample information*, written ENVSI(n) for short. It is obtained by taking into account the cost of inspecting n units, $K(n)$, which must be subtracted from the EVSI(n):

$$\text{ENVSI}(n) = \text{EVSI}(n) - K(n).$$

Further discussion of this result is presented in Biedermann et al. (2012a).

11.2.4   Decision after database searching

What action should be taken following the results of a database search? This question touches on a broad topic with various interrelated issues. As noted earlier in Section 7.10, one complication relates to inference based on two distinct items of information, given, on the one hand, by the observed correspondence between the profile of a trace and a person's profile found in the database, and, on the other hand, the observed differences with respect to the other database entries (e.g. in a setting in which there was exactly one database member found to correspond). In some wider sense, this can be seen as a problem of the assessment of the combined value of distinct observations (see also Chapter 8). Bayesian networks offer a way to handle this aspect coherently (Biedermann et al., 2012e, 2011a; Taroni et al., 2004). Another aspect relates to the question of what to decide about an individual found as a result of a database search, such as 'Should the person found as a result of a database search be considered as the source of the crime stain?' This question is sometimes referred to in this context as *individualization*. Influence diagrams represent an appropriate means to study the formal requirements of rationality for determining the answer to such a question. In particular, influence diagrams allow one to illustrate how to aggregate elements of inference networks constructed previously in Chapter 7 with decision theoretic considerations from Chapters 1 and 2.

Figure 11.11 shows a general structure of an influence diagram for decision analysis in a case where a database search has led to the observation that the profile of a crime stain corresponds to the profile of one of the n persons who have their profile stored in the database. Let the person with the corresponding profile be denoted without the loss of generality as 'person 1'. The probabilistic part of this network corresponds to the network described earlier in Figure 7.23 (Section 7.10). It is extended here by a decision node D and a node L for the loss function. The node D has two states d_1, 'Individualize person 1', and d_2, 'Do not individualize person 1'. This corresponds to the generic decision structure specified in Figure 11.1. It is augmented at this juncture solely with an additional node λ with numerical states that specify

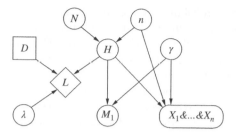

Figure 11.11 Influence diagram for a case in which a correspondence (M_1) is observed between the profile of a crime stain and that of a suspect found as a result of a search in a database with n entries. The size of the population of potential offenders is N, where n (with $n \leq N$) of them are on a database. The node H has three states: 'The suspect is the source of the crime stain' (H_1), 'One of the $n - 1$ other individuals in the database is the source of the crime stain' (H_{2_n}) and 'The source of the crime stain is amongst the $N - n$ individuals outside the database' (H_{n+1_N}). The population proportion of the corresponding genetic feature is γ. The node $X_2 \& ... \& X_n$ is binary and represents the proposition according to which the profiles of the $n - 1$ individuals in the database, other than the suspect, do *not* correspond to the crime stain. Node D represents the decision to individualize (d_1) and not individualize (d_2), respectively, the person with the corresponding profile. Node L provides the loss function and node λ particular values for the consequence of a missed individualization.

values – to be defined by the user – for the loss associated with the consequence of deciding d_2, not individualizing person 1, when H_1 is true, person 1 is the source of the crime stain. The loss associated with this particular consequence, that is, a missed individualization, might typically be different from the zero loss of correct conclusions and the maximum loss of 1 assigned for erroneous individualizations (as discussed in Chapter 1, Table 1.6). It may be subject to discussion, and for this reason, the possibility of modifying assumptions about its actual value through adjustments to the states of node λ may be a desirable feature of the model. Denote the proposition that person 1 is not the source by H_2. The loss function can then be summarized as follows:

$$L(d_i, H_j) = \begin{cases} 0, & i = j, \\ 1, & i = 1, \quad j = 2, \\ \lambda, & i = 2, \quad j = 1. \end{cases}$$

The following expression, developed according to discussion on utility theory in Chapter 1 (Section 1.4.2), can be used to assist with the choice for λ:

$$L(d_2, H_1) = L(d_1, H_1)(1 - \lambda) + L(d_1, H_2)\lambda,$$

The loss on the left-hand side of this expression is a weighted sum of the losses $L(d_1, H_1)$ and $L(d_1, H_2)$ where the weights are probabilities. This expresses the view that the loss associated with a missed individualization (i.e. deciding d_2 when H_1 is true) depends on the evaluator's probability λ of obtaining the worst consequence (here, an erroneous individualization) and the probability $(1 - \lambda)$ of obtaining the best consequence (i.e. a correct individualization). Given the $0 - 1$ loss function for correct conclusions defined above, $L(d_1, H_1)$ is zero so that the loss of interest $L(d_2, H_1)$ reduces to the probability λ of obtaining the worst consequence. Another way to express this result would be to say that the loss of a missed individualization is the fraction λ of the loss of an erroneous individualization. The loss λ intervenes in the calculus of the expected loss of decision d_2, as follows: $EL(d_2) = L(d_2, H_1)Pr(H_1) + L(d_2, H_2)Pr(H_2)$. The terms $Pr(H_i)$, for $i = \{1, 2\}$, are shorthand notation for the posterior probabilities of propositions H_i, given the observed correspondence between the profile of the trace and that of person 1 (event M_1) and the differences with respect to the other $n - 1$ profiles in the database (event $X_2 \& \ldots \& X_n$). Given that a correct non-individualization has a zero loss, the second term in the sum for $EL(d_2)$ is zero. Consequently, $EL(d_2)$ reduces to $\lambda Pr(H_1)$. See Section 1.4.2 for further details on utility theory and Section 1.4.4 for the loss function.

The expected loss of decision d_1, the individualization of person 1, is equal to $Pr(H_2)$ because in

$$EL(d_1) = L(d_1, H_1)Pr(H_1) + L(d_1, H_2)Pr(H_2),$$

the losses $L(d_1, H_1)$ and $L(d_1, H_2)$ are, respectively, 0 and 1. In this context, minimizing expected loss thus means to decide d_1 only if $Pr(H_2) < \lambda Pr(H_1)$. Developing this further, it can be shown that d_1 is the optimal decision only if $Pr(H_1) > 1/(1 + \lambda)$.

Example 11.9 (*Influence diagram for decision after a database search*) *Suppose a case in which the population of potential sources covers $N = 10^8$ individuals. The database contains the profiles of $n = 10^6$ persons. The findings consist of a correspondence (where the characteristic has population proportion $\gamma = 10^{-6}$) with the person 1 and no correspondence with the other $n = 2, \ldots, 10^6$ individuals (supposing error free analyses). These details can be specified at the nodes N, n, γ, M_1 (for the observed correspondence with person 1) and $X_2 \& \ldots \& X_n$ (for the observed differences with the $n - 1$ other profiles in the database) in the model shown in Figure 11.11. The posterior probability for proposition H_1, that is, $Pr(H_1|M_1, X_2 \& \ldots \& X_n)$, corresponding to such a case will be 0.01 (assuming a prior probability of $1/N$). Hence, the expected loss of decision d_1, individualizing person 1, will be $Pr(H_2|M_1, X_2 \& \ldots \& X_n) = 0.99$. This value will be greater than the expected loss for d_2, given by $\lambda Pr(H_1|M_1, X_2 \& \ldots \& X_n)$, because λ is defined as a value in the range between 0 and 1. This result, not to individualize person 1, is intuitively reasonable because the posterior probability for this person being the source of the crime stain is low. For a detailed discussion and analyses of alternative scenarios, see also Gittelson et al. (2012).*

11.3   Further readings

The question of whether or not to perform DNA analyses, as outlined in Section 11.2.1, represents an example for a particular instance of decision analysis in forensic science practice. As a natural extension to this, further questions may readily become of interest. The question 'How many markers to type?' is one such example. Taroni et al. (2007) considered this question by focusing on the gain in terms of utility by additional markers. Decision analysis for forensic DNA is also addressed in Mazumder (2010). More recently, Gittelson et al. (2014) approached the topic of genotype designation, that is, a decision problem marked by complications due to phenomena of drop-in and drop-out. More generally, literature on Bayesian decision networks for forensic science applications has remained rather scarce. Most reports focus on the generic analytic structure outlined in Figure 11.1, implementing the so-called one-stage decision problems. For an example of a staged decision analysis following the generic structure shown earlier in Figure 2.6 (Chapter 2), see the discussion on processing fingermarks presented in Gittelson et al. (2013b).

12

Object-oriented networks

12.1 Object orientation

In some areas of application in forensic science, inference and decision problems may be characterized by recurrent appearances of the same or similar components. Forensic kinship analyses (Section 7.9) provide an illustrative example for this. They typically involve multiple individuals, each being described in terms of their genotype. Graphical models for such applications can reflect this property in terms of repetitive sub-structures, also sometimes called *(local) network fragments* (for examples, see Figure 7.16). From a methodological point of view, decomposing a problem into smaller parts, in particular by recognizing generic structures and describing these in terms of local structures, can help to assure coherence in the overall model construction process. On a practical account, however, working with local structures can also bring complications. For example, when the analyst judges that a local structure needs definitional changes, then all instances where the structure of interest appears might also need a revision. More generally, the differences in focus on either local structures or aggregations of local structures within a larger context relate to the notion of hierarchy. This notion also reflects the way in which analysts tend to reason about problems. In fact, they commonly shift focus between different levels of abstractions, depending on the requirements of the situation at hand.

In order to support the construction and use of graphical models for potentially large problem domains, in particular where they involve recurrent aspects, more recent technological developments have focused on a framework referred to as *object-oriented probabilistic networks* (Section 2.2.5). The use of the notion of object orientation in the particular area of knowledge representation, processing and analysis using graphical models, covers two main aspects. One concerns the interpretation of network fragments as *objects*, where objects are instances of classes. Generally speaking, a class can be defined as a general model that can be reused as required in different contexts. A second major aspect is the idea of *inheritance*.

Bayesian Networks for Probabilistic Inference and Decision Analysis in Forensic Science, Second Edition.
Franco Taroni, Alex Biedermann, Silvia Bozza, Paolo Garbolino and Colin Aitken.
© 2014 John Wiley & Sons, Ltd. Published 2014 by John Wiley & Sons, Ltd.

This notion defines a relationship between classes because different classes may share common sub-structures or, so to speak, pass on structure to other classes.

In their object-oriented extension, probabilistic graphical models thus contain so-called instances nodes, in addition to the usual nodes dealt with so far in this book. The latter, usual nodes are also sometimes said to represent *basic* variables, because they do not represent an instance of a network class. In turn, an *instance node* is a node that represents an instantiation of a network class. Such a class represents a subnet, that is, a probabilistic graphical model with variables acting as interfaces and variables forming an internal hidden structural part. More generally, one can also think of an instance node as a copy of the network of which it is an instance. Further, it is important to note that an instance that is embedded in another network can itself contain instance nodes. Therefore, an object-oriented network can be considered a hierarchical description of a given domain under study.

When an instance node for one network is present within another network, then it connects to nodes in that other network through so-called interface nodes. These are basic variables in the sense defined above and can be divided into two categories, called *input* and *output* nodes. An input node is a placeholder for a node in another network in which the instance (or class) at hand is used. An output node has a different role: it can act as a parental node for a node in another network in which the instance is incorporated. It is useful to note that, usually, interface nodes represent only a subset of all the nodes that make up a network class. Therefore, an instance node will typically have a hidden structural part. This supports information hiding, which may be a property that is explicitly sought in large model constructions where the increasing number of nodes can tend to make it difficult to follow the visual representation (Fenton and Neil 2013).

It is useful to reflect these distinctions in definition in a visually appropriate way. In this chapter, following the usage of Section 2.2.5, input nodes will be specified with a dotted outer line. In turn, output nodes will be displayed in bold with a solid outer ring. This method of representation is found, for example, in some software packages as well as specialized forensic literature on the topic (e.g. Dawid et al. 2007). Note further that instance nodes can be displayed in two different ways. In the so-called expanded mode, the interface nodes are shown, whereas in the so-called collapsed mode, interface nodes are not shown. In the latter mode, instance nodes have a reduced size.

The description of object-oriented networks can be eased by adopting some notational conventions. Following standard literature in the field (Dawid et al. 2007), network classes will be indicated by **bold face** throughout this chapter. In turn, instance and regular nodes will be written in `teletype face`, which is a slight digression with respect to other chapters in this book where *simple italics* is used for node names in networks that do not contain instance nodes.

12.2   General elements of object-oriented networks

12.2.1   Static versus dynamic networks

For a considerable part of this book, Bayesian networks are considered for inference about target propositions based on particular observations. In one of its general forms, such probabilistic inference reduces to a two-node network fragment as illustrated in Figure 12.1(i). There are various ways in which the target (or hypothesis) variable and the conditioned information variable may be defined in such a basic model, depending on the context of the application. For the discussion here, consider again the problem of drawing an inference about the proportion of units in a large consignment that are of a certain kind (e.g. contain something illicit, such as an illicit drug), introduced earlier in Chapter 9 from the inspection of a subset of the consignment (Aitken 1999; Biedermann et al. 2008b). For such an application, more generally also known as *sampling*, the hypothesis variable, named Prop here, represents the proportion θ of positives in a consignment. The outcome variable, called Unit, represents the property of a particular unit drawn from the consignment: positive when the target substance (e.g. an illicit content) is present, and negative when the target substance is absent. Figure 12.1(i) represents inference about the proportion for a single observed unit, but this only gives an incomplete account of the reality of a sample inspection. In particular, it does not account for the fact that, in actual applications, scientists would typically inspect several distinct units. One way to deal with multiple variables that each add a distinct item of information was considered earlier in Chapter 9 and consists of considering distinct nodes for modelling the actual status (i.e. positive or negative) of each inspected unit. This is illustrated in Figure 12.1(ii) where an additional node Unit′ is introduced and which is conditionally independent of the first information variable Unit. In the same way, further information variables may be added. The main difference between Figure 12.1(i) and (ii) is that the former model is *static* in the sense that it does not consider the prevision of a positive (or negative) unit in a future trial, given knowledge about the state of a previously examined unit. In other words, it does not consider a temporal dimension in which one's belief in the current state of the variable Prop (i.e. the proportion θ) is updated through repeated observations. It also does not consider that a given state of belief about θ depends on what has been observed in the past. This is different for the model shown in Figure 12.1(ii), which is *dynamic*, because it accounts for temporal sequences. It models past observations but also covers prevision for future observations. To make these temporal sequences more explicit, Figure 12.1(iii) represents the current state of belief about the proportion θ through distinct nodes. This model clarifies that when taking into account information about a further examined unit (e.g. node Unit′), the prior belief about the proportion θ at node Prop′ is given by the posterior belief arrived at through previous observations (given at node Prop). The arc between Prop and Prop′ makes this dependency explicit. One could thus say that Figure 12.1(i) represents a time slice and Figure 12.1(iii) several connected time slices. As pointed out in further detail

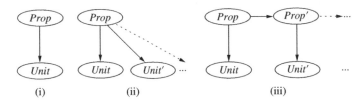

Figure 12.1   (i) Static Bayesian network for inference about a proportion θ of units in a large consignment that are of a certain kind (e.g. contain something illegal, node Prop), based on the information about the characteristic (e.g. the presence or absence of illegal content) of a single examined unit (node Unit), drawn from that consignment. (ii) Dynamic Bayesian network allowing to account for information about a further examined unit (node Unit′). (iii) Alternative representation of a dynamic Bayesian network for inference about a proportion θ using a distinct node Prop′ for inference based on the information about an additionally examined unit (node Unit′).

in the next section, such repetitive patterns of reasoning, representing a temporal dimension, are particularly well suited for modelling using object-oriented networks.

12.2.2   Dynamic Bayesian networks as object-oriented networks

Figure 12.1(ii) shows that a static Bayesian network, such as the time slice shown in Figure 12.1(i), can be extended to a dynamic Bayesian network by specifying multiple time slices and linking these in an appropriate way. Stated otherwise, Figure 12.1(ii) contains multiple instances of the time slice shown in Figure 12.1(i). This way of constructing a dynamic Bayesian network can be laborious, or inefficient, essentially because the various time slices have the same internal structure. For example, the transition probabilities for the node representing the proportion θ, that is, $Pr(Prop'|Prop)$, are the same in all time slices. In turn, whenever it is judged that the definition of the time slice requires a modification, all instances of this time slice may need to be updated as well.

To ease the construction of such time-sliced models, an approach based on so-called instance nodes may be used. Instead of explicitly constructing the same time slice multiple times, a generic network fragment – called a *class* (or class network) – may be defined and then invoked repeatedly. Invoking such a class network leads to a so-called instance of that class.

Figure 12.2(i) shows a class network for the time slice shown in the previous Figure 12.1(i). A particular feature of this class, named here **slice**, is the input node Prop*, shown with a dotted outer line. This node does not represent an actual variable but is given by the node Prop of the previous time slice. It can be thought of as a placeholder used for specifying the conditional probability table for the node Prop, shown with a bold solid outer ring (i.e. an output node). This output node Prop is to be bound to the input node of the next time slice.

Using two instances of the class network **slice** defined in Figure 12.2(i), one can construct the object-oriented network shown in Figure 12.2(ii). In this network, useable for inference about the proportion θ of units in a large consignment that are of a certain kind, the output node Prop of the first time slice is connected to the input node Prop* of the second time slice. In the same way, further time slices may be added as required. Note also that Figure 12.2(ii) shows that only the input and output nodes, the so-called interface nodes,

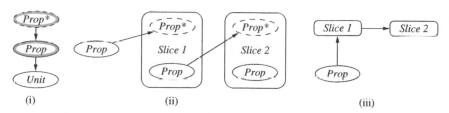

(i)   (ii)   (iii)

Figure 12.2   (i) Class network, named **slice**, for the time slice shown in Figure 12.1(i). Node Prop represents the proportion θ of a large consignment that is of a certain kind (e.g. contains something illegal). Node Unit represents the characteristic (e.g. the presence or absence of illegal content) of a unit drawn from the consignment. Node Prop* is a placeholder for a node Prop of a previous time slice. (ii) Object-oriented Bayesian network with two instances, named Slice 1 and Slice 2, of the class defined in (i). (iii) Representation of the object-oriented network with instance nodes shown in collapsed form.

are visible. The node Unit in each instance is part of the hidden structure (i.e. an encapsulated attribute). Figure 12.2(iii) shows the object-oriented network with instance nodes in collapsed form.

12.2.3   Refining internal class descriptions

A particular feature of object-oriented Bayesian networks is that they support a top-down construction process (Kjærulff and Madsen 2008). When constructing an object-oriented network, it may be sufficient to specify classes only on a coarse level of definition. A class is defined on a coarse level of detail if one restricts, in a first step, the scope of definition to the interface nodes, while leaving aside details of the internal structure (i.e. the encapsulated attributes). For the class network **slice** shown in Figure 12.2(i), for example, it would have been sufficient to specify only the input node Prop* and the output node Prop, because they represent the set of nodes that is relevant on the hierarchical construction level shown in Figure 12.2(ii). The level of abstraction of the hidden structural part may be refined at a later stage. Such changes will directly affect any instance of the class of interest in other networks, without the need of any further (i.e. manual) changes. This point is illustrated in Examples 12.1 and 12.2.

Example 12.1 (*Modified object-oriented network for inference about a proportion in a large consignment*) *Consider again the class network* **slice** *shown in Figure 12.2(i). In this class, the node* Unit *models the presence or absence of a target characteristic (e.g. illegal content) of a given examined unit (drawn from the consignment of units). This way of constructing the class does not, however, allow one to make a distinction between the true, but unknown, condition of a unit (i.e. containing or not containing an illegal substance) and what is observed in the course of an experiment designed to detect the presence or absence of that target feature. Stated otherwise, the mere observation of a positive (measurement) is equated with the examined unit being truly positive, which is a simplification of the underlying state of reality. Such distinctions are advocated, for example, in the context of DNA profiling analyses (Thompson et al. 2003), but are also challenged in other forensic disciplines (Saks and Koehler 2005).*

   In order to take this distinction into account, a binary observational variable Obs *with states 'positive' and 'negative' can be added as a direct descendant of the node* Unit. *Such a node* Obs *represents the observed result of a (diagnostic) test applied to the unit whose true state is represented by the node* Unit. *As shown in Figure 12.3(i), the directed arc from the latter to the former node indicates that the outcome of a test depends directly on the presence or absence of the target characteristic in the examined unit. This way of constructing the network allows one to account for the performance characteristics of the method used to analyse the unit of interest. These characteristics have two values: the probability of a positive result when the examined unit is truly positive,* $Pr(Obs = positive|Unit = positive)$, *and the probability of a negative result when the unit is actually negative,* $Pr(Obs = negative|Unit = negative)$, *the sensitivity and specificity of the method (see Section 9.3.3). Note that for a result to indicate the presence or absence of a unit's target characteristic with certainty, one would need to assume that* $Pr(Obs = positive|Unit = positive) = Pr(Obs = negative|Unit = negative) = 1$. *In practice, this may be an unrealistic assumption. Moreover, in a technical sense, it would render the additional node* Obs *superfluous because any instantiation made at this node would directly be copied to the node* Unit *(i.e. set to the same state).*

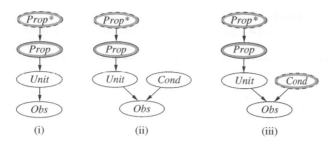

(i)   (ii)   (iii)

Figure 12.3 Modifications of the class network **slice** described earlier in Figure 12.2(i). Node Prop represents the proportion θ of a large consignment that is of a certain kind (e.g. contains something illegal). Node Unit represents the characteristic (e.g. presence or absence of illegal content) of a unit drawn from the consignment. Node Prop* is a placeholder for a node Prop of a previous time slice. Node Obs represents the observed result of a (diagnostic) test, whereas node Cond is binary, with states 'regular' and 'not regular', and represents the general conditions under which a unit of interest is analysed.

Example 12.2 *(Modified object-oriented network for inference about a proportion in a large consignment – continued) There may be circumstances shedding doubt on a result when examining a unit drawn from a consignment. Possible reasons for this may be that the examined unit is in a degraded condition or experimental settings were not optimal. Thus an examination may not always lead to a positive result when the unit of interest is truly positive, or it may lead to a positive result when in fact the examined unit is not positive. One can thus think of the result of conducting an examination (i.e. the performance characteristics of a method) as a function of the experimental conditions. The class networks shown in Figure 12.3(ii) and (iii) illustrate this in terms of an additional binary node called Cond, used as a parent for the observational variable Obs. The size of the conditional probability table of the latter node will double so that one can specify the method's sensitivity and specificity depending on whether or not the experimental conditions, represented by the node Cond, are optimal.*

It is worth noting that in Figure 12.3(ii), the node Cond is defined as a basic node that conditions the outcome of an examination of a given unit independently of the examination of another unit (i.e. represented by another instance of the same class). The probability for the experimental conditions being either optimal or not is, therefore, defined to be constant and equal for each examination. This is different for the class network shown in Figure 12.3(iii), where the node Cond is defined as an input node. This choice of definition allows one to introduce dependency between the assessment of results for distinct units, that is, between multiple, but connected, instances of this class network (see also Example 12.3).

Example 12.3 *(Object-oriented network for inference about a proportion in a large consignment: accounting for the potential of error and dependency in experimental conditions) Consider a case of inference about the proportion of a large consignment that is of a certain kind (e.g. the presence of a key feature). Suppose that four units drawn from the consignment show the target feature. Then, according to Aitken (1999), one can show that – assuming a uniform prior probability distribution for the proportion θ – that the probability that the proportion of positive units is greater than 0.5 is at least 0.95 (actually, 0.9689). This result can be reproduced, for example, by using an object-oriented network of the kind shown in Figure 12.2(ii), but with four connected slices. When instantiating the node Unit in each*

Table 12.1   Posterior probability that the proportion of units in a large consignment is greater than p (columns 3 and 4), given the result E (i.e. four inspected units all observed to be positive), for various initial assumptions about the experimental conditions (column 1).

		$Pr(Prop > p|E)$	
$Pr(Cond = optimal)$	$Pr(Cond = optimal|E)$	$p = 0.5$	$p = 0.9$
0.25	0.37	0.929	0.354
0.5	0.64	0.945	0.377
0.75	0.84	0.958	0.393
1	–	0.967	0.407
–	–	0.969[a]	0.410[a]

Column 2 lists the posterior probability for the state 'optimal' of node Cond, given the finding E. All results are obtained with the object-oriented network shown in Figure 12.4 (instantiations made at nodes Obs in each of the four instance nodes).

[a]Posterior probability that the proportion of units in a large consignment is greater than p given four positive units (i.e. assuming error-free analyses, specified by instantiating the nodes Unit in the instances of the class shown in Figure 12.3).

slice to 'positive', one can then find the posterior probability of interest in the node Prop (see also last row in Table 12.1). Notice that this evaluation supposes complete knowledge about the examined unit's actual properties. No distinction is introduced, as mentioned earlier in Example 12.1, between the observation of a test result and an examined unit's actual condition.

In order to account for results that cannot be assumed to be error free, instances of one of the class networks shown in Figure 12.3 may be used. Figure 12.4(i) shows such a network. It contains four instances of the class depicted in Figure 12.3(iii). For the purpose of illustration, suppose a value of 0.99 for the method's sensitivity and specificity under the assumption that the experimental conditions are 'optimal': $Pr(Obs = positive|Unit = positive, Cond = optimal) = Pr(Obs = negative|Unit = negative, Cond = optimal) = 0.99$. In turn, suppose a reduced value of 0.8 for settings in which the experimental conditions cannot be assumed to be optimal (i.e. node Cond is in state 'not optimal'). As a further difference with respect to the model defined earlier in Figure 12.2(ii), two summary nodes Prop>0.5? and Prop>0.9? are added. They have two possible states 'true' and 'false' and are amenable for evaluating cumulative probabilities.

Start by considering a distinction between the examined units' actual properties and the observations made during their examination. Suppose further that the experimental conditions are assumed to be optimal. This assumption can be specified by instantiating the node Cond to 'optimal'. A direct consequence of this is that inference about the actual condition of each unit (i.e. the nodes Unit in each slice) will depend only on the sensitivity and specificity value 0.99 as defined above. Note further that with this instantiation, there will be no dependency between the slices due to the node Cond (only through Prop). Thus, rather than instantiating the node Unit directly to 'positive' (as assumed at the beginning of this exercise), this model allows one to assess the state of the node Unit on the basis of 'positive' observations specified at the nodes Obs. In each slice, this inference is given by a likelihood ratio of 99 (i.e. sensitivity/(1 − specificity) = 0.99/0.01). This allowance for uncertainty about the actual condition of each unit has a weak overall impact on the target cumulative probability at the node Prop, as the probability changes less than 1% and is still greater than 95% (see also row four in Table 12.1).

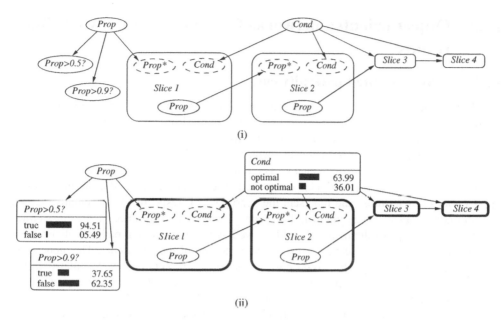

Figure 12.4   (i) Object-oriented Bayesian network with four instances of the class defined in Figure 12.3(iii) (see caption of Figure 12.3 for details of the definitions of the various nodes). Nodes Prop>0.5? and Prop>0.9? have two possible states 'true' and 'false' and are suitable for the evaluation of the cumulative probabilities for the population proportion θ. (ii) Partially expanded representation of the Bayesian network for evaluating a situation in which four units drawn from a large consignment are found to be positive. In each of the four instance nodes (marked with a bold border), the node Unit is set to 'positive'.

It is useful to note that the network shown in Figure 12.4, built on the basis of instances of the class shown in Figure 12.3(iii), is a general one. That is, with its top-level node Cond (common to all instances) instantiated, it produces the same results as an object-oriented network that consists of instances of the class shown in Figure 12.3(ii), with each slice having its own node Cond instantiated.

Next, consider an evaluation in which the experimental conditions are not assumed, by default, to be optimal in each trial. This can be achieved by leaving the node Cond uninstantiated. That is, initially, a prior probability distribution is defined over its possible states, which is then continually updated through inference made in each of the four instance nodes. Figure 12.1(ii) shows such an evaluation. For the sole purpose of illustration, equal prior probabilities were assigned to the states of the node Cond. The four instance nodes are shown with a bold border because their internal nodes Obs are instantiated (here to the state 'positive'). Specifying such a finding (i.e. four units observed to be positive) increases the probability of the experimental conditions to be optimal from 0.5 to 0.64. In turn, the posterior probability that the proportion θ is greater than 0.5 is no longer greater than 0.95 (actually 0.945) as in the previously discussed situation here above. The posterior probability that the proportion is greater than 0.9 is 0.377. Table 12.1 summarizes these results and compares them with the output for other initial assumptions about the experimental conditions (node Cond).

12.3   Object-oriented networks for evaluating DNA profiling results

12.3.1   Basic disputed paternity case

Recall the general paternity case scenario discussed earlier in Section 7.9.1 (Figure 7.15). It relates to an alleged family trio formed by a child, its undisputed mother and the putative father. The target proposition is whether the putative father is the true father, or whether the true father is some unknown individual, called *alternative father* here. Figure 12.5 shows an object-oriented network for this scenario. The rounded rectangles represent instance nodes. There are three different network classes invoked. The nodes pf, af and m are instances of the network class **founder** and represent the genotypes of, respectively, the putative father, the alternative father and the mother. The node c, which is an instance of the class **child**, models the genotype of the disputed child, as well as the transmission paths of alleles from the parents. In turn, the node tf represents the genotype of the true father. This node is an instance of the class **query** which is designed so as to copy the genotype of the putative father or an unknown individual, depending on the truth or otherwise of the principal proposition 'the putative father is the true father', represented by the basic node tf=pf?.

The forthcoming section will present further details on the network classes that are invoked in the object-oriented network shown in Figure 12.5. It is worth noting that this network represents the genotype of an unknown person (unrelated to the putative father), which assumed to be the true father under the alternative proposition, explicitly in terms of a separate instance node af. This was not the case for the model described earlier in Figure 7.15, where the probabilities for the true father's parental alleles were directly specified in the conditional node probability tables of nodes tfpg and tfmg (short for 'true father's paternal gene' and 'true father's material gene').

Figure 12.5(i) shows the object-oriented network in collapsed form, whereas Figure 12.5(ii) shows the same network with the instance nodes displayed in expanded form. This latter representation displays the interface nodes (i.e. input and output nodes) that

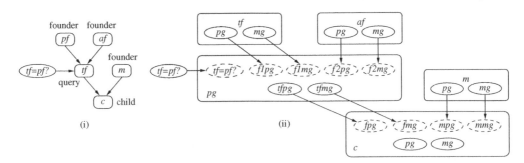

Figure 12.5   Object-oriented network for a basic disputed paternity case with instance nodes tf, pf, af, m and c, short for 'true father', 'putative father', 'alternative father', 'mother' and 'child', respectively, shown in (i) collapsed form and (ii) expanded form. Node tf=pf? is a basic variable and Boolean. It takes two values: 'true' if *tf* = *pf*, the true father is the putative father and 'false' if *tf* ≠ *pf*, the true father is not the putative father (i.e. an unknown person is the true father). Node names {f, f1, f2, tf, m}pg and {f, f1, f2, tf, m}mg denote, respectively, {father, father1, father2, true father, mother} paternal and maternal gene.

are used to ensure the connections between the various instance nodes and the single basic node tf=pf?. The class networks **founder**, **child** and **query** of which instances are used in the network shown in Figure 12.5 themselves contain instance nodes, but these are not shown at the level of representation given at this juncture. Section 12.3.2 covers further details on these hierarchical aspects and hidden structure.

12.3.2 Useful class networks for modelling kinship analyses

The class network **founder**, shown in Figure 12.6(i), contains two instance nodes pgin and mgin of the class **gene**. This latter class consists of a single output node as shown in Figure 12.6(iii) and models the different allelic forms that a marker (or gene) of interest may take, along with the respective allele probabilities. Here, pgin and mgin represent, respectively, the paternally and maternally inherited alleles of an individual. For the sole purpose of illustrating the different representational modes, Figure 12.6(i) shows the instance node pgin in collapsed form, whereas mgin is shown in expanded form. The latter two nodes feed two output nodes pg and mg that define the same scope of alleles as the class network **gene**. This distinct layer of output nodes is necessary because when using an instance of the class **founder** in another network, the single output nodes of the instances pgin and mgin cannot be linked directly to nodes in the other network as that would mean crossing more than one hierarchy (i.e. levels of encapsulation), which is not allowed in current software such as Hugin.

Besides **gene**, a further class network, called **genotype**, is invoked in the class **founder** [Figure 12.6(ii)]. The class **genotype** models an individual's genotype by defining observational nodes gtmin and gtmax as, respectively, the smaller and larger of the two parental alleles represented by pg and mg, respectively. This distinction is introduced because, in practice, one cannot tell which allele is inherited from which parent. The states of all the basic nodes in the class **genotype** correspond to the scope of alleles covered by the class network **gene**. The conditional probability table of the nodes gtmin and gtmax can be completed using, respectively, the following expressions (in Hugin syntax): min(pg,mg) and max(pg,mg).

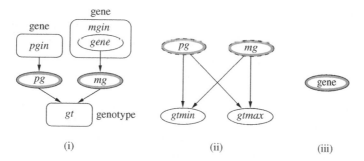

(i)   (ii)   (iii)

Figure 12.6 The class network (i), called **founder**, contains instance gt of the class **genotype**, shown in (ii), and instances pgin and mgin of the class **gene**, shown in (iii). The nodes pg and mg represent, respectively, paternally and maternally inherited alleles. Nodes gtmin and gtmax are defined as, respectively, the minimum and the maximum of the conditioning input nodes pg and mg.

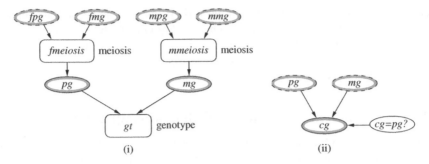

Figure 12.7   Class networks **child** (i) and **meiosis** (ii). Two instances, denoted `fmeisosis` and `mmeiosis`, of the class **meiosis** are used in **child**, along with an instance `gt` of the class **genotype** [Figure 12.6(i)]. Nodes `pg` and `mg` denote, respectively, the child's paternal and maternal alleles. Nodes `fpg`, `fmg`, `mpg` and `mmg` denote, respectively, the father's and mother's paternal and maternal alleles. Node `cg=pg?` is Boolean with possible states 'true' and 'false' and regulates whether a child's allele, node `cg` received from a given parent, will be that parent's paternal or maternal allele.

The sink node `c` in the paternity network shown in Figure 12.5 is an instance of the class **child**, displayed in some further detail in Figure 12.7(i). This class is similar to **founder** in the sense that it contains an instance `gt` of the class **genotype** with entering edges from two output nodes `pg` and `mg`. The latter two nodes represent, respectively, the child's paternally and maternally inherited alleles. Since each parent can only transmit one of the two alleles that make up their genotype, some means is required to model the process by which alleles are passed on. This is operated here through the use of the nodes `fmeiosis` and `mmeiosis`, which are instances of the class **meiosis** [Figure 12.7(ii)] which models the process by which a parent transmits one of its two alleles to offspring. This class has two input nodes `pg` and `mg` which represent, respectively, the paternal and maternal alleles of a parent. In case of the instance node `fmeiosis`, its internal input nodes `pg` and `mg` connect to the nodes representing the father's paternal and maternal alleles (i.e. nodes `fpg` and `fmg`), and similarly, for the maternal inheritance line, modelled by the instance node `mmeiosis`. In the class **meiosis**, the selection of one of the two parental alleles is regulated by a Boolean node `cg=pg?` and by the assignation of equal probabilities to its two states 'true' and 'false'. If node `cg=pg?` is in state 'true', then the output node `cg`, representing the allele inherited by the child, will copy the current state of the node `pg`, otherwise it will copy that of `mg`. It is worth noting that modelling parent to child allele transmission in terms of a distinct class **meiosis** is useful because possible extensions to further aspects, in particular mutation, can be readily implemented within this class. Any network that will contain instances of such a modified class network will thus be updated immediately, without the need of further changes.

Yet another class network used in the paternity network shown in Figure 12.5 is **query**. Its internal structure is shown in Figure 12.8. It has two output nodes `tfpg` and `tfmg` that model the true father's paternal and maternal alleles, respectively. Besides, there are also input nodes `f1pg`, `f2pg`, `f1mg` and `f2mg` that model, respectively, the paternal and maternal alleles of two candidate people, called here generically 'father 1' and 'father 2'. If the Boolean input node `tf=f1?` is in state 'true', then `tfpg` copies the current state of `f1pg`, otherwise it copies that of `f2pg`, with an analogous association for the true father's maternal allele, `tfmg`,

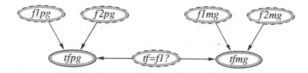

Figure 12.8 Internal structure of the class network **query**. The output nodes tfpg and tfmg represent, respectively, the true father's paternal and maternal alleles. The input nodes f1pg, f2pg, f1mg and f2mg represent, respectively, the paternal and maternal alleles of two distinct individuals, called 'father 1' and 'father 2'. Node tf=f1? is a Boolean variable and takes two values: 'true' if the true father is 'father 1' and 'false' if the true father is 'father 2'.

which is a function of f1mg and f2mg. Note that the input node tf=f1? connects to a node with the same name that models the target propositions regarding paternity in the main network shown in Figure 12.5.

Class networks can also be developed to deal with additional aspects and complicating phenomena of forensic DNA profiling analyses, such as silent or missed alleles. Examples are given in the works of Dawid et al. (2005, 2007).

12.3.3   Object-oriented networks for kinship analyses

The description of class networks in Section 12.3.2 was generic in the sense that no reference was made to a particular marker system. For example, the class network **gene** was said to cover the different allelic forms that a marker (or gene) of interest may take, along with the respective allele probabilities. Example 12.4 seeks to provide a full numerically specified illustration using population data from published literature. Example 12.5 points out the ease and flexibility with which the object-oriented modelling mechanism allows one to tackle variations to standard paternity case settings.

Example 12.4 *(Basic paternity case, results for a single marker) Consider again the classic paternity case scenario recalled at the beginning of Section 12.3.1 where an alleged family trio covers a child, its undisputed mother and the putative father. The main proposition of interest is whether the putative father is the true father or whether an unknown individual, called alternative father, is the true father. Figure 12.5 shows an object-oriented network for this scenario. In order to use this network for the evaluation of profiling results for a particular marker, it is necessary to specify the definition of the class network **gene** according to features of the marker of interest. For the purpose of illustration, suppose that profiling results are available for the marker CSF1PO. Suppose further that the data reported in Butler et al. (2003) for US Caucasians are deemed to be relevant for the case at hand. This dataset lists the set of alleles* {8, 9, 10, 11, 12, 13, 14} *for marker CSF1PO with the following population proportions:* {0.005, 0.012, 0.217, 0.301, 0.361, 0.096, 0.008}. *To ensure that the resulting network will be able to cope with any genotypes observed amongst the members of the alleged family trio, one should define a separate numbered state in the class **gene** for each of the CSF1PO alleles* {8, 9, ... , 14}. *The probability table can then be filled using the relative frequencies mentioned above, accepting the simplification of interpreting these values directly as allele proportions. Specifying the class **gene** in that way will directly affect all other classes that contain an instance of the class **gene**. Suppose next the following profiling results:* {10, 11} *for the* *child, $\{10, 13\}$ for the mother and $\{11, 12\}$ for the putative father. The numerator of the likelihood ratio in this case thus is $Pr(c = \{10, 11\}|m = \{10, 13\}, pf = \{11, 12\}, H_p, I) = 0.25$, and the denominator $Pr(c = \{10, 11\}|m = \{10, 13\}, H_d, I) = 0.5 \times \gamma_{11}$. With $\gamma_{11} = 0.301$, this leads to a likelihood ratio of $0.25/(0.5 \times 0.301) = 1.661130$. Note that this result is given here with more decimals than would be retained in practical applications. The purpose is to compare the result with the output of the object-oriented network, to demonstrate the strict agreement between the two ways of approaching this case. Assuming a network structure as defined by Figure 12.5, with instances of the class* **gene** *as defined here above, profiling results of the child, the mother and the putative father can be specified in the nodes* c, m *and* pf, *respectively. From the node* tf=pf?, *one can then find the following posterior odds ratio $0.62421973/0.37578027 = 1.661130$. This result agrees here with the likelihood ratio because of the special assumption of equal prior odds.*

The result of Example 12.4 is elementary in the sense that it can also be obtained in a rather simple algebraic approach. Notwithstanding, it serves the purpose of illustrating the principles of object-oriented Bayesian network modelling for disputed kinship cases in a very general setting. It can be used as a starting point for introducing means to deal with particular events that may occur during the process of allele transmission between generations (e.g. mutation) or with profiling results of further individuals in possibly complex familial relationships. Extensions to account for effects of allelic dependencies in paternity analyses can also be included in such networks (Hepler and Weir 2008).

Example 12.5 *(Disputed maternity case, putative mother unavailable) Consider again the case of questioned maternity proposed as part of the relationship testing workshop 2010 of the English Speaking Working Group of the International Society for Forensic Genetics (ISFG), described earlier in Section 7.9.3 (pedigree shown in Figure 7.20). The alleged mother is dead and not available for typing. As a slight modification to the original scenario, suppose that the grandparents of the alleged mother are not available for typing, only an undisputed full sister of the alleged mother is available. Besides, typing results are also available for the child. The target question is whether the alleged mother is the true mother. A pedigree for this scenario is shown in Figure 12.9(i). In terms of an object-oriented Bayesian network, this scenario can be approached in a way that closely resembles the paternity network shown in Figure 12.5. The nodes dealing with the true father* tf *and the alternative father* af *can be interpreted here as representations of the genotypes of the true mother* tm *and the alternative mother* am. *In turn, the node for the undisputed mother* m *in the paternity network takes here the role of the undisputed father* f. *The instance node* c *obviously remains the same. The node for the disputed parent, however, requires a minor comment. In the paternity network (Figure 12.5), the role of the putative father was modelled in terms of an instance of the class* **founder** *because all that was needed for further analysis was the parental alleles of this individual (defined as output nodes). The case of the disputed mother here is slightly different. The genotype of this person is not available but needs to be inferred from knowledge of the genotype of a full sister* s. *Both the putative mother* pm *and her sister* s *share the same biological parents, denoted here* gf *(grandfather) and* gm *(grandmother). The two nodes for* gf *and* gm *are instances of the class* **founder**. *The node for the target propositions of interest is a basic node named* tm=pm?. *It is Boolean with states 'true' and 'false'. Figure 12.9(ii) summarizes the overall network structure with instance nodes shown in collapsed form. This network is useable for evaluating profiling results for a single marker, depending on the way in which the class* **gene** *is defined (i.e. the locus and associated allelic proportions).*

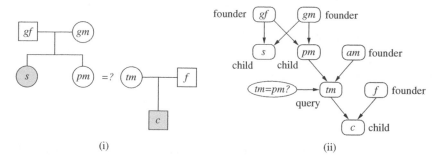

(i)   (ii)

Figure 12.9 (i) Pedigree for a case of questioned maternity with the putative mother *pm* unavailable. Individuals for which typing results are available are shown in grey-shaded nodes. Squares represent males and circles females. (ii) Object-oriented network with instance nodes tm, pm, om, gm, gf, s, f and c, short for, respectively, 'true mother', 'putative mother', 'alternative mother', 'grandmother', 'grandfather', 'sister', 'father' and 'child'. Node tm=pm? is a basic variable and Boolean. It takes two values: 'true' if *tf − pf*, the true mother is the putative mother and 'false' if *tm ≠ pm*, the true mother is not the putative mother (i.e. an unknown female is the true mother).

12.3.4   Object-oriented networks for inference of source

The scope of application of the class networks described in the previous sections is not restricted to kinship analyses. The generic patterns of reasoning that they present are also relevant for inference problems that pertain to the source of DNA detected in various forms of biological stains or trace material. More generally, there is no neat separation between forensic inference of source and analyses of familial relationships. This is illustrated by the fact that inference of source may be coupled with questions about familial relationships. For example, in a case where a putative source of a biological trace is not available for DNA profiling analyses, considerations may be extended to a sibling or a parent of the putative source. The two examples presented below illustrate this point.

Example 12.6 (*Inference of source for a DNA stain from a single donor*) *In one of its most general forms, source inference relates to single donor stains. Again, the generality of such settings may not require a graphical modelling approach, but their restricted scope of considerations is helpful to illustrate the main principles of object-oriented modelling for such problems. Consider thus a scenario involving a crime stain that is assumed to come from a single source. Suppose further that DNA profiling results are available for an individual who is considered as a potential source of that crime stain. The alternative source-level proposition is that an unknown individual, unrelated to the suspect, is the source of the crime stain. Figure 12.10 shows an object-oriented Bayesian network for this scenario. The instance nodes* s *and* o *are of type* **founder** *and represent, respectively, the allelic constitution of the potential source and an unknown individual. The instance node* t *is of type* **query** *and models the genotype of the trace (i.e. the true source of the trace). The labels of the class* **query** *as used here differ slightly with respect to the description given earlier in Section 12.3.2, although the overall logic of their terminology remains the same. Recall that the class* **query** *used for paternity analyses represents the genotype of the true father as that of the putative father or an unknown person, depending on the truth or otherwise of the main proposition of paternity.*

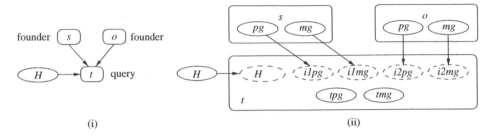

Figure 12.10 Collapsed (i) and expanded (ii) representation of an object-oriented Bayesian network for a basic case of inference of source. Node H is Boolean and in state 'true' if the potential source s is the source of the crime stain t, and in state 'false' if an unknown person o is the source. Nodes {i1,i2,t}pg and {i1,i2,t}mg represent, respectively, the paternal and maternal alleles of the generic sources 'individual 1', 'individual 2' and the true source of the trace.

*This logic is used here again for the genotype of the crime stain. The instance node t in Figure 12.10 ensures that the parental allele nodes tpg and tmg, representing the true source of the DNA trace, copy the states of the parental allele nodes i1pg and i1mg of the potential source, called here generically individual 1, if the Boolean node H is in state 'true' (i.e. the suspect is the source of the crime stain). In turn, if the node H is in state 'false' (i.e. the stain comes from an unknown person), then the parental allele nodes tpg and tmg for the crime stain copy the states of the parental allele nodes i2pg and i2mg of the unknown person, referred to here as individual 2. As with all the networks presented so far in this chapter, the model shown in Figure 12.10 copes with profiling results for a single genetic marker, depending on the way in which the allele nodes are specified (e.g. pg and mg). Their numerical states define the range of alleles taken into account. The allelic proportions of the genetic marker of interest are specified in the node table of the class **gene**. As noted earlier in Section 12.3.2, instances of this class are used in the class **founder**, but they are not displayed in the instances s and o of class **founder** shown in Figure 12.10, because they are part of the hidden structure.*

Example 12.7 *(Inference of source for a DNA stain from a single donor, suspect not available)*
*Consider again a case involving a crime stain originating from a single source, as introduced above in Example 12.6, but with the difference that no DNA profiling results are available for the suspect. This may occur, for example, if the suspect is temporarily not available to give control material for analysis. In such a case, control material from close relatives of the suspect may be analysed, and based on their genotypes, inferences are made about the suspect's genotype. Typically, parents, siblings or children could be individuals to be investigated but more distant relationships may also be considered. For illustrative purposes, the object-oriented network shown in Figure 12.11 covers nodes for the parents, a brother and a child of the suspect. In order to extend considerations to these individuals, the instance node s (of type **founder**) in Figure 12.10 needs to be replaced by an instance node of the class **child**. As the partially expanded network in Figure 12.11(ii) shows, this change is needed in order to construct a connection to the instances nodes gf and gm, representing the suspect's parents. Let us recall that a suspect's parental allele, for example, his/her paternal allele (represented here by node pg in instance node s), may be one of the two parental alleles pg or mg of his/her father (instance node gf). This process of selection is operated by an instance of the*

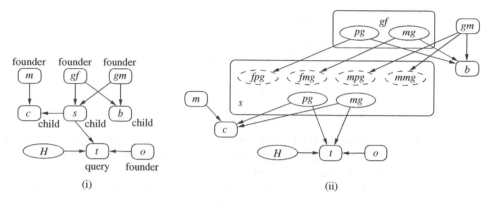

Figure 12.11 Collapsed (i) and expanded (ii) representation of an object-oriented Bayesian network for inference of source in a case where the genotype of the suspect may potentially be missing. Node H is Boolean and in state 'true' if the potential source s is the source of the crime stain t and in state 'false' if an unknown person o is the source. Instance nodes gf, gm, b and c represent, respectively, the suspect's father, mother, brother and child. Nodes {f, m}pg and {f, m}mg represent, respectively, the paternal and maternal alleles of the suspect's father and mother.

class **meiosis**, *which is part of the hidden structure of the class* **child** *(see also Figure 12.7). Using the logic of their terminology, two further instances of the class* **child**, *denoted* b *and* c, *are added to represent, respectively, a full brother and a child of the suspect. The use of this object-oriented network is obviously not limited to cases in which the suspect is not available. If the suspect's genotype is available, it can be specified in the instance node* s *but with the consequence that knowledge of the genotype of other close relatives would become irrelevant.*

12.3.5 Refining internal class descriptions and further considerations

The general idea of refining internal class descriptions introduced earlier in Section 12.2.3 can also be illustrated usefully in the context of inference modelling for DNA profiling results. That is, based on a given class description, even on a coarse level of definition, details of the internal structure can be revised and augmented as required, depending on the question that is to be addressed. Example 12.8 discusses this aspect on the basis of the class **founder** defined earlier in Section 12.3.2.

Example 12.8 *(Accounting for population structure) So far in this chapter, the paternally and maternally inherited alleles of an individual are assigned probabilities according to their rarity in a particular population. There may be uncertainty about the relevant population, but such uncertainty can readily be accounted for in Bayesian networks. To illustrate this point, consider the class* **founder** *(Section 12.3.2). This class represents an individual's paternally and maternally inherited alleles, at a given locus, in terms of the two nodes* pg *and* mg, *which are copies of the internal nodes* gene *of, respectively, the instance nodes* pgin *and* mgin. *These two nodes are themselves instances of the class* **gene** *[Figure 12.6(iii)], given by a single output node that is defined with respect to a particular population. The probabilities that are assigned to the various states (representing allele numbers) of the node* gene *relate to a*

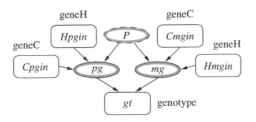

Figure 12.12 Class **founderHET** contains instance gt of the class **genotype** [Figure 12.6(ii)], instances Cpgin and Cmgin of the class **geneC** and instances Hpgin and Hmgin of the class **geneH**. The classes **geneC** and **geneH** represent the scope of alleles at a given genetic marker for, respectively, Caucasians and Hispanics. The nodes pg and mg represent, respectively, paternally and maternally inherited alleles. The node P, with states C and H, regulates the way in which the nodes pg and mg are supplied with allele proportions for Caucasians and Hispanics.

particular population (e.g. Caucasians, as in Example 12.4). In situations in which there is a need to account for more than one population, because there is uncertainty about the relevant population, this structure can be augmented by defining a distinct class of the kind **gene**, *for another population, and then using instances of that class to input the nodes* pg *and* mg. *Figure 12.12 shows such a modification of the class* **founder**, *renamed* **founderHet**. *Its interface nodes* pg *and* mg *are connected to instances of the classes* **geneC** *and* **geneH**. *These two classes are given by single output nodes whose states represent the scope of alleles at a given genetic marker but with different allele proportions. Here, the class* **geneC** *is assigned allele proportions for Caucasians, whereas the class* **geneH** *is assigned allele proportions for Hispanics. The node P, with states C and H, regulates the way in which the nodes* pg *and* mg *are supplied with allele proportions. If P is in state C, then the nodes* pg *and* mg *will copy the allele proportions of the instances of the class* **geneC**, *otherwise they will copy the allele proportions of the instances of the class* **geneH**. *The probabilities assigned to the states C and H of the node P can be viewed as a mixing weight for the synthetic population composed of Caucasians and Hispanics. Note, however, that if P is not defined as an interface node, then this mixing weight will be constant and fixed across markers. In turn, if P is defined as an input node, it can be related to a single node P in a higher hierarchical level. This will induce dependency between loci (Green and Mortera 2009).*

To illustrate the effect of uncertainty about the relevant population, consider the genotype {8, 9} *at the locus CSF1PO. Using data from Butler et al. (2003), the genotype probability for Caucasians can be written as* $2\gamma_8\gamma_9 = 2 \times 0.005 \times 0.012 = 0.00012$. *For Hispanics,* $2\gamma_8\gamma_9 = 2 \times 0.060 \times 0.037 = 0.00444$. *Specify probabilities of 0.8 for Caucasians and 0.2 for Hispanics in the table for node* P, *then the genotype probabilities are weighted according to these values and*

$$Pr(\{8,9\}) = Pr(\{8,9\}|P = Caucasian)Pr(P = Caucasian)$$
$$+ Pr(\{8,9\}|P = Hispanic)Pr(P = Hispanic)$$
$$= 2 \times 0.005 \times 0.012 \times 0.8 + 2 \times 0.060 \times 0.037 \times 0.2$$
$$= 0.000984.$$

Besides population heterogeneity, inference modelling for cases involving DNA profiling results may also need to account for relatedness between individuals involved in a particular scenario. The use of object-oriented networks for such aspects, as well as their joint consideration, is discussed in Green and Mortera (2009). For an application of object-oriented networks to DNA profiling results for markers other than classical short tandem repeats, in the context of unbalanced mixtures, see Cereda et al. (2014).

13

Qualitative, sensitivity and conflict analyses

Bayesian networks are presented and proposed in this book as a general framework for the representation and sensible evaluation of uncertainty in knowledge. It has been pointed out that the graphical structure is amongst a network's most robust constituents. It conveys, on a rather general level of detail, a transparent and clear idea of the factors considered as well as their assumed relationships. When information is available on the strength of the relationships amongst the specified variables, models may be considered at a numerical level and used for inference, that is, a quantitative revision of personal beliefs based on new information.

However, it is sometimes argued or felt that the quantitative specification is too problematic a task for the use of Bayesian networks in forensic science applications. Even though data may be available from various sources such as literature, surveys, databases, or elicited from experts, forensic scientists may be reluctant to provide particular numerical values for all components of interest in the model. One of the reasons for this is that forensically relevant items and traces usually come into existence under particular configurations of real-world circumstances. This may make it difficult to perceive given cases as instances of a series of other cases that occurred under comparable circumstances. This is also one of the reasons why notions such as 'frequency' or 'relative frequency' may be of little help in the context of discussing probability assignment for node tables. Moreover, the unavailability of case-tailored experimental data may make it hard for scientists to justify quantitative assessments of factors relevant to a given scenario. This is reinforced by the fact that Bayesian networks are sometimes thought of as an essentially numerical method that requires 'exact' numbers with 'high accuracy' for their implementation. However, this view is unnecessarily restrictive. Although it is certainly the case that there may be network components that have a substantial bearing on a particular line of reasoning, it is also important to emphasize that not all parts of a network may have the same impact, and indeed some parts may even have no substantial impact. It is for this reason that it is crucial to draw attention to additional concepts, such as sensitivity

Bayesian Networks for Probabilistic Inference and Decision Analysis in Forensic Science, Second Edition.
Franco Taroni, Alex Biedermann, Silvia Bozza, Paolo Garbolino and Colin Aitken.
© 2014 John Wiley & Sons, Ltd. Published 2014 by John Wiley & Sons, Ltd.

analyses, that can help scientists inspect models and recognize those components that require the most careful specification.

The present chapter describes formalisms that can help scientists deal with scenarios in which there is a lack of numerical data. Section 13.1 presents qualitative probabilistic networks (QPNs) and Section 13.2 deals with sensitivity analyses. A further concept, conflict analysis, is presented in Section 13.3. Informally, QPN are abstractions of Bayesian networks in which numerical relationships are replaced by qualitative probabilistic relationships. In such models, inferences can still be made according to the laws of probability theory. The result is an indication of the change, if any, in the direction of belief. Sensitivity analysis requires the use of techniques of 'data investigation' to evaluate an inferential model or a sequence of procedures for decisions. Sensitivity analyses consist of varying the probabilities assigned to one or more variables of interest and comparing the results, for example, but analyses may also focus on the effect of different observations evaluated using a fixed model. Conflict analyses relate to the investigation of multiple observations in the context of particular modes of inference. In combination, these auxiliary aspects provide assistance in the examination of a model's property. This is consistent with the general idea pursued in this book that the use of Bayesian networks is not restricted to a quantitative use with the focus solely on probability updating. It is equally important to gain further insight into particular problems of inference and recognize those aspects that exhibit a major influence. This, in turn, can provide guidance in the allocation of resources for gathering data in a more focused way.

13.1   Qualitative probability models

Wellman (1990a) formally defined a QPN as a pair $G = (V, Q)$, where V is a set of variables, also called *vertices*, of the graph and Q is a set of qualitative relationships. In analogy to the definition of Bayesian networks (Section 2.1.2), a QPN G is required to be acyclic. The vertices of G are discrete and either binary or multiple valued. The qualitative relationships Q cover qualitative influences and synergies. A qualitative influence defines the sign of the direct influence one variable has on another, whereas qualitative synergies are used to describe interactions amongst influences. These concepts will be developed in further detail in Sections 13.1.1–13.1.5 with their relevance for reasoning in forensic contexts illustrated through various examples.

13.1.1   Qualitative influence

13.1.1.1   The binary case

The notion of qualitative influence between two variables of a graph is an expression of how the state of one variable influences the state of the other variable. Imagine two binary variables Y and Z connected so that $Y \to Z$. Then, given a possible value x of X (i.e. an assignment to other ancestors of Z), Y is said to have a positive influence on Z if and only if

$$Pr(Z|Y, X = x) \geq Pr(Z|\bar{Y}, X = x). \tag{13.1}$$

Negative and zero qualitative influences can be defined analogously by substituting $\geq$ in (13.1) by $\leq$ and $=$, respectively. A fourth sign of qualitative influence is '?' and denotes ambiguity. It applies when the effect of the influence is not known. Verbally expressed, a positive qualitative influence between two variables Y and Z is, for example, a statement of the form 'the presence of Y makes Z more likely' (Wellman 1990a). Notice that qualitative influences adhere to the property of symmetry so that one is also allowed to state that 'Z makes Y more likely'. Another way to express this would be to say 'the occurrence of event Z favours hypothesis Y over $\bar{Y}$' (Schum 1994). See also Garbolino (2001) for a discussion on the use of such expressions in judicial contexts.

In forensic science, qualitative reasoning is quite common. For example, forensic scientists often follow a procedure that goes from the general to the particular. In the context of evaluating forensic findings, this means that scientists would start by asking questions of the kind 'is this finding what I would expect to observe if the prosecution's case were true?' and, 'to what degree would I expect to see this finding occurring if the defence case were true?' Initially, the scientist would frame these questions on a general level and think about the possible outcomes in essentially qualitative terms. In a second stage, more detailed data may then be sought and used to sustain a particular conclusion.

Example 13.1 *(Qualitative reasoning in range of firing evaluations) Consider again the evaluation of a distance of firing, that is, an inferential problem discussed at some length in Section 8.4.4. For a given distance, D, of firing, the expected number of gunshot residues (GSR) Y has been approximated by a Normal distribution. Note that this approach has solely been concerned with the amount of visualized GSR. The major aim was to demonstrate principally that GSR evidence is amenable to quantification and probabilistic evaluation. In practice, however, scientists may also use information other than GSR. For example, experts may assess the presence or absence of particular wound characteristics, damage to textile clothing in the area of the entrance hole (e.g. melted fibres) and so on. When the scientist considers these items of information as a whole, then the general pattern of reasoning involved can be described on a purely qualitative level. For example, the scientist would need to evaluate whether the observations are more, less or equally likely, given one's propositions of interest, rather than the stated alternative. Such judgements may be considered as the most general level at which scientists may be required to characterize their findings. If the scientist is unable to tell whether he/she believes the findings to be more, less or equally likely under each of the specified propositions, then his/her finding is not suitable for discrimination amongst those propositions. In the latter situation, the uninformative '?' sign is applicable, the notation chosen to illustrate its intuitive meaning.*

Notice that the concept of qualitative influence allows one to describe the relation between two variables on the basis of two conditional probabilities. The revision of belief can be made according to the laws of probability even though no particular numerical probabilities are provided.

13.1.1.2   The non-binary case

Take again two variables Y and Z and let them both have more than two states. These states are indexed with i, so that a particular state of Y would be y_i, for example. Also, an ordering from the highest to lowest value is assumed so that for all $i < j$, one has, for example, $y_i \geq y_j$. Then, the variable Y may be said to have a positive qualitative influence on a variable Z, if and only if for all values y_j and y_k of Y, with $y_j > y_k$, all values z_i of Z, and a given state x of X (i.e. other ancestors of Z),

$$Pr(Z \geq z_i | Y = y_j, X = x) \geq Pr(Z \geq z_i | Y = y_k, X = x) . \tag{13.2}$$

The relationship $\geq$ in the centre of (13.2) can be substituted with $\leq$ and $=$ to define negative and zero qualitative influences, respectively. A verbal expression of a qualitative influence between two non-binary variables may be, for example, 'higher values of Y make higher values of Z more likely', assuming the influence under investigation to be a positive one.

The direction of influence between two variables can be evaluated by comparing cumulative probability distribution functions (CDFs) (Wellman 1990a). The CDF of a variable Z with states $z_1 < \cdots < z_n$, and $n \geq 1$, is given by the function $F_Z(z_i) = Pr(z_1 \lor z_2 \lor \cdots \lor z_i)$, noted $Pr(Z \leq z_i)$ for short. Cumulative conditional probability distribution functions are defined analogously and are denoted as $F_{Z|Y=y_j}(z) = Pr(Z \leq z | Y = y_j)$, or, more shortly, as $F_{z|y_j}$. For example, $F_{z_i|y_j}$ denotes the CDF computed in correspondence of $Z = z_i$, given the observed value y_j.

Consider the comparison of CDFs in the context of a positive qualitative influence between the non-binary variables Y and Z, connected so that $Y \to Z$. In order for the statement 'higher values of Y make higher values of Z more likely' to be valid, the following must hold for all values y_i, y_j with $y_i > y_j$:

$$F_{z|y_i}(z_i) \leq F_{z|y_j}(z_i) \text{ for all values } z_i \text{ of } Z.$$

Example 13.2 (*Relationship between extent of contact and amount of transferred material*) *Consider the relation between the variables C, short for 'degree (or, extent) of contact', and Q, short for 'amount of transferred fibres recovered from a garment' as described by Cook et al. (1999). The variable C has the three states 'heavy', 'medium' and 'light', abbreviated as c_1, c_2 and c_3, respectively. Three states are also proposed for the variable Q, Namely, 'none', 'few' and 'many', abbreviated as q_1, q_2 and q_3, respectively. From the values specified in Table 13.1, the corresponding cumulative conditional probability distributions thus are*

$$Pr(Q \leq q_1|c_1) = 0.02, Pr(Q \leq q_1|c_2) = 0.10, Pr(Q \leq q_1|c_3) = 0.20$$

$$Pr(Q \leq q_2|c_1) = 0.30, Pr(Q \leq q_2|c_2) = 0.50, Pr(Q \leq q_2|c_3) = 0.80$$

$$Pr(Q \leq q_3|c_1) = 1.00, Pr(Q \leq q_3|c_2) = 1.00, Pr(Q \leq q_3|c_3) = 1.00.$$

From this listing of values, it may be seen that for each c_i of C, with $c_1 > c_2 > c_3$, the cumulative conditional probability distributions of Q given c_i obey to

$$F_{q|c_1}(q_i) \leq F_{q|c_2}(q_i) \leq F_{q|c_3}(q_i) \text{ for all } q_i \text{ of } Q.$$

It may be concluded from this analysis that higher values of Q are more probable for higher values of C. In the context of fibre transfer, this means that the higher the degree of contact,

Table 13.1 Conditional probabilities for the quantity Q of transferred fibres recovered from a garment given different degrees of contact C.

Intensity of contact (C)		Heavy (c_1)	Medium (c_2)	Light (c_3)
Quantity (Q):	None (q_1)	0.02	0.1	0.2
	Few (q_2)	0.28	0.4	0.6
	Many (q_3)	0.70	0.5	0.2

Values are adopted from Cook et al. (1999).

the greater the amount of transferred fibres recovered from a garment. An important aspect of this result is that the assumed relationship between the extent of contact and the quantity of recovered fibres is handled in a probabilistically rigorous way whilst particular numerical values need not necessarily be available.

Example 13.3 *(Qualitative reasoning in comparative handwriting examinations) In Section 8.4.1, a handwriting scenario with two variables S and E has been described. S is binary, defined as 'Mr. Adams wrote the signature', and E is non-binary, referring to the results of comparative handwriting analyses. The possible states for E are few (e_1), some (e_2) and many (e_3) similarities. Recall, also the conditional probabilities specified for each outcome of E given S_p and S_d, Mr. Adams being or not being the source of the signature, respectively:*

$$Pr(E = e_1|S_p) = 0.60, \ Pr(E = e_1|S_d) = 0.78$$

$$Pr(E = e_2|S_p) = 0.27, \ Pr(E = e_2|S_d) = 0.20$$

$$Pr(E = e_3|S_p) = 0.13, \ Pr(E = e_3|S_d) = 0.02.$$

The corresponding cumulative conditional probability distributions thus are as follows:

$$Pr(E \leq e_1|S_p) = 0.60, \ Pr(E \leq e_1|S_d) = 0.78$$

$$Pr(E \leq e_2|S_p) = 0.87, \ Pr(E \leq e_2|S_d) = 0.98$$

$$Pr(E \leq e_3|S_p) = 1.00, \ Pr(E \leq e_3|S_d) = 1.00.$$

For each e_i of E, the cumulative conditional probability distribution is smaller given S_p than given S_d. The underlying qualitative probabilistic statement thus is that higher values for e are more probable under S_p than under S_d. Another way to express this would be to say that more similarities are more likely to be found when Mr. Adams wrote the signature than if someone else wrote the signature.

13.1.2   Additive synergy

'Synergy' is a term used when studying interactions amongst influences. This is typically the case when two nodes share a common child. Additive synergy is a concept applicable for evaluating the direction of interaction between two variables in their influence on a third (Wellman 1990b). Note that this property of additive synergy holds regardless of any other direct influence on that third variable.

Additive synergy is described here for the binary case. Let Y, W and Z be binary variables, connected in a converging connection: $Y \to Z \leftarrow W$. Y and W are said to exhibit a positive additive synergy on their common descendant Z if

$$Pr(Z|Y, W, X = x) + Pr(Z|\bar{Y}, \bar{W}, X = x)$$
$$\geq Pr(Z|Y, \bar{W}, X = x) + Pr(Z|\bar{Y}, W, X = x), \qquad (13.3)$$

where x represents an assignment to other ancestors X of Y. Again, negative and zero additive synergies are defined by substituting $\geq$ by $\leq$ and $=$, respectively.

Another way to express an additive synergy, for example, a positive one, is to say 'the effect of Y on Z is greater when W is true' (Wellman 1990b). This can be illustrated by rearranging the terms in (13.3) to obtain the following:

$$Pr(Z|Y, W, X = x) - Pr(Z|\bar{Y}, W, X = x)$$

$$\geq Pr(Z|Y, \bar{W}) - Pr(Z|\bar{Y}, \bar{W}, X = x).$$

The left-hand side describes the changes in Z due to changes in Y whilst W is true. The right-hand side describes the changes in Z due to changes in Y whilst $\bar{W}$ is true. The relative magnitude of these changes is compared and expressed in terms of the signs '+', '−', '0' and '?'. In forensic contexts, patterns of reasoning exist that comply with the definition of additive synergy. Example 13.4 proposes an illustration in the context of particle detection.

Example 13.4 (*Qualitative probabilistic reasoning for results of a particle detection apparatus) Imagine the use of some sort of detection apparatus, such as an ion mobility Spectrometer (IMS). Such devices have been developed and marketed for the detection of explosives and illicit drugs, for example. Surfaces of interest such as tables, luggage, clothing and so on may be vacuumed, and particles thus collected are analysed. When a detectable quantity of a target substance is present, then the IMS would indicate that event by a sonar signal, which is given with a certain reliability. Consider the three variables D, 'presence of a detectable quantity of drugs'; P, 'performance of the device', and S, 'sonar signal given by the device'. It is thought that the probability of obtaining a sonar signal is dependent on whether a detectable quantity of drugs is present and on the apparatus' performance. An appropriate representation thus is P → S ← D. The variables are binary and their states are defined as follows: 'present' and 'absent' (variable D), 'good' and 'poor' (variable P), 'yes' and 'no' (variable S). The descriptor 'performance' is used here, roughly speaking, as an expression of one's belief in the apparatus' capacity to do what it is supposed to do. So, if the apparatus' performance is good and a detectable quantity of drugs is present, then a sonar signal may be expected with a reasonably high probability. In the absence of a detectable quantity of drugs, but still assuming a good performance, one should be able to expect a signal to occur with a low probability. This argument allows one to express general qualitative beliefs satisfying $Pr(S_1|P_1, D_1) > Pr(S_1|P_1, D_2)$. For the purpose of illustration, let $Pr(S_1|P_1, D_1)$ be 0.99 and $Pr(S_1|P_1, D_2)$ be 0.01. In turn, when the performance of the detection apparatus is poor, then the probability of obtaining a sonar signal may be less influenced by the actual presence or absence of drugs. Qualitatively speaking, $Pr(S_1|P_2, D_1) \approx Pr(S_1|P_2, D_2)$ or even $Pr(S_1|P_2, D_1) = Pr(S_1|P_2, D_2)$ could be justified. For the purpose of illustration, let these last two probabilities be 0.5.*

All the necessary conditional probabilities of the variable S are now defined. They can be used to examine the interaction amongst the influences that both P and D exert on S. The relative magnitude of the target probabilities is such that the following holds:

$$Pr(S_1|D_1, P_1) - Pr(S_1|D_2, P_1) > Pr(S_1|D_1, P_2) - Pr(S_1|D_2, P_2).$$

The two variables P and D thus exhibit a positive additive synergy on S. Verbally expressed, the presence of drugs is thought to have a stronger effect on the probability of obtaining a sonar signal when the apparatus is a viable device rather than when its performance is poor.

13.1.3 Product synergy

One can distinguish between two types of synergies, one of which, additive synergy, has been introduced in the previous Section. A second type of interaction is product synergy. It regroups the sign of conditional dependence between a pair of immediate predecessors of a node that has been observed or has indirect evidential support (Druzdzel and Henrion 1993a). As for additive synergy, the notion of product synergy applies to scenarios involving converging connections. Product synergy has been proposed by Henrion and Druzdzel (1991) and has further been studied by Wellman and Henrion (1993).

Consider a converging connection where X and Y have direct influences on a third variable Z, that is, $X \rightarrow Z \leftarrow Y$. The variables X and Y are said to exhibit a negative product synergy with respect to the outcome Z of Z, if and only if

$$Pr(Z|X, Y) \times Pr(Z|\bar{X}, \bar{Y}) \leq Pr(Z|X, \bar{Y}) \times Pr(Z|\bar{X}, Y). \tag{13.4}$$

Again, positive and zero product synergies are defined by replacing $\leq$ by $\geq$ and $=$, respectively. Notice that there are as many product synergies as there are possible outcomes for Z. So, for the binary case, there would be a product synergy for both Z and $\bar{Z}$. Note also that the above definition of product synergy assumes that Z either has no parents other than X and Y or if there are parents, they are all instantiated. See Druzdzel and Henrion (1993c) for an extension to situations involving additional, unobserved predecessors of Z.

The negative product synergy between X and Y as expressed by (13.4) means that, given Z, confirmation of X tends to decrease the probability of Y. It is also useful to note that given Z, X and Y are no longer d-separated. Product synergy can thus be taken as a concept describing the interactions between two parental variables who become dependent upon receiving information on the truthstate of their common child.

Hereafter, product synergy will be used to study two particular aspects of reasoning amongst variables sharing a common descendant (Wellman and Henrion 1993). One of these aspects relates to a very commonly encountered pattern of reasoning that became known as *explaining away*. It applies when the confirmation of a believed event reduces the need to invoke alternative events. A second aspect relates to the opposite of 'explaining away', that is, a mode of reasoning applied when the confirmation of one event increases the belief in another.

13.1.3.1 Explaining away

Discussions on explaining away may be found, for example, in Kim and Pearl (1983), Henrion (1986) or Pearl (1988). More general conditions for this particular form of reasoning have been formulated by Henrion and Druzdzel (1991) and Wellman and Henrion (1993). The latter authors have shown that negative product synergy is a general probabilistic criterion that precisely justifies explaining away.

Example 13.5 (*Residual quantities of flammable liquids*) *Imagine a case in which an individual suspected of arson is arrested a few hours after a house had been set on fire, a fire which was aided by spilt inflammable liquid. Forensic scientists examine the suspect's hands and clothing for the presence of residual quantities of a flammable liquid (e.g. its vapours). Assume that analyses indicate the presence of considerable quantities of such a substance. Represent this event by P. When evaluating this finding, the scientist may consider various, not necessarily mutually exclusive, events that may lead to the detection of a flammable substance* *on the hands and the clothing of a person of interest. For the purpose of illustration, consider two potential sources defined as follows. A first variable is B, termed* background contamination, *and refers to the accidental presence of a flammable liquid, acquired, for example, whilst visiting or working in an environment where such a substance is prevalent. A second variable is T and refers to a recent handling of a flammable substance, eventually resulting in a primary transfer. It shall be ignored here, however, whether that transfer (or spill) is crime related or not. Both B and T may be realized simultaneously, so they are modelled as distinct parental nodes of P. For the purpose of this example, let the conditional probabilities for P be close to 1 in cases where T is true, such as, $Pr(P|T, B) = 0.99$ and $Pr(P|T, \bar{B}) = 0.97$. When T is false, but B is true, the conditional probability for P is reduced: $Pr(P|\bar{T}, B) = 0.7$. When neither B nor T is true, then it appears reasonable to hold a low probability for P, such as $Pr(P|\bar{B}, \bar{T}) = 0.01$.*

The above values reflect a scientist's point of view according to which a positive analytical result is (i) highly probable whenever a transfer occurred (T) and a background contamination (B) is present; (ii) a fairly probable event when either T only is true and B is not, and vice-versa, and (iii) an event with very low probability if neither a transfer (T) occurred nor a background contamination (B) is present.

Note that the absolute values specified above are not of primary interest here. It is sufficient here to consider them as rough indications of the beliefs held by a reasoner. A closer examination of these beliefs shows that they satisfy the following:

$$\frac{Pr(P|T, B)}{Pr(P|T, \bar{B})} \leq \frac{Pr(P|\bar{T}, B)}{Pr(P|\bar{T}, \bar{B})} . \tag{13.5}$$

Verbally, this relation can be taken to mean that the proportional increase of the probability of P, the occurrence of a positive analytical result, due to B, the accidental background presence of a flammable substance, is smaller given T, the presence of a transferred flammable liquid, than given $\bar{T}$, the absence of such substance. Alternatively, one could also say that the incremental effect of one of the two possible sources, for example, B, is less than it would be if the other, T, were absent.

For the above statements to be valid, one's beliefs must comply with the formal requirement of negative product synergy. *This can be explained by the fact that (13.5) is obtained from (13.4) by rearranging the terms. A practical consequence of this is that, given a positive analytical result, the confirmation of one potential 'cause' would decrease one's belief in the alternative event that may lead to the observed result. It is such modes of reasoning that are referred to as* explaining away. *Further applications of Bayesian networks for evaluating the analytical results concerning residual quantities of flammable liquids are described in, for example, Biedermann et al. (2005a, b).*

13.1.3.2   The opposite of explaining away

When there are two events that may both lead to the same consequence, then these events may not only compete with one another in providing an account for the observed effect but might also complement each other (Wellman and Henrion 1993). The latter situation is one in which 'explaining away' does not appear to be appropriate, that is, when the confirmation of one event, given its consequence, tends to *increase* the probability of another event that may lead to the same effect (rather than decrease as it would be the case for explaining away).

Example 13.6 (*Qualitative probabilistic reasoning for particle detection, continued*) *Consider again a scenario discussed earlier in Section 13.1.2. An analytical device is used in order to detect the presence of traces of an illicit substance (e.g. a drug). The general pattern of reasoning involved when obtaining a positive sonar signal could be as follows: given the apparatus' sonar signal* (S_1), *one's suspicion of traces of drugs being present in a detectable quantity* (D_1) *would be higher the more one believes the apparatus' performance to be good* (P_1), *that is, the apparatus is sensitive to drugs. This sort of reasoning applies when the beliefs about the various outcomes of S, given P and D, satisfy*

$$\frac{Pr(S_1|P_1,D_1)}{Pr(S_1|P_1,D_2)} \geq \frac{Pr(S_1|P_2,D_1)}{Pr(S_1|P_2,D_2)}. \tag{13.6}$$

The opposite of explaining away can thus be considered in situations where a positive product synergy applies. The condition defined by (13.6) states that the proportional increase in the probability of S_1 *(the apparatus giving a sonar signal) when changing from* D_2 *(no detectable quantity of drugs) to* D_1 *(the presence of a detectable quantity of drugs) is greater, given* P_1 *(the apparatus' performance being good), than* P_2 *(the apparatus' performance being poor).*

13.1.4   Properties of qualitative relationships

So far, qualitative influences have been considered as direct relationships between two adjacent variables of a network, that is, a concept associated with each of a network's arcs. However, influences amongst nodes may not only be evaluated along arcs. It is also possible to consider indirect influences between separated variables. To this end, Wellman (1990a) has proposed an approach based on network transformations where all trails – alternating sequences of directed links and nodes such that each link starts with the node ending the preceding link – between two nodes of interest are collapsed into a single arc. The sign of this resulting arc is determined from the signs associated with the collapsed arcs. Two operators are available for computing the sign of a resulting arc, the sign product operator and the sign addition operator, as given in Table 13.2.

The sign product operator $\otimes$ is used for computing the sign of qualitative influence between two separated variables based on the signs of the qualitative influences that are associated with each arc of the trail connecting two variables (transitivity). Suppose a QPN involving the three variables X, Y and Z with $X \to Y$ and $Y \to Z$ forming a trail from X to Z [Figure 13.1(i)]. Let δ_1 and δ_2 denote the signs of qualitative influences associated with $X \to Y$ and $Y \to Z$, respectively. The sign of qualitative influence of X on Z along the trail $X \to Y \to Z$ can be evaluated by collapsing the two arcs $X \to Y$ and $Y \to Z$ into a

Table 13.2   Sign product ($\otimes$) and sign addition ($\oplus$) operators as defined by Wellman (1990a).

$\otimes$	+	−	0	?	$\oplus$	+	−	0	?
+	+	−	0	?	+	+	?	+	?
−	−	+	0	?	−	?	−	−	?
0	0	0	0	0	0	+	−	0	?
?	?	?	0	?	?	?	?	?	?

single arc $X \to Z$. Denote the sign of that single arc by δ_3. It is obtained by applying the sign product operator to δ_1 and δ_2. For example, if $\delta_1 = +$ and $\delta_2 = -$, then δ_3 is given by $\delta_1 \otimes \delta_2 = + \otimes - = -$. It is assumed here that no direct influence exists between X and Z. However, as noted by Wellman (1990b), even if there were such a direct link, it still makes sense to say that X influences Z with the sign δ_i along the particular trail considered, that is, $X \to Y \to Z$. Thus, it is possible to maintain the original graph and use the sign product operator for defining the sign of qualitative influence along a trail in a graph.

The sign addition operator $\oplus$ is used to combine parallel influences (composition). Consider again a QPN with a trail from X to Z through Y. In addition, assume a direct link from X to Z as shown in Figure 13.1(ii). A sign δ_i of qualitative influence is associated with each link. The sign of qualitative influence that X exerts on Z can be evaluated in two steps. First, the sign of qualitative influence along the trail $X \to Y \to Z$ is evaluated. For this purpose, the sign product operator is applicable as mentioned above: $\delta_1 \otimes \delta_2$. Second, the sign associated with this trail is combined with the sign associated with the direct link $X \to Z$. This step involves a combination of parallel influences that requires the application of the sign addition operator. For the network shown in Figure 13.1(ii), the sign addition operator is applied as follows: $(\delta_1 \otimes \delta_2) \oplus \delta_3$.

Generally, note that the operators $\otimes$ and $\oplus$ commute, associate and distribute as ordinary multiplication and addition do. Besides transitivity and composition, qualitative relationships also adhere to the property of *symmetry*. In the case of a qualitative influence (Section 13.1.1), the property of symmetry states that if a node Y exerts a qualitative influence on another node Z, then, inversely, Z exerts a qualitative influence of the same sign on Y. Notice, however, that the term symmetry is restricted to the sign of influence. The magnitude of the influence of a variable Y on a variable Z can be arbitrarily different from the magnitude of the influence of Z on Y (Druzdzel and Henrion 1993b). The property of symmetry also applies to additive and product synergy, respectively. In this context, symmetry refers to the property that the two nodes exhibiting a synergy are interchangeable.

Example 13.7 *(Trace quantities of illegal drugs) Consider a scenario introduced earlier in Section 5.1.5 where a person of interest is suspected of trafficking of illegal drugs and a Bayesian network with a structure as shown in Figure 5.2. Let H be the major proposition with H_p defined as 'the suspect is involved in illegal drug trafficking, reselling and activities related thereto' and H_d defined as 'the suspect has nothing to do with such activities'. Material collected from the surface of the suspect's clothing is analysed by an IMS. A positive result is obtained for cocaine. This result is then confirmed by a second analytical procedure, such as gas chromatography/mass spectrometry (GC/MS). Let these findings be denoted by a variable E. This is a binary variable that relates to the proposition that the instrumental analysis yields a positive result for cocaine. The two states are E and $\bar{E}$, denoting the truth and falsity of the*

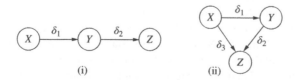

Figure 13.1 Qualitative probabilistic networks over three variables, X, Y and Z. The δ_i designate the signs of qualitative influence ($\delta_i \in \{+, -, 0, ?\}$).

proposition, respectively. An argument is constructed to connect to the variable H by defining two intermediate propositions, C and T. The outcome C (of variable C) states that the suspect has been in contact with a primary source of illegal drugs (i.e. cocaine), and $\bar{C}$ states that the suspect has never been in contact with a primary source of illegal drugs. Variable T refers to the event of transfer (primary, secondary, tertiary or a combination of these) of a certain quantity of illegal drugs, and $\bar{T}$ to the event of no transfer. Recall that an illegal drug, in pow-der form or as a compressed bloc, has been referred to as a primary source *earlier in Section 5.1.5. It provides an instance for a possible (primary) transfer to an individual, which, in turn, may initiate secondary and tertiary transfers.*

Figure 13.2 reflects the assumed relationships amongst the variables. The symbols δ_i, for $i = 1, \ldots ,4$, indicate the sign of qualitative influence. Depending on the scenario, the signs of qualitative influence may change (e.g. due to varying case-related circumstantial information or the strategy chosen by the defence). Each time an assessment changes, the aim is to re-evaluate the direction of influence that knowledge of E exerts on H.

Imagine a scenario in which a fairly high-intensity signal is recorded by the IMS during analysis of trace material collected on the suspect. Usually, confirmative analyses by GC/MS would not involve a quantification. Nevertheless, the intensity of the signal obtained through IMS can provide an idea as to the extent of the contamination. Let the proposition proposed by the defence, H_d, be that the suspect has never been involved in any way in activities of which he/she is accused by the prosecution, that is, the trafficking of illegal drugs, reselling and activities related thereto. An inference from E to H requires a series of assessments to be made, as explained below.

If the suspect engages in trafficking of illegal drugs, their reselling and activities such as their packaging (H_p), it appears reasonable to assign a high probability for the event of direct contact with a primary source of illicit drugs. Thus, let $Pr(C|H_p)$ assume a value of 0.9, for example. If the suspect is innocent (H_d), then the probability of a contact with a primary source can be considered impossible: $Pr(C|H_d) = 0$. This expresses the assumption that there is no possibility for a legitimate contact with a primary source of illegal drugs.

If the suspect has been in contact with a primary source (C), then one may take it for certain that a transfer has occurred, independently of whether H_p or H_d is true: $Pr(T|C,H_p) = Pr(T|C,H_d) = 1$. If the suspect was not in contact with a primary source ($\bar{C}$),

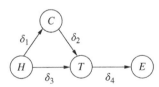

Figure 13.2 Qualitative probabilistic network for evaluating recovered trace material, given activity-level propositions. Node C has two states, C ('The suspect has been in contact with a primary source of illegal drugs') and $\bar{C}$ ('The suspect has never been in contact with a primary source of illegal drugs'). Node T has two states, T ('There has been a transfer of a certain quantity of illegal drugs') and $\bar{T}$ ('No transfer has occurred'). Node H has two states, H_p ('The suspect is involved in illegal drug trafficking, reselling and activities related thereto') and H_d ('The suspect has nothing to do with such activities'). Binary node E represents the findings of the IMS/GC/MS analyses: state E represents 'A positive result for an illicit drug is obtained' (e.g. cocaine) and state $\bar{E}$ represents 'A negative result is obtained'.

then the probability of transfer is reduced but not necessarily zero. For example, if the suspect is involved in drug trafficking (H_p), but did not himself handle any drugs, then secondary or tertiary transfers may eventually occur due to the exposure to a contaminated environment (e.g. a car used for the transport of drugs or a room used for the storage of drugs). For the purpose of illustration, consider a value of 0.4 for $Pr(T|\bar{C}, H_p)$. If the suspect is not involved in any of the alleged activities (H_d), and he/she did not have contact with a primary source, then the probability of a transfer can be set to zero or close to zero: $Pr(T|\bar{C}, H_d) = 0$. Note that the actual value of this probability is not of primary importance here, rather its relative value when compared to $Pr(T|\bar{C}, H_p)$: in particular, $Pr(T|\bar{C}, H_p) > Pr(T|\bar{C}, H_d)$ is assumed.

Assuming that the event of a positive analytical result is much more probable when there was a transfer of trace quantities of drugs than when there was no such a transfer leads to the consideration of $Pr(E|T) \gg Pr(E|\bar{T})$. Exemplary values could be, for instance, 0.99 and 0.01.

These qualitative expressions of beliefs about the various probabilistic relationships can now be used to extract the signs of qualitative influence associated with each arc of the QPN shown in Figure 13.2. Start with the relationship between H and C (δ_1). Here, the assumption $Pr(C|H_p) > Pr(C|H_d)$ implies that H positively ($\delta_1 = +$) influences C along $H \to C$.

Next, consider the relationship between C and T (δ_2). Given H_p or H_d, the probability of transfer given contact is considered to be higher than given no contact. Formally, this may be written as $Pr(T|C, H_p) > Pr(T|\bar{C}, H_p)$ and $Pr(T|C, H_d) > Pr(T|\bar{C}, H_d)$. This implies that $Pr(T|C, x) > Pr(T|\bar{C}, x)$ for any value of x of other parents of T (i.e. other than C). The conclusion thus is that the variable C positively ($\delta_2 = +$) influences T along $C \to T$.

As a third relationship, consider the link that goes from H to T (δ_3). Here, the changes in the probability of T due to varying H need to be considered for each state of C. If C is true, one has $Pr(T|H_p, C) = Pr(T|H_d, C)$. If $\bar{C}$ is true, one has $Pr(T|H_p, \bar{C}) > Pr(T|H_d, \bar{C})$. The overall relationship is thus described by $Pr(T|H_p, x) \geq Pr(T|H_d, x)$. Thus, H positively ($\delta_3 = +$) influences T along $H \to T$.

Finally, consider the relationship between T and E (δ_4). From the above assumption $Pr(E|T) > Pr(E|\bar{T})$, one can retain a positive ($\delta_4 = +$) qualitative influence of T on E.

The signs of qualitative influence in the QPN are now defined and can be used for inference. Attention is drawn here to the effect that knowledge about E may have on the truthstate of H. Let δ_5 denote the indirect influence between H and E. The direction of this influence depends, on the one hand, on the graphical structure of the QPN, that is, the various trails between H and E, and, on the other hand, on the signs associated with each arc of the trails connecting H and E. In the case at hand, δ_5 derived from the QPN shown in Figure 13.2 is

$$\delta_5 = [(\delta_1 \otimes \delta_2) \oplus \delta_3] \otimes \delta_4 . \tag{13.7}$$

Following the definition of the sign product and sign addition operators (Table 13.2), one can find that $\delta_5 = +$ for $\delta_1 = \delta_2 = \delta_3 = \delta_4 = +$. The result of the qualitative analysis is that under the stated assumptions, knowledge of E tends to increase the probability of H.

The computation of the sign of overall influence between H and E in Example 13.7, based on (13.7), may be more attractive, in some way, than the development of a formal likelihood ratio because the contribution of each of the underlying qualitative influences is shown in (13.7). In addition, whenever one or more of the δ_i ($i = 1, \ldots, 4$) change, (13.7) can readily be updated. This is illustrated in the scenarios discussed in Examples 13.8 and 13.9.

Example 13.8 *(Trace quantities of illegal drugs, continued) Consider a scenario slightly different from the one discussed in Example 13.7. Imagine that the position of the defence is still to argue that the suspect is not involved in drug trafficking. However, information is offered by the defence according to which the suspect has regularly been exposed to a potentially contaminated environment. This may be the case, for example, if accomplices of the suspect have already been proven to be involved in drug trafficking. It may then be a question for consideration whether this newly acquired circumstantial information should change the previously derived direction of inference between E and H. The additional information offered by the defence seems relevant for assessing the probabilities of transfer, as discussed below.*

If there were a contact with a primary source (C), then an event transfer is assumed to occur with very high probability or it may even be taken for certain (in a practical sense): $Pr(T|C, H_p) = Pr(T|C, H_d) = 1$. If there were no contact but the suspect is involved in drug trafficking, a non-zero transfer probability, denoted $Pr(T|\bar{C}, H_p) = t$, may be appropriate. As noted earlier, secondary or tertiary transfers may occur in such circumstances. These assessments are all analogous to Example 13.7. In turn, if there was no contact and the suspect was not involved in drug trafficking, there was no possibility for a transfer: $Pr(T|\bar{C}, H_d) = 0$. In the scenario considered here, information offered by the defence sheds new light on this assessment. It may appear reasonable to argue that there could be a secondary or tertiary transfer in the same way as for someone who is involved in drug trafficking. So, if $Pr(T|\bar{C}, H_d)$ is written more shortly as t', probabilities could be assessed so that $t \approx t'$ or $t = t'$.

Because of changes in the probabilities associated with the node T, the signs δ_2 and δ_3 (Figure 13.2) need to be reviewed. To assess the influence associated with $C \rightarrow T$ (δ_2), one needs to consider that $Pr(T|C, H_p) > Pr(T|\bar{C}, H_p)$ and $Pr(T|C, H_d) > Pr(T|\bar{C}, H_d)$. This observation implies that C exerts a positive qualitative influence on T: $\delta_2 = +$. The influence associated with $H \rightarrow T$ (δ_3) is a consideration of $Pr(T|H_p, C) = Pr(T|H_d, C)$ and $t = t'$, which implies $\delta_3 = 0$. The signs thus determined can now be entered in (13.7). The sign of overall influence between H and E is given by

$$\delta_5 = [(+ \otimes +) \oplus 0] \otimes + = + \,.$$

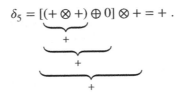

The result of this qualitative analysis is that the finding E still supports H_p over H_d despite the new circumstantial information proposed by the defence. This raises the question of whether the additional information has any effect. The answer to this question depends on one's point of view. As far as the effect of E on H is concerned, the direction of the overall inference remains unchanged. Effects may, however, be observed more locally, notably at the arc $H \rightarrow T$. Here, information relating to a possible secondary or tertiary transfer has 'neutralized' the qualitative influence that H exerts on T: the '+' sign has changed to a '0' sign. Notice that the assumption $Pr(T|H_p, C) = Pr(T|H_d, C)$ and $t = t'$ is equivalent to say that the edge $H \rightarrow T$ is uninformative and could be omitted, as long as the assumption holds.

The qualitative analysis is also capable of clarifying the reasons why there is still a positive effect of E on H. A close inspection of the assessments for the probability table of the node T is helpful for this. Notably, it has been assumed that for H_p and H_d, $Pr(T|C) > Pr(T|\bar{C})$. So, even though there may be a possibility for a secondary or tertiary transfer, a contact with a primary *source (C) provides a better account of an event transfer (T). Consequently, support for T due to E favours C over C̄, which in turn supports H_p over H_d. This scenario is considered further in Example 13.9.*

Example 13.9 *(Trace quantities of illegal drugs, continued) As a modification to the previous two Examples 13.7 and 13.8, imagine a scenario where the analytical result is negative. The definition of the variable E could then be changed in order to refer to such an event. Doing so, the conditional probabilities of E given T need to be reviewed. Clearly, a negative result is considered less probable given T than T̄. Consequently, $Pr(E|T) < Pr(E|\bar{T})$ and $\delta_4 = -$. Assuming the remaining signs to be the same as in Example 13.7, the sign of overall influence between E and H is given by*

$$\delta_5 = \underbrace{[\underbrace{(+ \otimes +)}_{+} \oplus +] \otimes -}_{+} = - \, .$$

The result of this analysis is that E, a negative result, is a finding in support of the non-occurrence of event T, that is, T̄. An increased probability of T̄ is better accounted for by an event of no contact (C̄) and the event that the suspect is not the offender (H_d), a result that appears perfectly reasonable.

13.1.5   Implications of qualitative graphical models

Notice that in the examples discussed in the previous section, no probabilities have been assigned to the root variable H, because they are irrelevant for the kind of qualitative analyses studied here. In some way, this seems a convenient property as it can help forensic scientists focus on the probability of the findings whilst avoiding the provision of an opinion on an issue, such as the proposition H, that lies outside their area of competence.

The qualitative notions presented here are neither intended nor suggested for use in written reports, or for presentation before trial. The qualitative framework is primarily intended as an aid for probabilistic reasoning in situations where there is a lack of explicit numerical data but where some informal expert knowledge is available. The aim is to show that such models can offer logical guidance in evaluative processes. Scientists can be provided with an approach that allows them to discuss, lay down and articulate the foundations of their inferences and beliefs in order to justify the coherence and credibility of their conclusions.

The elegance of the qualitative framework lies in its abstraction from numbers. The relative magnitude of probabilistic assessments can be sufficient to process uncertainty about target factors, in the light of expert knowledge, according to the accepted rules for inference. Also, literature (Druzdzel and Henrion 1993a; Druzdzel 1996; Druzdzel 1997; Wellman and Henrion 1993) has noted that it may be advantageous to maintain qualitative distinctions even when numerical information is available. Also, a qualitative specification might often be obtained with much less effort than a quantitative one. QPNs can thus supplement Bayesian networks where numerical values are either not available or not necessary for the questions of interest. However, if a numerically specified network is already available, the qualitative dependencies may readily be determined from the formal definitions of the various QPN properties. Thus, along with Bayesian networks, QPNs extend the spectrum of possible levels of precision for the specification of probabilistic inference models. Notably, the level of detail can be made dependent on the extent of available information, allowing maximum efficiency with the least possible effort. The results of such qualitative analyses provide a workable basis for the generation of verbal explanations of probabilistic reasoning. Explanations of this kind may be easier to follow than explanations using numerical probabilities.

Generally, the understanding of both the assumptions, and the reasoning of probabilistic expert systems, are prerequisites for a successful collaboration between these systems and their user (Henrion and Druzdzel 1991). In this context, qualitative verbal explanations extracted from QPNs are of particular interest. A distinction may be made between the explanation of assumptions and the explanations of reasoning (Druzdzel 1996; Henrion and Druzdzel 1991). The explanation of assumptions is concerned with the communication of static knowledge, such as a graph structure used as a representation of the main entities of a real-world problem. The explanation of reasoning is a more dynamic property as it focuses on extracting conclusions from assumptions encoded in the original model and the findings and observations. For the purpose of inference in QPNs, the interested reader may find it useful to consider further readings in a topic concerned with a method known as *qualitative belief propagation* (Druzdzel and Henrion 1993b; Henrion and Druzdzel 1991). Its underlying algorithm traces the qualitative effect of particular findings from one variable to the next and thus provides an illustrative basis for evaluating the effect of newly acquired knowledge and its propagation through the network. However, there may be practical limitations to this because of generated '?' signs, which may spread through a network, limiting the informative value of the overall output. Also, the QPN formalism appears to be rather rarely used currently because of the widespread availability of computationally powerful systems that can process even large numerically specified networks almost instantly.

13.2   Sensitivity analyses

13.2.1   Preliminaries

When a scientist has set up a graphical model for structuring a given inferential task, the model typically conveys structural assumptions, that is, relationships believed to hold true amongst particular variables of interest. In Bayesian networks, probability theory is used as a concept to characterize the nature of the relationships assumed to hold amongst variables. Most often, however, a given model is considered only as a tentative approach for a given problem at hand and represents, as such, a starting point for further investigations. In particular, it is important for scientists to study the properties of a given model. This leads to a topic broadly called *sensitivity analysis*, a term that brings together several related aspects.

One important aspect concerns sensitivity to findings. In specialized literature on Bayesian networks, this is also referred to as *evidence sensitivity analysis* [e.g. Kjærulff and Madsen (2008)]. This kind of analysis focuses on question of the following kind: 'What is the support offered by a given item of information on the belief in a given hypothesis of interest – versus the belief in another?', 'Which item of information (in a large set of findings) provides the best discrimination between the hypotheses of interest?' Such sensitivity analyses may help understanding the conclusions reached when using a given model for inference. The provision of answers to questions as mentioned above is crucial in a process of deciding whether or not a model is acceptable for the purpose for which it has been designed. In this context, it has been pointed out that

> […] we should be able to obtain useful insights about the relative importance to our conclusions of the various assumptions, decisions, uncertainties, and disagreements in the inputs. These can help us decide whether it is likely to be worthwhile gathering more information, making more careful uncertainty assessments, or refining the model, and which of these could most reduce the uncertainty in the conclusions. (Morgan and Henrion 1990)

The likelihood ratio is a very common measure for evaluating the impact of findings, particularly in the area of forensic science. When H denotes a selected proposition of interest, E denotes particular findings and I denotes circumstantial information, then $Pr(E|H_p, I)/Pr(E|H_d, I)$ measures the impact that findings E have on the propositions H or, alternatively speaking, the degree to which findings E discriminate H_p from H_d. This measure has extensively been studied throughout this book. A crucial role of the likelihood ratio was encountered, for example, in Chapter 10, where the expression was used to assess the value of potential findings before conducting analyses, in order to inform about the usefulness of particular investigations. Sensitivity to findings is thus not pursued in this chapter in further detail. Attention is drawn to another topic in sensitivity analysis, as mentioned below.

Usually, the impact of distinct findings on other variables of interest is considered for a model with a fixed structure. However, the impact on a line of reasoning may also vary due to features of the model at hand, in particular with respect to the numerical specification, as given by the (conditional) probabilities specified in the various node tables. In specialized literature on Bayesian networks e.g. Jensen and Nielsen (2007), such node probabilities are also sometimes called *parameters* and the study of the implications of their variation is known as *parameter sensitivity analysis*. It is widely held that disciplines that rely on quantitative analyses can benefit from examining variations in inference as a result of varying key assessments within a reasonable range (Watson and Buede 1987). This topic is pursued below in various perspectives. Sections 13.2.2 and 13.2.3 focus on what are known as one- and two-way sensitivity analyses due to changes in assignments in node probability tables. Section 13.2.4 focuses on sensitivity to the prior distribution in a continuous network.

13.2.2 Sensitivity to a single probability assignment

Consider the Bayesian network described in Section 7.11.1 (Figure 7.24), which may be used for the evaluation of the potential of error associated with DNA profiling results (Biedermann and Taroni 2004). There are three nodes H, M and R that are serially connected: $H \rightarrow M \rightarrow R$. The network is chosen here for the purpose of illustrating the idea of sensitivity analysis as applied to a single probability assignment, also known as *one-way sensitivity analysis*. The model is well suited for this purpose because it contains, on the one hand, logical probabilities of 0 and 1 and, on the other hand, epistemic probabilities such as γ and *fpp*. The model involves only a few variables, but it may be thought of as a local part of a larger network. Proposition H is binary and refers to the event according to which a crime stain comes from the suspect. Denote the truth and falsity of H more succinctly as H_p and H_d. The variable M refers to the event that there is a true correspondence between the profile of the crime stain and the profile of the control material taken from the suspect. Recall that it was assumed that when the suspect is in fact the source of the crime stain (i.e. H_p is true), then certainly there is a correspondence: $Pr(M|H_p) = 1$. Notice that M may also be true even though H is false. The probability of such an event has been described in terms of γ, the conditional genotype probability. The variable R is the scientist's report of a correspondence between the profile of the crime stain and that of the material known to originate from the suspect. It is assumed that if there is a correspondence (i.e. M is true), then the scientist would report a correspondence: $Pr(R|M) = 1$. This probability could be taken to be less than 1, but this is not an aspect that is investigated at this point. In turn, the probability with which the scientist would falsely declare a correspondence, $Pr(R|\bar{M})$, is denoted x (in Section 7.11.1 written fpp to denote the false positive probability).

Imagine now a scientist's report of an observed correspondence. Such information would be communicated to the network by instantiating the node R. Let the target probability be denoted by $y = Pr(H_p|R)$, that is, the probability that the suspect is the source of the crime stain, given the report provided by the scientist. Let $x = Pr(R|\bar{M})$, the so-called false positive probability (Thompson et al. 2003). This latter probability assignment for the node table of R is chosen here for sensitivity analysis. The question of interest thus is how changes in the value taken by x affect the probability y. To conduct such a sensitivity analysis, the functional relation between y and x is investigated. Start by rewriting y as

$$y = Pr(H_p|R) = \frac{Pr(R|H_p)Pr(H_p)}{Pr(R)} .$$

In its extended form, the numerator is

$$[Pr(R|M)Pr(M|H_p) + Pr(R|\bar{M})Pr(\bar{M}|H_p)]Pr(H_p).$$

Consider this as a linear function in x:

$$ax + b,$$

where both $a = Pr(\bar{M}|H_p)Pr(H_p)$ and $b = Pr(R|M)Pr(M|H_p)Pr(H_p)$ will be considered fixed, whilst x is as defined above. For the denominator, one can write

$$Pr(R) = \sum_M \sum_H Pr(R|M)Pr(M|H)Pr(H).$$

The denominator can also be considered as a linear function in x, notably,

$$Pr(R) = cx + d,$$

where c is given by $Pr(\bar{M}|H_p)Pr(H_p) + Pr(\bar{M}|H_d)Pr(H_d)$ and d is given by $Pr(R|M)[Pr(M|H_p)Pr(H_p) + Pr(M|H_d)Pr(H_d)]$ and will be considered fixed (as a and b above).

As may be seen from these transformations, the numerator and denominator of y relate linearly to the conditional probability x. The probability of interest, y, is thus expressed as a quotient of two linear functions of x. A more formal statement of this property can be found in specialized literature e.g. Coupé and Van der Gaag (1998).

Recall from above that $Pr(M|H_p)$ and $Pr(R|M)$ have both been set to 1. For H, default probabilities of $Pr(H_p) = Pr(H_d) = 0.5$ are used. This assignment of prior probabilities is one of a purely technical nature. In practice, it assignment is informed by case-related background information that lies outside the province of the forensic scientist. Based on these assessments, it can be shown that for the scenario discussed here,

$$y = \frac{ax+b}{cx+d} = \frac{1}{\gamma(1-x)+1+x} \, .$$

Example 13.10 (*One-way sensitivity analysis for the potential of error*) *Consider a numerical illustration for the result obtained above by specifying a value* $\gamma = 10^{-5}$. *The probability of* H_p *given a reported correspondence, R, can then be represented as shown in Figure 13.3. This plot shows that as the probability of a false positive (x) tends to zero, then the probability of interest,* $y = Pr(H_p|R)$, *tends to* $1/[1 + Pr(M|H_d)] = 1/(1 + \gamma)$. *This corresponds to a situation in which the likelihood ratio* $V = Pr(R|H_p, I)/Pr(R|H_d)$ *is* $1/\gamma$. *On the other hand, when x reaches one, then the probability of interest y falls back to the prior probability of 0.5 as specified initially. This situation corresponds to a likelihood ratio V of 1, independently of how low the value* γ *is. The observation that the target probability is 'dominated' by increased values of x is sometimes raised to argue that attention ought to be drawn on the elicitation of assignments for x, the false positive probability, rather than for* γ, *the expression of the rarity of the corresponding features.*

13.2.3 Sensitivity to two probability assignments

A so-called *two-way sensitivity analysis* may be considered when information is needed about the joint effect of changes in two probability assignments. To illustrate this topic through an example, consider again a situation where a scientist's report of a correspondence between the profile of a crime stain and that of comparison material from a suspect is available. As outlined in Section 13.2.2, such a setting can be approached through a Bayesian network with the structure $H \to M \to R$. Suppose further that there is additional information according to which

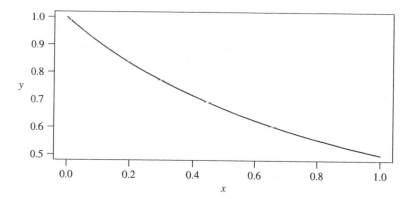

Figure 13.3 Representation of the conditional probability $y = Pr(H_p|R)$ as a function of $x = Pr(R|\bar{M})$, the false positive probability. Variable R is the report of a correspondence between the profile of a crime stain and that of control material from a suspect. State $\bar{M}$ denotes the event that there is no correspondence between the profile of a crime stain and that of control material from the suspect. Proposition H_p states that the suspect is the source of the crime stain.

the suspect has been selected on the basis of a search of the crime stain's DNA profile against a relevant database (containing N profiles), with no other correspondences being observed besides that with the suspect. Following the arguments presented in Section 7.10, one way to deal with this issue is to extend the above Bayesian network with a binary variable D, defined as 'the other $(N - 1)$ profiles of the database do not match'. D is, besides M, a further child variable of H. So, when the available observations and results consists of both R and D, then the likelihood ratio is given by

$$V = \frac{Pr(R, D | H_p)}{Pr(R, D | H_d)} = \frac{1}{\gamma + [fpp + (1 - \gamma)]} \times \frac{1}{1 - \phi}, \tag{13.8}$$

where ϕ is the probability that the source of the crime stain is to be found amongst the other $(N - 1)$ suspects, fpp is the false positive probability and γ the relevant profile probability.

In Section 13.2.2, the sensitivity analysis concentrated on the probability y taken by a particular node state, given an observation (i.e. finding), whilst varying a single conditional probability assignment x in a node table. The values observed for the target probability y depended on the specified prior probabilities, that is, in the scenario of interest, the prior probability of the event that the suspect is the source of the crime stain, $Pr(H_p)$. This may be perceived as a drawback as one may prefer to work with as few assumptions as possible. To avoid such an assumption about prior probabilities, one can consider a likelihood ratio. It draws attention to the value of the findings whilst having the same potential for analysing variation in values of probabilities assigned in node tables. Example 13.11 illustrates this for the above scenario on a reported correspondence and a database search result. In this context, a two-way sensitivity analysis amounts to an investigation of the changes in the likelihood ratio V, as formulated above, due to variation in two probability assignments. This can be represented graphically in the form of the so-called contour lines. Such lines connect combinations of values for the two probabilities of interest that result in the same value for the likelihood ratio.

Example 13.11 (*Two-way sensitivity analysis for a reported correspondence and the result of a database search*) *Consider the likelihood ratio V for a reported correspondence R ('The scientist reports a correspondence between the profile of a crime stain and the control material from the suspect') and the database search result D ('No correspondence with any database member other than the suspect'), with respect to the propositions H_p ('The suspect is the source of the crime stain') and H_d ('An unknown person is the source of the crime stain'). The likelihood ratio V as given by (13.8) involves several probability assignments that may be varied in the course of a sensitivity analysis. The discussion here focuses on γ, the profile probability, and ϕ, the probability that the source of the crime stain is to be found amongst the other $(N - 1)$ suspects, whilst keeping fpp (the false positive probability) fixed, for example, at 0.001.*

Figure 13.4 shows contour lines for $\log_{10}$ likelihood ratios of 0.25, 0.5, 1, 1.5, 2 and 3. It can be observed that the distances between the contour lines are greater in the lower right part of the graph than in other parts of the graph. This indicates that the likelihood ratio is relatively insensitive for high values of γ and low values of ϕ. Whenever γ decreases and/or ϕ increases, the contour lines become more close, that is, changes in the values of the target probabilities tend to provoke greater changes in the value of V.

If it is felt that the scientist can consider what may be called plausible ranges of values for both γ and ϕ, then this can be marked in the graph in terms of parallel lines. Illustrative regions are shown in Figure 13.4, using dashed lines. Their intersection is visualized by the

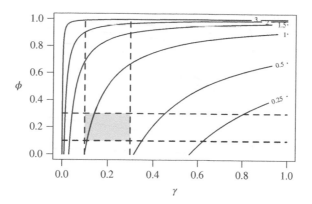

Figure 13.4   Two-way sensitivity analysis of the likelihood ratio, (13.8), in a case where a correspondence is reported between the profile of a crime stain and that of a suspect, and the suspect was selected as result of a search in a database of size N (Example 13.11). The values assigned for γ, the genotype probability, and ϕ, probability that the source of the crime stain is to be found amongst the other $(N-1)$ suspects, are varied simultaneously. Each contour line represents a $\log_{10}$ likelihood ratio of a particular value (i.e. 0.25, 0.5, 1, 1.5, 2 and 3). The dashed parallel lines delimit illustrative plausible intervals.

grey-shaded area. The contour lines that pass through this shaded area correspond to the $\log_{10}$ likelihood ratios that can result from the combination of values defined by the plausible ranges. Note further that Figure 13.4 also illustrates that depending on the region of values considered, the likelihood ratio may vary considerably even though only a very narrow range of values is considered. Conversely, there may also be configurations with broadly defined regions for γ and ϕ, but where the likelihood ratio varies only to a minor extent.

Example 13.12 *(Two-way sensitivity analysis for a case of material recovered on a crime scene) Consider again a scenario introduced earlier in Section 3.3.3, involving k offenders and a single stain found on a crime scene. A likelihood ratio development for such a setting is given by Evett (1993). Section 6.1 translates this approach into a Bayesian network. Suppose that the single blood stain is found at the crime scene in a position where it may have been left by one of the offenders, but there may be disagreement about the extent to which the recovered stain should be considered relevant. A suspect is found and he/she gives blood for comparative analyses. The suspect's blood and the crime stain are found to have the same profile. This profile is supposed to be shared by a proportion γ of the relevant population from which the criminals have come. In turn, γ' denotes is the profile probability amongst the (innocent) people who may have left the stain. Further, denote by p the probability that the stain would have been left by the suspect even though being innocent of the offence. The main propositions of interest are 'The suspect is one of the k offenders' (H_p), and 'The suspect is not one of the k offenders' (H_d). With these notational definitions, one can derive (6.10) (Evett 1993).*

 Suppose that it is interest to study the impact that varying assessments for the relevance factor r and the probability of innocent acquisition p have on the value of the likelihood ratio, so as to gain an idea as to the conditions under which the recovered stain and the observed correspondence with the profile of the suspect can be of guidance to a court. For this purpose, suppose a setting proposed in Evett (1993), with k = 1 and γ = γ' = 0.001. In Evett (1993), plots of $\log_{10}$ likelihood ratios are given for selected values of r, that is, r = (0.25, 0.5, 0.75, 1),

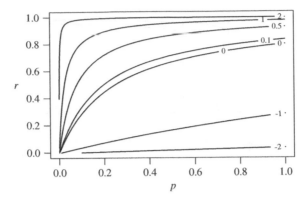

Figure 13.5 Two-way sensitivity analysis of the likelihood ratio, (6.10), in a case where a profile of a crime stain corresponds to the suspect. The main propositions are 'The suspect is one of the k offenders' (H_p) and 'The suspect is not one of the k offenders' (Example 13.12), and the number of offenders is taken to be $k = 4$. Each contour line represents a $\log_{10}$ likelihood ratio of a particular value (i.e. -2, -1, 0, 0.1, 0.5, 1, 2, and 3), as obtained by a particular combination of values for r, the relevance of the crime stain, and p, the probability of innocent acquisition.

whilst varying p over the full range of values between 0 and 1. Another way to convey a sensitivity analysis consists of using a contour plot, as in Example 13.11. For illustration, consider contour lines for $\log_{10}$ likelihood ratios of -2, -1, 0, 0.1, 0.5, 1, 2 and 3 and variation of r and p over the full range of values between 0 and 1. This leads to the plot shown in Figure 13.5.

13.2.4   Sensitivity to prior distribution

Bayes' theorem does not specify which prior distribution should be chosen. Though in the original version it was proposed in a context where data followed a binomial distribution and the parameter of interest was a proportion for which a uniform prior was chosen, the theorem is far more general and a wide variety of probability models and prior distributions can be adopted. Probability and prior distributions may be specified in various different ways, and sensitivity analysis may be performed with the objective of determining the effect of alternative likelihood and prior specifications on posterior inferences. In this section, attention is confined to the sensitivity to prior specification.

A simple way for specifying a prior distribution is to consider a few summary statistics, such as the mean and the standard deviation, and check whether the shape of the resulting prior distribution looks close to one's prior knowledge, in the sense that the probability is reasonably spread over the range of values in which one believes the parameter to lie. Otherwise, the prior distribution needs to be revised (O'Hagan 1994).

Still, even if the prior elicitation process is very accurate and the prior distribution reflects properly the currently available prior knowledge, it can be agreed that it is not always possible to make an arbitrarily precise prior elicitation. Consider Example 9.3 where a prior distribution was introduced to model the uncertainty about the unknown value of alcohol in blood. As there was confidence that a unimodal distribution represented an appropriate reflection of the available prior knowledge about θ, a Normal distribution was specified with a prior mean μ equal to 0.7 g kg^{-1} and a prior standard deviation τ equal to 0.1 g kg^{-1}. In this way, prior beliefs were given by a 95% interval area (i.e. values outside the interval 0.5–0.9 were considered unlikely in that the overall probability of a value lying outside that interval was thought to be about 0.05 or 1 in 20). However, values of the prior mean close to 0.7 would be more or less equally acceptable, and this is valid also for the variance. If the posterior inference is only slightly affected by a different (but still plausible) specification of the prior parameters, then it can be said they are robust to an inappropriate specification of the prior parameters. Otherwise, they are sensitive to the prior parameters, and it would be advisable to improve the prior specification.

Consider a sensitivity analysis of the posterior mean. The posterior mean represents a natural candidate as a summary of the location of the posterior distribution and as an estimate of the parameter of interest. Take, for instance, the alcohol-test scenario described in Example 9.3. Practitioners might be interested to focus on the posterior mean, compare this value to the legal limit and take it as a basis to conclude whether a given item is positive or not e.g. Taroni et al. (2010). This is clearly not the only way to summarize a posterior distribution, though elements of Bayesian decision theory can provide an answer to justify this choice. Bayesian statistical inference can be seen as a decision problem where a decision maker is asked to consider the expected loss of alternative decisions, which correspond to reporting a probability distribution for the unknown quantity of interest and to choose the optimal decision that is the one which minimizes the expected loss. It can be verified that whenever the undesirability of the consequences of available actions can be represented by a quadratic loss function, the posterior mean represents the quantity that minimizes the posterior expected loss, that is, the posterior mean is the optimal decision.

Suppose that the prior uncertainty about the parameter of interest θ is modelled by a symmetric unimodal prior distribution, say $\theta \sim N(\mu, \tau^2)$ and the number of observations in the sample for the estimation of the likelihood is n. Suppose that the prior mean μ is changed by an amount ξ, that is, the prior mean is taken equal to $\mu + \xi$. The posterior mean can be obtained by substituting the amount μ with $\mu + \xi$ in (9.15). The effect of this change on the posterior mean can be easily computed and is equal to

$$\Delta\mu(\xi) = \left(\frac{\sigma^2/n}{\sigma^2/n + \tau^2}(\mu + \xi) + \frac{\tau^2}{\sigma^2/n + \tau^2}\bar{x} \right) - \left(\frac{\sigma^2/n}{\sigma^2/n + \tau^2}\mu + \frac{\tau^2}{\sigma^2/n + \tau^2}\bar{x} \right)$$

$$= \frac{\xi\sigma^2/n}{\sigma^2/n + \tau^2}. \tag{13.9}$$

This is a linear function of ξ and is represented in Figure 13.6 for $\sigma^2 = 0.005$, $\tau^2 = 0.01$ and different number of observations.

This sensitivity analysis allows one to quantify the effect of a different specification of the prior mean to the posterior mean. If it is found that a slightly different specification of the prior distribution could substantially affect the decision process, then it would be advisable to reflect more carefully on the specification or, if possible, to increase the number of observations in the sample. The greater the number of observations in the sample, the less the posterior will be influenced by the prior (i.e. the prior is said to be *dominated* by the likelihood). Also, one can observe that if the prior variance τ^2 is sufficiently large relative to the variance of observations σ^2, the posterior mean will be robust to inappropriate specification of the prior mean μ in (13.9). In the same way, if the prior variance τ^2 is changed by an amount δ, the

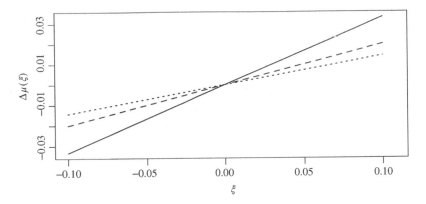

Figure 13.6   The effect of a change ξ in the prior mean to the posterior mean, for $\sigma^2 = 0.005$, $\tau^2 = 0.01$ and $n = 1$ (continuous line), $n = 2$ (dashed line) and $n = 3$ (dotted line).

effect on the posterior mean can be obtained by substituting τ^2 with the amount $\tau^2 + \delta$ in (9.15) and quantified in

$$\frac{(\sigma^2/n)\delta(\bar{x} - \mu)}{(\sigma^2/n + \tau^2 + \delta)(\sigma^2/n + \tau^2)}.$$

Example 13.13 *(Blood alcohol, sensitivity) Consider again the case illustrated in Example 9.3 where a person was stopped with the suspicion of driving under the influence of alcohol. Blood was taken from the individual and submitted to a forensic laboratory, and a value of 0.82 g kg^{-1} was measured. Assuming that the person is considered to be under the influence of alcohol if the alcohol level in blood is greater than 0.8 g kg^{-1}, this implies that a posterior mean larger than 0.8 could be an indication of that. A Normal prior distribution for the true but unknown level of alcohol in blood was chosen, $N(\mu = 0.7, \tau = 0.1)$, reflecting an a priori belief that values less than 0.5 g kg^{-1} and greater than 0.9 g kg^{-1} were considered unlikely. The posterior distribution is represented in Figure 9.5(ii), with a posterior mean equal to 0.78, that is, below the threshold of 0.8. It might be agreed that a slightly different prior specification could still represent the prior knowledge appropriately. The Bayesian network to model this situation is represented in Figure 13.7(i). Suppose now that the prior distribution is taken as $N(0.75, 0.1)$, with a prior mean changed by an amount of $\xi = 0.05$. The posterior mean increases to 0.7967, as it can be seen in Figure 13.7(ii), which is still below the legal limit.*

13.3   Conflict analysis

Given an inferential domain of interest, scientists can develop expert systems that use Bayesian networks as models, so as to support selected reasoning tasks. As noted earlier in Section 3.5.2, it is not realistic to expect that particular models will cover all aspects of a problem of interest. Proposed Bayesian networks are most often approximations of the problem domain. They are tentative in the sense that they rely on given assumptions. In practice, it can happen that one observes findings (whose probative value ought to be assessed) that are very unlikely given the problem at hand. The observation of a low probability of some findings does not

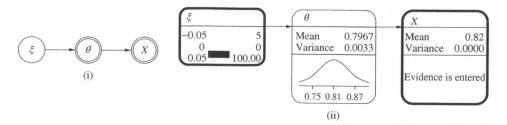

Figure 13.7 (i) A Bayesian network with two continuous nodes, θ and X, where θ represents the true level of alcohol in blood and X the measured alcohol level. Node θ depends on a discrete node ξ that allows one to change the prior mean by an amount ξ. (ii) Posterior distribution of the true value θ of blood alcohol concentration given a measured value X of blood alcohol concentration of 0.82 g kg^{-1} and a prior mean equal to $\mu + \xi = 0.7 + 0.05 = 0.75$. The instantiation of the nodes X and ξ is shown with a bold border.

necessarily disclose the presence of a conflict between the findings and the model; in fact, it may be justified whenever it is related to a rare case.

As a simple example, consider the case where two independent laboratory analyses can be performed to detect the presence or the absence of a given target substance in trace material. Denote by E_1 and E_2 the random variables measuring the quantities of target substance using the first procedure and the second procedure, respectively. Suppose further that the two variables are discrete with states 'large' (e_{i1}), 'medium' (e_{i2}), 'small' (e_{i3}) and 'absent' (e_{i4}), where the first subscript denotes the ith procedure, $i = 1, 2$. Imagine now that the first procedure reveals a large quantity of the target substance, whilst the second procedure reveals a low quantity of it. This result may go against one's expectations so that one may determine a low probability for these findings. Notwithstanding this result, there can be circumstances under which such a finding can arise, such as the fact that the second procedure is less accurate than the first and that in peculiar conditions, it might not work reliably. The distinction between these situations is not obvious, in the sense that it may be very difficult to tell whether the findings derive from a rare case that the proposed network is fully equipped to handle, or from kinds of data that make the background assumptions no longer valid. For this reason, it is of interest for analysts to have a device that can alert them about the possible presence of a so-called conflict and that the model may be weak. Note that the presence of a conflict does not prevent the network from producing a probability distribution for the variables of interest. The problem is that if a model is used under circumstances that do not conform to the underlying assumptions, then results can be unreliable and a low probability for observations can be an indication of that failure to conform to the assumptions.

13.3.1 Conflict detection

One strategy for the detection of a possible conflict is to consider a so-called straw model in which all variables are independent of each other [e.g. Jensen (1996)]. The straw model is structurally poor, since all relationships amongst variables, some of which may be relevant, are removed. However, the straw model can give better results with respect to the original model whenever, in the latter, there is a data conflict that is removed by the elimination of all dependencies. A conflict measure can thus be computed by starting from the assumption that observations (i.e. findings to be evaluated) are positively correlated; that is, for any variables

E_i and E_j in a network, $Pr(E_i = e_i | E_j = e_j) > Pr(E_i = e_i)$ holds, or in a more compact form $Pr(e_i | e_j) > Pr(e_i)$.

Consider the two laboratory tests for the detection of a given target substance introduced at the beginning of the section. It is fully reasonable to assume that the detection of a large presence of the target substance with one procedure will lead one to infer an increased probability for the event of the detection of a large presence of the substance with the alternative procedure, that is, $Pr(E_2 = large | E_1 = large) > Pr(E_2 = large)$. Recalling the definition of conditional probability, that is,

$$Pr(e_i | e_j) = \frac{Pr(e_i, e_j)}{Pr(e_j)} \quad \text{or} \quad Pr(e_i, e_j) = Pr(e_i | e_j) Pr(e_j),$$

and starting from the assumption that distinct observations are positively correlated, then one may expect to observe

$$Pr(e_i, e_j) > Pr(e_i) Pr(e_j).$$

The observation of a joint probability $Pr(e_i, e_j)$ of the observations which is less than the product of the marginal probabilities of the observations can thus indicate a possible conflict. By rearranging terms, the presence of a possible conflict can be detected whenever

$$\frac{Pr(e_i) Pr(e_j)}{Pr(e_i, e_j)} > 1 \quad \text{or} \quad \log \frac{Pr(e_i) Pr(e_j)}{Pr(e_i, e_j)} > 0.$$

Given n distinct observations $e_1, \ldots, e_n$ over n variables denoted by $E = \{E_1, E_2, \ldots, E_n\}$, the general conflict measure is defined as

$$\text{conf}(E) = \text{conf}(\{E_1, \ldots, E_n\}) = \log \frac{\prod_{i=1}^{n} Pr(e_i)}{Pr(e_1, \ldots, e_n)}. \tag{13.10}$$

This implies that a positive value of the conflict measure in (13.10) suggests the presence of a conflict.

Example 13.14 *(Examination of handwriting and fingermarks) Consider a scenario described earlier in Section 8.4.1, involving a suspect, Mr. Adams, arrested in a shop trying to pass a stolen cheque. Suppose that the issue is whether or not Mr. Adams wrote the signature on the stolen cheque. In a first approach to this case, let the forensic examinations cover two aspects (Example 13.15 will consider further forensic examinations), comparative handwriting examinations and detection techniques for fingermarks potentially present on the cheque, which may be close to the signature in question and in a position where the author of the signature would be expected to leave fingermarks. In particular, suppose that in this case the findings consist of many similarities between the handwriting of Mr. Adams and the signature in question and of detected fingermarks that do not correspond to the fingerprints obtained from Mr. Adams under controlled conditions. Such a starting point could be considered as conflicting in the sense that if Mr. Adams is the author of the questioned signature, then one may expect to find agreement with respect to both types of findings, handwriting and fingermarks. In turn, if Mr. Adams were not the author of the signature, then both handwriting and fingermarks may be expected to show discrepancies with respect to comparison material obtained from Mr. Adams.*

To study the concept of conflict in this scenario, consider the Bayesian network shown previously in Figure 8.2(iii), reproduced here in Figure 13.8. This model covers a main proposition defined in terms of the binary node S with states 'The suspect is the author of the

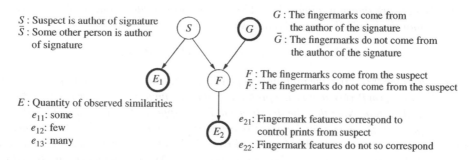

Figure 13.8 Bayesian network for the questioned signature case discussed in Examples 13.14–13.16. Findings are available for nodes E_1 and E_2 whilst node G represents a proposition under which the conflict in the findings is reconsidered; these nodes are highlighted with a bold circle. Definitions of node states are indicated besides each node.

questioned signature (S)' and 'The suspect is not the author of the questioned signature ($\bar{S}$)'. Node S conditions node E_1 which represents the findings of the handwriting examinations, defined by three states, the finding of few (e_{11}), some (e_{12}) and many (e_{13}) similarities. Thus, connection $S \rightarrow E_1$ holds. Node S also conditions the binary node F, covering the states 'The suspect is the source of the fingermarks (F)' and 'The suspect is not the source of the fingermarks ($\bar{F}$)'. Node G is the relevance term and has two states, 'The fingermarks have been left by the author of the questioned signature (G)' and 'The fingermarks have not been left by the author of the questioned signature ($\bar{G}$)'. Node G acts as further parent of F, which leads to the structure $S \rightarrow F \leftarrow G$. Finally, there is a node E_2, conditioned on node F, and has two states, 'There are correspondences between the fingermarks on the questioned document and the fingerprints of the suspect (e_{21})' and 'There are no correspondences between the fingermarks on the questioned document and the fingerprints of the suspect (e_{22})'.

Given the above model, the overall findings 'many similarities in handwriting' (e_{13}) and 'no correspondence in fingermark features' (e_{22}) can thus be summarized as $E = \{E_1 = e_{13}, E_2 = e_{22}\}$, written more shortly as $E = \{e_{13}, e_{22}\}$. The conflict measure to be computed thus is $\mathrm{conf}(E) = \mathrm{conf}(\{e_{13}, e_{22}\})$, written in further detail as

$$\mathrm{conf}(E) = \log \frac{Pr(e_{13})Pr(e_{22})}{Pr(e_{13}, e_{22})}.$$

Assuming that the forensic scientific findings $E = \{e_{13}, e_{22}\}$ are positively correlated, then a positive value of the conflict measure detects a possible conflict. Start by computing the marginal probability $Pr(e_{13})$, given by $Pr(e_{13}|S)Pr(S) + Pr(e_{13}|\bar{S})Pr(4\bar{S})$. Using the likelihoods $Pr(e_{13}|S) = 0.13$ and $Pr(e_{13}|\bar{S}) = 0.02$ defined earlier in Section 8.4.1, and equal prior probabilities for the states of the node S, the marginal probability for e_{13} can be found to be $Pr(e_{13}) = 0.075$.

The computation of the marginal $Pr(e_{22})$ is slightly more demanding because there are intermediate nodes F and G between node E_2 and node S. Full details of the probabilistic underpinnings of this structure have been previously outlined in Section 6.1.2, in particular the factorization

$$Pr(e_{22}) = \sum_S \sum_G \sum_F Pr(e_{22}|F)Pr(F|S, G)Pr(S)Pr(G).$$

given by the network shown in Figure 13.8. Using these results allows one to observe, with a little effort, that the required marginal reduces to

$$Pr(e_{22}) = (1 - \gamma)(1 - r)Pr(S) + (1 - \gamma)Pr(\bar{S}),$$

where γ is short for $Pr(e_{21}|\bar{F})$, the probability of a correspondence in fingermark features, and r is the relevance term $Pr(G)$, that is, the probability that the fingermarks have been left by the author of the signature. Note that the development here supposes that there is no possibility that the suspect left the marks for innocent reasons. Hence, the term $Pr(F|\bar{S}, \bar{G})$, defined in Section 6.1.2 as factor p, is set to 0. Moreover, $Pr(e_{21}|F)$ is set to 1 as explained previously in Section 8.4.1. The use of illustrative values $\gamma = 0.01$ and $r = 0.9$, and equal prior probabilities for the states of the node S, gives $Pr(e_{22}) = 0.5445$.

As a third step, the joint probability $Pr(e_{13}, e_{22})$ is required. Recalling the definition of conditional probability, one can write the joint probability as $Pr(e_{13}|e_{22})Pr(e_{22})$. The conditional probability in this product can be obtained by replacing $Pr(S)$ and $Pr(\bar{S})$ by, respectively, the posterior probabilities $Pr(S|e_{22})$ and $Pr(\bar{S}|e_{22})$ in the equation for the marginal $Pr(e_{13})$ computed in the first step above. The posterior probabilities for node S, given $E_2 = e_{22}$, can be found through Bayes' theorem. Using these posterior probabilities allows one to determine the conditional probability $Pr(e_{13}|e_{22}) = 0.03$. Then, multiplication with the marginal $Pr(e_{22})$ leads to the required joint probability $Pr(e_{13}, e_{22})$, which is found to be 0.01634. Based on the above results, the conflict measure can now be computed as

$$\text{conf}(E) = \log \frac{Pr(e_{13})Pr(e_{22})}{Pr(e_{13}, e_{22})}$$

$$= \log \frac{0.075 \times 0.5445}{0.01634} = \log \frac{0.04084}{0.01634} = 0.916.$$

Because this result is greater than zero (indicating that the joint probability of e_{13} and e_{22} is less than the product of the marginal probabilities of e_{12} and e_{22}), $\text{conf}(E)$ indicates a possible conflict in the group of findings $E = \{e_{13}, e_{22}\}$. Note that a positive value for the conflict measure does not indicate that there is necessarily a misspecification in the model. It serves as an indication for the user to inspect modelling assumptions with respect to the observations in the case at hand.

13.3.2   Tracing a conflict

Once a possible conflict has been detected, further investigation of the model should focus on locating the possible conflict, that is, tracing its origin. This can be achieved by performing the conflict measure for different subsets of the findings E. That is, for each pair of findings $\{e_i, e_j\}$ on variables $\{E_i, E_j\}$, one can compute the partial conflicts as

$$\text{conf}(E_i, E_j) = \log \frac{Pr(e_i)Pr(e_j)}{Pr(e_i, e_j)}, \qquad \forall \ i, j = 1, \dots, n; \ i \neq j,$$

where n is the number of distinct findings comprised in E. The computation of the partial conflicts shows for which pair of the findings there exists a conflict. In Example 13.14, there were only two distinct scientific results available, so the concept of partial conflicts and the tracing of conflicts does not apply to this situation. However, the concepts can be explored by considering a further forensic result, as is illustrated in Example 13.15.

Example 13.15 *(Conflict tracing) Consider again the handwriting scenario introduced in Example 13.14, with findings 'many similarities in handwriting' (e_{13}) and 'no correspondence in fingermark features' (e_{22}) (see also Figure 13.8). Suppose now a third item of information (supposed to be positively correlated with the previous observations), E_3, defined as 'Results of the writing ink comparison between the questioned signature and the ink of a writing instrument (pen) found in possession of the suspect'. Let this variable be binary with two states 'The ink of the questioned signature cannot be distinguished (with the techniques used by the scientist) from the ink of the suspect's pen (e_{31})' and 'The ink of the questioned signature is found to be different from the ink of the suspect's pen (e_{32})'. This is a very coarse description of the findings, accepted here for the sole purpose of illustrating the general concept of conflict tracing. In practical applications, more refined variable definitions may be retained according to the case circumstances. Let variable E_3 be directly dependant on the proposition S, defined in terms of the binary states 'The suspect is the author of the questioned signature (S)' and 'The suspect is not the author of the questioned signature (S̄)'. This leads to the network extension $S \rightarrow E_3$ (not explicitly displayed in Figure 13.8). Further suppose that the illustrative likelihoods $Pr(E_3 = e_{31}|S) = 1$ and $Pr(E_3 = e_{31}|\bar{S}) = 0.1$ are considered.*

For the scenario considered here, assume that the analysis of the writing ink of the questioned signature reveals no difference with respect to the composition of the suspect's pen. Let this finding be denoted $E_3 = e_{31}$. In the case here, there are three findings, given by $E = \{E_1 = e_{13}, E_2 = e_{22}, E_3 = e_{31}\}$, or $E = \{e_{13}, e_{22}, e_{31}\}$ for short. The following are three partial conflicts that can, therefore, be computed:

$$\text{conf}(\{E_1, E_2\}) = \log \frac{Pr(e_{13})Pr(e_{22})}{Pr(e_{13}, e_{22})} = \log \frac{0.075 \times 0.5445}{0.01634} = 0.916$$

$$\text{conf}(\{E_1, E_3\}) = \log \frac{Pr(e_{13})Pr(e_{31})}{Pr(e_{13}, e_{31})} = \log \frac{0.075 \times 0.55}{0.066} = -0.47$$

$$\text{conf}(\{E_2, E_3\}) = \log \frac{Pr(e_{22})Pr(e_{31})}{Pr(e_{22}, e_{31})} = \log \frac{0.5445 \times 0.55}{0.099} = 1.107.$$

Note that the first conflict measure is already available from Example 13.14. This example also provides the marginal probabilities $Pr(e_{13})$ and $Pr(e_{22})$ needed for the computation of the second and third conflict measures above. The introduction of the additional observation $E_3 = e_{31}$ in this example only requires one to compute three additional values, that is, the marginal $Pr(e_{31})$, $Pr(e_{13}, e_{31})$ and $Pr(e_{22}, e_{31})$, which can be obtained by standard calculations as explained previously, or more directly with the help of a Bayesian network software; these calculations are, therefore, left as an exercise.

The results of the partial conflicts computed above show that positive values are obtained for subsets that contain the finding $E_2 = e_{22}$ (i.e. 'No correspondence in fingermark features'), that is, subsets $\{e_{13}, e_{22}\}$ and $\{e_{22}, e_{31}\}$. A negative conflict value is obtained in the one setting that does not contain the result $E_2 = e_{22}$. In conclusion, the origin of the conflict can be associated with $E_2 = e_{22}$.

13.3.3   Conflict resolution

As noted in the introduction of this section, a conflict may arise in a rare case, which gives rise to a low probability for the findings. Let H be a proposition under which such findings may occur, so that in some more informal discussion H may also be referred to as an *explanation*.

The term explanation is, however, avoided in the discussion here because, as noted earlier at the end of Section 2.1.7, this term has an ambiguous connotation in forensic contexts in the sense that it may be introduced for the sole purpose of the construction of an alternative scenario, without regard as to whether such an alternative scenario can be reasonably justified within the circumstances of the case. For the current discussion, consider H thus as a proposition which, when introduced, may resolve the conflict by leading to a conflict measure $\mathrm{conf}(E, H)$ with a negative sign. According to the definition in (13.10), the conflict measure $\mathrm{conf}(E, H)$ can be written as

$$\mathrm{conf}(\{E_1, \ldots, E_n, H\}) = \log \frac{Pr(e_1) \cdot \cdots \cdot Pr(e_n)Pr(h)}{Pr(e_1, \ldots, e_n, h)}$$

$$= \log \left(\frac{Pr(e_1) \cdot \cdots \cdot Pr(e_n)}{Pr(e_1, \ldots, e_n)} \frac{Pr(h)}{Pr(h|e_1, \ldots, e_n)} \right)$$

$$= \mathrm{conf}(E) + \log \frac{Pr(h)}{Pr(h|e_1, \ldots, e_n)},$$

where h is a possible state of proposition H that could resolve the conflict. Therefore, whenever

$$\mathrm{conf}(E) + \log \frac{Pr(h)}{Pr(h|e_1, \ldots, e_n)} < 0,$$

the hypothesis H can explain away the conflict. Rearranging terms, one obtains

$$\mathrm{conf}(E) < \log \frac{Pr(h|e_1, \ldots, e_n)}{Pr(h)}.$$

Example 13.16 (*Conflict resolution*) *Consider again the questioned signature case discussed in Example 13.14, focusing again on only two findings, $E_1 = e_{13}$ and $E_2 = e_{22}$. The finding $E_1 = e_{13}$, defined as many observed similarities between the questioned signature and writings obtained from the suspect under controlled conditions, provides support for the proposition S ('The suspect is the author of the signature'), compared to the proposition $\bar{S}$ ('Some other person is the author of the signature'). The second finding, differences in ridge skin features between the suspect's fingers and the fingermarks found on the questioned document (e_{22}), supports the proposition $\bar{F}$ that the suspect is not the source of the fingermarks rather than F that the suspect is the source of the fingermarks. This support for $\bar{F}$ tends to favour the proposition $\bar{S}$ over S. This effect is strengthened by the fact that, initially, there is a predominant belief that the fingermarks are those of the author of the signature (the marginal probability of proposition G is 0.9). The findings $E = \{e_{13}, e_{22}\}$ thus point in different directions with respect to the main pair of propositions S and $\bar{S}$, and the conflict measure developed in Example 13.14 provides a value greater than 0.*

This starting point raises the question as to whether there is a proposition that, if included in the computation, would lead to a negative value for the conflict measure. Stated otherwise, the question is whether one can think of a proposition that, if known to be true, would not make one expect to obtain a conflict. In the example considered here, the proposition $\bar{G}$, defined as 'The fingermarks do not come from the author of the signature', is one such proposition. It is the negation of G, known also as the relevance term. Indeed, in Example 13.14, the marginal probabilities $Pr(G) = r = 0.9$ and $Pr(\bar{G}) = 1 - r = 0.1$ were assigned. Choosing $\bar{G}$

as a proposition that may resolve the conflict, the required computation is $\text{conf}(E \cup \{\bar{G}\})$, *defined as*

$$\text{conf}(E \cup \{\bar{G}\}) = \text{conf}(E) + \log \frac{Pr(\bar{G})}{Pr(\bar{G}|E)}$$

$$= \text{conf}(\{e_{13}, e_{22}\}) + \log \frac{Pr(\bar{G})}{Pr(\bar{G}|\{e_{13}, e_{22}\})}.$$

Proposition $\bar{G}$ could be said to resolve the conflict if the above expression has a value < 0. The result for $\text{conf}(\{e_{13}, e_{22}\})$ *is already available from Example 13.14 and is 0.916. The marginal probability $Pr(\bar{G})$ is $1 - r = 0.1$ as mentioned above. All that remains is the computation of the posterior probability $Pr(\bar{G}|\{e_{13}, e_{22}\})$. This term can be evaluated by applying Bayes' theorem in the form of*

$$[Pr(\{e_{13}, e_{22}\}|\bar{G})Pr(\bar{G})]/Pr(\{e_{13}, e_{22}\}).$$

The denominator in this ratio has already been calculated during the computation of the conflict measure (Example 13.14) and was found to be 0.01634. The first term in the numerator can be factorized as $Pr(e_{22}|e_{13}, \bar{G})Pr(e_{13})$, and be shown to be equal to $(1 - \gamma) \times Pr(e_{13})$, where γ is short for $Pr(e_{21}|\bar{F})$, the rarity of a correspondence in fingermark features (previously assigned the value 0.01). The marginal $Pr(e_{13})$ is also available from the previous computation of the conflict measure and takes the value 0.075. Bayes' theorem for calculation of the posterior probability $Pr(\bar{G}|\{e_{13}, e_{22}\})$ can now be applied as follows: $[(1 - 0.01) \times 0.075 \times (1 - 0.9)]/0.01634 = 0.45455$.

On the basis of the above results, the term $\log(Pr(\bar{G})/Pr(\bar{G}|E))$ can now be computed as $\log(0.1/0.45455) = -1.514$. Substitution of this result in the expression of the conflict measure leads to

$$\text{conf}(E \cup \{\bar{G}\}) = \text{conf}(\{e_{13}, e_{22}\}) + \log \frac{Pr(\bar{G})}{Pr(\bar{G}|\{e_{13}, e_{22}\})}$$

$$= 0.916 - 1.514 = -0.598.$$

This negative result for the conflict measure indicates that under the assumption of the fingermarks not being relevant to the case under investigation (i.e. they do not come from the author of the signature), that is, proposition $\bar{G}$, one would not expect a conflict for the findings e_{13} (many corresponding handwriting features) and e_{22} (no correspondence in ridge skin features between the fingermark and the suspect's control prints).

For the purpose of exploring the concept, suppose now proposition G is known to be true, rather than $\bar{G}$. Proposition G states that the fingermarks come from the author of the questioned signature. In such a situation, one would expect the conflict to be strengthened. This can be verified by computing the expression:

$$\text{conf}(E \cup \{G\}) = \text{conf}(\{e_{13}, e_{22}\}) + \log \frac{Pr(G)}{Pr(G|\{e_{13}, e_{22}\})}$$

$$= 0.916 + \log \frac{0.9}{0.54545}$$

$$= 0.916 + 0.501 = 1.417.$$

The value 0.916 of the conflict measure found in Example 13.14 for findings e_{13} and e_{22} is found to be increased if, in addition, the fingermarks are considered relevant (i.e. proposition G is assumed to be true). Note that, in the above computation, the probabilities $Pr(G)$ and $Pr(G|\{e_{13}, e_{22}\})$ are given by, respectively, $1 - Pr(\bar{G})$ and $1 - Pr(\bar{G}|\{e_{13}, e_{22}\})$, because proposition G is binary.

The approach to conflict analysis pursued throughout this section has the advantage of requiring relatively standard terms only, which makes it attractive and the necessary computations can readily be performed. It is worth noting, however, that this method may not be able to detect conflicts when the assumption of positive (negative) correlation between findings does not hold. Other methods have been proposed that are based on more complex straw (Kim and Valtorta 1995).

References

Aitken CGG (1995) *Statistics and the Evaluation of Evidence for Forensic Scientists*. John Wiley & Sons, Ltd, Chichester.

Aitken CGG (1999) Sampling - how big a sample? *Journal of Forensic Sciences* **44**, 750–760.

Aitken CGG (2005) Sampling issues in the administration of justice. *Significance* **2**, 24–27.

Aitken CGG, Connolly T, Gammerman A, Zhang G, Bailey D, Gordon R and Oldfield R (1996a) Statistical modelling in specific case analysis. *Science & Justice* **36**, 245–255.

Aitken CGG, Gammerman A, Zhang G, Connolly T, Bailey D, Gordon R and Oldfield R (1996b) Bayesian belief networks with an application in specifc case analysis In *Computational Learning and Probabilistic Reasoning* (ed. Gammerman A) John Wiley & Sons, Ltd, Chichester pp. 169–184.

Aitken CGG and Gammerman A (1989) Probabilistic reasoning in evidential assessment. *Journal of the Forensic Science Society* **29**, 303–316.

Aitken CGG and Gold E (2013) Evidence evaluation for discrete data. *Forensic Science International* **230**, 147–155.

Aitken CGG and Taroni F (1997) A contribution to the discussion on 'Bayesian analysis of the deoxyribonucleic acid profiling data in forensic identification applications', foreman et al. *Journal of the Royal Statistical Society, Series A* **160**, 463.

Aitken CGG and Taroni F (2004) *Statistics and the Evaluation of Evidence for Forensic Scientists* second edn. John Wiley & Sons, Ltd, Chichester.

Aitken CGG, Taroni F and Garbolino P (2003) A graphical model for the evaluation of cross-transfer evidence in DNA profiles. *Theoretical Population Biology* **63**, 179–190.

Anderson T, Schum DA and Twining W (2005) *Analysis of Evidence Law in Context* second edn. Cambridge University Press, Cambridge.

Anderson T and Twining W (1998) *Analysis of Evidence: How to do Things with Facts Based on Wigmore's Science of Judicial Proof*. Northwestern University Press, Evanston, IL.

Andrasko J and Maehly AC (1977) Detection of GSR on hands by Scanning Electron Microscopy. *Journal of Forensic Sciences* **22**, 279–287.

Balding DJ (2000) Interpreting DNA evidence: can probability theory help? In *Statistical Science in the Courtroom* (ed. Gastwirth JL) Statistics for Social Science and Public Policy Springer, New York, pp. 51–70.

Balding DJ (2005) *Weight-of-Evidence for Forensic DNA Profiles*. John Wiley & Sons, Ltd, Chichester.

Bayesian Networks for Probabilistic Inference and Decision Analysis in Forensic Science, Second Edition.
Franco Taroni, Alex Biedermann, Silvia Bozza, Paolo Garbolino and Colin Aitken.
© 2014 John Wiley & Sons, Ltd. Published 2014 by John Wiley & Sons, Ltd.

Balding DJ and Donnelly P (1995) Inference in forensic identification. *Journal of the Royal Statistical Society, Series A* **158**, 21–53.

Balding DJ and Donnelly P (1996) Evaluating DNA profile evidence when the suspect is identified through a database search. *Journal of Forensic Sciences* **41**, 603–607.

Balding DJ, Donnelly P and Nichols RA (1994) Some causes for concern about DNA profiles - comment. *Statistical Science* **9**, 248–251.

Balding DJ and Nichols RA (1994) DNA profile match probability calculation: how to allow for population stratification, relatedness, database selection and single bands. *Forensic Science International* **64**, 125–140.

Balding DJ and Nichols RA (1997) Significant genetic correlations among Caucasians at forensic DNA loci. *Hereditiy* **78**, 583–589.

Bangsø O and Wuillemin PH (2000) Object Oriented Bayesian networks: a framework for top-down specification of large Bayesian networks with repetitive structures. Technical report, Hewlett-Packard Laboratory for Normative Systems, Aalborg University, Aalborg.

Bayes T (1763) An essay towards solving a problem in the doctrine of chances In *Studies in the History of Statistics and Probability (1970)* (ed. Pearson ES and Kendall MG) vol. **1** Griffin, London pp. 134–153.

Berger JO (1985) *Statistical Decision Theory and Bayesian Analysis* second edn. Springer, New York.

Berger JO (1988) *Statistical Decision Theory and Bayesian Analysis* second edn. Springer, New York.

Berk RE, Rochowicz SA, Wong M and Kopina MA (2007) Gunshot residue in Chicago police vehicles and facilities: an empirical study. *Journal of Forensic Sciences* **52**, 838–841.

Bernardo JM and Smith AFM (1994) *Bayesian Theory*. John Wiley & Sons, Ltd, Chichester.

Bernardo JM and Smith AFM (2000) *Bayesian Theory* second edn. John Wiley & Sons, Ltd, Chichester.

Berry DA (1994) [DNA fingerprinting: a review of the controversy]: comment. *Statistical Science* **9**, 252–255.

Berzuini C, Dawid, P and Bernardinelli L (2012) *Causality: Statistical Perspectives and Applications*. John Wiley & Sons, Ltd, Chichester.

Biasotti AA (1959) A statistical study of the individual characteristics of fired bullets. *Journal of Forensic Sciences* **4**, 34–50.

Biasotti AA and Murdock J (1997) Firearms and toolmark identification: the scientific basis of firearms and toolmark identification In *Modern Scientific Evidence: The Law and Science of Expert Testimony* (ed. Faigman L, Kaye DH, Saks MJ and Sanders J) West Publishing, St.Paul, MN pp. 144–150.

Biedermann A, Bozza S and Taroni F (2008a) Decision theoretic properties of forensic identification: underlying logic and argumentative implications. *Forensic Science International* **177**, 120–132.

Biedermann A, Taroni F, Bozza S and Aitken CGG (2008b) Analysis of sampling issues using Bayesian networks. *Law, Probability and Risk* **7**, 35–60.

Biedermann A, Bozza S and Taroni F (2009) Implementing statistical learning methods through Bayesian networks (Part I): a guide to Bayesian parameter estimation using forensic science data. *Forensic Science International* **193**, 63–71.

Biedermann A, Bozza S, Garbolino P and Taroni F (2012a) Decision-theoretic analysis of forensic sampling criteria using Bayesian decision networks. *Forensic Science International* **223**, 217–227.

Biedermann A, Bozza S, Konis K and Taroni F (2012b) Inference about the number of contributors to a DNA mixture: comparative analyses of a Bayesian network approach and the maximum allele count method. *Forensic Science International: Genetics* **6**, 689–696.

Biedermann A, Taroni F and Champod C (2012c) How to assign a likelihood ratio in a footwear mark case: an analysis and discussion in the light of R v T. *Law, Probability and Risk* **11**, 259–277.

Biedermann A, Voisard R and Taroni F (2012d) Learning about Bayesian networks for forensic interpretation: an example based on the 'the problem of multiple propositions'. *Science & Justice* **52**, 191–198.

Biedermann A, Vuille J and Taroni F (2012e) A Bayesian network approach to the database search problem in criminal proceedings. *Investigative Genetics* **3**, 1–16.

Biedermann A, Garbolino P and Taroni F (2013) The subjectivist interpretation of probability and the problem of individualisation in forensic science. *Science & Justice* **53**, 192–200.

Biedermann A, Gittelson S and Taroni F (2011a) Recent misconceptions about the 'database search problem': a probabilistic analysis using Bayesian networks. *Forensic Science International* **212**, 51–60.

Biedermann A, Taroni F and Thompson WC (2011b) Using graphical probability analysis (Bayes nets) to evaluate a conditional DNA inclusion. *Law, Probability and Risk* **10**, 89–121.

Biedermann A and Taroni F (2004) Bayesian networks and the assessment of scientific evidence: the use of qualitative probabilistic networks and sensitivity analyses Proceedings of the 14th Interpol Forensic Science Symposium, Lyon, France.

Biedermann A and Taroni F (2006) A probabilistic approach to the joint evaluation of firearm evidence and gunshot residues. *Forensic Science International* **163**, 18–33.

Biedermann A and Taroni F (2011) Evidential relevance in scene to offender transfer cases: development and analysis of a likelihood ratio for offence level propositions. *Law, Probability and Risk* **10**, 277–301.

Biedermann A, Taroni F, Delémont O, Semadeni C and Davison AC (2005a) The evaluation of evidence in the forensic investigation of fire incidents (Part I): an approach using Bayesian networks. *Forensic Science International* **147**, 49–57.

Biedermann A, Taroni F, Delémont O, Semadeni C and Davison AC (2005b) The evaluation of evidence in the forensic investigation of fire incidents (Part II): practical examples of the use of Bayesian networks. *Forensic Science International* **147**, 56–69.

Biedermann A, Taroni F and Garbolino P (2007) Equal prior probabilities: can one do any better? *Forensic Science International* **172**, 85–93.

Booth G, Johnston F and Jackson G (2002) Case assessment and interpretation - application to a drugs supply case. *Science & Justice* **42**, 123–125.

Bovens L, Fitelson B, Hartmann S and Snyder J (2002) Too odd (not) to be true? A reply to Olsson. *British Journal for the Philosophy of Science* **53**, 539–563.

Bovens L and Hartmann S (2003) *Bayesian Epistemology*. Clarendon Press, Oxford.

Box GEP and Draper NR (1987) *Empirical Model Building and Response Surfaces*. John Wiley & Sons, Inc., New York.

Brier GW (1950) Verification of forecasts expressed in terms of probability. *Monthly Weather Review* **78**, 1–3.

Bruijning-van Dongen CJ, Slooten K, Burgers W and Wiegerinck W (2009) Bayesian networks for victim identification on the basis of DNA profiles *Forensic Science International: Genetics Supplement Series* **2**, 466–468. Progress in Forensic Genetics 13 - Proceedings of the 23rd International ISFG Congress.

Buckleton JS, Evett IW and Weir BS (1998) Setting bounds for the likelihood ratio when multiple hypotheses are postulated. *Science & Justice* **38**, 23–26.

Buckleton JS and Triggs CM (2005) Relatedness and DNA: are we taking it seriously enough? *Forensic Science International* **152**, 115–119.

Buckleton JS, Triggs CM and Champod C (2006) An extended likelihood ratio framework for interpreting evidence. *Science & Justice* **46**, 69–78.

Buckleton JS, Triggs CM and Walsh SJ (2005) *Forensic DNA Evidence Interpretation*. CRC Press, Boca Raton, FL.

Bunch SG (2000) Consecutive matching striation criteria: a general critique. *Journal of Forensic Sciences* **45**, 955–962.

Butler JM, Schoske R, Vallone PM, Redman JW and Kline MC (2003) Allele frequencies of 15 autosomal STR loci on U.S. Caucasian, African American and Hispanic populations. *Journal of Forensic Sciences* **48**, 908–911.

Cardinetti B, Ciampini C, Abate S, Marchetti C, Ferrari F, Di Tullio D, D'Onofrio C, Orlando G, Gravina L, Torresi L and Saporita G (2006) A proposal for statistical evaluation of the detection of gunshot residues on a suspect. *Scanning* **28**, 142–147.

Castillo E, Gutierrez JM and Hadi S (1997) *Expert Systems and Probabilistic Network Models*. Springer, New York.

Cavallini D and Corradi F (2006) Forensic identification of relatives of individuals included in a database of DNA profiles. *Biometrika* **93**, 525–536.

Cereda G, Biedermann A, Hall D and Taroni F (2014) Object-oriented Bayesian networks for evaluating DIP-STR profiling results from unbalanced DNA mixtures. *Forensic Science International: Genetics* **8**, 118–133.

Champod C (2010) The use of probabilistic networks in the area of fingerprints Impression and Pattern Evidence Symposium, Clearwater, FL.

Champod C and Evett IW (2001) A probabilistic approach to fingerprint evidence. *Journal of Forensic Identification* **51**, 101–122.

Champod C and Jackson G (2000) European fibres group workshop: case assessment and Bayesian interpretation of fibres evidence Proceedings of the 8th Meeting of European Fibres Group, pp. 33–45, Krakow.

Champod C, Lennard C, Margot P and Stoilovic M (2004) *Fingerprints and Other Ridge Skin Impressions*. CRC Press, Boca Raton, FL.

Champod C and Taroni F (1999) The Bayesian approach In *Forensic Examination of Fibres* (ed. Robertson J and Grieve M) Taylor and Francis, London pp. 379–398.

Charniak E (1991) Bayesian networks without tears. *AI Magazine* **12**, 50–63.

Cohen JL (1977) *The Probable and the Provable*. Clarendon Press, Oxford.

Cohen JL (1988) The difficulty about conjunction in forensic proof. *Statistician* **37**, 415–416.

Conan Doyle SA (1953) *The Complete Sherlock Holmes*. Doubleday & Company, Inc., Garden City.

Cook R, Evett IW, Jackson G, Jones PJ and Lambert JA (1998a) A hierarchy of propositions: deciding which level to address in casework. *Science & Justice* **38**, 231–239.

Cook R, Evett IW, Jackson G, Jones PJ and Lambert JA (1998b) A model for case assessment and interpretation. *Science & Justice* **38**, 151–156.

Cook R, Evett IW, Jackson G, Jones PJ and Lambert JA (1999) Case pre-assessment and review in a two-way transfer case. *Science & Justice* **39**, 103–111.

Cooper GF (1999) An overview of the representation and discovery of causal relationships using Bayesian networks In *Computation, Causation, and Discovery* (ed. Glymour C and Cooper GF) The MIT Press, Cambridge, MA pp. 3–62.

Corradi F, Lago G and Stefanini FM (2003) The evaluation of DNA evidence in pedigrees requiring population inference. *Journal of the Royal Statistical Society, Series A* **166**, 425–440.

Corradi F and Ricciardi F (2013) Evaluation of kinship identification systems based on short tandem repeat DNA profiles. *Applied Statistics* **62**, 649–668.

Coupé VMH and Van der Gaag LC (1998) Practicable sensitivity analysis of Bayesian belief networks In *Prague Stochastics '98 - Proceedings of the Joint Session of the 6th Prague Symposium of Asymptotic Statistics and the 13th Prague Conference on Information Theory, Statistical Decision Functions and Random Processes* (ed. Hušková M, Lachout P and Víšek JA) Union of Czech Mathematicians and Physicists, Prague, pp. 81–86.

Cowell RG (2003) FINEX: a probabilistic expert system for forensic identification. *Forensic Science International* **134**, 196–206.

Cowell RG, Dawid AP, Lauritzen SL and Spiegelhalter DJ (1999) *Probabilistic Networks and Expert Systems*. Springer, New York.

Cowell RG, Lauritzen SL and Mortera J (2007a) A Gamma model for DNA mixture analyses. *Bayesian Analysis* **2**, 333–348.

Cowell RG, Lauritzen SL and Mortera J (2007b) Identification and separation of DNA mixtures using peak area information. *Forensic Science International* **166**, 28–34.

Cowell RG, Lauritzen SL and Mortera J (2008) Probabilistic modelling for DNA mixture analysis. *Forensic Science International: Genetics Supplement Series* **1**, 640–642. Progress in Forensic Genetics 12 - Proceedings of the 22nd International ISFG Congress.

Cowell RG, Lauritzen SL and Mortera J (2011) Probabilistic expert systems for handling artifacts in complex DNA mixtures. *Forensic Science International: Genetics* **5**, 202–209.

Curran JM, Hicks TN and Buckleton JS (2000) *Forensic Interpretation of Glass Evidence*. CRC Press, Boca Raton, FL.

Curran JM, Triggs CM, Buckleton J, Walsh KAJ and Hicks T (1998) Assessing transfer probabilities in a Bayesian interpretation of forensic glass evidence. *Science & Justice* **38**, 15–21.

D'Agostini GD (2003) *Bayesian Reasoning in Data Analysis: A Critical Introduction*. World Scientific, New York.

Darboux JG, Appell PE and Poincaré JH (1908) *Examen critique des divers systèmes ou études graphologiques auxquels a donné lieu le bordereau*. In L'affaire Dreyfus - La révision du procès de Rennes - Enquête de la chambre criminelle de la Cour de Cassation Ligue française des droits de l'homme et du citoyen Paris.

Daston L (1988) *Classical Probability in the Enlightment*. Princeton University Press, Princeton, NJ.

Dawid AP (1987) The difficulty about conjunction. *Statistician* **36**, 91 97.

Dawid AP (2000) Causal inference without counterfactuals (with discussion). *Journal of the American Statistical Association* **95**, 407–448.

Dawid AP (2001) Comment on Stockmarr's 'Likelihood ratios for evaluating DNA evidence, when the suspect is found through a database search'. *Biometrics* **57**, 976–978.

Dawid AP (2002) Influence diagrams for causal modelling and inference. *International Statistical Review* **70**, 161–189.

Dawid AP (2003a) An object-oriented Bayesian network for estimating mutation rates. Proceedings of the Ninth International Workshop on Artificial Intelligence and Statistics, Key West, Florida (FL).

Dawid AP (2003b) An object-oriented Bayesian network for estimating mutation rates. Technical Report 226, Department of Statistical Science, University College London.

Dawid AP, van Boxel DW, Mortera J and Pascali VL (1999) Inference about disputed paternity from an incomplete pedigree using a probabilistic expert system. *Bulletin of the International Statistical Institute* **58** (Book 1), 241–242.

Dawid AP and Evett IW (1997) Using a graphical method to assist the evaluation of complicated patterns of evidence. *Journal of Forensic Sciences* **42**, 226–231.

Dawid AP and Mortera J (1996) Coherent analysis of forensic identification evidence. *Journal of the Royal Statistical Society, Series B* **58**, 425–443.

Dawid AP, Mortera J, Pascali VL and van Boxel D (2002) Probabilistic expert systems for forensic inference from genetic markers. *Scandinavian Journal of Statistics* **29**, 577–595.

Dawid AP, Mortera J and Vicard P (2005) Object-oriented Bayesian networks for complex forensic DNA profiling problems. Technical Report 256, Departement of Statistical Science, University College London.

Dawid AP, Mortera J and Vicard P (2007) Object-oriented Bayesian networks for complex forensic DNA profiling problems. *Forensic Science International* **169**, 195–205.

Dawid AP and Vovk V (1999) Prequential probability: principles and properties. *Bernoulli* **5**, 125–162.

DeGroot MH (1970) *Optimal Statistical Decisions* Wiley Classics Library Edition Published 2004. John Wiley & Sons, Inc., Hoboken, NJ.

DeGroot MH and Fienberg SE (1982) The comparison and evaluation of forecasters. *Journal of the Royal Statistical Society. Series D (The Statistician)* **32**, 12–22.

Diaconis P and Freedman D (1981) The persistence of cognitive illusions. *Behavioural and Brain Sciences* **4**, 333–334.

Diaconis P and Zabell S (1982) Updating subjective probability. *Journal of the American Statistical Association* **77**, 882–830.

Dillon JH (1990) The Modified Griess Test: a chemically specific chromophoric test for nitrite compounds in gunshot residues. *AFTE Journal* **22**, 243–250.

Domotor Z, Zanotti M and Graves H (1980) Probability kinematics. *Synthese* **44**, 421–442.

Donkin WF (1851) On certain questions relating to the theory of probability. *The London, Edinburgh and Dublin Philosophical Magazine and Journal of Science, Series 4* **1**, 353–368.

Donnelly P and Friedman RD (1999) DNA database searches and the legal consumption of scientific evidence. *Michigan Law Review* **97**, 931–984.

Druzdzel MJ (1996) Explanation in probabilistic systems: is it feasible? Will it work? Proceedings of the Workshop Intelligent Information Systems V, pp. 1–13, Deblin, Poland.

Druzdzel MJ (1997) Five useful properties of probabilistic knowledge representations from the point of view of intelligent systems. *Fundamenta Informaticae* **30**, 241–254. Special Issue on Knowledge Representation and Machine Learning.

Druzdzel MJ and Henrion M (1993a) Belief propagation in qualitative probabilistic networks In *Qualitative Reasoning and Decision Technologies* (ed. Carrete NP and Singh MG) CIMNE, Barcelona pp. 451–460.

Druzdzel MJ and Henrion M (1993b) Efficient reasoning in qualitative probabilistic networks Proceedings of the 11th Annual Conference on Artificial Intelligence (AAAI-93), pp. 548–553, Washington, DC.

Druzdzel MJ and Henrion M (1993c) Intercausal reasoning with uninstantiated ancestor nodes Proceedings of the Ninth Annual Conference on Uncertainty in Artificial Intelligence (UAI-93), pp. 317–325, Washington, DC.

Earman J (1992) *Bayes or Bust?* The MIT Press, Cambridge, MA.

Edwards W (1954) The theory of decision making. *Psychological Bulletin* **51**, 380–417.

Edwards AWF (1987) *Pascal's Arithmetical Triangle.* Griffin & Oxford University Press, London, New York.

Edwards W (1991) Influence diagrams, Bayesian imperialism, and the *Collins* case: an appeal to reason. *Cardozo Law Review* **13**, 1025–1074.

ENFSI Drugs Working Group (2004) Guidelines on representative drug sampling Den Haag.

Evett IW (1984) A quantitative theory for interpreting transfer evidence in criminal cases. *Applied Statistics* **33**, 25–32.

Evett IW (1987) On meaningful questions: a two-trace transfer problem. *Journal of the Forensic Science Society* **27**, 375–381.

Evett IW (1992) Evaluating DNA profiles in a case where the defence is 'It was my brother'. *Journal of the Forensic Science Society* **32**, 5–14.

Evett IW (1993) Establishing the evidential value of a small quantity of material found at a crime scene. *Journal of the Forensic Science Society* **33**, 83–86.

Evett IW, Foreman LA and Weir BS (2000a) Letter to the editor. *Biometrics* **56**, 1274–1277.

Evett IW, Jackson G and Lambert JA (2000b) More on the hierarchy of propositions: exploring the distinction between explanations and propositions. *Science & Justice* **40**, 3–10.

Evett IW, Jackson G, Lambert JA and McCrossan S (2000c) The impact of the principles of evidence interpretation and the structure and content of statements. *Science & Justice* **40**, 233–239.

Evett IW, Gill PD, Jackson G, Whitaker J and Champod C (2002) Interpreting small quantities of DNA: the hierarchy of propositions and the use of Bayesian networks. *Journal of Forensic Sciences* **47**, 520–530.

Evett IW, Gill P and Lambert JA (1998a) Taking accout of peak areas when interpreting mixed DNA profiles. *Journal of Forensic Sciences* **43**, 62–69.

Evett IW, Lambert JA and Buckleton JS (1995) Further observations on glass evidence interpretation. *Science & Justice* **35**, 283–289.

Evett IW, Lambert JA and Buckleton JS (1998b) A Bayesian approach to interpreting footwear marks in forensic casework. *Science & Justice* **38**, 241–247.

Evett IW and Weir BS (1998) *Interpreting DNA Evidence*. Sinauer Associates Inc., Sunderland.

Fenton N, Berger D, Lagnado D, Neil M and Hsu A (2013) When 'neutral' evidence still has probative value (with implications from the Barry George Case). *Science & Justice*, in press.

Fenton N and Neil M (2011) Avoiding legal fallacies in practice using Bayesian networks. *Australian Journal of Legal Philosophy* **36**, 114–151.

Fenton N and Neil M (2013) *Risk Assessment and Decision Analysis with Bayesian Networks*. CRC Press, Boca Raton, FL.

de Finetti B (1930a) Fondamenti logici del ragionamento probabilistico. *Bollettino della Unione matematica Italiana* **9**, 258–261.

de Finetti B (1930b) Funzione caratteristica di un fenomeno aleatorio. *Memorie della Regia Accademia dei Lincei* **4**, 86–133.

de Finetti B (1937) La prévision: ses lois logiques, ses sources subjectives. *Annales de l'Institut Henri Poincaré* **7**, 1–68 (english translation) In *Studies in Subjective Probability (1980)* (ed. Kyburg HE and Smokler HE) second edn. Dover, New York pp. 93–158.

de Finetti B (1962) Does it make sense to speak of 'Good Probability Appraisers'? In *The Scientist Speculates: An Anthology of Partly-Baked Ideas* (ed. Good IJ) Basic Books, New York pp. 357–364.

de Finetti B (1968) Probability: the subjective approach In *La Philosophie Contemporaine* (ed. Klibansky R) La Nuova Firenze, Firenze pp. 45–53.

de Finetti B (1972) *Probability, Induction and Statistics, The Art of Guessing*. John Wiley & Sons, Inc., New York.

de Finetti B (1975) *Theory of Probability, A Critical Introductory Treatment*. vol. 2. John Wiley & Sons, Inc., New York.

de Finetti B (1982) The proper approach to probability In *Exchangeability in Probability and Statistics* (ed. Koch G and Spizzichino K) North-Holland Publishing Company, Amsterdam pp. 1–6.

de Finetti B (1989) La probabilità: guardarsi dalle contraffazioni! In *La logica dell'incerto* (ed. Mondadori M) Il Saggiatore, Milano pp. 149–188.

Finkelstein MO and Fairley WB (1970) A Bayesian approach to identification evidence. *Harvard Law Review* **83**, 489–517.

van Fraassen BC (1986) A demonstration of Jeffrey conditionalization rule. *Erkenntnis* **24**, 17–24.

van Fraassen BC (1989) *Laws and Symmetry*. Clarendon Press, Oxford.

Friedman RD (1986a) A close look at probative value. *Boston University Law Review* **66**, 733–759.

Friedman RD (1986b) A diagrammatic approach to evidence. *Boston University Law Review* **66**, 571–622.

Friedman RD (1996) Assessing evidence. *Michigan Law Review* **94**, 1810–1838.

Friedman N and Goldszmidt M (1996) Learning Bayesian networks with local structure In *Uncertainty in Artificial Intelligence: Proceedings of the Twelfth Conference* (ed. Horvitz E and Jensen FV) Morgan Kaufmann, San Mateo, CA pp. 252–262.

Galavotti MC (2005) *Philosophical Introduction to Probability*. CSLI Publications, Stanford, CA.

Garbolino P (2001) Explaining relevance. *Cardozo Law Review* **22**, 1503–1521.

Garbolino P and Taroni F (2002) Evaluation of scientific evidence using Bayesian networks. *Forensic Science International* **125**, 149–155.

Gärdenfors P, Hansson B and Sahlin NE (1983) *Evidentiary Value: Philosophical, Judicial and Psychological Aspects of a Theory: Essays Dedicated to Sören Halldén on His Sixtieth Birthday.* Gleerups, Lund.

Geiger D and Heckerman D (1991) Advances in probabilistic reasoning In *Uncertainty in Artificial Intelligence: Proceedings of the Seventh Conference* (ed. Bonissone P, D'Ambrosio B and Smets P) Morgan Kaufmann, Los Angeles, pp. 118–126.

Gill P, Puch-Solis R and Curran J (2009) The *low-template-DNA* (stochastic) threshold - its determination relative to risk analysis for national DNA databases. *Forensic Science International: Genetics* **3**, 104–111.

Gittelson S, Biedermann A, Bozza S and Taroni F (2012) The database search problem: a question of rational decision making. *Forensic Science International* **222**, 186–199.

Gittelson S, Biedermann A, Bozza S and Taroni F (2013a) Modeling the forensic two-trace problem with Bayesian networks. *Artificial Intelligence and Law* **21**, 221–252.

Gittelson S, Bozza S, Biedermann A and Taroni F (2013b) Decision-theoretic reflections on processing a fingermark. *Forensic Science International* **226**, e42–e47.

Gittelson S, Biedermann A, Bozza S and Taroni F (2014) Decision analysis for the genotype designation in low-template-DNA profiles. *Forensic Science International: Genetics* **9**, 118–133.

Good IJ (1950) *Probability and the Weighing of Evidence.* Griffin, London.

Good IJ (1952) Rational decisions. *Journal of the Royal Statistical Society. Series B (Methodological)* **14**, 107–114.

Good IJ (1985) Weight of evidence: a brief survey (with discussion) In *Bayesian Statistics 2* (ed. Bernardo JM, DeGroot MH, Lindley DV and Smith AFM) Elsevier Science Publishers B.V., North Holland pp. 249–270.

Green PJ and Mortera J (2009) Sensitivity of inferences in forensic genetics to assumptions about founding genes. *Annals of Applied Statistics* **3**, 731–763.

Haag LC (2006) *Shooting Incident Reconstruction.* Elsevier, Amsterdam.

Hacking I (1975) *The Emergence of Probability.* Cambridge University Press, Cambridge.

Halliwell J, Keppens J and Shen Q (2003) Linguistic Bayesian networks for reasoning with subjective probabilities in forensic statistics 9th International Conference on Artificial Intelligence and Law (ICAIL 2003), pp. 42–50, Edinburgh.

Harbison SA and Buckleton JS (1998) Applications and extensions of subpopulation theory: a caseworkers guide. *Science & Justice* **38**, 249–254.

Harman G (1965) The inference to the best explanation. *Philosophical Review* **77**, 88–95.

Hatsch D, Keyser C, Hienne R and Bertrand L (2007) Resolving paternity relationships using X-chromosome STRs and Bayesian networks. *Journal of Forensic Sciences* **52**, 895–897.

Hempel CG (1965) *Aspects of Scientific Explanation and Other Essays in the Philosophy of Science.* The Free Press, New York.

Henrion M (1986) Uncertainty in artificial intelligence: is probability epistemically and heuristically adequate? In *Expert Judgement and Expert Systems (NATO ASI Series F)* (ed. Mumpower J, Phillips LD, Renn O and Uppuluri VRR) vol. **35** Springer Berlin pp. 105–130.

Henrion M and Druzdzel MJ (1991) Qualitative propagation and scenario-based schemes for explaining probabilistic reasoning In *Uncertainty in Artificial Intelligence 6* (ed. Bonissone PP, Henrion M, Kanal LN and Lemmer JF) Elsevier, North Holland pp. 17–32.

Hepler A, Dawid AP and Leucari V (2007) Object-oriented graphical representations of complex patterns of evidence. *Law Probability and Risk* **6**, 275–293.

Hepler A and Weir BS (2007) Object-oriented Bayesian networks for paternity cases with allelic dependencies. Technical Report 281, Departement of Statistical Science, University College London, London.

Hepler A and Weir BS (2008) Object-oriented Bayesian networks for paternity cases with allelic dependencies. *Forensic Science International: Genetics* **2**, 166–175.

Hitchcock C (2012) Probabilistic causation In *Stanford Encyclopedia of Philosophy* (ed. Zalta EN) http://plato.stanford.edu/entries/causation-probabilistic/ (last accessed May 2014).

Højsgaard S, Edwards D and Lauritzen S (2012) *Graphical Models with R*. Springer, Boston, MA.

Horwich P (1982) *Probability and Evidence*. Cambridge University Press, Cambridge.

Howard R and Matheson J (1984) Influence diagrams In *Readings on the Principles and Applications of Decision Analysis* (ed. Howard RA and Matheson JE) Vol. **2** Strategic Decisions Group, Menlo Park, CA pp. 719–762.

Howson C and Urbach P (1996) *Scientific Reasoning: The Bayesian Approach* second edn. Open Court, La Salle.

Hummel K (1971) Biostatistical opinion of parentage based upon the results of blood group tests In *Biostatistische Abstammungsbegutachtung mit Blutgruppenbefunden* (ed. Schmidt P) Gustav Fisher, Stuttgart.

Hummel K (1983) Selection of gene frequency tables In *Inclusion Probabilities in Parentage Testing* (ed. Walker R) American Association of Blook Banks, Arlington, TX.

Jackson G (2000) The scientist and the scales of justice. *Science & Justice* **40**, 81–85.

Jalanti T, Henchoz P, Gallusser Λ and Bonfanti MS (1999) The persistence of gunshot residue on shooters' hands. *Science & Justice* **39**, 48–52.

Jaynes ET (1983) Where do we stand on Maximum Entropy? In *Papers on Probability, Statistics and Statistical Physics* (ed. Rosenkrantz RD) Reidel Publishing, Dordrecht pp. 211–314.

Jaynes ET (2003) *Probability Theory: The Logic of Science*. Cambridge University Press, Cambridge.

Jeffrey RC (1983) *The Logic of Decision* second edn. University of Chicago Press, Chicago.

Jeffrey RC (1992) *Probability and the Art of Judgment*. Cambridge University Press, Cambridge.

Jeffrey RC (2004) *Subjective Probability, The Real Thing*. Cambridge University Press, Cambridge.

Jensen FV (1996) *An Introduction to Bayesian Networks*. University College London Press, London.

Jensen FV (2001) *Bayesian Networks and Decision Graphs*. Springer, New York.

Jensen FV, Lauritzen SL and Olesen KG (1990) Bayesian updating in causal probabilistic networks by local computation. *Computational Statistics Quarterly* **4**, 269–282.

Jensen FV and Nielsen TD (2007) *Bayesian Networks and Decision Graphs* second edn. Springer, New York.

Juchli P, Biedermann A and Taroni F (2012) Graphical probabilistic analysis of the combination of items of evidence. *Law, Probability and Risk* **11**, 51–84.

Kadane JB and Schum DA (1996) *A Probabilistic Analysis of the Sacco and Vanzetti Evidence*. John Wiley & Sons, Inc., New York.

Katterwe H (2003) True or false. *Information Bulletin for Shoeprint/Toolmark Examiners* **9**, 18–25.

Kaye DH (1987) The validity of tests: caveant omnes. *Jurimetrics Journal* **27**, 349–361.

Kaye DH (1988) What is Bayesianism? In *Probability and Inference in the Law of Evidence, The Uses and Limits of Bayesianism (Boston Studies in the Philosophy of Science)* (ed. Tillers P and Green ED) Springer, Dordrecht pp. 1–19.

Kaye DH (2009) Rounding up the usual suspects: a logical and legal analysis of DNA trawling cases. *North Carolina Law Review* **87**, 425–503.

Kilty JW (1975) Activity after shooting and its effect on the retention of primer GSR. *Journal of Forensic Sciences* **20**, 219–230.

Kim JH and Pearl J (1983) A computational model for causal and diagnostic reasoning in inference systems Proceedings of IJCAI-83, Karlsruhe, West Germany, pp. 190–193. Morgan Kaufmann, Los Altos, CA.

Kim Y and Valtorta M (1995) On the detection of conflicts in diagnostic Bayesian networks using abstraction Proceedings of the Eleventh Conference on Uncertainty in Artificial Intelligence, pp. 362–367.

Kind SS (1994) Crime investigation and the criminal trial: a three chapter paradigm of evidence. *Journal of the Forensic Science Society* **35**, 155–164.

Kirk PL and Kingston CR (1964) Evidence evaluation and problems in general criminalistics 16th Annual Meeting of the American Academy of Forensic Sciences, Chicago.

Kjærulff UB and Madsen AL (2008) *Bayesian Networks and Influence Diagrams, A Guide to Construction and Analysis*. Springer, New York.

Knechtle P and Gallusser A (1996) La persistance des résidus de tir sur les mains selon l'activité du tireur. *Revue Internationale de Criminologie et de Police Technique* **49**, 228–246.

Köller N, Nissen K, Riess M and Sadorf E (2004) *Probability Conclusions in Expert Opinions on Handwriting. Substantiation and Standardization of Probability Statements in Expert Opinions*. Luchterhand, München.

Koller D and Pfeffer A (1997) Object-Oriented Bayesian networks Proceedings of the Thirteenth Annual Conference on Uncertainty in Artificial Intelligence (UAI-97), pp. 302–313.

Korb KB and Nicholson AE (2011) *Bayesian Artificial Intelligence* second edn. Chapman & Hall/CRC, Boca Raton, FL.

Kullback S (1959) *Information Theory and Statistics*. John Wiley & Sons, New York.

Laskey KB and Mahoney SM (1997) Network fragments: representing knowledge for constructing probabilistic models Proceedings of the Thirteenth Annual Conference on Uncertainty in Artificial Intelligence (UAI-97), pp. 334–341.

Lauritzen SL and Mortera J (2002) Bounding the number of contributors to mixed DNA stains. *Forensic Science International* **130**, 125–126.

Lauritzen SL and Nilsson D (2001) Representing and solving decision problems with limited information. *Management Science* **47**, 1238–1251.

Lauritzen SL and Spiegelhalter DJ (1988) Local computations with probabilities on graphical structures and their application to expert systems. *Journal of the Royal Statistical Society, Series B* **50**, 157–224.

Lee PM (2004) *Bayesian Statistics, An Introduction* third edn. Arnold, London.

Leibniz GW (1930) *Sämtliche Schriften und Briefe* vol. **6**. Otto Reichl, Darmstadt.

Lemmer JF and Barth SW (1982) Efficient minimum information updating for Bayesian inferencing in expert systems Proceedings of the 2nd American National Conference on AI, pp. 424–427, Pittsburgh.

Lempert RO (1977) Modeling relevance. *Michigan Law Review* **75**, 1021–1057.

Levitt TS and Blackmond Laskey K (2001) Computational inference for evidential reasoning in support of judicial proof. *Cardozo Law Review* **22**, 1691–1731.

Lichtenberg W (1990) Methods for the determination of shooting distance. *Forensic Science Review* **2**, 37–62.

Lindley D (1965) *Introduction to Probability and Statistics from a Bayesian Viewpoint*. Cambridge University Press, Cambridge.

Lindley DV 1975 Probabilities and the Law. In: *Utility, Probability and Human Decision Making* Wendt/Vlek pp 223–232. D.Reidel Publishing Company, Dordrecht.

Lindley D (1985) *Making Decisions* second edn. John Wiley & Sons, Ltd, Chichester.

Lindley DV (1990) The 1988 Wald Memorial Lectures: the present position in Bayesian statistics. *Statistical Science* **5**, 44–89.

Lindley DV (1991) Probability In *The Use of Statistics in Forensic Science* (ed. Aitken CGG and Stoney DA) Ellis Horwood, New York pp. 27–50.

Lindley DV (2000) The philosophy of statistics. *Statistician* **49**, 293–337.

Lindley DV (2004) Foreword In *Statistics and the Evaluation of Evidence for Forensic Scientists* second edn (ed. Aitken CGG and Taroni F) John Wiley & Sons, Inc., New York.

Lindley DV (2006) *Understanding Uncertainty*. John Wiley & Sons, Inc., Hoboken, NJ.

Luce RD and Raiffa H (1958) *Games and Decisions: Introduction and Critical Survey*. John Wiley & Sons, Inc., New York.

Mackie JL (1974) *The Cement of the Universe*. Clarendon Press, Oxford.

Madsen AL and Jensen FV (1999) Lazy propagation: a junction tree inference algorithm based on lazy evaluation. *Artificial Intelligence* **113**, 203–245.

Mahoney SM and Laskey KB (1999) Representing and combining partially specified CPTs In *Uncertainty in Artificial Intelligence: Proceedings of the Fifteenth Conference* (ed. Laskey KB and Prade H) Morgan Kaufmann, San Francisco, CA pp. 391–400.

Marschak J (1950) Rational behavior, uncertain prospects, and measurable utility. *Econometrica* **18**, 111–141.

Mazumder A (2010) Planning in Forensic DNA Identification Using Probabilistic Expert Systems PhD thesis University of Oxford, Department of Statistics Oxford.

Morgan MG and Henrion M (1990) *Uncertainty: A Guide to Dealing with Uncertainty in Quantitative Risk and Policiy Analysis*. Cambridge Unviersity Press, Cambridge.

Mortera J (2003) Analysis of DNA mixtures using probabilistic expert systems In *Highly Structured Stochastic Systems* (ed. Green PL, Hjort NL and Richardson S) Oxford University Press Oxford, Oxford.

Mortera J, Dawid AP and Lauritzen SL (2003) Probabilistic expert systems for DNA mixture profiling. *Theoretical Population Biology* **63**, 191–205.

Mueller C and Kirkpatrick L (1988) *Federal Rules of Evidence*. Little, Brown, Boston, MA.

Nagarajan R, Scutari M and Lèbre S (2013) *Bayesian Networks in R with Applications in Systems Biology* Use R! Series. Springer, New York.

National Research Council (1992) *NRC I - National Research Council Committee on DNA Technology in Forensic Science*. National Academy Press, Washington, DC.

Neil M, Fenton N and Nielsen L (2000) Building large-scale Bayesian networks. *Knowledge Engineering Review* **15**, 257–284.

Nesbitt RS, Wessel JE and Jones PF (1976) Detection of GSR by use of the Scanning Electron Microscope. *Journal of Forensic Sciences* **21**, 595–610.

von Neumann J and Morgenstern O (1953) *Theory of Games and Economic Behavior* third edn. Princeton University Press, Princeton, NJ.

Nichols RG (1997) Firearms and toolmark identification criteria: a review of the literature. *Journal of Forensic Sciences* **42**, 466–474.

Nichols RG (2003) Firearms and toolmark identification criteria: a review of the literature, Part II. *Journal of Forensic Sciences* **48**, 318–327.

O'Hagan A (1988) *Probability: Methods and Measurements*. Chapman and Hall, London.

O'Hagan A (1994) *Kendall's Advanced Theory of Statistics, Vol. 2B, Bayesian Inference*. Edward Arnold, London.

Olsson EJ (2002) Corroborating testimony, probability and surprise. *British Journal for the Philosophy of Science* **53**, 273–288.

Pearl J (1988) *Probabilistic Reasoning in Intelligent Systems: Networks of Plausible Inference*. Morgan Kaufmann Publishers, Inc., San Mateo, CA.

Pearl J (1999) Graphs, structural models, and causality In *Computation, Causation, and Discovery* (ed. Glymour C and Cooper GF) The MIT Press, Cambridge, MA pp. 95–138.

Pearl J (2000) *Causality: Models, Reasoning, and Inference*. Cambridge University Press, Cambridge.

Pearl J (2009) *Causality: Models, Reasoning, and Inference* second edn. Cambridge University Press, Cambridge.

Pearl J and Verma T (1987) The logic of representing dependencies by directed graphs *Proceedings of the National Conference on AI (AAAI-87)*. AAAI Press/MIT Press, Menlo Park, CA pp. 374–379.

Pourret O, Naïm P and Marcot B (2008) *Bayesian Networks: A Practical Guide to Applications*. John Wiley & Sons, Ltd, Chichester.

Press SJ (1982) *Applied Multivariate Analysis: Using Bayesian and Frequentist Methods of Inference* second edn. Robert E. Krieger Publishing Company, Malabar.

Press SJ (1989) *Bayesian Statistics: Principles, Models, and Applications*. John Wiley & Sons, Ltd, New York.

Press SJ (2003) *Subjective and Objective Bayesian Statistics: Principles, Models, and Applications* second edn. Wiley-Interscience, Hoboken, NJ.

R Core Team (2013) *R: A Language and Environment for Statistical Computing*. R Foundation for Statistical Computing Vienna, Austria.

Raiffa H (1968) *Decision Analysis, Introductory Lectures on Choices under Uncertainty*. Addison-Wesley, Reading, MA.

Ramsey FP (1931) Truth and probability In *The Foundations of Mathematics and Other Logical Essays* (ed. Braithwaite RB) Routledge & Kegan Paul Ltd, London pp. 156–198. Reprinted in Kyburg HE and Smokler HE (eds) (1980) *Studies in Subjective Probability* second edn. Dover, New York pp. 61–92.

Ribaux O and Margot P (2003) Case based reasoning in criminal intelligence using forensic case data. *Science & Justice* **43**, 135–143.

Robert CP (2007) *The Bayesian Choice, A Decision-Theoretic Motivation* second edn. Springer-Verlag, New York.

Robertson B and Vignaux GA (1993) Taking fact analysis seriously. *Michigan Law Review* **91**, 1442–1464.

Robertson B and Vignaux GA (1995) *Interpreting Evidence: Evaluating Forensic Science in the Courtroom*. John Wiley & Sons, Ltd, Chichester.

Romolo FS, Christopher ME, Donghi M, Ripani L, Jeynes C, Webb RP, Ward NI, Kirkby KJ and Bailey MJ (2013) Integrated Ion Beam Analysis (IBA) in gunshot residue (GSR) characterisation. *Forensic Science International* **231**, 219–228.

Romolo FS and Margot P (2001) Identification of gunshot residue: a critical review. *Forensic Science International* **119**, 195–211.

Rowe WF (2000a) Range In *Encyclopedia of Forensic Science* (ed. Siegel JH, Saukko PJ and Knupfer GC) vol. **2** Academic Press, San Diego, CA pp. 949–953.

Rowe WF (2000b) Residues In *Encyclopedia of Forensic Science* (ed. Siegel JH, Saukko PJ and Knupfer GC) vol. **2** Academic Press, San Diego, CA pp. 953–961.

Saks MJ and Koehler JJ (2005) The coming paradigm shift in forensic identification science. *Science* **309**, 892–895.

Salmon WC (1990) Rationality and objectivity in science or Tom Kuhn meets Tom Bayes In *Scientific Theories, Minnesota Studies in the Philosophy of Science* (ed. Savage CW) vol. **14** University of Minnesota Press, Minneapolis, MN pp. 175–204.

Salmon WC (1998) *Causality and Explanation*. Oxford University Press, New York, Oxford.

Salmon WC (1999) Scientific explanation In *Introduction to the Philosophy of Science* (ed. Salmon MH, Earman J, Glymour C, Lennox JG, Machamer P, McGuire JE, Norton JD, Salmon WC and Schaffner KF) second edn. Hackett Publishing Co., Indianapolis, IN pp. 7–41.

Salmon W, Jeffrey RC and Greeno JG (1971) *Statistical Explanation and Statistical Relevance*. University of Pittsburgh Press, Pittsburgh, PA.

Savage LJ (1951) Theory of statistical decision. *Journal of the American Statistical Association* **46**, 55–67.

Savage LJ (1954) *The Foundations of Statistics*. John Wiley & Sons, Inc., New York.

Savage LJ (1972) *The Foundations of Statistics* second edn. Dover, New York.

Scheines R (1997) An introduction to causal inference In *Causality in Crisis?* (ed. McKim VR and Turner S) University of Notre Dame Press, Notre Dame pp. 185–199.

Schervish MJ (1995) *Theory of Statistics*. Springer-Verlag, Berlin, New York.

Schum DA (1994) *Evidential Foundations of Probabilistic Reasoning*. John Wiley & Sons, Inc., New York.

Shachter RD (1986) Evaluating influence diagrams. *Operations Research* **34**, 871–882.

Shafer G and Vovk V (2001) *Probability and Finance. It's Only a Game!*. John Wiley & Sons, Inc., New York.

Shenoy PP and Shafer G (1990) Axioms for probability and belief-function propagation In *Uncertainty in Artificial Intelligence, Proceedings of the Fifth Conference* (ed. Shachter RD, Levitt TS, Kanal LN and Lemmer JF) North Holland, Amsterdam pp. 169–198.

Sjerps M and Kloosterman AD (1999) On the consequences of DNA profile mismatches for close relatives of an excluded suspect. *International Journal of Legal Medicine* **112**, 176–180.

Smith JQ (1988) *Decision Analysis: A Bayesian Approach*. Chapman and Hall, London.

Spirtes P, Glymour C and Scheines R (2001) *Causation, Prediction and Search* second edn. The MIT Press, Cambridge, MA.

Stockton A and Day S (2001) Bayes, handwriting and science Proceedings of the 59th Annual ASQDE Meeting - Handwriting & Technology: at the Crossroads, pp. 1–10, Des Moines, IA.

Stoney DA (1991) Transfer evidence In *The Use of Statistics in Forensic Science* (ed. Aitken CGG and Stoney DA) Ellis Horwood, New York pp. 107–138.

Stoney DA (1994) Relaxation of the assumption of relevance and an application to one-trace and two-trace problems. *Journal of the Forensic Science Society* **34**, 17–21.

Swinburne R (2002) *Bayes's Theorem*. Published for the British Academy by Oxford University Press, Oxford.

Taroni F and Aitken CGG (1998) Probabilités et preuve par ADN dans les affaires civiles et criminelles. Questions de la cour et réponses fallacieuses des experts. *Revue Pénale Suisse* **116**, 291–313.

Taroni F, Aitken CGG and Garbolino P (2001) De Finetti's subjectivism, the assessment of probabilities and the evaluation of evidence: a commentary for forensic scientists. *Science & Justice* **41**, 145–150.

Taroni F, Aitken CGG, Garbolino G and Biedermann A (2006a) *Bayesian Networks and Probabilistic Inference in Forensic Science*. John Wiley & Sons, Ltd, Chichester.

Taroni F, Bozza S and Biedermann A (2006b) Two items of evidence, no putative source: an inference problem in forensic intelligence. *Journal of Forensic Sciences* **51**, 1350–1361.

Taroni F and Biedermann A (2005) Inadequacies of posterior probabilities for the assessment of scientific evidence. *Law, Probability and Risk* **4**, 89–114.

Taroni F, Biedermann A, Bozza S, Comte J and Garbolino P (2012) Uncertainty about the true source - a note on the likelihood ratio at the activity level. *Forensic Science International* **220**, 173–179.

Taroni F, Biedermann A, Garbolino P and Aitken CGG (2004) A general approach to Bayesian networks for the interpretation of evidence. *Forensic Science International* **139**, 5–16.

Taroni F, Biedermann A, Vuille J and Morling N (2013) Whose DNA is this? How relevant a question? (a note for forensic scientists). *Forensic Science International: Genetics* **7**, 467–470.

Taroni F, Bozza S and Aitken CGG (2005) Decision analysis in forensic science. *Journal of Forensic Sciences* **50**, 894–905.

Taroni F, Bozza S, Bernard M and Champod C (2007) Value DNA tests: a decision perspective. *Journal of Forensic Sciences* **52**, 31–39.

Taroni F, Bozza S, Biedermann A, Garbolino G and Aitken CGG (2010) *Data Analysis in Forensic Science: A Bayesian Decision Perspective*. John Wiley & Sons, Ltd, Chichester.

Taroni F, Champod C and Margot P (1998) Forerunners of Bayesianism in early forensic science. *Jurimetrics Journal* **38**, 183–200.

Thagart P (1988) *Computational Philosophy of Science*. The MIT Press, Cambridge, MA.

Thagart P (2003) Why wasn't O.J. convicted? Emotional coherence and legal inference. *Cognition and Emotion* **17**, 361–383.

Thompson WC (1995) Subjective interpretation, laboratory error and the value of DNA evidence: three case studies. *Genetica* **96**, 153–168.

Thompson WC and Ford S (1991) The meaning of a match: sources of ambiguity in the interpretation of DNA prints In *Forensic DNA Technology* (ed. Farley M and Harrington J) CRC Press, Inc., New York pp. 93–152.

Thompson WC, Taroni F and Aitken CGG (2003) How the probability of a false positive affects the value of DNA evidence. *Journal of Forensic Sciences* **48**, 47–54.

Tikochinsky Y, Tishby NZ and Levine RD (1984) Consistent inference of probabilities for reproducible experiments. *Physical Review Letters* **52**, 1357–1360.

Tillman WL (1987) Automated GSR particle search and characterization. *Journal of Forensic Sciences* **32**, 62–71.

Twining W (1994) *Rethinking Evidence*. Northwestern University Press, Evanston, IL.

Unites Nations Office on Drugs and Crime (2009) *Guidelines on Representative Drug Sampling*. UNO, New York.

Vanderweele TJ and Staudt N (2011) Causal diagrams for empirical legal research: a methodology for identifying causation, avoiding bias and interpreting results. *Law, Probability and Risk* **10**, 329–354.

Vicard P, Dawid AP, Mortera J and Lauritzen SL (2008) Estimating mutation rates from paternity casework. *Forensic Science International: Genetics* **2**, 9–18.

Vuille J, Biedermann A and Taroni F (2013) The importance of having a logical framework for expert conclusions in forensic DNA profiling: illustrations from the Amanda Knox case In *Wrongful Convictions and Miscarriages of Justice: Causes and Remedies in North American and European Criminal Justice Systems* (ed. Huff RC and Killias M) Routledge Chapman & Hall, New York pp. 137–159.

Watson SR and Buede DM (1987) *Decision Synthesis: The Principles and Practice of Decision Analysis*. Cambridge University Press, Cambridge.

Weir BS (1996) *Genetic Data Analysis II*. Sinauer Associates, Sunderland, MA.

Weir BS (2000) Statistical analysis In *Encyclopedia of Forensic Sciences* (ed. Siegel JA, Saukko PJ and Knupfer GC) Academic Press, San Diego, CA pp. 545–550.

Wellman MP (1990a) Fundamental concepts of qualitative probabilistic networks. *Artificial Intelligence* **44**, 257–303.

Wellman MP (1990b) Qualitative probabilistic networks for planning under uncertainty In *Readings in Uncertain Reasoning* (ed. Shafer G and Pearl J) Morgan Kaufmann, San Mateo, CA pp. 711–722.

Wellman MP and Henrion M (1993) Explaining 'explaining away'. *IEEE Transactions on Pattern Analysis and Machine Intelligence* **15**, 287–292.

Whittaker J (1990) *Graphical Models in Applied Multivariate Statistics*. John Wiley & Sons, Ltd, Chichester.

Wigmore JH (1913) The problem of proof. *Illinois Law Review* **8**, 77–103.

Wigmore JH (1937) *The Science of Judicial Proof: As Given by Logic, Psychology, and General Experience and Illustrated in Judicial Trials* third edn. Little, Brown, Boston, MA.

Williams PM (1980) Bayesian conditionalisation and the principle of Minimum information. *British Journal for the Philosophy of Science* **31**, 131–144.

Williamson J (2004) *Bayesian Nets and Causality: Philosophical and Computational Foundations*. Clarendon Press, Oxford.

Zeichner A and Levin N (1995) Casework experience of GSR detection in Israel on samples from hands, hair and clothing using an Autosearch SEM/EDX System. *Journal of Forensic Sciences* **40**, 1082–1085.

Author index

Bayesian Networks for Probabilistic Inference and Decision Analysis in Forensic Science, Second Edition.
Franco Taroni, Alex Biedermann, Silvia Bozza, Paolo Garbolino and Colin Aitken.
© 2014 John Wiley & Sons, Ltd. Published 2014 by John Wiley & Sons, Ltd.

Subject index

Bayesian Networks for Probabilistic Inference and Decision Analysis in Forensic Science, Second Edition.
Franco Taroni, Alex Biedermann, Silvia Bozza, Paolo Garbolino and Colin Aitken.
© 2014 John Wiley & Sons, Ltd. Published 2014 by John Wiley & Sons, Ltd.

Statistics in Practice

Human and Biological Sciences

Senn – Cross-over Trials in Clinical Research, Second Edition

Senn – Statistical Issues in Drug Development, Second Edition

Spiegelhalter, Abrams and Myles – Bayesian Approaches to Clinical Trials and Health-Care Evaluation

Walters – Quality of Life Outcomes in Clinical Trials and Health-Care Evaluation

Welton, Sutton, Cooper and Ades – Evidence Synthesis for Decision Making in Healthcare

Whitehead – Design and Analysis of Sequential Clinical Trials, Revised Second Edition

Whitehead – Meta-Analysis of Controlled Clinical Trials

Willan and Briggs – Statistical Analysis of Cost Effectiveness Data

Winkel and Zhang – Statistical Development of Quality in Medicine

Earth and Environmental Sciences

Buck, Cavanagh and Litton – Bayesian Approach to Interpreting Archaeological Data

Chandler and Scott – Statistical Methods for Trend Detection and Analysis in the Environmental Statistics

Christie, Cliffe, Dawid and Senn (Eds.) – Simplicity, Complexity and Modelling

Gibbons, Bhaumik and Aryal – Statistical Methods for Groundwater Monitoring, 2nd Edition

Haas – Improving Natural Resource Management: Ecological and Political Models

Haas – Introduction to Probability and Statistics for Ecosystem Managers

Helsel – Nondetects and Data Analysis: Statistics for Censored Environmental Data

Illian, Penttinen, Stoyan and Stoyan – Statistical Analysis and Modelling of Spatial Point Patterns

Mateu and Muller (Eds) – Spatio-Temporal Design: Advances in Efficient Data Acquisition

McBride – Using Statistical Methods for Water Quality Management

Okabe and Sugihara – Spatial Analysis Along Networks: Statistical and Computational Methods

Webster and Oliver – Geostatistics for Environmental Scientists, Second Edition

Wymer (Ed.) – Statistical Framework for RecreationalWater Quality Criteria and Monitoring

Industry, Commerce and Finance

Aitken – Statistics and the Evaluation of Evidence for Forensic Scientists, Second Edition

Balding – Weight-of-evidence for Forensic DNA Profiles

Brandimarte – Numerical Methods in Finance and Economics: A MATLAB-Based Introduction, Second Edition

Brandimarte and Zotteri – Introduction to Distribution Logistics

Chan – Simulation Techniques in Financial Risk Management

Coleman, Greenfield, Stewardson and Montgomery (Eds) – Statistical Practice in Business and Industry

Frisen (Ed.) – Financial Surveillance

Fung and Hu – Statistical DNA Forensics

Gusti Ngurah Agung – Time Series Data Analysis Using EViews

Jank and Shmueli – Modeling Online Auctions

Jank and Shmueli (Ed.) – Statistical Methods in e-Commerce Research

Lloyd – Data Driven Business Decisions

Kenett (Ed.) – Operational Risk Management: A Practical Approach to Intelligent Data Analysis

Kenett (Ed.) – Modern Analysis of Customer Surveys: With Applications using R

Kenett and Zacks – Modern Industrial Statistics: With Applications in R, MINITAB and JMP, Second Edition

Kruger and Xie – Statistical Monitoring of Complex Multivariate Processes: With Applications in Industrial Process Control

Lehtonen and Pahkinen – Practical Methods for Design and Analysis of Complex Surveys, Second Edition

Mallick, Gold, and Baladandayuthapani – Bayesian Analysis of Gene Expression Data

Ohser and Mücklich – Statistical Analysis of Microstructures in Materials Science

Pasiouras (Ed.) – Efficiency and Productivity Growth: Modelling in the Financial Services Industry

Pfaff – Financial Risk Modelling and Portfolio Optimization with R

Pourret, Naim and Marcot (Eds) – Bayesian Networks: A Practical Guide to Applications

Rausand – Risk Assessment: Theory, Methods, and Applications

Ruggeri, Kenett and Faltin – Encyclopedia of Statistics and Reliability

Taroni, Biedermann, Bozza, Garbolino and Aitken – Bayesian Networks for Probabilistic Inference and Decision Analysis in Forensic Science, Second Edition

Taroni, Bozza, Biedermann, Garbolino and Aitken – Data Analysis in Forensic Science

Printed and bound by CPI Group (UK) Ltd, Croydon, CR0 4YY

27/10/2024

14580217-0005